VLADIMIR I. ARNOLD

Collected Works

Springer

Vladimir I. Arnold, 1993
Photo by Boris Khesin

VLADIMIR I. ARNOLD

Collected Works

VLADIMIR I. ARNOLD

Collected Works

VOLUME VI

Dynamics, Combinatorics, and Invariants
of Knots, Curves, and Wave Fronts
1992–1995

VLADIMIR I. ARNOLD

Collected Works

VOLUME VI

Dynamics, Combinatorics, and Invariants
of Knots, Curves, and Wave Fronts
1992–1995

Edited by

Alexander B. Givental
Boris A. Khesin
Mikhail B. Sevryuk
Victor A. Vassiliev
Oleg Ya. Viro

Springer

Vladimir I. Arnold
June 12, 1937–June 3, 2010

Editors

Alexander B. Givental
Department of Mathematics
University of California
Berkeley, CA, USA

Boris A. Khesin
Department of Mathematics
University of Toronto
Toronto, ON, Canada

Mikhail B. Sevryuk
N.N. Semënov Federal Research Center
of Chemical Physics
Russian Academy of Sciences
Moscow, Russia

Victor A. Vassiliev
Steklov Mathematical Institute
Russian Academy of Sciences
Moscow, Russia

Oleg Ya. Viro
Institute for Mathematical Sciences
Stony Brook University
Stony Brooks, NY, USA

ISBN 978-3-031-04800-5

Mathematics Subject Classification: 35A18, 70S05, 53D05, 53D12, 37C05, 37C99, 00A35, 97A40

This Springer imprint is published by the registered company Springer Nature Switzerland AG
The registered company address is: Gewerbestrasse 11, 6330 Cham, Switzerland

Preface

The present volume of the *Collected Works* includes papers written by V.I. Arnold during the period from 1992 to 1995. Arnold's work of this period ranges from combinatorics of Bernoulli and Euler numbers to contact geometry of wave fronts, invariants and perestroikas of plane curves, and Vassiliev's knot invariants. The interested reader will certainly enjoy Arnold's unique style in his paper on mathematical problems in physics written in 1994, which opens with a by-now famous phrase "Mathematics is the name for those domains of theoretical physics that are temporarily unfashionable."

The articles specifically translated for this volume cover problems in dynamical systems and a discussion of problem solvability and insolvability, a paper on symplectic geometry of caustics, comments on two problems of A.D. Sakharov, and a very unusual paper on projective topology, from a popular math-phys magazine for school students. We would like to draw the reader's attention to translations of two interviews given by Arnold to the French and Spanish media. One can see how worried he was about the fate of Russian and world mathematics and science in general. His grim predictions are still very relevant today.

This is the first volume related to the work of Arnold written after the fall of the Iron Curtain: free travel and research exchanges became a part of life in the beginning of the 90s. As a byproduct, more of his papers were originally published in English and French rather than translated from Russian, while the quality of journal translations themselves had also substantially improved.

May 2021

Alexander Givental
Boris Khesin
Mikhail Sevryuk
Victor Vassiliev
Oleg Viro

Acknowledgements

The Editors thank the Göttingen State and University Library, the Mathematics Department Library of Heidelberg University, and the Mathematisches Forschungsinstitut Oberwolfach for providing the original articles for this edition. They are also grateful to Emmanuel Ferrand, Nikolai Kruzhilin, and Daniel Peralta-Salas for translating several papers in this volume, and to Ágata Timón García-Longoria for helping to locate the El País interview. The Editors are particularly indebted to Natalia Ustalova and Martin Peters for their extensive help and support with this project.

Contents

Books by V.I. Arnold written in 1992–1995 but not included in the "Collected Works"

1. Topological Invariants of Plane Curves and Caustics. Amer. Math. Soc., Providence, RI, 1994, vii+60 pp. (Dean Jacqueline B. Lewis Memorial Lectures presented at Rutgers University, New Brunswick, NJ. University Lecture Series, vol. 5).
2. Lektsii po uravneniyam s chastnymi proizvodnymi (Lectures on Partial Differential Equations) (in Russian). Publishing House of the Mathematical College of the Independent Moscow University, Moscow, 1995, 147 pp. The second (expanded and corrected) Russian edition: Lektsii ob uravneniyakh s chastnymi proizvodnymi (Lectures on Partial Differential Equations). FAZIS (PHASIS), Moscow, 1997, xi+175 pp. (the series "A Mathematical Student's Library", vol. 2). Two subsequent unrevised Russian editions: by FAZIS (1999) and by MTsNMO (Moscow, 2017, 182 pp.). The English translation: Lectures on Partial Differential Equations. Translated from the second Russian edition by R. Cooke. Springer, Berlin; PHASIS, Moscow, 2004, x+157 pp. The German translation: Vorlesungen über partielle Differentialgleichungen. Übersetzt aus dem Russischen von T. Damm. Springer, Berlin, 2004, ix+174 S.

During this period 1992–1995 Arnold also wrote paper #42 in Volume II.

Bernoulli–Euler updown numbers associated with function singularities, their combinatorics and arithmetics

V.I. Arnold

The numbers we shall study below are the numbers of topologically different maximal morsifications of critical points of smooth functions. In the simplest case of functions of one variable they are closely related to the Bernoulli and Euler numbers. But they are defined for any critical point of function. Since the classification of the critical points of functions is related to that of simple Lie algebras and to that of finite Euclidean Coxeter reflection groups, our construction associates the generalized Bernoulli-Euler numbers to the simple Lie algebras or to the Coxeter groups.

These numbers have remarkable combinatorial and arithmetical properties; some of these may be new even for the classical case.

1. Introduction. The *bifurcation diagram of functions* is the hypersurface in the space of functions, consisting of those functions which have either a degenerate critical point or two coinciding critical values.

The bifurcation diagram separates the components of its complement in the neighborhood of a given function. The representatives of these components will be called the morsifications of the given function.

A morsification (a component) is called maximal (for short, *M-morsification*), if the number of real nondegenerate critical points, born from the initial degenerate critical point, is maximal (equal to the multiplicity of the critical point).

Similar definitions can be applied to the critical points of real smooth functions of finite multiplicity (see [1], [2] for the details).

Example. Let us consider the morsifications of the function x^{n+1} of one variable. The M-morsifications are equivalent to those polynomials

$$x^{n+1} + a_1 x^{n-1} + \cdots + a_n$$

which have n different real critical values. The number K_n of nonequivalent M-morsifications is equal to the number of connected components, into which the bifurcation diagram cuts the above n-dimensional space of polynomials. The numbers K_n form the *updown sequence*

Originally published in: *Duke Math. Journal* **63**(2), 537–555 (1991), © 1991 Duke University Press.

n	0	1	2	3	4	5	6	7	8	$\cdots$
K_n	1	1	1	2	5	16	61	272	1385	$\cdots$

The exponential generating function of the updown sequence is equal (an exercise) to

$$\sum_{n=0}^{\infty} K_n \frac{t^n}{n!} = \sec t + \tan t.$$

Hence K_{2r} is equal to the Euler number E_r, and $K_{2r-1} = T_r$ is proportional to the Bernoulli number B_r with well-known coefficient (see [3]):

$$T_r = \frac{4^r(4^r - 1)}{2r} B_r.$$

Definition. The *Euler-Bernoulli number* of a critical point of a function is the number of its M-morsifications.

Thus, for the series A_n of critical points (of type x^{n+1}) the Euler-Bernoulli numbers form the updown sequence K_n. Since the A_n singularity corresponds to the Lie algebra $\mathrm{SU}(n+1)$ and to the Coxeter group of symmetries of a simplex in R^n, the Euler-Bernoulli numbers of the special unitary groups and of the Coxeter groups of symmetries of the simplices form the same classical sequence.

In this article the Euler-Bernoulli numbers of the next two series B_n, C_n are calculated. They correspond to the functions

$$B_n = x^n + y^2 \qquad \text{and} \qquad C_n = xy + y^n$$

on the plane with boundary $x = 0$, to the Lie algebras

$$B_n \sim \mathrm{O}(2n+1), \qquad C_n \sim \mathrm{Sp}(n),$$

and to the Coxeter group of the symmetries of an n-cube.

Unexpectedly, the answer is the same updown sequence, conveniently shifted:

$$K(B_n) = K(C_n) = K_{n+1}.$$

This fact implies a strange purely combinatorial identity, relating the usual updown (Bernoulli-Euler) numbers:

THEOREM 1. *The sum $\sum V_i$ of the variations*

$$V_i = |i_0 - i_1| + |i_1 - i_2| + \cdots + |i_n - i_{n+1}|$$

of all the K_n sequences

$$i_0 < i_1 > i_2 < \cdots i_{n+1}, \qquad \{i_1, \ldots, i_n\} = \{1, \ldots, n\}$$
$$i_0 = 0, \qquad i_{n+1} = \begin{cases} 0 & \text{if } n \text{ is odd} \\ n+1 & \text{if } n \text{ is even} \end{cases} \tag{1}$$

is equal to the number K_{n+2} of the sequences

$$i_0 < i_1 > i_2 \cdots i_{n+1}, \qquad \text{where } \{i_0, \ldots, i_{n+1}\} = \{0, \ldots, n+1\}. \tag{2}$$

Examples. Let $n = 2$. The only sequence (1) is (0, 2, 1, 3). Its variation is equal to 5 = the number K_4 of sequences (2) with 4 elements.

Let $n = 3$. There exist two sequences (1), namely, (0, 2, 1, 3, 0) and (0, 3, 1, 2, 0). Their variations are equal to 8. The sum of variations is equal to $16 = K_5$.

Some other combinatorial identities, suggested by the critical point theory, may also be new. For instance, the following strange identities are proved in §3:

THEOREM 2. (A) *The number of sequences*

$$i_0 < i_1 > i_2 < \cdots i_n, \qquad \{i_0, i_1, \ldots, i_n\} = \{0, 1, \ldots, n\} \tag{3}$$

is equal to the number of sequences

$$j_0 < j_1 > j_2 < \cdots j_n, \qquad j_s \equiv s \ (\mathrm{mod}\ 2), \qquad |j_s| \leqslant s. \tag{4}$$

(B) *The number of sequences* (3), *for which $i_1 - i_0 = k + 1$, is equal to the number of sequences* (4), *for which $j_n = (-1)^n(2k - n)$.*

Examples. (A) Let $n = 2$. The sequences (3) are (0, 2, 1) and (1, 2, 0). The sequences (4) are (0, 1, 0) and (0, 1, −2).

(B) Let $n = 3$, $k = 0$. The sequences (3) are (1, 2, 0, 3) and (2, 3, 0, 1). The sequences (4) are (0, 1, 0, 3) and (0, 1, −2, 3).

The updown sequence is closely related to the following Bernoulli-Euler triangle

```
                    Bernoulli    1     Euler
                             1     0
                          0     1     1
                       2     2     1     0
                    0     2     4     5     5
                16    16    14    10     5     0
             0    16    32    46    56    61    61
         272   272   256   224   178   122    61     0
      0    272   544   800  1024  1202  1324  1385  1385
                              .....
```

Each number in this Pascal-like triangle is equal to the number of the *shuttle ways* connecting it to the top vertex. A shuttle way starts at the top vertex and descends one row down at each step. At the first step it goes to the left, at the second to the right, at the third to the left, and so on.

Each shuttle way intersects the row above the last one and is uniquely defined by the truncated shuttle way above the intersection point and by the end point. Hence each number in the Bernoulli-Euler triangle is equal to the sum of the numbers in the preceding row, situated to the left of the required number if its row number is even and to the right if the row number is odd (the rows are numbered as in the Pascal triangle: the top vertex lives in the 0th row, the nth row contains $n + 1$ elements).

The left side of this triangle provides the Bernoulli part of the updown sequence K_n (n being odd), while the right side provides the Euler part (the first step of any shuttle way being in the Bernoulli direction).

The powers of 2 occurring in the triangle suggest the mod 2^k calculation. Let us replace each number $a = 2^b \pmod{2c + 1}$ in the triangle by b. We obtain the following ("Von Staudt") triangle

```
                                 0
                               0   *
                             *   0   0
                           1   1   0   *
                         *   1   2   0   0
                       4   4   1   1   0   *
                     *   4   5   1   3   0   0
                   4   4   8   5   1   1   0   *
                 *   4   5   5  10   1   2   0   0
               8   8   4   4   4   4   1   1   0   *
             *   8   9   4   6   4   5   1   4   0   0
           9   9   8   8   4   4   5   6   1   1   0   *
         *   9  10   8   8   4   5   6   8   1   2   0   0
      12  12   9   9   8   8   4   4   4   4   1   1   0   *
     *  12  13   9  11   8   9   4   7   4   5   1   3   0   0
  11      11  11   9   9   8   8   4   4   6   5   1   1   0   *
        12  11  13   9  10   8  11   4   5   5   8   1   2   0   0.
```

Some of the laws relating these numbers to each other are "evident" (say, the series of 1^s, of 4^s, of 9^s on the lines at distances 4, 8, 12 from the Euler side), but the proofs are not so simple (see §6 below).

The power of 2, which divides the elements of the Bernoulli-Euler (BE) triangle, grows with the distance from the Euler side. Hence the BE triangle, reduced mod 2^b, is almost filled by the zeroes. The nonzero mod 2^b entries belong only to the first s lines, parallel to the Euler side. The first values of s are given by the table

b	0	1	2	3	4	5	6	7	8	9	10	11	12	13
s	2	4	4	4	4	8	8	8	8	10	12	12	16	16

The behavior of the residues mod n, where $n \neq 2^b$, is quite different. With the exception of a finite neighborhood of the Euler side and of the Bernoulli side, the lines of the triangle are formed by the segments of periodic sequences.

Example. The mod 3 and mod 5 triangles can easily be continued, departing from the fragments

```
               Bernoulli    1    Euler
                          1   0
                        0   1   1
                      2   2   1   0
                    0   2   1   2   2
                  1   1   2   1   2   0
                0   1   2   1   2   1   1
              2   2   1   2   1   2   1   0
            0   2   1   2   1   2   1   2   2
          1   1   2   1   2   1   2   1   2   0,
```

```
               Bernoulli    1    Euler
                          1   0
                        0   1   1
                      2   2   1   0
                    0   2   4   0   0
                  1   1   4   0   0   0
                0   1   2   1   1   1   1
              2   2   1   4   3   2   1   0
            0   2   4   0   4   2   4   0   0
          1   1   4   0   0   1   4   0   0   0
        0   1   2   1   1   1   2   1   1   1   1
      2   2   1   4   3   2   1   4   3   2   1   0.
```

The horizontal lines of the BE triangle define finite mass distributions. These distributions converge to the sine function on $[0, \pi/2]$, when normalized conveniently.

Among the different applications of the updown numbers is the following theorem, communicated to me by F. Catanese. Let us call a complex polynomial of degree $m+1$ in one variable *generic*, if the values of its modulus at the m critical points are all different. The number of the connected components of the space of generic polynomials is then equal to K_m (which is also the number of topologically different $|f|$ functions and the number of the nonhomeomorphic configurations of the generalized critical lemniscates).

Example. There exists two topologically different configurations of the three critical level lines of the modulus function of a generic polynomial of degree 4. The highest critical level is an 8-shaped lemniscate. The two other critical points may be situated either in the same disc, bounded by the lemniscate, or in both discs; $K_3 = 2$.

Remark. The numbers of topologically different generic complex polynomials of degree n and of generic real polynomials $x^n + \cdots$ whose critical values are all real coincide. But I see no natural bijection between these sets of classes of generic complex and real polynomials.

2. The Bernoulli-Euler triangle. We shall denote the elements of the BE triangle like the numbers of the Pascal triangle:

$$K_0^0 = 1$$

$$K_1^0 = 1 \qquad K_1^1 = 0$$

$$K_2^0 = 0 \qquad K_2^1 = 1 \qquad K_2^2 = 1$$

$$\cdots\cdots$$

Definition 1. A *shuttle sequence* is a finite sequence of integers

$$i_0 \geqslant i_1 < i_2 \geqslant i_3 < i_4 \geqslant \cdots i_n, \qquad 0 \leqslant i_s \leqslant s \quad \text{for all } s.$$

Remark. This sequence defines an itinerary of a shuttle way, connecting K_0^0 to $K_n^{i_n}$.

Example. There exist 5 shuttle sequences ending at $i_4 = 3$, namely, (0, 0, 1, 0, 3), (0, 0, 2, 0, 3), (0, 0, 1, 1, 3), (0, 0, 2, 1, 3), and (0, 0, 2, 2, 3).

Definition 2. The element K_n^p of the BE triangle is the number of shuttle sequences, ending with $i_n = p$.

Example. $K_4^3 = 5$.

LEMMA 1. *K_n^p is equal to the sum of the numbers of the preceding row to the left (to the right) of it if n is even (odd):*

$$K_{2m}^p = K_{2m-1}^0 + \cdots + K_{2m+1}^{p-1}, \qquad K_{2m+1}^p = K_{2m}^p + K_{2m}^{p+1} + \cdots + K_{2m}^{2m}.$$

Proof. A shuttle way intersects the preceding row once and is defined by the truncated way, provided that the end point (p) is fixed.

COROLLARY 1. $K_{2m}^p = K_{2m}^{p-1} + K_{2m-1}^p$, $K_{2m+1}^p = K_{2m+1}^{p+1} + K_{2m}^{p-1}$.

These formulae provide an easy algorithm for the calculation of the BE triangle.

COROLLARY 2. *The total number $K_n = K_{n-1}^0 + \cdots + K_{n-1}^{n-1}$ of shuttle sequences with n elements is equal to K_n^n if n is even and to K_n^0 if n is odd:*

$$K_{2m} = K_{2m}^{2m} = K_{2m}^{2m-1} = K_{2m+1}^{2m}, \qquad K_{2m+1} = K_{2m+1}^0 = K_{2m+1}^1 = K_{2m+2}^1.$$

Remark. Lemma 1 implies the fast convergence of the distributions K_n^p for $n \to \infty$ to the sine distribution. Indeed, the formulae of Lemma 1 in the continuous limit define the transformation, sending a function f, defined on a segment $[0, r]$ to the function g on the same segment, defined by the equation

$$g(y) = \int_0^y f(r - x)\, dx.$$

We normalize the functions, fixing the value 1 for the total integral over the whole segment. Now f is sent to

$$h(y) = c^{-1} g(y), \qquad \text{where } c = \int_0^r g(y)\, dy.$$

The mapping, sending f to h, is a contraction, and its fixed point is $f(x) = \sin x$, provided that we wisely choose $r = \pi/2$.

Hence the normalized distributions converge to the distribution with density $\sin x$.

Remark. The same continuous version of the BE triangle appears in Kolmogorov's proof of the inequalities between the derivatives of the smooth functions, where the Euler numbers are also crucial ingredients.

3. The updown triangle.

Definition 1. An *updown sequence* of length n is a sequence of integers

$$i_1 < i_2 > i_3 < \cdots i_n,$$

such that $\{i_1, \ldots, i_n\} = \{1, \ldots, n\}$.

Definition 2. The *updown number* L_n is the number of the updown sequences of length n.

Example. $L_3 = 2$, the sequences being (1, 3, 2) and (2, 3, 1).

Remark. We shall prove soon, that the updown numbers L_n coincide with the shuttle numbers K_n of §2.

Definition 3. The *updown triangle* is formed by the numbers L_n^k = the number of the updown sequences of length $n + 1$, such that the interval between the first two elements contains exactly k other elements of the sequence:

$$i_2 - i_1 = k + 1. \tag{5}$$

Example. The initial fragment of the updown triangle is

$$L_0^0 = 1$$

$$L_1^0 = 1 \qquad L_1^1 = 0$$

$$L_2^0 = 1 \qquad L_2^1 = 1 \qquad L_2^2 = 0$$

$$L_3^0 = 2 \qquad L_3^1 = 2 \qquad L_3^2 = 1 \qquad L_3^3 = 0$$

$$\cdots\cdots$$

(We define *formally* $L_0^0 = 1$.)

The astonishing property of these numbers is the following

THEOREM 2′. *Let $s + k \leqslant n$. Then the number of the updown sequences of length $n + 2$, such that the interval between the first two elements contains exactly s other elements and that the sequence contains exactly k elements, which are greater than the second element,*

$$i_2 - i_1 = s + 1, \qquad i_2 + k = n + 2, \tag{6}$$

is equal to L_n^k (and hence does not depend on s).

Proof. We define a mapping, sending the updown sequences of length $n + 2$, verifying condition (6), to the updown sequences of length $n + 1$, verifying condition (5), and then we prove that it is a bijection, constructing the inverse mapping.

(A) Let us start from an $n + 2$ updown sequence, verifying the conditions (6). Among the elements $i_4, \ldots, i_{n+2}$ there exist exactly m elements, $i_{r_m} > \cdots > i_{r_1}$ which are greater than i_3. Their number m is at least k, since k elements are greater than i_2. Let us define $j = i_{r_k} + 1/2$. The down-up $(n + 1)$-sequence

$$j > i_3 < i_4 > \cdots$$

contains exactly k elements in the interval between j and i_3. It is easy to transform it into the required updown sequence verifying condition (5) by an orientation reversing diffeomorphism of the axis where i and j live.

(B) Let us define the inverse mapping. We start from an $n + 1$ down-up sequence $j_1 > j_2 < j_3 > \cdots$, where

$$\{j_1, \ldots, j_{n+1}\} = \{1, \ldots, n + 1\} \qquad \text{and} \qquad j_1 - j_2 = k + 1.$$

Among the elements $(j_3, \ldots, j_{n+1})$ there exist exactly u elements $j_{r_1} > \cdots > j_{r_u}$, which are greater than j_2. Their number is at least k, since $j_1 - j_2 = k + 1$. Let us choose

$i_2 = j_{r_k} - \frac{1}{2}$. Then $i_2 > j_2$ (it may be greater or smaller than j_1). Among the n elements $(j_2, \ldots, j_{n+1})$ there exist exactly v elements $j_{q_v} < \cdots < j_{q_1}$, which are smaller than i_2. Their number v is at least s, since the number of elements above i_2 is equal to k and $n - k \geqslant s$. We choose $i_1 = j_{q_s} - \frac{1}{2}$. The $(n+2)$-sequence $i_1 < i_2 > j_2 < j_3 > \cdots$ contains exactly s elements between i_1 and i_2 and exactly k elements are higher than i_2. To transform it into the required updown sequence, verifying conditions (6), we use a diffeomorphism of the axis of i and j.

(C) The operations (A) and (B) are obviously inverse to each other.

Corollary 1. *$L^s_{n+1} = L^0_n + \cdots + L^{n-s}_n$, provided that $s \leqslant n$.*

Proof. An updown sequence $i_1 < i_2 > \cdots$ of length $n+2$, such that $i_2 - i_1 = s + 1$, has k elements higher than i_2, where $0 \leqslant k \leqslant n - s$. The summation over k gives the required corollary, since the number of updown $(n+2)$-sequences with any given k and s is equal to L^k_n.

Corollary 2. *The updown triangle and the BE triangle differ only in the orderings of the even rows:*

$$L^p_{2n-1} = K^p_{2n-1}, \qquad L^p_{2n} = K^{2n-p}_{2n}.$$

Proof. The formula of Corollary 1 and that of Lemma 1 of §2 provide isomorphic calculations of K and of L, since the boundary conditions

$$L^{2n-1}_{2n-1} = K^0_{2n-1} = 0, \qquad L^0_{2n} = K^0_{2n} = 0 \quad \text{for } n > 0, \qquad L^0_0 = K^0_0 = 1$$

are isomorphic.

Corollary 3. *The updown numbers L_n coincide with the shuttle numbers K_n of §2.*

We have thus proved Theorem 2 of §1.

4. The calculation of $K(B_n)$. The singularity B_n is a boundary singularity of the function x^n on the line $\{x\}$, equipped with the "boundary" $x = 0$ (see [1], [2]). The n "critical values" in this case are the $n - 1$ usual critical values and the value at the origin. The standard methods of singularity theory provide the following information.

Lemma 1. *The number of maximal boundary morsifications $K(B_n)$ is equal to the number of components of the space of real polynomials*

$$f = x^n + a_1 x^{n-1} + \cdots + a_{n-1} x, \tag{7}$$

whose critical points are all real and whose n critical values are all different. These components are contractible and each of them is smoothly parametrized by the $n - 1$ basic symmetric functions of the $n - 1$ nonzero critical values.

Example 1. Let $n = 2$. The polynomials $x^2 + ax$ form a real line. The critical point is always real. The critical values are 0 (the boundary level) and $-a^2/4$. The complement to the bifurcation diagram has two components ($a > 0$, $a < 0$). Hence, $K(B_2) = 2$.

Example 2. Let $n = 3$. The polynomials $x^3 + ax^2 + bx$ form a real plane. The critical points are real if and only if $b \leqslant a^2/3$. Two critical values coincide if either $b = a^2/3$, or the polynomial has a double root, in which case either $b = a^2/4$ or $b = 0$. The complement to the bifurcation diagram $b(b - a^2/4) = 0$ in the M-morsifications domain $b < a^2/3$ consists of five components. Hence $K(B_3) = 5$.

These 5 components are shown in Figure 1. The graph of the polynomial f, corresponding to each component, is shown in Figure 1 together with the position of the origin ($x = 0, f(x) = 0$) on the graph.

The whole graph is cut into 5 pieces by the 2 critical level sets.

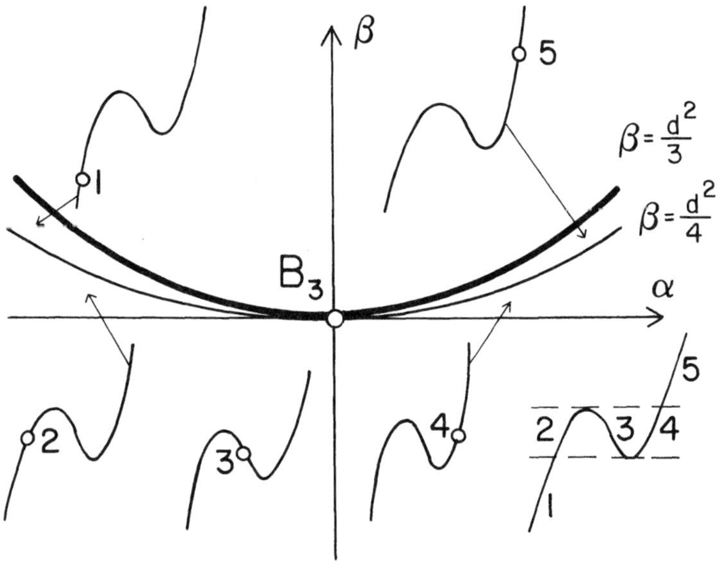

FIGURE 1

Definition. The number of connected components, into which the critical level sets decompose a graph of a function, is called the *variation* of the function.

Example. The variation of the polynomial $x^3 - x$ is equal to 5. The variation of a polynomial of degree 4 with 3 real critical values is equal to 8.

Let us associate to a real polynomial $x^n + \cdots$ having $n - 1$ different real critical values an updown sequence. We use the decreasing ordering $x_1 > x_2 > \cdots > x_{n-1}$ of the critical points. We obtain the updown sequence

$$y_1 < y_2 > y_3 < \cdots y_{n-1}$$

of the critical values. We transform the set $\{y_1, \ldots, y_{n-1}\}$ into $\{1, \ldots, n-1\}$ by an orientation preserving diffeomorphism of the y-axis. We denote by i_k the image of the critical value y_k.

LEMMA 2. *The variation of our polynomial is equal to the variation*

$$|i_0 - i_1| + |i_1 - i_2| + \cdots + |i_{n-1} - i_n|$$

of the down-up sequence $i_0 > i_1 < \cdots i_n$, where $i_0 = n$ and $i_n = n$ if n is even, and $i_n = 0$ if n is odd.

Indeed, the part of the graph between x_s and x_{s+1} $(s = 1, \ldots, n-2)$ is cut into the intervals whose number is $|i_{s+1} - i_2|$. To take into account the rays $x \geqslant x_1$ and $x < x_{n-1}$, we put $x_0 = +\infty$, $x_n = -\infty$. Then $y_0 = +\infty$, $y_n = (-1)^n\infty$. Transforming finally $y = +\infty$ into $i = n$ and $y = -\infty$ into $i = 0$, we prove Lemma 2.

Remark. The formula of Lemma 2 holds for any function, behaving as a polynomial at infinity. The variation of a polynomial does not change along a component of the space of polynomials with $n - 1$ critical values. Each component contains a polynomial with critical values $1, \ldots, n-1$. This polynomial is unique up to the translations along the x axis. These components are classified by the $(n-1)$-updown sequences of the critical values of the above polynomials, each sequence defining such a component. Finally, we obtain (by the usual technique of singularity theory) the following corollary.

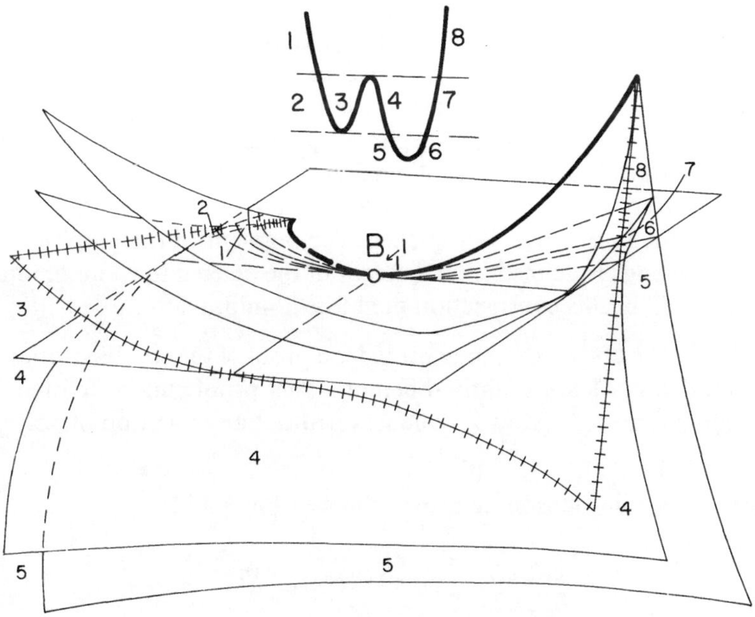

FIGURE 2

Corollary. *The number $K(B_n)$ of the components of the space of polynomials (7) whose n critical values are all different is equal to the sum of the variations $\sum |i_{k+1} - i_k|$ of all the $(n+1)$-sequences $i_0 > i_1 < \cdots i_n$, where $\{i_1, i_2, \ldots, i_{n-1}\} = \{1, \ldots, n-1\}$, $i_0 = n$, $i_n = n$ if n us even, $i_n = 0$ if n is odd.*

Example. Let $n = 4$. Then there exist two possible 3-sequences $(i_1, i_2, i_3) = (1, 3, 2)$ or $(2, 3, 1)$. The corresponding 5-sequences are $(4, 1, 3, 2, 4)$ and $(4, 2, 3, 1, 4)$. Each has variation 8; hence $K(B_4) = 16$. The 16 components are represented in Figure 2 together with the corresponding graphs.

5. Proof of Theorem 1. According to §3 and §4, Theorem 1 is equivalent to the following

Lemma. $K(B_n) = K(A_{n+1})$.

Proof. Let us start with a representative of a component, counted by $K(B_n)$. It is a function with $n-1$ critical values and a point on its graph corresponding to the noncritical value. Let us add a δ-shaped function, concentrated near the above noncritical point. The sum will have two critical points more than the initial function: a local minimum and a local maximum. The value at the local minimum belongs to the same interval between the old critical levels, to which the critical point belongs. The value at the local maximum is higher than all the old local maxima values. Hence the topological type (the $(n+1)$-sequence) of the new function is well defined by the topological type of the initial sequence (by its $(n-1)$-updown sequence) and by the segment of the initial graph containing the initially chosen point of noncritical level.

To reconstruct the initial combinatorial data from the $(n+1)$-sequence (which may be an arbitrary updown sequence), we consider the higher maximum H and choose the highest h of the (at most two) neighboring local minima (h, h'). The segment of the graph between the lowest of the neighboring minima (h') and the local maximum (H'), separated from the highest local maximum (H) by the highest local minimum (h) can be replaced by a graph of a monotonic function (Figure 3). Thus we have reconstructed the initial $(n-1)$-sequence. The segment of the graph to which the initial point belongs is defined by the height of the deleted local minimum (h).

Points h' or H' in this construction might be at infinity.

Remark 1. The above reasoning provides a proof of the combinatorial Theorem 1 (the notations in §1 are slightly different). This proof can be written in purely combinational terms of updown sequences without any mention of polynomial or functions.

Remark 2. Similar reasoning proves the well-known identity

$$2K_{n+1} = \sum_{i=0}^{n} C_n^i K_i K_{n-i} \qquad (n > 0).$$

Indeed, let us push to $+\infty$ the highest local maxima (or to $-\infty$ the lowest local

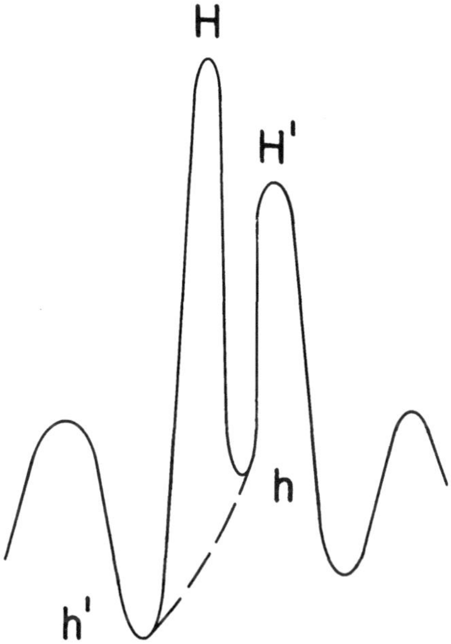

FIGURE 3

minima) of a function with $n+1$ critical points with different critical values and with the usual conditions at infinity. Then the function will be decomposed into two parts: the left part with i critical points and the right part with $n-i$ critical points. The number of the ordered pairs of types is $K_i K_{n-i}$. To reconstruct the type of the initial function, we need to know which of the n critical levels correspond to the left, and which to the right part. This gives the factor C_n^i.

Finally, in our pairs the initial function is represented twice: once we count the decomposition obtained pushing to infinity the highest maximum, and once that corresponding to the lowest minimum. The identity is thus proved.

Remark 3. A similar reasoning counts the lemniscates configurations (see §1): we decompose the configuration in two parts, according to the two discs bounded by the outer lemniscate. Then we count separately the configurations with i critical points in the "first" and $n-1$ in the "second" disc. Finally, we find that each configuration was counted twice, since the two discs are interchangeable by a diffeomorphism of the plane.

In the terms of the exponential generating function $K(t)=\sum_{n=0}^{\infty} K_n t^n/n!$ the identity of Remark 2 takes the form of a differential equation $2K'=1+K^2$. Taking into account that $K(0)=1$, we obtain the classical formula $K(t)=\sec t+\tan t$.

Remark 4. The formula

$$d\tan\frac{t}{dt}=\sec^2 t$$

implies the identity

$$K_{2m+1} = \sum C_{2m}^{i} K_i K_{2m-i}, \qquad \text{where } i \text{ even}.$$

The combinatorial proof: push the lowest local minimum of $x^{2m+2} + \cdots$ to minus infinity. If we push to plus infinity the highest local maximum, we find

$$K_{2m+1} = \sum C_{2m}^{i} K_i K_{2m-i}, \qquad \text{where } i \text{ odd},$$

which corresponds to the formula

$$d \tan t/dt = 1 + \tan^2 t.$$

Similarly, pushing the maxima or minima of $x^{2m+1} + \cdots$, we find the identity

$$K_{2m} = \sum C_{2m-1}^{i} K_i K_{2m-i}, \qquad (i \text{ even}) = \sum C_{2m-1}^{i} K_i K_{2m-i}, \qquad (i \text{ odd}),$$

which corresponds to the formula

$$d \sec t/dt = \tan t \cdot \sec t.$$

The last identity is a direct formal corollary of the identity of Remark 2, while the preceding identities yield some formally new information.

Remark 5. The numbers K_n count the types of functions on the circle with n nondegenerate critical values, provided that n is even (the equivalence being defined by the orientation preserving diffeomorphisms of the circle and of the line of values of functions). If n is odd, the number K_n counts the types of the sections of the Möbius bundle over the circle.

Remark 6. The fact that the Bernoulli-Euler numbers K_n form a single sequence provides a method of construction of identities. The identity (see [3])

$$E_n = (-1)^n + \sum (-1)^s C_{2n}^{2s+1} T_{n-s}, \qquad 0 \leqslant s \leqslant n,$$

takes in our notation the form

$$K_m = \sum (-1)^s C_m^{2s+1} K_{m-2s-1}, \qquad m \text{ even},$$

(where $C_m^{m+1} K_{-1} = 1$).

We guess that the identity holds for all m's:

$$K_{2n-1} = \sum (-1)^s C_{2n+1}^{2s+1} K_{2n-2s}, \qquad 0 \leqslant s \leqslant n.$$

This identity is true and seems to be "new". In the notation of [3] it takes the form

$$T_{n+1} = \sum (-1)^s C_{2n+1}^{2s-1} E_{n-s}, \qquad 0 \leqslant s \leqslant n.$$

6. Arithmetics of the Bernoulli-Euler triangle. The Bernoulli-Euler triangle mod 2^b is filled by zeroes with the exception of a few lines parallel to the Euler side.

Perhaps, $2mK_n^{n-j} = 4^{2m-1}(2c+1)$, $n = 4m-1$, like in the Von Staudt Theorem (see [3]). The relation $K_n^{n-j} = 0 \bmod 2^b$ implies the periodicities of the numbers $K_n^{n-i} \bmod 2^b$, $i < j$, with respect to n.

Example. $E_n \bmod 32$ is periodic with period of length 16, namely

$$(1, 5, 29, 9, 25, 13, 21, 17, 16+1, 16+5, \ldots).$$

Let p be an odd prime number. (See [4], [5].)

THEOREM 3. $K_n^i \equiv K_n^{i+(p-1)} \bmod p$, $0 < i < n-p+1$.

Let us denote Δ^k the k-th difference operation, the shift of the argument i being $p-1$. For instance, $(\Delta K)_n^i = K_n^{i+(p-1)} - K_n^i$.

THEOREM 4. $(\Delta^r K)_n^i \equiv 0 \bmod p^r$, *if* $r \leqslant i \leqslant n - rp$.

Example. $K_n^i - 2K_n^{i+12} + K_n^{i+24} \equiv 0 \bmod 169$.

Similar congruences hold for the lines parallel to the Euler or to the Bernoulli sides, provided that we change the signs of those horizontal lines, which contain $4c$ or $4c-1$ elements:

$$G_{ij} = (-1)^s K_{i+j}^i, \qquad (s = 1 \text{ for } i+j = 4c-1, 4c-2, s = 0 \text{ otherwise}).$$

(This change of notation was suggested to me by Ira Gessel; see [4]).

The exponential generating function

$$U(x, y) = \sum G_{ij} x^i/i!\, y^j/j! \tag{1}$$

is then equal to $e^x/\mathrm{ch}(x+y)$. Indeed, the inductive definition of the BE triangle takes the form

$$\partial U/\partial x - \partial U/\partial y = U,$$

and we know the boundary conditions $U(x, 0) = 1 + \mathrm{th}\, x$, $U(0, y) = 1/\mathrm{ch}\, y$ from Remark 3 of §5.

We shall consider the ring of series of the form (1) with any integer coefficients G_{ij}, and we call a series divisible by an integer q if all these coefficients G_{ij} are multiples of q. In this notation Theorem 4 takes the form

$$\Delta^r U \equiv 0 \bmod p^r \tag{2}$$

where $\Delta = D_y D_x^p - D_x D_y^p$ ($D_x = \partial/\partial x$, $D_y = \partial/\partial y$).

THEOREM 5. *The congruence* (2) *holds for any differential operator* $\Delta = \Delta_{k,l} = D_y^l D_x^{kp} - D_x^k D_y^{lp}$, $k \geqslant 0$, $l \geqslant 0$.

Example. For $(k, l) = (1, 0), (0, 1), (1, 1)$ we obtain the congruences mod p^r for the BE numbers forming a line, parallel to the Bernoulli side, to the Euler side and to the horizontal line respectively.

Example. $T_n + 2T_{n+3} + T_{n+6} \equiv 0(49)$, if $n > 1$; $E_n + 2E_{n+3} + E_{n+6} \equiv 0(49)$, if $n > 0$

$$T_n - 2T_{n+6} + T_{n+12} \equiv 0(169), \quad \text{if } n > 1;\ E_n - 2E_{n+6} + E_{n+12} \equiv 0(169), \quad \text{if } n > 0.$$

Theorem 5 implies that for any fixed i and j the number $G_{i+k(p-1),j+l(p-1)}$ is congruent mod p^r to the value of a polynomial in k, l, whose degree is smaller, than r (at least, for sufficiently large values of k and l).

Example. Let $r = 1$. Then the above polynomial is a constant. Hence G_{ij} mod p depends only on i mod$(p - 1)$, j mod$(p - 1)$, provided that $i > 0$, $j > 0$.

The coefficients of the polynomials seem to be integers:

$$G_{2k+1,2l+1} \equiv (2 + 6l) \bmod 9, \text{ if } k > 0; \qquad G_{2k,2l} \equiv (1 + 3l)(9), \text{ if } k > 0, 1 > 0.$$

$$G_{4k,4l} \equiv 24 \bmod 25, \text{ if } k > 0, l > 0; \qquad G_{4k,4l+1} \equiv (21 - 5(k - l))(25), \text{ if } k > 0.$$

$$G_{2k,2l} \equiv [27(kl^2 + k^2 l) - 9l^3] + (9kl - 18k^2 - 27l^2) + (9k + 30l) + 1(81), \quad k, l > 1,$$

$$G_{2k,2l+1} \equiv [27(kl^2 + k^2 l) - 9l^3] + (36kl - 45k^2) - 24l - 2(81), \qquad k, l > 1.$$

Proof of Theorem 5. We start with an obvious corollary of the Leibnitz formula.

LEMMA F (FOR FROBENIUS). D^p *mod* p *is a derivation for any prime number* p *and any derivation* D *of any commutative ring*

$$D^p(ab) = aD^p b + bD^p a + pf[a, b]$$

where f *is a bilinear combination of* $D^i a$, $D^j(b)$ ($i > 0$, $j > 0$).

Now let $R = \mathbb{Z}[A]$ be the ring of integer polynomials in $A = \operatorname{th} z$ and let $D = d/dz$. Then $DA = 1 - A^2$; hence D is a derivation of R. Let p be an odd prime number.

LEMMA 1. *$D^pA = DA \bmod p$; i.e., $D^pA = 1 - A^2 + pf, f \in R$.*

Proof. Let $C = \operatorname{ch} z, S = \operatorname{sh} z$. We apply Lemma F to the product $S = AC$. Since p is odd, we find

$$C = D^pS = CD^pA + AS + pCh, \qquad h \in R,$$

since $D^pC = S$, $D^iA = f_i$, $D^jC = g_jC$ $(f_i, g_j \in R)$.

LEMMA 2. *For any $a \in R$ there exists $f[a] \in R$, such that*

$$(D^p - D)a = pf[a].$$

Proof. $f[ab] = af[b] + bf[a] + f[a, b] \in R$, according to Lemma F.

LEMMA 3. *For any $a \in R$ there exists $f_r[a] \in R$, such that*

$$(D^p - D)^r a = p^r f_r[a].$$

Proof. $f_{r+1}[a] = f[f_r[a]]$.

Example. $(D^p - D)^r A = p^r f_r[1]$. The Bernoulli side congruences are thus proved.

Now let $B = 1/C = \operatorname{sech} z$; hence $DB = -AB$.

LEMMA 4. *$D^pB = DB \bmod p$; that is $(D^p - D)B = phB, h \in R$.*

Proof. We apply Lemma F to the product $1 = BC$. We obtain

$$0 = (D^pB)C + BS - ph, \qquad h \in R,$$

since $D^pC = S$, $D^iB = h_iB$, $D^jC = g_jC$ $(h_i, g_j \in R)$. Thus $D^pB = -BS/C + phB$, as required.

LEMMA 5. *For any $a \in R$ there exists $h[a] \in R$, such that*

$$(D^p - D)(aB) = ph[a]B.$$

Proof. We apply Lemma F to the product aB. We obtain

$$D^p(aB) = (D^pa)B + a(D^pB) + pH[a]B, \qquad H[a] \in R,$$

since $D^iB = h_iB, h_i \in R, D^ja \in R$. We find $h[a] = f[a] + ah + H[a]$ from Lemmas 2 and 4.

LEMMA 6. *For any $a \in R$ there exists $h_r[a] \in R$, such that*

$$(D^p - D)^r(aB) = p^r h_r[a]B.$$

Proof. $h_{r+1}[a] = h[h_r[a]]$.

Example. $(D^p - D)^r B = p^r h_r[1]B$. The Euler side congruences are thus proved.

Now let $A = \text{th}(x+y)$, $B = \text{sech}(x+y)$, $E = e^x$, $U = BE$. The derivations D_x and D_y act on $R = \mathbb{Z}[A]$ as D, while $D_x E = E$, $D_y E = 0$.

LEMMA 7. $(D_x^p - D_x)U = pvU$, $(D_y^p - D_y)U = phU$, $v, h \in R$.

Proof. We apply Lemma F to the product BE. We obtain

$$D_x^p U = EB + ED_x^p B + pFU, \qquad F \in R,$$

since $D_x^i E = E$ and $D_x^j B = h_j B$, $h_j \in R$. Substructing $D_x U = EB + ED_x B$ and replacing $(D_x^p - D_x)B$ by its value from Lemma 4, we find $v = h + F \in R$, and similarly for D_y.

LEMMA 8. *For any $a \in R$ there exists $x[a]$, $y[a] \in R$, such that*

$$(D_x^p - D_x)(aU) = px[a]U, \qquad (D_y^p - D_y)(aU) = py[a]U.$$

Proof. Since $D_x U = (1-A)U$, $D_y U = -AU$, we find $D_x^i U = X_i U$, $D_y^j U = Y_j U$, where X_i, $Y_j \in R$. We apply Lemma F to the product aU. We obtain $D_x^p(aU) = (D_x^p a)U + a(D_x^p U) + pv[a]U$, $v[a] \in R$. Substituting $D_x^p a = D_x a + pf[a]$ from Lemma 2 and $D_x^p U = D_x U + pvU$ from Lemma 7, we find $D_x^p(aU) = (D_x a)U + a(D_x U) + pf[a]U + pavU + pv[a]U$; hence $x[a] = f[a] + av + v[a]$. Similarly, $D_y^p(aU) = (D_y^p a)U + a(D_y^p U) + pw[a]U$; hence $y[a] = f[a] + ah + w[a] \in R$.

LEMMA 9. *Let $\Delta = D_y D_x^p - D_x D_y^p$. Then for any $a \in R$ there exists $F[a] \in R$, such that $\Delta(aU) = pF[a]U$.*

Proof. From Lemma 8, $\Delta(aU) = D_y(px[a]U) - D_x(py[a]U)$. Since $D_x U = XU$, $D_y U = YU$, where X, $Y \in R$, we obtain $F[a] = D_y x[a] - D_x y[a] + x[a]Y - y[a]X \in R$.

LEMMA 10. *For any $a \in R$ there exists $F_r[a] \in R$, such that $\Delta^r(aU) = p^r F_r[a]U$.*

Proof. $F_{r+1}[a] = F[F_r[a]]$.

Example. $\Delta^r U = p^r F_r[1]U$. The horizontal lines congruences are thus proved.

Let k and l be nonnegative integer numbers, and let

$$\Delta_{k,l} = D_y^l D_x^{kp} - D_x^k D_y^{lp}.$$

LEMMA 11. *For any $a \in R$ there exists $F_{k,l}[a] \in R$, such that $\Delta_{k,l}(aU) = pF_{k,l}[a]U$.*

Proof. $\Delta_{k+1,l}(aU) = D_x \Delta_{k,l}(aU) + (D_x^p - D_x)(D_y^l D_x^{kp}(aU))$. $D_y^l D_x^{kp}(aU) = b_{k,l}U$, $b_{k,l} \in R$, since $D_x U = XU$, $D_y U = YU (X, Y \in R)$. From Lemma 5 we obtain $\Delta_{k+1,l}(aU) = p(D_x(F_{k,l}[a]U) + x[b_{k,l}]U)$. Hence $F_{k+1,l}[a] = D_x(F_{k,l}[a] + F_{k,l}[a]X + x[b_{k,l}]) \in R$ and similarly for $F_{k,l+1}[a]$.

LEMMA 12. *For any $a \in R$ there exists $F_{k,l,r}[a] \in R$, such that $\Delta^r_{k,l}(aU) = p^r F_{k,l,r}[a] U$.*

Proof. $F_{k,l,r+1} = F_{k,l}[F_{k,l,r}[a]]$.

Example. $\Delta^r_{k,l} U = p^r F_{k,l,r}[1] U$. Theorem 5 is thus proven.

Acknowledgement. This work was mainly written while the author was Eilenberg Professor of Mathematics at Columbia University and was completed while visiting the University of North Carolina at Chapel Hill. I would like to thank those departments of mathematics for their hospitality, Prof. Ira Gessel and P. Deligne for useful comments, Mrs. Elaine Jackson for quickly typing a difficult manuscript, and Ya. Minski for the computer experimentations.

REFERENCES

1. V. I. ARNOLD, S. M. GUSEIN-ZADE, AND A. N. VARCHENKO, *Singularities of Differentiable Maps*, Vol. 1, 1985, Birkhauser, Boston, 304; Vol. 2, 1988, 492.
2. V. I. ARNOLD, V. A. VASSILIEV, V. V. GORYUNOV, AND O. V. LYASHKO, *Singularities* 1 (Dyn. Syst. 6, Springer Encyclopedia of Math. Sciences, vol. 6), and *Singularities* 2 (Dyn. Syst. 8, Springer Encyclopedia of Math. Sciences, vol. 39). Russian versions: Itogi Nauki i Tekhn. VINITI, Sovrem. Probl. Math., Fund, Napravl., vol. 6, Moscow 1985, 256, vol. 39, Moscow 1989, 256.
3. N. NIELSEN, *Traité elementaire des nombres de Bernoulli*, Gauthier-Villars, Paris, 1923.
4. I. GESSEL, *Congruences for Bell and tangent numbers*, Fibonacci Quart. **19** (1981), 137–143.
5. D. E. KNUTH AND T. J. BUCKHOLTZ, *Computation of tangent, Euler, and Bernoulli numbers*, Math. Comp. **21** (1967), 663–688.

STEKLOV MATHEMATICAL INSTITUTE, VAVILOVA 42, 117966, MOSCOW GSP-1, UNION OF SOVIET SOCIALIST REPUBLICS

Congruences for Euler, Bernoulli and Springer numbers of Coxeter groups

V.I. Arnol'd

Abstract. The classical and generalized Euler numbers, reduced with respect to an odd modulus, are represented as sums of exponentials. From this representation there follow congruences modulo powers of an odd prime p between elements of the Euler-Bernoulli triangles and the values of certain polynomials in two variables on sublattices with step $p-1$.

§1. Congruences for Euler numbers and tangent numbers

Consider the sequences of integers $f(n)$ and $g(n)$ $(n \geq 0)$ defined by the expansions

$$\sum f(n)z^n/n! = 1 + \tanh z, \qquad \sum g(n)z^n/n! = 1/\cosh z.$$

Let $p > 1$ be an odd number (not necessarily prime).

Theorem 1.

$$1^n + 2^n - 3^n - 4^n + + - \cdots - (p-1)^n + p^n \equiv f(n) \operatorname{mod}(p = 4s+1),$$
$$1^n + 2^n - 3^n - 4^n + + - \cdots + (p-1)^n - p^n \equiv g(n) \operatorname{mod}(p = 4s+3),$$
$$1^n - 2^n - 3^n + 4^n + - - \cdots + (p-1)^n + p^n \equiv g(n) \operatorname{mod}(p = 4s+1),$$
$$1^n - 2^n - 3^n + 4^n + - - \cdots - (p-1)^n - p^n \equiv -f(n) \operatorname{mod}(p = 4s+3).$$

The iterated differences of the numbers $f(n)$ form the (twisted) Euler-Bernoulli triangle $F_{k,l}$ (see [1], [2], and [3]):

$$F(x, y) = \sum F_{k,l}(x^k/k!)(y^l/l!) = e^x/\cosh(x+y).$$

Suppose p is an odd prime. Consider the sublattice $\{(k_i = a + i(p-1),\ l_j = b + j(p-1))\}$ with step $p-1$.

Theorem 2. *The coefficients F_{k_i, l_j} at the points of the sublattice are congruent modulo p^r to the values of some polynomial in i and j of degree less than r (with integral coefficients when $r \leq p$) under the condition $a, b \geq r$ (this polynomial depends on a, b, p, and r).*

The *Springer number* of a group generated by reflections is the maximal number of Weyl chambers (the connected domains into which the mirrors divide the Euclidean space) in the connected domain bounded by the extensions of the mirrors forming the walls of one Weyl chamber (see [2] and [4]). The Springer numbers a_n of the

Originally published in: *Izvestiya RAN, Ser. Mat.* **56**(5), 1129–1133 (1992), © 1992 Russ. Acad. Sciences
English translation: *Russ. Acad. Sci. Izvestiya Math.* **41**(2), 389–393 (1993),

groups A_n of symmetries of n-dimensional simplexes (i.e., the Weyl groups of the Lie algebras $\mathrm{su}(n+1)$) form the sequence

$$(a_{-1}, a_0, a_1, \dots) = (1, 1, 1, 2, 5, 16, 61, 272, 1385, \dots).$$

The tangent numbers a_{2m} and the Euler numbers a_{2m-1} occurring in this sequence differ from our numbers f and g only in their signs:

$$a_{2m} = (-1)^m f(2m+1), \qquad a_{2m-1} = (-1)^m g(2m).$$

The combinatorial significance of the elements of the Euler-Bernoulli triangle was discussed in [1]–[3]. Theorem 2 was formulated in [1] as a conjecture and was partially proved there (not proved was the fact about the coefficients of the polynomial being integral).

§2. Congruences for the Springer numbers of the series B and D

The series B and D are the Weyl groups of the orthogonal groups.

The corresponding numbers are defined by the exponential generating functions

$$\sum f^B(n)z^n/n! = \sinh z/\cosh 2z, \qquad \sum g^B(n)z^n/n! = \cosh z/\cosh 2z,$$

$$\sum f^{B+D}(n)z^n/n! = 1 + (\sinh 2z/\cosh 2z), \qquad \sum g^{B+D}(n)z^n/n! = 1/\cosh 2z,$$

where $f^{B+D} = f^B + f^D$, $g^{B+D} = g^B + g^D$, and $n \geq 0$.

Example. The first few numbers $f^B, \dots, g^D$ are given by the table

n	0	1	2	3	4	5
f^B	0	1	0	−11	0	361
g^B	1	0	−3	0	57	0
f^D	1	1	0	−5	0	151
g^D	0	0	−1	0	23	0

Theorem 3. *The numbers $f^{\cdot}$ and $g^{\cdot}$ satisfy the following congruences modulo an odd $p > 1$:*

$$f^{\cdot}(n) \equiv \varepsilon_1 1^n + \varepsilon_2 2^n + \dots + \varepsilon_p p^n, \qquad g^{\cdot}(n) \equiv \varepsilon_3 1^n + \varepsilon_4 2^n + \dots + \varepsilon_{p+2} p^n,$$

where the coefficients ε_k depend only on the remainders of the division of k and p by 8, and $\varepsilon_k + \varepsilon_{k+4} = 0$.

The values of $\varepsilon_k = 0, 1, -1$ for $\cdot = B$ and $\cdot = B + D$ are given in the following tables:

	p	$k=$ 1	2	3	4
$\cdot = B$	1	0	+	+	0
	3	−	+	0	0
	5	0	−	−	0
	7	−	−	0	0

	p	$k=$ 1	2	3	4
$\cdot = B + D$	1	+	+	+	+
	3	+	−	+	+
	5	−	+	−	+
	7	−	−	−	+

Example.

$$f^B(n) \equiv 2^n + 3^n - 6^n - 7^n + 10^n + 11^n - \cdots - (p-3)^n - (p-2)^n (p = 8r+1),$$
$$g^B(n) \equiv 1^n - 4^n - 5^n + 8^n + 9^n - \cdots - (p-1)^n + p^n (p = 8r+1),$$
$$g^D(n) \equiv 2^n - 3^n - 6^n + 7^n \pmod 9, \qquad 23 \equiv 2^4 - 3^4 - 6^4 + 7^4 \ (9).$$

§3. Congruences for the elements of triangles B and D

With each series of Coxeter groups B and D there are associated two generalized Euler-Bernoulli triangles: a left and a right (see [2] and [3]). The corresponding exponential generating functions $*F^{\cdot}(x, y) = \sum *F^{\cdot}_{k,l}(x^k/k!)(y^l/l!)$ are given by

$$LF^B = e^{-y}\sinh z/\cosh 2z, \qquad RF^B = e^x \cosh z/\cosh 2z,$$
$$LF^{B+D} = e^{-y}e^{2z}/\cosh 2z, \qquad RF^{B+D} = e^x/\cosh 2z,$$

where $z = x + y$ and $F^{B+D} = F^B + F^D$ $(\cdot = B, D, B+D;\ * = L, R)$.

Theorem 4. *The elements of the generalized Euler-Bernoulli triangles satisfy the following congruences modulo an odd number* $p > 1$:

$$LF^{\cdot}_{k,l} \equiv \varepsilon_1 1^k 0^l + \varepsilon_2 2^k 1^l + \cdots + \varepsilon_p p^k (p-1)^l,$$
$$RF^{\cdot}_{k,l} \equiv \varepsilon_{p+2} 1^k 0^l + \varepsilon_3 2^k 1^l + \cdots + \varepsilon_{p+1} p^k (p-1)^l,$$

where the coefficients ε_i *are indicated in the tables of Theorem* 3 $(\cdot = B, D, B+D$; *here* 0^0 *means* $1)$.

Remark. The congruences for the elements of the triangle $F_{k,l}$ in Theorem 2 are analogous to the congruence for $LF^{\cdot}_{k,l}$, but $\varepsilon_{i+4} = \varepsilon_i$, and for $p = 4s+1$ we have $(\varepsilon_1, \dots, \varepsilon_4) = (++--)$, while for $p = 4s+3$ we have $(\varepsilon_1, \dots, \varepsilon_4) = (-++-)$.

Theorem 5. *The assertion of Theorem 2 about congruences modulo* p^r, *where* p *is an odd prime, is also valid for the elements of each of the generalized triangles* $*F^{\cdot}$.

§4. Proofs of Theorems 1, 3, and 4

The main difficulty is to guess that f and g are congruent to sums of exponentials with certain coefficients. Once one somehow gets a feeling for the coefficients, a concise proof can be given. The following short proof of Theorem 1 was communicated to me by Dov Zagier.

Let us sum $k^n z^n/n!$ over n from 0 to ∞. The first congruence in Theorem 1 follows from the identity

$$e^z + e^{2z} - e^{3z} - e^{4z} + \cdots - e^{(p-1)z} + e^{pz} - (1 - \tanh z)$$
$$= (e^{pz} - 1)e^z(e^z - 1)/(e^{2z} + 1),$$

since the coefficients of $z^n/n!$ in the first parenthesis are divisible by p. The other three congruences in Theorem 1 can be proved analogously.

Theorem 3 can be proved in the same way.

Theorem 4 follows from Theorem 3.

Remark. It is impossible to *think up* these theorems in this way: they were discovered as a result of computer experiments in attempts to understand why the polynomials in Theorem 2, calculated in [1], have integral coefficients. The search for $f(n) \bmod p$

as a sum of exponentials with unknown coefficients $\varepsilon_1, \dots, \varepsilon_p$ leads to an expression of all elements of the Euler-Bernoulli triangle as sums of exponentials. Knowing the null boundary conditions

$$f(2m) = g(2m-1) = 0 \quad \text{for } m > 0,$$

we can find the system of equations

$$\varepsilon_i + \varepsilon_{p-i} = 0 \quad (i = 1, \dots, p-1), \qquad \varepsilon_i = \varepsilon_{p+2-i} \quad (i = 2, 3, \dots, p).$$

The product of the two involutions $i \leftrightarrow p-i$ and $i \leftrightarrow p+2-i \pmod p$ is a rotation. Therefore the p-periodically continued sequence ε has the form $\varepsilon_{2p-2r} = (-1)^r \varepsilon_{2p}$ $(r = 0, 1, \dots, p-1)$, and then $\varepsilon_p = 1$ can be found from the condition $f_0 = 1$.

The proofs of Theorems 3 and 4 can be reduced to an investigation of pairs of involutions of a circle by the same method. In the case $B + D$ we obtain equations for the p-periodically continued sequence of coefficients ε_i of the form

$$\varepsilon_i + \varepsilon_{p-i} = 0 \quad (i = 1, \dots, p-1), \qquad \varepsilon_i = \varepsilon_{p+4-i} \quad \text{for all } i.$$

In case B we obtain the equations

$$\varepsilon_i + \varepsilon_{p-i} = 0 \quad \text{for all } i, \qquad \varepsilon_i = \varepsilon_{p+4-i} \quad \text{for } i = 4, 5, \dots, p.$$

Moreover, the conditions at the vertices of the triangles yield $\varepsilon_3 = 1 + \varepsilon_1$.

Accordingly, we find in the casc $B + D$ that

$$\varepsilon_{4p-4r} = (-1)^r \varepsilon_{4p} \qquad (r = 0, 1, \dots, p-1).$$

In case B, the solution of the system has the form $\varepsilon_{4r} = (-1)^r \delta$, where $\delta = (-1)^{(p^2-9)/8}(\varepsilon_3 - \varepsilon_1)$ when $p < 4r < 3p$, $\delta = 0$ when $4r < p$ or $4r > 3p$ $(r = 1, \dots, p)$.

§5. Proofs of the congruences modulo p^r

Theorem 1 represents the triangle $\{F_{k,l} \pmod{p^r}\}$ as an array of iterated differences of sums of exponentials. Therefore it suffices to prove Theorem 2 for an array of differences of the one exponential $F_{k,0} = s^k$. This array has the form $F_{k,l} = s^k(s-1)^l$. Substituting k_i and l_j from the sublattice, we obtain $F_{k,l} = CU^iV^j$, where $U = s^{p-1}$, $V = (s-1)^{p-1}$; and $C = s^a(s-1)^b$. If s or $s-1$ is divisible by p, then $F_{k,l} \equiv 0 \pmod{p^r}$ when $k, l \geq r$. If, however, neither s nor $s-1$ is divisible by p, then, by Fermat's Little Theorem, $U = 1 + up$ and $V = 1 + vp$; hence, by the binomial formula,

$$\begin{aligned} U^iV^j &= \left(1 + iup + \binom{i}{2}u^2p^2 + \cdots\right)\left(1 + jvp + \binom{j}{2}v^2p^2 + \cdots\right) \\ &= 1 + (iu_jv)p + \left[\binom{i}{2}u^2 + ijuv + \binom{j}{2}v^2\right]p^2 + \cdots. \end{aligned}$$

Modulo p^r this product is congruent to the value of a polynomial in i and j of degree less than r with integral coefficients when $r \leq p$, which proves Theorem 2.

Theorem 5 can be deduced from Theorem 3 in the same way as Theorem 2 was from Theorem 1.

This paper was begun at the Université Paris-Sud (Orsay) and completed at the Max Planck Institute for Mathematics in Bonn. The author thanks the University and the Institute for their support, J.-P. Serre and especially D. Zagier for useful discussions that helped to simplify the proof, and I. V. Arnol'd for conducting the computer experiments.

Bibliography

1. V. I. Arnol′d, *Bernoulli-Euler updown numbers associated with function singularities, their combinatorics and arithmetics*, Duke Math. J. **63** (1991), 537–555.

2. ——, *Springer numbers and Morsification spaces*, J. Algebraic Geom. **1** (1992), 197–214.

3. ——, *The calculus of snakes and the combinatorics of Bernoulli, Euler, and Springer numbers of Coxeter groups*, Uspekhi Mat. Nauk **47** (1992), no. 1 (283), 3–45; English transl. in Russian Math. Surveys **47** (1992).

4. T. A. Springer, *Remarks on a combinatorial problem*, Nieuw Arch. Wisk. (3) **19** (1971), 30–36.

Steklov Mathematical Institute, Russian Academy of Sciences

Received 5/MAY/92

Translated by G. A. Kandall

The calculus of snakes and the combinatorics of Bernoulli, Euler and Springer numbers of Coxeter groups

V.I. Arnol'd

Contents

Introduction

Euler and Bernoulli numbers arise in many parts of classical analysis: from number theory to optimal control theory, from the Euler–Maclaurin asymptotic summation formula to the calculation of the coefficients in the power series expansions of tan and sec, and from Kummer congruences to the classification of Milnor spheres.

It was recently discovered that the same numbers control the topology of bifurcation diagrams and the classification of morsifications of simple

Originally published in: *Uspekhi Mat. Nauk* **47**(1), 3–45 (1992), © 1992 Russ. Acad. Sciences
English translation: *Russ. Math. Surveys* **47**(1), 1–51 (1992), © 1992 Russ. Acad. Sciences and London Math. Soc.

singularities of smooth functions on the one hand, and the combinatorics of the mirrors of reflection groups on the other ([1], [2], [3]).

The present article describes these unexpected connections between combinatorics and singularity theory, providing, in particular, homeomorphisms between complicated real bifurcation diagrams of singularity theory, the so-called caustics and Maxwell strata, and polyhedral cones bounded by mirrors of Euclidean reflection groups, and their subdivisions into Weyl chambers.

The investigation of these topological aspects is based on purely combinatorial and essentially elementary constructions. This combinatorics of permutations of a special kind (the "snakes" connected with reflection groups) is presented below independently of applications and connections with singularity theory or the theory of reflection groups, since the corresponding combinatorial results may have quite different applications. In spite of the natural and elementary nature of the enumerative combinatorial problems considered here, they have apparently not been considered previously, since they occur most naturally in connection with the theory of reflection groups. The combinatorial results obtained in this way are unexpected and non-trivial, and their bijective proofs are not simple.

The simplest example illustrating the combinatorial results under discussion is the following.

Definition. An *A_n-snake* is a permutation of the numbers $\{0, 1, ..., n\}$ for which the system of inequalities $(x_0 < x_1 > x_2 < ... x_n)$ holds. A sequence satisfying these inequalities is also called an *updown*-sequence.

Example. The five A_3-snakes are shown in Fig. 1. Let a_n denote the number of A_n-snakes, so that $a_3 = 5$. The sequence of numbers a_n begins as follows:

n	−1	0	1	2	3	4	5	6	7	...
a_n	1	1	1	2	5	16	61	272	1385	...

(it will become clear from what follows why we need to take $a_{-1} = 1$).

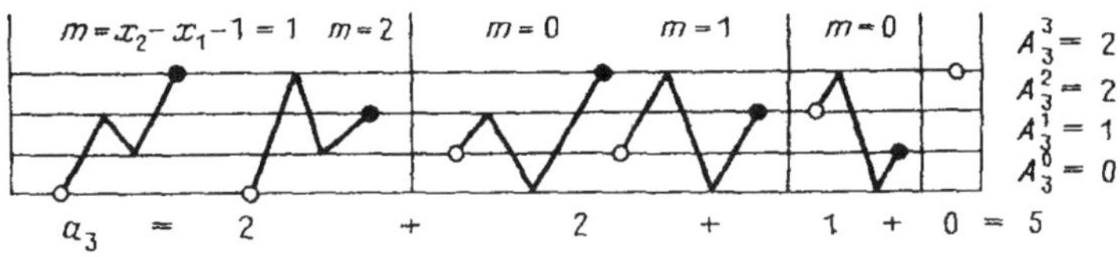

Fig. 1. The classification of A_3-snakes.

A classical result of updown-permutation theory is the following.

Let us form the exponential generating function for the number of A_n-snakes:

$$A(t) = \sum_{n=0}^{\infty} a_{n-1} t^n / n!.$$

***Theorem* 1.** *The exponential generating function for the number of A_n-snakes is*

$$A(t) = \sec t + \tan t. \tag{1}$$

The combinatorial theory presented below is a far-reaching generalization of this formula.

One can see from (1) that the numbers a_n with even n are the coefficients of the Taylor expansion of sec at 0, and those with odd n are the analogous coefficients for tan. The numbers E_i occurring in the expansion of sec are called *Euler numbers*:

i	0	1	2	3	4
E_i	1	1	5	61	1385

The numbers T_i occurring in the expansion of tan are called *tangent numbers*:

i	1	2	3	4	...
T_i	1	2	16	272	...

The tangent numbers are connected with the Bernoulli numbers B_n by the relation

$$T_n = B_n \frac{4^n(4^n - 1)}{2n}.$$

Analogous numbers will be connected below (§1) with an arbitrary finite group generated by reflections in Euclidean space. The Euler and tangent numbers a_n are thereby connected with the symmetry groups of simplexes, that is, with the Weyl groups of the simple Lie algebras of the SU$(n+1)$. Other series of reflection groups (the Weyl groups of the Lie algebras of the orthogonal and symplectic groups B_n and D_n) give rise to other sequences of numbers, which we call Springer numbers b_n, d_n:

n	0	1	2	3	4	5	6	...
b_n	1	1	3	11	57	361	2763	...
d_n	1	1	1	5	23	151	1141	...

(see the definition in §1).

The corresponding exponential generating functions were found by Springer in [1] (see §6 below).

To calculate the Euler and Bernoulli numbers quickly, it is convenient to use the classical Euler – Bernoulli triangle, similar to the Pascal triangle:

Euler side   Bernoulli side

1
0 1
1 1 0
0 1 2 2
5 5 4 2 0
0 5 10 14 16 16
61 61 56 46 32 16 0
. . . .

The sequence a_n consists of the numbers on the sides of the triangle.

An algorithm for calculating the Euler–Bernoulli triangle is as follows: each number in the triangle is equal to the sum of the numbers in the preceding row situated to the left of the required number if the number of terms in the row containing the required number is even, and to the right if that number is odd.

It is clear that the rows have zero at the beginning and end alternately. Subsequent terms are easily calculated by shuttling to the right in the rows of even length and to the left in the rows of odd length (such a movement always starts from zero):

$0 + 2 = 2,\ 2 + 2 = 4,\ 4 + 1 = 5,\ 5 + 0 = 5$
$0 + 5 = 5,\ 5 + 5 = 10,\ 10 + 4 = 14,\ 14 + 2 = 16,\ 16 + 0 = 16$

and so on.

Each number occurring in the Euler–Bernoulli triangle has its own combinatorial meaning. The row consisting of $n+1$ elements corresponds to a classification of A_n-snakes according to the last element.

Example. In Fig. 1 two A_3-snakes end with the number 3, two end with 2, and one ends with 1, and $a_3 = 2+2+1+0 = 5$.

The Springer numbers b_n and d_n are also connected with Euler–Bernoulli-like triangles, and more precisely with pairs of triangles. These pairs of triangles have the following form:

$L(b)$	$R(b)$
0	1
1 1	1 0
0 1 2	2 3 3
11 11 10 8	8 6 3 0
0 11 22 32 40	40 48 54 57 57
. . . . .	. . . . .

$L(d)$	$R(d)$
0	1
0 0	0 1
1 1 1	1 1 0
0 1 2 3	3 4 5 5
23 23 22 20 17	17 14 10 5 0
. . . . . .	. . . . . .

The algorithm for calculating pairs of triangles is the same as in the Euler–Bernoulli triangle: each number is equal to the sum of all the numbers in the preceding horizontal row lying to the left or to the right (depending on whether the number of terms in the row containing the required number is even or odd).

Thus the numbers on the sides of the triangles facing one another are the same in each row. The algorithm fills both the triangles $L(b)$ and $R(b)$ simultaneously:

$0 + 3 = 3,\ 3 + 3 = 6, 6 + 2 = 8,\ 8 + 2 = 10,\ 10 + 1 = 11,\ 11 + 0 = 11.$

The triangles $L(d)$ and $R(d)$ are filled in the same way, but the direction of movement in the even and odd rows is reversed:

$$0+1=1,\ 1+1=2,\ 2+1=3,\ 3+1=4,\ 4+1=5,\ 5+0=5.$$

The calculus of snakes described below is the combinatorics of the different objects enumerated by these triangles and their combinations.

As an example we present here the following result.

Theorem 2. *The number of sequences of integers satisfying the conditions*

$$(x_1 < x_2 > x_3 < \dots x_n),\ 0 < |x_i| \leqslant n,\ |x_i| \neq |x_j|$$

and beginning with a fixed negative number $x_1 = -m$ *is equal to the sum of the numbers of sequences of integers satisfying the conditions*

$$(y_1 < y_2 > y_3 < \dots y_n),\ 0 < |y_i| \leqslant n,\ |y_i| \neq |y_j|,$$

and beginning with $y_1 = k$ *or* $y_1 = -k$, *where* $k = n+1-m$, *for which* $(-1)^n y_n > 0$.

Example. The snakes (x_i) and (y_i) corresponding to $n = 3$ are shown in Fig. 2. Their total number is equal to $11 = 3+4+4$, in accordance with $m = 1, 2, 3$. The reader may attempt to find a proof of Theorem 2 by establishing a natural one-to-one correspondence between the snakes (x_i) and (y_i). This correspondence exists (see §8), but it is not easy to find!

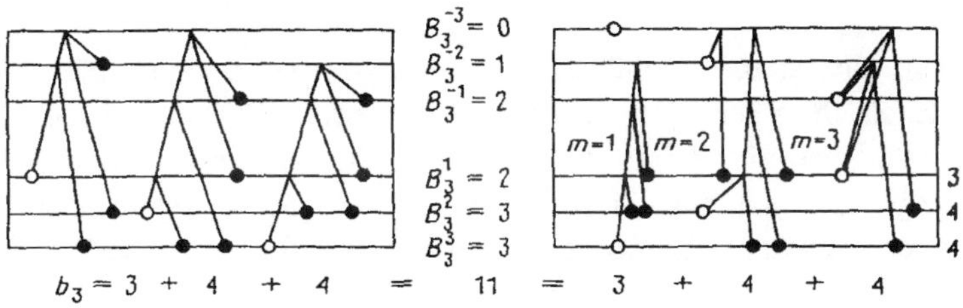

Fig. 2. On Theorem 2.

The numbers described in Theorem 2 form another remarkable triangle:

```
                1
           2         1
      4         4         4
 16        16        14        11
80    80        76        68        57
           . . . . . . .
```

It can be obtained from the triangles $L(b)$ and $R(b)$ by adding their opposite elements:

$$11+0=11,\ 11+3=14,\ 10+6=16,\ 8+8=16.$$

The same numbers can be obtained by adding the triangles $R(d)$ and $L(b)$:

$$3+8=11,\ 4+10=14,\ 5+11=16,\ 5+11=16$$

or $L(d)$ and $R(b)$:

$$23+57=80,\ 22+54=76,\ 20+48=68,\ 17+40=57.$$

All these relations are non-trivial theorems of enumerative combinatorics, similar to Theorem 2. To formulate and prove them, we need first of all to give combinatorial (not recursive) definitions of snake triangles. A motivation for these definitions is given by the connection with singularity theory (where snakes are realized as graphs of functions) or with the theory of Coxeter groups (where they represent Weyl chambers). Therefore we preface the combinatorial definitions of our triangles by their description in terms of Coxeter groups.

CHAPTER I

THE ENUMERATIVE COMBINATORICS OF SNAKES

§1. Combinatorics of Springer cones

Let us consider a finite group generated by reflections in Euclidean space of dimension n. Hyperplanes, reflections in which are included in the group, are called *mirrors*.

The mirrors divide the Euclidean space into simplicial cones called Weyl chambers. Each chamber is bounded by n mirrors. These n mirrors divide the space into 2^n orthants (simplicial cones). We shall call them *Springer cones* (although Springer himself stated that he did not consider them). Two of the Springer cones are chambers (the initial and the antipodal). The rest are divided by mirrors into chambers.

Definiton. The principal Springer cone is that consisting of the largest number of Weyl chambers. This number is called the *Springer number* of the given reflection group.

Example 1. The minimal angle between axes of symmetry of a regular p-gon is equal to π/p. Thus the Springer numbers of the symmetry groups $I_2(p)$ of the regular p-gons are

$$i_2(p)=p-1,\ a_2=2,\ b_2=3,\ g_2=5$$

(including $A_2 = I_2(3)$, $B_2 = I_2(4)$, $G_2 = I_2(6)$).

Example 2. Two of the three Weyl chamber walls are orthogonal in $\mathbb{R}^3$. They divide the space into 4 congruent parts, each of which consists of $|W|/4$ chambers, where $|W|$ is the total number of chambers (that is, the order of the group). Therefore the Springer number of each reflection group in the three-dimensional space is equal to

$$s_3 = |W|/4 - 1.$$

In particular,

$$a_3 = 4!/4 - 1 = 5, \ b_3 = 8 \cdot 3!/4 - 1 = 11, \ h_3 = 120/4 - 1 = 29.$$

It is convenient to show configurations of Springer cones and mirrors in a projective plane (Fig. 3).

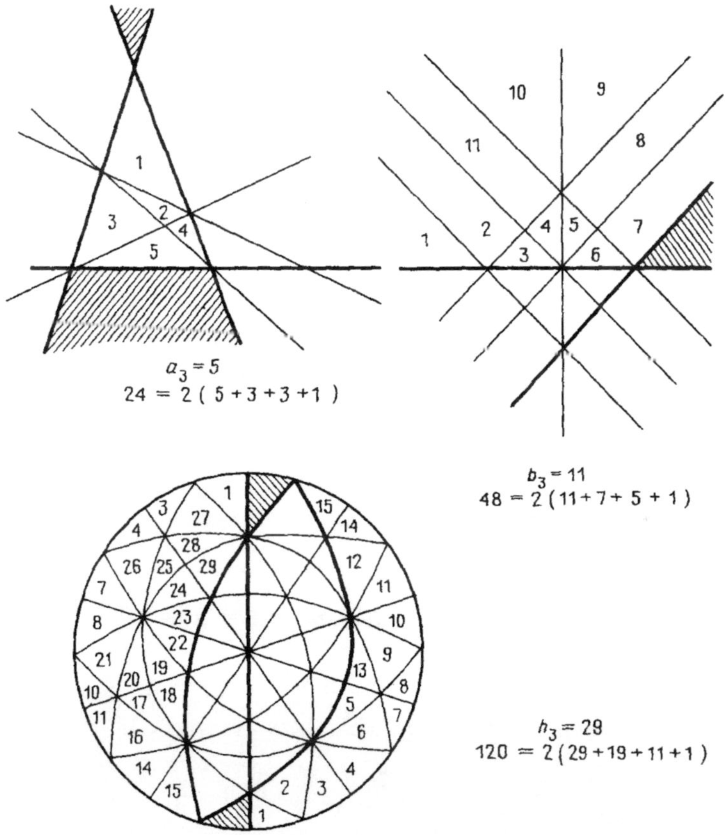

Fig. 3. The Springer cones of the groups A_3, B_3 and H_3.

Each small triangle shows two opposite chambers. Springer cones are shown by large triangles bounded by bold lines (one of which in the case b_3 is infinitely distant). The initial chamber is shaded.

Remark. The numbers of chambers in Springer cones of the symmetry group of the icosahedron H_3

$$(1, 11, 19, 29)$$

are related in a strange way to the degrees of invariants of the symmetry group of the hypericosahedron H_4 (these are equal to (2, 12, 20, 30)). In the same way the numbers of chambers for A_3 (1, 3, 3, 5) generate the degrees (2, 4, 4, 6) of the invariants for D_4, and the numbers of chambers for B_3 (1, 5, 7, 11) generate the degrees of invariants of the reflection group F_4 (2, 6, 8, 12).

Example 3. The area (and hence the number of chambers) of a spherical triangle is expressed by a well-known formula in terms of the angles between corresponding hyperplanes. This formula is obtained by the inclusion-exclusion method from a formula for the area of a lune (proportional to the angle). Repeating this argument, it is not difficult to calculate the numbers of chambers in Springer cones for higher dimensions. For example, for H_4 one gets

$$|W| = 14400 = 2\,(2281 + 1681 + 1199 + 719 + 601 + 599 + 119 + 1),$$

hence $h_4 = 2281$.

Springer calculated his numbers in this way. For the exceptional groups he obtained

$$e_6 = 2663,\ e_7 = 94121,\ e_8 = 14422081,\ f_4 = 169.$$

For the series A_n, B_n, D_n Springer explicitly indicated the following main cones.

Let us consider in Euclidean space $\mathbb{R}^n$ the reflection groups A_n, B_n, D_n with standard mirrors and a standard Weyl chamber given by the usual Dynkin diagram (see [4]).

***Theorem* 3.** *The principal Springer cone of a reflection group A_n (respectively B_n, D_n) is given by the inequalities*

for $A_n : (x_0 < x_1 > x_2 < \dots x_n)$
(where $\mathbb{R}^n = \{x : x_0 + \dots + x_n = 0\}$, mirrors: $x_i = x_j$);
for $B_n : (0 < x_1 > x_2 < \dots x_n)$ (mirrors: $x_i = 0$, $x_i = \pm x_j$);
for $D_n : (-x_2 < x_1 < x_2 > x_3 < \dots x_n)$ (mirrors: $x_i = \pm x_j$).

Remark. The corresponding initial Weyl chambers are given by the following inequalities:

for $A_n : (x_0 < x_1 < x_2 < \dots < x_n)$,
for $B_n : (0 < x_1 < x_2 < \dots < x_n)$,
for $D_n : (-x_2 < x_1 < x_2 < x_3 < \dots < x_n)$.

§2. Numbers and triangles of snakes

Definition. A snake of type A_n (respectively B_n, D_n) is a sequence of integers x_i satisfying the conditions:

for $A_n : x_0 < x_1 > x_2 < \dots x_n,\ x_i \neq x_j$;
for $B_n : 0 < x_1 > x_2 < \dots x_n,\ x_i \neq \pm x_j$;
for $D_n : -x_2 < x_1 < x_2 > x_3 < \dots x_n,\ x_i \neq \pm x_j$;
where $0 \leqslant x_i \leqslant n$ for A_n, $1 \leqslant |x_i| \leqslant n$ for B_n, $0 \leqslant |x_i| < n$ for D_n.

Example. The number of A_3-snakes is equal to 5 (Fig. 1). The number of B_3-snakes is equal to 11 (Fig. 2). The number of D_4-snakes is equal to 23 (Fig. 4).

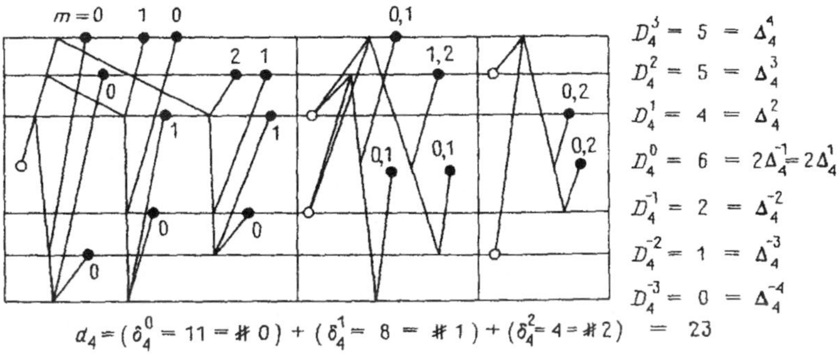

Fig. 4. The classification of D_4-snakes.

Theorem 3 implies the following result.

Corollary. *The numbers a_n, b_n, d_n of Weyl chambers in the principal Springer cones of the Coxeter groups A_n, B_n, D_n are equal respectively to the numbers of all possible snakes of corresponding types.*

Remark. The corollary can be deduced from Theorem 3 in the following way: B_n- and D_n-snakes lie in the corrresponding Springer cones. Each Weyl chamber in a Springer cone contains one and only one snake (as is easy to check). In the case A_n the snake gets into the chamber after shifting all x_i to their arithmetic mean. For shifted A_n-snakes the same argument is applied as in the B and D cases; there is exactly one shifted snake in each chamber of the Springer cone.

To obtain Pascal-like triangles, we consider a classification of snakes from their tails.

Definition. We denote by A_n^m (respectively, B_n^m, D_n^m) the number of A_n-snakes (respectively, B_n-snakes, D_n-snakes) ending with the number $x_n = m$.

Example. The numbers A_3^m, A_4^m, B_3^m, B_4^m, D_3^m, D_4^m are given by the following tables (partially following from Figs. 1, 2, 4):

m	0	1	2	3
A_3^m	0	1	2	2

m	0	1	2	3	4
A_4^m	5	5	4	2	0

m	−3	−2	−1	1	2	3
B_3^m	0	1	2	2	3	3

m	−4	−3	−2	−1	1	2	3	4
B_4^m	11	11	10	8	8	6	3	0

m	−2	−1	0	1	2
D_3^m	1	1	2	1	0

m	−3	−2	−1	0	1	2	3
D_4^m	0	1	2	6	4	5	5

Let us form Pascal-like triangles from these numbers, writing out the number $D_n^0/2$ twice instead of D_n^0.

Definition. Pascal-like triangles of snakes

$$\begin{array}{ccccc} & & a_{0,0} & & \\ & a_{1,0} & & a_{0,1} & \\ a_{2,0} & & a_{1,1} & & a_{0,2} \\ & & \cdots & & \end{array} \qquad \begin{array}{ccccc} & & L_{0,0} & & \\ & L_{1,0} & & L_{0,1} & \\ L_{2,0} & & L_{1,1} & & L_{0,2} \\ & & \cdots & & \end{array} \qquad \begin{array}{ccccc} & & R_{0,0} & & \\ & R_{1,0} & & R_{0,1} & \\ R_{2,0} & & R_{1,1} & & R_{0,2} \\ & & \cdots & & \end{array}$$

are defined by the relations

$$a_{i,j} = A_{i+j}^j,$$
$$L_{i,j}(b) = B_{i+j+1}^{-i-1}, \quad R_{i,j}(b) = B_{i+j+1}^{j+1},$$
$$L_{i,j}(d) = d_{i+j+1}^{-i}, \quad R_{i,j}(d) = d_{i+j+1}^{j}.$$

Here $d_n^m = D_n^m$ if $m \neq 0$, $d_n^0 = D_n^0/2$ for $n > 1$. In addition we put formally $L_{0,0}(d) = 0$, $R_{0,0}(d) = 1$.

Example. The initial parts of all these triangles of snakes are given in the introduction. For example, the initial part of a triangle $L(b)$ is

$$L(b) = \begin{array}{ccccc} & & L_{0,0}(b) & & \\ & L_{1,0}(b) & & L_{0,1}(b) & \\ L_{2,0}(b) & & L_{1,1}(b) & & L_{0,2}(b) \\ & & \cdots & & \end{array} = \begin{array}{ccccc} & & B_1^{-1} & & \\ & B_2^2 & & B_2^1 & \\ B_3^{-3} & & B_3^{-2} & & B_3^{-1} \\ & & \cdots & & \end{array} = \begin{array}{ccccc} & & 0 & & \\ & 1 & & 1 & \\ 0 & & 1 & & 2 \\ & & \cdots & & \end{array}$$

Besides b- and d-triangles, corresponding to a classification of snakes according to their tails, we shall define two more triangles, corresponding to a classification of snakes according to their heads.

Definition. A β_n*-snake* is a sequence of integers x_i satisfying the conditions

$$(x_1 < x_2 > x_3 \ldots x_n), \quad x_i \neq \pm x_j,$$

where $1 \leqslant |x_i| \leqslant n$.

The number of β_n-snakes beginning with $x_1 = m$ is denoted by β_n^m.

Example. The numbers β_2^m, β_3^m (Fig. 5) are given by the tables

m	-2	-1	1	2
β_2^m	1	1	1	0

m	-3	-2	-1	1	2	3
β_3^m	4	4	3	3	2	0

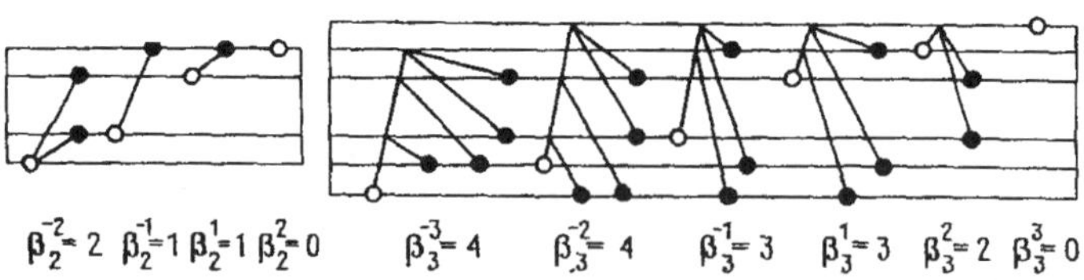

Fig. 5. The classification of β_2- and β_3-snakes.

From the rows β_n^m we form, as usual, two Pascal-like triangles, orienting the rows alternately by zero to the left and zero to the right.

$$L(\beta) = \begin{array}{ccccccc} & & & 1 & & & \\ & & 2 & & 1 & & \\ & 0 & & 2 & & 3 & \\ 16 & & 16 & & 14 & & 11 \\ \cdots & & \cdots & & \cdots & & \cdots \end{array} \qquad \begin{array}{ccccccc} & & & 1 & & & \\ & & 1 & & 0 & & \\ & 3 & & 4 & & 4 & \\ 11 & & 8 & & 4 & & 0 \\ \cdots & & \cdots & & \cdots & & \cdots \end{array} = R(\beta)$$

In other words we introduce the following definition.

Definition. The *triangles of snakes* $L(\beta)$, $R(\beta)$ are defined by the conditions

$$L_{i,j}(\beta) = \begin{cases} \beta_{i+j+1}^{i+1} & \text{if } i+j \text{ is even,} \\ \beta_{i+j+1}^{-i-1} & \text{if } i+j \text{ is odd;} \end{cases}$$

$$R_{i,j}(\beta) = \begin{cases} \beta_{i+j+1}^{-j-1} & \text{if } i+j \text{ is even } >0, \\ \beta_{i+j+1}^{j+1} & \text{if } i+j \text{ is odd;} \end{cases}$$

$$R_{0,0}(\beta) = 1 \text{ by definition.}$$

§3. Combinatorial identities for triangles of snakes

The triangles of snakes defined in §2 possess the following properties.

Theorem 4. *Each array of a, b, d, β is calculated by the algorithm described in the introduction.*

In other words, the following identities hold:

$$X_{i,j} = X_{i+1,j-1} + X_{i,j-1},$$

or

$$X_{i,j} = X_{i-1,j+1} + X_{i-1,j}.$$

For $X = a$, $L(d)$ and $R(d)$

the first case holds for even $i+j$, and the second one for odd $i+j$.

For $X = L(b)$, $R(b)$, $L(\beta)$ and $R(\beta)$

the first case holds for odd $i+j$, and the second one for even $i+j$.

In addition, the following boundary conditions are satisfied:

case A:

$$a_{2k-1,0} = 0,\ a_{0,2k} = 0 \ \text{for} \ k \geqslant 1,\ a_{0,0} = 1;$$

case B:

$$L_{2k,0}(b) = 0,\ R_{0,2k-1}(b) = 0,$$

$$L_{0,j}(b) = R_{j,0}(b) \ \text{for} \ j > 1,\ L_{0,0}(b) = 0,\ R_{0,0}(b) = 1;$$

case D:

$$L_{2k-1,0}(d) = 0,\ R_{0,2k}(d) = 0 \ \textit{for}\ \ k > 1,$$

$$L_{0,j}(d) = R_{j,0}(d) \ \textit{for}\ \ j > 1,\ L_{0,0}(d) = 0,\ R_{0,0}(d) = 1;$$

case β:

$$L_{0,2k-1}(\beta) = 0,\ L_{2k,0}(\beta) = 0 \ \textit{for}\ \ k > 0,$$

$$L_{0,j}(\beta) = R_{j,0}(\beta),\ L_{0,0}(\beta) = R_{0,0}(\beta) = 1.$$

There are several relations between the triangles of snakes b, d, and β. Every such relation is a non-trivial combinatorial theorem similar to Theorem 2 presented in the introduction, and has a direct bijective proof. But since there are many of these relations, I formulate them formally by means of the notations previously introduced and those that follow.

Definition. A *bar* above a symbol of a triangle of snakes denotes the triangle reflected in a bisector, that is, $\overline{X}_{i,j} = X_{j,i}$.

The operation ε changes the signs of those sides of the triangle with even length: $(\varepsilon X)_{i,j} = (-1)^{i+j}X_{i,j}$.

Theorem 5. *The triangles of snakes introduced above satisfy the following relations*:

$$L(b) + \varepsilon\overline{R}(b) = R(\beta),\ R(b) - \varepsilon\overline{L}(b) = L(\beta),$$

$$\overline{L}(d) + R(b) = R(\beta),\ \overline{R}(d) + L(b) = L(\beta).$$

The relations of Theorem 5 were found experimentally and seem astonishing.

Example. The third relation in extended form is

$$\begin{array}{ccccccccc} &&&&0&&&& \\ &&&0&&0&&& \\ &&1&&1&&1&& \\ &3&&2&&1&&0& \\ 17&&20&&22&&23&&23 \\ &&&&\cdots&&&& \end{array} \quad + \quad \begin{array}{ccccccccc} &&&&1&&&& \\ &&&1&&0&&& \\ &&2&&3&&3&& \\ &8&&6&&3&&0& \\ 40&&48&&54&&57&&57 \\ &&&&\cdots&&&& \end{array} \quad = \quad \begin{array}{ccccccccc} &&&&1&&&& \\ &&&1&&0&&& \\ &&3&&4&&4&& \\ &11&&8&&4&&0& \\ 57&&68&&76&&80&&80 \\ &&&&\cdots&&&& \end{array}$$

Here, for example, the equality $3+8 = 11$ means the following quite astonishing combinatorial statement.

The number β_n^{-1} ($= 11$) of quadruples of integers satisfying the conditions

$$(x_1 < x_2 > x_3 < x_4),\ x_i \neq \pm x_j,\ 1 \leqslant |x_i| \leqslant 4,\ x_1 = -1,$$

is equal to the sum of the number B_4^{-1} ($= 8$) of quadruples of integers satisfying the conditions

$$(0 < y_1 > y_2 < y_3 > y_4),\ y_i \neq \pm y_j,\ 1 \leqslant |y_i| \leqslant 4,\ y_4 = -1,$$

and half the number D_4^0 (= 6) of quadruples of integers satisfying the conditions

$$(-z_2 < z_1 < z_2 > z_3 < z_4),\ z_i \neq \pm z_j,\ 0 \leqslant |z_i| < 4,\ z_4 = 0.$$

The corresponding snakes are shown in Fig. 6.

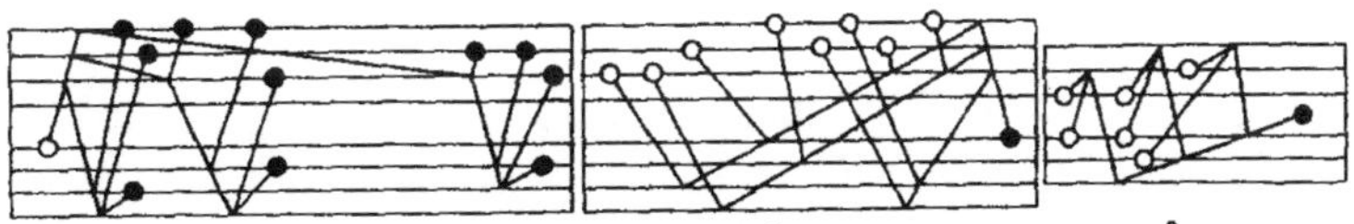

Fig. 6. On Theorem 5.

The first relations of Theorem 5 contain Theorem 2. Theorem 5 contains in a concentrated form many other astonishing statements in enumerative combinatorics analogous to Theorem 2 which, however, do not reduce to Theorem 2 and are proved in another way, as for example the far from obvious earlier equality $\beta_4^{-1} = B_4^{-1} + D_4^0/2$ illustrated above.

There follow several combinatorial theorems, also involving numbers of triangles of snakes.

§4. Combinatorial identities for numbers of *A*-snakes

***Theorem* 6** (see [2]). *The number of integer sequences*

$$(x_0 < x_1 > \ldots x_n),\ 0 \leqslant x_i \leqslant n,\ x_i \neq x_j,$$

such that $x_1 - x_0 = m+1$ *is equal to the number* A_n^m *of integer sequences*

$$(y_0 < y_1 > \ldots y_n),\ 0 \leqslant y_i \leqslant n,\ y_i \neq y_j,$$

ending with $y_n = m$ *if* n *is even, or to the number* A_n^{n-m} *of sequences ending with* $y_n = n-m$ *if* n *is odd.*

Example. It is seen from Figs. 1 and 7 that the number of A_n-snakes for which the interval strictly between x_1 and x_0 contains $m = 0, 1, \ldots$ integers is equal to

2, 2, 1, 0 for $n = 3$,
5, 5, 4, 2, 0 for $n = 4$.

Definition. The *variation* of a sequence is the sum of the moduli of the differences of its consecutive terms:

$$V(x) = \Sigma\ |x_i - x_{i+1}|.$$

***Theorem* 7** (see [2]). *The sum of the variations of all the* a_{n-1} *snakes*

$$(x_0 < x_1 > x_2 < \ldots x_{n-1}),\ \{x_1, \ldots, x_n\} = \{1, \ldots, n\},$$

with boundary conditions

$$x_0 = 0, \; x_{n+1} = 0 \quad \textit{if } n \textit{ is odd},$$
$$x_0 = 0, \; x_{n+1} = n + 1 \quad \textit{if } n \textit{ is even},$$

is equal to the number a_{n+1} *of all* A_{n+1}*-snakes*

$$(y_0 < y_1 > y_2 < \ldots y_{n+1}), \; \{y_0, \ldots, y_{n+1}\} = \{0, \ldots, n+1\}.$$

Example. For $n = 2$ the only snake x is $0 < 2 > 1 < 3$. Its variation is $5 = a_3$. For $n = 3$ there are two snakes x, namely $0 < 2 > 1 < 3 > 0$ and $0 < 3 > 1 < 2 > 0$ (Fig. 8). The variation of each one is equal to 8. The sum of the variations is $16 = a_4$.

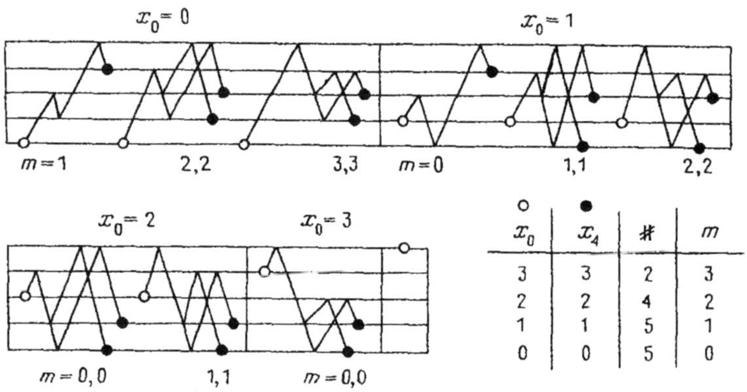

Fig. 7. The classification of A_4-snakes.

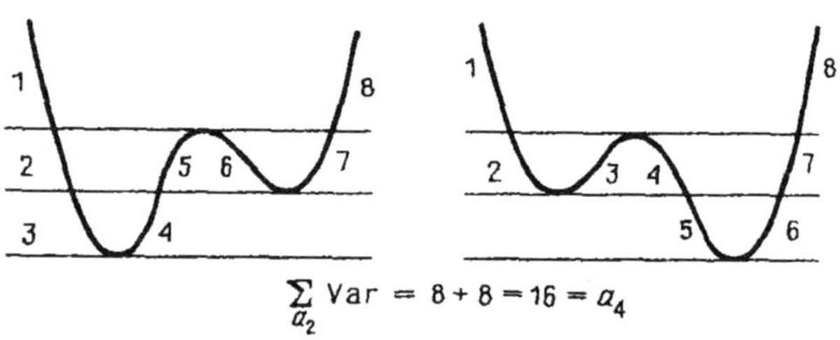

Fig. 8. The sum of variations of A_2-snakes.

***Theorem* 8** (see [2]). *The number of* A_n*-snakes*

$$(x_0 < x_1 > x_2 < \ldots x_n), \; \{x_0, \ldots, x_n\} = \{0, \ldots, n\},$$

is equal to the number of integer sequences

$$(y_0 < y_1 > y_2 < \ldots y_n), \; y_s \equiv s \bmod 2, \; |y_s| \leqslant s.$$

Here the number of snakes ending with $x_n = m$ is equal to the number of sequences for which $y_n = 2m - n$.

Example. For $n = 2$ the snakes are $(0 < 2 > 1)$ and $(1 < 2 > 0)$, and the sequences are $(0, 1, 0)$ and $(0, 1, -2)$. For $n = 3$, $m = 3$ the snakes are $(1 < 2 > 0 < 3)$, $(0 < 2 > 1 < 3)$, and the sequences are $(0, 1, 0, 3)$ and $(0, 1, -2, 3)$.

Theorem 9. *The number of permutations $(x_0, ..., x_n)$ of numbers $\{0, ..., n\}$ such that in the sequence of signs $x_{i+1}-x_i$ there is no more than one pair of similar signs one after the other is equal to the number a_{n+1} of snakes $(y_0 < y_1 > ... y_{n+1})$, where $\{y_0, y_1, ..., y_{n+1}\} = \{0, ..., n+1\}$.*

In other words, $a_{n+1} = a_1 + ... + a_n$, where the dots denote the numbers of Weyl chambers in the gallery of Springer cones of the group A_n, obtained from the principal Springer cone if one crosses in succession the first, the second, and so on, walls until the cone opposite to the initial main cone is reached.

Example. $16 = 5+3+3+5$.

Extreme 5-tuples are the Weyl chamber numbers in the principal Springer cones of the group A_3 (in Fig. 3 to them there correspond five small triangles in one large one, bounded by a bold line). The triples in the middle are the numbers of small triangles in two other triangles, bounded by bold lines (not shaded). The whole gallery consists of four Springer cones.

§5. Combinatorial identities for numbers of D-snakes and B-snakes

Theorem 10. *The number of D_n-snakes, that is, integer sequences $(-x_2 < x_1 < x_2 > x_3 < ... x_n)$, $x_i \neq \pm x_j$, $|x_i| < n$, for which $|x_1| = m$, is equal to the number of D_n-snakes for which there are exactly m terms of the sequence in the interval between x_1 and x_2.*

Example. D_n-snakes with $|x_1| = 0$, 1 and 2 are shown in Fig. 4. There are 11, 8 and 4 of them. The number m of terms between x_1 and x_2 is written at the end of each snake. The numbers $m = 0$, 1 and 2 occur 11 times, 8 times and 4 times. We denote the number of D_n-snakes with $|x_1| = m$ by δ_n^m. For $n = 4$ the numbers δ_n^m are equal to

m	0	1	2	3
δ_4^m	11	8	4	0

For $n = 3$ we get in the same way

m	0	1	2
δ_3^m	3	2	0

Theorem 11. *The number δ_n^m of integer sequences*

$$(-x_2 < x_1 < x_2 > x_3 \dots x_n),\ x_i \neq \pm x_j,\ |x_i| < n,$$

for which $|x_1| = m$, *is equal to the number of integer sequences*

$$(y_1 < y_2 > y_3 \ldots y_n),\ y_i \neq \pm y_j,\ 1 \leqslant |y_i| \leqslant n,$$

with $y_1 = m+1$. *In the notation introduced above*: $\delta_n^m = \beta_n^{m+1}$.

Example. $\delta_4^2 = 4$, as is seen from Fig. 4. $\beta_4^3 = 4$, as is seen from Fig. 9.

Let us now consider the sides of triangles $L(b)$ and $R(b)$ facing each other.

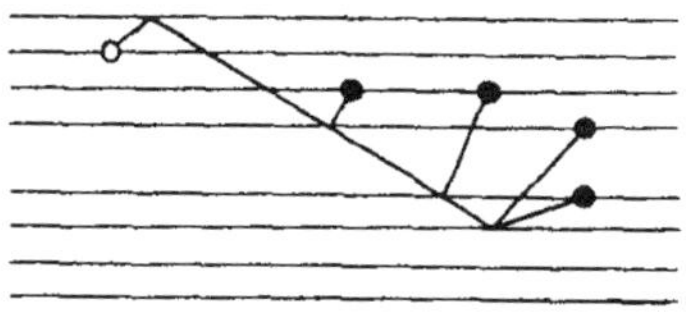

$\beta_4^3 = 4$

Fig. 9. The classification of β_4^3-snakes.

***Theorem* 12.** *The number* $B_n^1 = B_n^{-1}$ *of integer sequences*

$$(0 < x_1 > x_2 < \ldots x_n),\ x_i \neq \pm x_j,\ 1 \leqslant |x_i| \leqslant n,\ n > 1,$$

ending with the number $x_n = 1$ (*or* $x_n = -1$) *is divisible by* 2^{n-2}, *and the quotient is equal to the number* a_{n-2} *of permutations*

$$(y_0 < y_1 > \ldots y_{n-2}),\ \{y_0, \ldots, y_{n-2}\} = \{0, \ldots, n-2\}.$$

Example. $1 = 2^0 \cdot 1$, $2 = 2^1 \cdot 1$, $8 = 2^2 \cdot 2$, $40 = 2^3 \cdot 5$; $B_6^1 = B_6^{-1} = 256$.

§6. Exponential generating functions of triangles of snakes

Definition. The exponential generating function of the triangle $C_{i,j}$ is the function

$$f(x, y) = \sum_{i \geqslant 0,\, j \geqslant 0} c_{i,j} \frac{x^i}{i!} \frac{y^j}{j!},$$

where $c_{i,j} = C_{i,j}\varepsilon(i+j)$, and $\varepsilon(k) = (-1)^{k(k-1)/2}$.

In other words, we change the signs of those horizontal rows such that the number of their elements is congruent to 2 or 3 modulo 4, and then form the usual exponential generating function.

The multiplication by ε, proposed by Deligne and Gessel, is extremely useful, because after this the difference in the behaviour of the algorithm for calculating our triangles in even and odd rows vanishes.

***Theorem* 13.** *The exponential generating function of a triangle* $a_{i,j}$ *is*

$$A(x, y) = \frac{e^y}{\cosh(x+y)}.$$

Proof (see the details in [2]). Theorem 4 in terms of the generating function transforms into the partial differential equation

$$A_y - A_x = A$$

with boundary conditions

$$A\,(x,\,0) = \operatorname{sech} x,\; A\,(0,\,y) = 1 + \tanh y$$

following from Theorem 1. This boundary-value problem is easy to solve by the method of characteristics.

Remark. Theorem 1 itself follows from the fact that the function $A(t)$ satisfies the ordinary differential equation $2A' = 1+A^2$. This equation is equivalent to the identity

$$2a_n = \sum_{i=0}^{n} \binom{n}{i} a_{i-1}a_{n-i-1} \qquad (n > 0),$$

which has a simple combinatorial proof (see [2]). Namely, let us delete from the snake $(x_0 < x_1 > \dots x_n)$ either the largest term x_i, or the smallest one. In each case the snake breaks into two shorter snakes $(x_0 < x_1 > \dots x_{i-1})$ and $(x_{i+1} < x_{i+2} > \dots x_n)$ (or $(x_{i+1} > x_{i+2} < \dots x_n)$). The binomial coefficient takes into account the variants of mutual disposition of the sets $\{x_0, \dots, x_{i-1}\}$, $(x_{i+1}, \dots, x_n\}$, and the number two takes into account the difference between maximum and minimum.

***Theorem* 14.** *The exponential generating functions of triangles $L(b)$, ..., $R(\beta)$ are given by the formulae*

$$b: \quad L\,(x,\,y) = \frac{\sinh z}{\cosh 2z} e^{-y}, \qquad R\,(x,\,y) = \frac{\cosh z}{\cosh 2z} e^{x};$$

$$d: \quad L\,(x,\,y) = \frac{1 - \cosh z}{\cosh 2z} e^{y}, \quad R\,(x,\,y) = \frac{e^{2z} - \sinh z}{\cosh 2z} e^{-x};$$

$$\beta: \quad L\,(x,\,y) = \frac{e^{x+z}}{\cosh 2z}, \qquad R\,(x,\,y) = \frac{e^{x}}{\cosh 2z};$$

where $z = x+y$.

Proof. Theorem 4 gives a partial differential equation

$$b,\, \beta\colon f_x - f_y = f, \quad d\colon\; f_y - f_x = f.$$

The boundary conditions in case b are given by Theorem 12 (see the details in [3], where Theorem 14 is proved for case b).

The cases d and β are reduced to case b by means of Theorem 5. Indeed, we find from it that

$$\bar{L}\,(d) + R\,(b) = L\,(b) + \varepsilon\bar{R}\,(b),$$
$$\bar{R}\,(d) + L\,(b) - R\,(b) \quad c\bar{L}\,(b)$$

and hence we get expressions for the generating functions for the arrays $L(d)$ and $R(d)$ in terms of $L(b)$ and $R(b)$.

Remark. The even and odd parts

$$f_{ev}(x, y) = \frac{1}{2}(f(x, y) + f(-x, -y)),$$
$$f_{odd}(x, y) = \frac{1}{2}(f(x, y) - f(-x, -y))$$

of our generating functions are:

$$L_{ev}(b) = -\frac{\sinh y \sinh z}{\cosh 2z}, \qquad L_{odd}(b) = \frac{\cosh y \sinh z}{\cosh 2z},$$

$$R_{ev}(b) = \frac{\cosh x \cosh z}{\cosh 2z}, \qquad R_{odd}(b) = \frac{\sinh x \cosh z}{\cosh 2z},$$

$$L_{ev}(d) = \frac{1 - \cosh z}{\cosh 2z}\cosh y, \qquad L_{odd}(d) = \frac{1 - \cosh z}{\cosh 2z}\sinh y,$$

$$R_{ev}(d) = \frac{\cosh(y + z) + \sinh x \sinh z}{\cosh 2z}, \quad R_{odd}(d) = \frac{\sinh(y + z) - \cosh x \sinh z}{\cosh 2z}.$$

Therefore, the identities in Theorem 5 are addition theorems for hyperbolic functions. In other words, the combinatorial identities in Theorems 5 – 12 can be deduced from the explicit formulae in Theorems 13 and 14 for exponential generating functions. The explicit formulae of Theorem 14 can be deduced from the recurrence relation of Theorem 4, using the boundary conditions $L(x, 0)$ (or $R(0, y)$). These boundary conditions were calculated by Springer in [1] (in the formula for d_{2n} on p. 36 of that article one should change $x/2$ to $x^2/2$).

Below, however, the proofs of Theorems 4 – 12 are given independently of explicit formulae. These proofs contain additional information about the combinatorics of snakes. Namely, not only the equal cardinality of many sets of snakes (those mentioned above and others) will be proved, but also the fact that these mappings of one of the comparable sets onto the other are bijective.

Remark. Each line of our triangle a or a pair of triangles b (or d, or β) approximates (after an appropriate normalization) the function $\sin x$ on the interval $[0, \pi/2]$. Let us define a mapping that associates with the function f with integral 1 on this interval a new function

$$h(y) = c^{-1}g(y), \quad g(y) = \int_0^y f(\pi/2 - x)\,dx, \quad c = \int_0^{\pi/2} g(y)\,dy.$$

This contracting mapping, which is the continual analogue of our recurrence relation, has the fixed point $f = \sin x$. From this the convergence of distributions follows.

CHAPTER II

PROOFS OF CONBINATORIAL THEOREMS

§7. Proof of recurrence relations

Here the relations of Theorem 4 for triangles a, b, d, β are proved.

Case A. Let us discard from the A_n-snake

$$(x_0 < x_1 > x_2 < \dots x_n),\ \{x_0, \dots, x_n\} = \{0, \dots, n\},$$

the last term $x_n = m$ and assign to the integer points x of the segment $[0, n]$, excluding m, the values y from the segment $[0, n-1]$, keeping the order. The rest $(x_0 < x_1 > \dots x_{n-1})$ of the initial A_n-snake becomes the A_{n-1}-snake $(y_0 < y_1 > \dots y_{n-1})$, $\{y_0, \dots, y_{n-1}\} = \{0, \dots, n-1\}$. Depending on the parity of n, the value x_{n-1} will belong to one of the parts into which m divides the segment $[0, n]$. Thus for y_{n-1} we obtain the restriction $y_{n-1} < m$ if n is odd, and $y_{n-1} \geqslant m$ if n is even. It is easy to check that the A_{n-1}-snake y and the number m satisfying these relations uniquely define the initial A_n-snake x. Hence,

$$A^m_{2k+1} = A^0_{2k} + \dots + A^{m-1}_{2k},\quad A^m_{2k} = A^{2k-1}_{2k-1} + \dots + A^m_{2k-1},$$

which was to be proved. The relations $A^0_{2k+1} = A^{2k}_{2k} = 0$, except for $A^0_0 = 1$, follow from these recurrence relations.

Case B. The proof is completely analogous, except that together with $x_n = i$ one needs to exclude also $-i$. As a result we obtain for $m > 0$ the required relations:

$$B^m_{2k+1} = B^{-2k}_{2k} + \dots + B^{m-1}_{2k},\quad B^{-m}_{2k+1} = B^{-2k}_{2k} + \dots + B^{-m}_{2k},$$
$$B^m_{2k} = B^{2k-1}_{2k-1} + \dots + B^m_{2k-1},\quad B^{-m}_{2k} = B^{2k-1}_{2k-1} + \dots + B^{-(m-1)}_{2k-1},$$

where formally one takes $B^0_n = 0$.

In particular, from these relations it follows that $B^{-1}_n = B^1_n$ for $n > 1$. This is clear beforehand: changing the sign of an element of a B-snake whose modulus is equal to 1 takes a B-snake into a B-snake so long as this element was not the first one. The relations $B^{2k}_{2k} = 0$ and $B^{1-2k}_{2k-1} = 0$ also follow from the relations proved above. Obviously, $B^1_1 = 1$.

Case D. The proof is analogous to the proof for case B if $x_n \neq 0$ is crossed out. Thus we find for $m > 0$ that

$$D^m_{2k} = D^{-(2k-2)}_{2k-1} + \dots + D^{m-1}_{2k-1},\quad D^{-m}_{2k} = D^{-(2k-2)}_{2k-1} + \dots + D^{-m}_{2k-1},$$
$$D^m_{2k+1} = D^{2k-1}_{2k} + \dots + D^m_{2k},\quad D^{-m}_{2k+1} = D^{2k-1}_{2k} + \dots + D^{-(m-1)}_{2k}$$

(even and odd n change places in comparison with the B-case, because $x_1 > x_2$ for B-snakes). In particular, $D^{1-2k}_{2k} = 0$, $D^{2k}_{2k+1} = 0$ for $k > 0$.

If $x_n = 0$, then on crossing it out from the D_n-snake we shall decrease the moduli of all the remaining elements of the snake by 1, keeping their signs. As a result we get the D_{n-1}-snake y. The sign of its last number y_{n-1}, if it is different from 0, is defined by the parity of n. In other respects this D_{n-1}-snake is arbitrary.

The initial D_n-snake x is reconstructed from the snake y in a two-valued way if $y_{n-1} \neq 0$. Namely, the sign of the element x_i that has the modulus one remains completely undefined. All the other elements are reconstructed uniquely. If $y_{n-1} = 0$, the original D_n-snake x is reconstructed uniquely. For in this case $|x_{n-1}| = 1$. Therefore only the sign of x_{n-1} remains in doubt. But the sign of x_{n-1} is fixed by the parity of n, since $x_n = 0$. Thus

$$D^0_{2k} = D^0_{2k-1} + 2\,(D^{-(2k-2)}_{2k-1} + \ldots + D^{-1}_{2k-1}) \quad \text{for} \quad k > 1,$$
$$D^0_{2k+1} = D^0_{2k} + 2\,(D^{2k-1}_{2k} + \ldots + D^1_{2k}).$$

These relations, together with the previous ones referring to the case $x_n \neq 0$, are equivalent to the desired recurrence relations of Theorem 4 for triangles $L(d)$ and $R(d)$.

Remark. The values $L_{0,0} = 0$, $R_{0,0} = 1$ are chosen so that these last relations hold formally also for $i+j = 1$.

In accordance with the combinatorial definition of numbers of snakes ($D^0_1 = 1$, $D^0_2 = 0$, $D^0_3 = 2$) we must have $L_{0,0} = R_{0,0} = 1/2$.

Case β. Let us cross out $x_1 = i$ in the β-snake $(x_1 < x_2 > \ldots x_n)$, $1 \leqslant |x_k| \leqslant n$, $|x_k| \neq |x_l|$. We change the moduli of x_k $(k > 2)$, decreasing by one those of them that exceed $|i|$. We change all the signs. We obtain the β_{n-1}-snake $(z_1 < z_2 > \ldots)$, $1 \leqslant |z_k| \leqslant n-1$, $|z_k| \neq |z_l|$.

Let $|i| = m$. If $x_1 = m$, then $-z_1 = x_2 - 1 \geqslant m$, that is, $z_1 \leqslant -m$. If $x_1 = -m$, then $-z_1 \geqslant -(m-1)$, that is, $z_1 \leqslant m-1$. All the β_{n-1}-snakes with such z_1 are realized. Conversely, the initial n-snake is uniquely reconstructed from the resulting $n-1$ snake and the number i. Thus,

$$\beta^m_n = \beta^{-(n-1)}_{n-1} + \ldots + \beta^{-m}_{n-1}, \quad \beta^{-m}_n = \beta^{-(n-1)}_{n-1} + \ldots + \beta^{m-1}_{n-1},$$

which was to be proved.

§8. Proofs of the relations between triangles b and β

Here we prove the first two relations of Theorem 5, connecting enumerations of B-snakes with their tails and heads. Theorem 2 from the introduction is also proved. First I call to mind the following definition.

Definition. A sequence of n integers x_i is called a β_n*-snake* if

$$(x_1 < x_2 > x_3 < \ldots x_n), \quad 0 < |x_i| \leqslant n, \quad |x_i| \neq |x_j|.$$

The number of β_n-snakes with $x_1 = i$ is denoted by β^i_n.

Example. The numbers β_n^i with $n \leqslant 5$ form the table

1	1
2 1	1 0
4 4 3	3 2 0
16 16 14 11	11 8 4 0
80 80 76 68 57	57 46 32 16 0

In the left part of the n-th row i changes from $-n$ to -1, in the right part from 1 to n. All 16 β_3-snakes are shown in Fig. 5.

Remark. $\beta_n^1 = \beta_n^{-1}$, since the replacement of $x_1 = 1$ by $x_1 = -1$ takes a β_n-snake into a β_n-snake.

Definition. A β_n-snake is called a γ_n*-snake* if it remains a snake after adding 0 on the right, that is, if $(-1)^n x_n > 0$.

We denote the number of γ_n-snakes with $x_1 = i$ by γ_n^i.

Example. The numbers γ_n^i for $n \leqslant 5$ are given by the table

1	0
1 1	1 0
3 3 2	2 1 0
11 11 10 8	8 6 3 0
57 57 54 48 40	40 32 22 11 0

The numbering by the index i in the n-th row is the same as above: from $-n$ to -1 on the left, from 1 to n on the right.

All 11 γ_3-snakes are shown in Fig. 2 on the right.

Remark. $\gamma_n = \gamma_n^{-1}$ when $n > 1$, since the replacement of $x_1 = 1$ by $x_1 = -1$ (like the reverse replacement) takes a γ_n-snake into a γ_n-snake (when $n = 1$ this is not true).

Remark. The numbers β_n^i and γ_n^i with increasing i (non-strictly) decrease. This is easy to foresee: the lower a snake begins, the more possibilities there are for it to continue. This monotonicity property follows from the recurrence relations (Theorem 4) or from Theorem 15, which will be proved below.

Definition. The points m and w are called *conjugate* if they are symmetric with respect to the mid-point of the segment $\pm[1, n]$, to which both of them belong: if $m > 0$, then

$$w = n + 1 - m \quad \text{is the conjugate point for } m;$$

$$-w = m - (n + 1) \quad \text{is the conjugate point for } -m.$$

Theorem 15. *The number of β_n-snakes with fixed negative beginning is equal to the sum of the numbers of γ_n-snakes with conjugate and anticonjugate beginning. The number of β_n-snakes with fixed positive beginning is equal to the difference between the numbers of γ_n-snakes with anticonjugate and conjugate beginning.*

For $m > 0$, in the previous notation,

$$\beta_n^{-m} = \gamma_n^{-w} + \gamma_n^{w},$$
$$\beta_n^{m} = \gamma_n^{-w} - \gamma_n^{w}.$$

Example. $68 = 57+11$, $46 = 57-11$ ($n = 5$, $m = 2$, $w = 4$).

Remark. The signs ($+$ for β_n^{-m}, $-$ for β_n^{m}) and the order ($-w$ before w) are reconstructed from considerations of monotonicity:

$$\beta_n^{-m} > \beta_n^{m}, \quad \gamma_n^{-m} > \gamma_n^{m}.$$

Remark. Theorem 15 differs from the first half of Theorem 5 only in the notation.

Proposition. *The relations $\gamma_n^i = B_i^i$ if n is even, $\gamma_n^i = B_n^{-i}$ if n is odd, are valid.*

Proof of the proposition. Let us turn the γ_n-snake back to front. We obtain the snake $(0 \ldots < x_2 > x_1)$. If n is even, the rearranged snake $(0 < x_n > x_{n-1} \ldots)$ will be a B_n-snake with the beginning equal to the end of the original snake. If n is odd, the rearranged snake $(0 > x_n < x_{n-1} \ldots)$ will become a B_n-snake after changing the sign. The resulting B_n-snake has a beginning which is opposite to the end of the initial γ_n-snake. The proposition is proved.

Changing γ_n^i to $B_n^{\pm i}$ in accordance with this proposition and using the definitions

$$B_n^{-m} = L_{m-1,\, n-m}(b), \quad B_n^{m} = R_{n-m,\, m-1}(b) \qquad (m > 0),$$

we bring the relations of Theorem 15 to the form indicated in Theorem 5.

Beginning of the proof of Theorem 15. Let us associate with each β_n-snake

$$(x_1 = \varepsilon_1 i_1, \ldots, x_n = \varepsilon_n i_n), \quad i_k = |x_k|, \quad \{i_1, \ldots, i_n\} = \{1, \ldots, n\},$$

a permutation of the moduli $(i_1, \ldots, i_n)$ and a sequence of signs $(\varepsilon_1, \ldots, \varepsilon_n)$. Let us fix the permutation of moduli. We call it *increasing* if $i_1 < i_2$, and *decreasing* if $i_1 > i_2$. We define the *conjugate permutation* $(j_1, \ldots, j_n)$ by the symmetry condition with respect to the middle: $i_k + j_k = n+1$. Obviously the permutation conjugate to an increasing one decreases, and conversely.

For a fixed permutation of moduli Δ let us call a sequence of signs a β_n-(γ_n-)*admissible assignment* if the resulting sequence $x_k = \varepsilon_k i_k$ is a β_n-(γ_n-)snake. Suppose also that $n > 1$.

***Lemma* 1.** *The numbers $\beta^+(\Delta)$, $\gamma^+(\Delta)$ of β^+-(γ^+-)admissible assignments of the given permutation moduli beginning with $\varepsilon_1 = +1$ are equal to the numbers $\beta^-(\Delta)$, $\gamma^-(\Delta)$ of its assignments beginning with $\varepsilon_1 = -1$ if the permutation is increasing, and are equal to zero if the permutation is decreasing.*

Proof of Lemma 1. In the first case $|x_1| < |x_2|$, $x_1 < x_2$. Therefore $|x_1| < x_2$, and changing the sign of x_1 does not disturb the admissible assignment. In the second case $|x_1| > |x_2|$ and $x_1 < x_2$, therefore $x_1 < 0$, in spite of $\varepsilon_1 = +1$.

Lemma 2. *The number of β_n-assignments beginning with $\varepsilon_1 = -1$ for a given permutation Δ is equal to the number of β_n-assignments ending with $\varepsilon_n = (-1)^n$ for the conjugate permutation ∇:*

$$\beta^-(\Delta) = \gamma(\nabla) = \gamma^+(\nabla) + \gamma^-(\nabla).$$

The proof of Lemma 2 is given separately below.

Deduction of Theorem 15 *from Lemmas* 1 *and* 2.

Proposition. *For conjugate permutations Δ and ∇ the following relations are valid:*

$$\beta^+(\Delta) = \gamma^-(\nabla) - \gamma^+(\nabla),$$
$$\beta^-(\Delta) = \gamma^-(\nabla) + \gamma^+(\nabla).$$

Proof. If Δ increases, then ∇ decreases. By Lemma 1, $\gamma^+(\nabla) = 0$, $\beta^+(\Delta) = \beta^-(\Delta)$. By Lemma 2, $\beta^-(\Delta) = \gamma^+(\nabla)+\gamma^-(\nabla) = \gamma^-(\nabla)$, as required.

If Δ decreases, then ∇ increases. By Lemma 1, $\gamma^+(\nabla) = \gamma^-(\nabla)$, $\beta^+(\Delta) = 0$. By Lemma 2, $\beta^-(\Delta) = \gamma^+(\nabla)+\gamma^-(\nabla) = 2\gamma^-(\nabla)$, as required. The proposition is proved.

End of the proof of Theorem 15. Let us sum the formulae of the proposition over all permutations beginning with $i_1 = m$. We obtain the formulae of Theorem 15, which is thus proved.

Proof of Lemma 2. Consider the γ_n-snake $(y_1 < y_2 > \dots y_n = (-1)^n|y_n|)$, obtained by assigning the permutation conjugate to Δ. The symmetric sequence $(z_1, \dots, z_n) = (-1)^{n+1}(y_n, \dots, y_1)$ is a β_n-snake with negative first term.

The graph of the permutation of moduli corresponding to this snake is centrally symmetric to the graph of the moduli of the initial permutation Δ, since it is obtained by reflecting in two axes (by conjugation and symmetry). Thus Lemma 2 follows from the following statement.

Lemma 3. *The numbers of β_n-assignments beginning with $\varepsilon_1 = -1$ for permutations with graphs centrally symmetric to each other are identical.*

Proof of Lemma 3. This lemma follows from the explicit formula given below for the number of $\beta^-(\Delta)$ assignments beginning with $\varepsilon_1 = -1$.

§9. Numbers of assignments

Definition. A *β-assignment of the permutation* $(i_1, ..., i_n)$ *of the numbers* $\{1, ..., n\}$ is a choice of signs $\varepsilon_k = \pm 1$ such that the numbers $x_k = \varepsilon_k i_k$ satisfy the inequalities $(x_1 < x_2 > x_3 < ... x_n)$.

Definition. A *falling segment* in a sequence $\Delta = (i_1, ..., i_n)$ is a maximal segment in which the sequence decreases:

$$(i_{a-1} <)\ i_a > i_{a+1} > \dots > i_b\ (< i_{b+1})$$

(the cases $a = 1$ and $b = n$ are also admissible).

Remark. The numbers of falling segments in permutations whose graphs are centrally symmetric to each other are identical. Indeed, a central symmetry takes falling segments into falling ones.

Theorem 16. *The number of β-assignments beginnng with* $\varepsilon_1 = -1$ *for a given permutation of moduli* Δ *is equal to* 2^r, *where r is the number of falling segments in* Δ:

$$\beta^-(\Delta) = 2^{r(\Delta)}.$$

Example. Let $n = 3$. The graph of a permutation is one of four types. All possible β-assignments for each type are shown in Fig. 10 and the case $\varepsilon_1 = -1$ is singled out. The two extreme types are self-dual, and the middle ones are dual to each other.

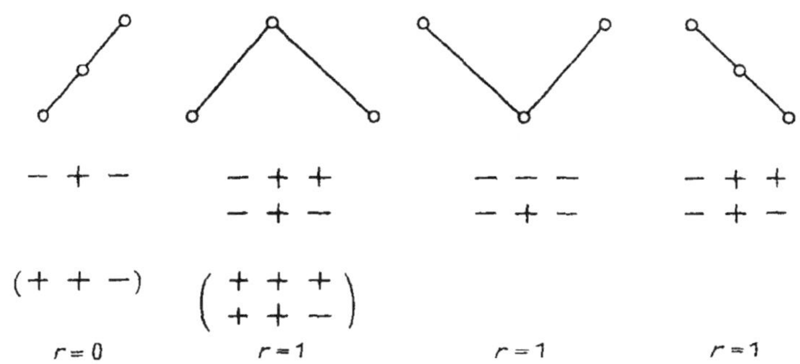

Fig. 10. The β_3-assignments.

Proof of Theorem 16. Consider three positive numbers $i \neq j \neq k$ and three signs $\alpha, \beta, \gamma = \pm 1$.

Definition. The triple (x, y, z) *alternates* if $x < y > z$ or $x > y < z$. It is easy to check the following lemma.

***Lemma* 1.** *The triple (αi, βj, γk) alternates if and only if one of the following holds:*

$$\begin{aligned} &i<j<k, \quad \gamma=-\beta; \\ &i<j>k; \\ &i>j<k, \quad \gamma=\alpha; \\ &i>j>k, \quad \alpha=-\beta. \end{aligned}$$

Let us adjoin to the sequence $\Delta = (i_1, \ldots, i_n)$ the element $i_0 = 1/2 < i_1$ on the left and choose the sign $\varepsilon_0 = 1$. Now to prove Theorem 16 it is necessary to count the number of all possible choices of signs of sequences $x_0 > x_1 < x_2 \ldots x_n$ with given moduli i_k and with fixed signs $\varepsilon_0 = 1, \varepsilon_1 = -1$.

Consider the restrictions imposed on the sequence of signs $(\varepsilon_0, \ldots, \varepsilon_n)$ by Lemma 1. We divide the whole sequence i_k into rising and falling segments, joining alternating maxima and minima. The restrictions of Lemma 1 consist in the following. Along each rising or falling segment, the signs alternate except for the minimum points. The signs of the minimum points are arbitrary (excluding $\varepsilon_0 = 1$). The signs of the neighbours of each minimum point are the same.

These rules together with the initial condition $\varepsilon_0 = 1, \varepsilon_1 = -1$ uniquely define all the signs, except the signs of the minimum points other than i_0, which are arbitrary. So $\beta^-(\Delta) = 2^r$, where r is the number of minima other than i_0. This number r is equal to the number of falling segments, the right-hand ends of which are the minimum points mentioned above. This completes the proof of Theorem 16.

It also completes the proof of Lemma 3 (and hence of Lemma 2) for Theorem 15, and so the proof of this theorem is also completed.

Remark. Let $p_n(r)$ denote the number of permutations of n elements with r falling segments. From the above we get

$$\sum_r p_n(r) = n!, \qquad \sum_r p_n(r) 2^r = b_n.$$

Example. $n = 4$: $p_4(0) = 1,\ p_4(1) = 18,\ p_4(2) = 5;$

$1 + 18 + 5 = 24, \qquad 1 \cdot 1 + 18 \cdot 2 + 5 \cdot 4 = 57.$

$n = 5$: $p_5(0) = 1,\ p_5(1) = 58,\ p_5(2) = 61;$

$1 + 58 + 61 = 120, \quad 1 \cdot 1 + 58 \cdot 2 + 61 \cdot 4 = 361.$

I do not know a general formula for $p_n(r)$, but the number $61 = a_5$ does not appear here accidentally.

***Theorem* 17.** *The number of permutations of $2r$ elements with r falling segments is equal to the number a_{2r-1} of permutations ($x_1 < x_2 > \ldots$) of $2r$ elements $\{1, \ldots, 2r\}$.*

The number of permutations of $2r-1$ elements with $r-1$ falling segments is equal to the number of permutations of $2r$ elements with r falling segments.

Proof. The first statement is obvious, since r segments have $2r$ ends, and hence the permutation is $(y_1 > y_2 < \ldots)$.

Suppose that in a permutation of $2r-1$ elements there are $r-1$ falling segments. The ends of these segments give $2r-2$ vertices of the graph of the permutation. The last vertex lies either on one of the segments (falling or rising), or to the left and below the left vertex, or to the right and above the right one.

Let us construct a function that behaves like x at infinity and has $2r-2$ critical points, so that the points of the function graph corresponding to them are the above-mentioned $2r-2$ vertices of the graph of the permutation. Between these vertices the function is monotonic. We can choose the function in such a way that its graph passes through the last vertex of the permutation graph. At this point we can make the first and second derivatives of the function equal to zero, and the third non-zero. There are no other critical points of this function.

Let us deform the function in the neighbourhood of a cubic singular point so that this singularity splits into a maximum and a minimum. We then make this maximum very high (leaving the previous minimum fixed and changing the function only in the neighbourhood of the maximum point). The resulting function has $2r$ non-degenerate critical points. Its critical values form a sequence $(y_1 > y_2 < \ldots)$ of $2r$ elements.

Conversely, from any such sequence we construct a function with $2r$ critical points. We find the highest local maximum. We compress it to the higher of the neighbouring local minima (not changing the function elsewhere nor the heights of these two minima). We get a function with one cubic singularity. We construct a permutation from its $2r-1$ critical points and values. It has $r-1$ falling segments.

Indeed, the initial sequence of $2r$ elements had r falling segments (each of length 1). One of them connected the highest maximum with one of the neighbouring minima. If this neighbouring minimum is the higher of the two, we have constructed this falling segment. If this neighbouring minimum is the lower, then the cubic singularity arising from the compression combines two falling segments of the original permutation into one falling segment of length 2 in the new permutation.

Of course, all this may be described without functions, in purely combinatorial language. We have constructed two mappings: from permutations of $2r-1$ elements with $r-1$ falling segments into permutations $(y_1 > y_2 < \ldots)$ of $2r$ elements and conversely. These mappings are mutually inverse (as is easy to check). Hence they are one-to-one, which proves Theorem 17.

§10. Proof of the relations between triangles b and d

We recall that D_n^i is the number of D_n-snakes, that is, the number of integer sequences

$$(-x_2 < x_1 < x_2 > x_3 < \dots x_n),\quad |x_k| < n,\quad |x_k| \neq |x_j|,$$

ending with $x_n = i$ $(n > 1)$. In what follows it will be convenient to use the following notation:

$$\begin{aligned} \Delta_n^{-m} &= D_n^{-(m-1)},\quad \Delta_n^m = D_n^{m-1} && \text{for}\quad m > 1,\\ \Delta_n^{-1} &= \Delta_n^1 = D_n^0/2 && \text{for}\quad n > 1,\\ \Delta_1^{-1} &= 0,\ \Delta_1^1 = 1. \end{aligned}$$

Thus, for a fixed n, the sequence Δ_n^i is a row of $2n$ numbers in the triangles $L(d)$, $R(d)$ (in this order), enumerated by index i from $-n$ to -1 in $L(d)$ and from 1 to n in $R(d)$. This is the same enumeration as in b-triangles.

Example. For $n = 4$ (see Fig. 4)

$$\Delta_4^{-4} = 0,\ \Delta_4^{-3} = 1,\ \Delta_4^{-2} = 2,\ \Delta_4^{-1} = 3 = \Delta_4^1,\ \Delta_4^2 = 4,\ \Delta_4^3 = 5,\ \Delta_4^4 = 5.$$

Remark. The total number of D_n-snakes is

$$d_n = \sum D_n^i = \sum \Delta_n^i.$$

Example. $d_4 = 1+2+3+3+4+5+5 = 23$ (see Fig. 4).

In this notation the second part of Theorem 5 has the following form.

Theorem 18. *The following relations hold:*

$$\begin{aligned} \Delta_n^i &= \beta_n^i - \gamma_n^i && \textit{for odd } n,\\ \Delta_n^i &= \beta_n^{-i} - \gamma_n^{-i} && \textit{for even } n \end{aligned}$$

(where the numbers β_n^i and γ_n^i are as defined in §8).

Proof of Theorem 18. Consider a β_n-snake $(z_1 < z_2 > \dots z_n)$ that is not a γ_n-snake (so that after adjoining the zero on the right, the sequence ceases to alternate: $(-1)^n z_n < 0$. In this case, $|z_{n-1}| > |z_n|$ and the signs of z_{n-1} and z_n are identical. We decrease the modulus of each z_k by 1 without changing the signs. We get a snake

$$(y_1 < y_2 > \dots y_n),\quad |y_k| < n,\quad |y_k| \neq |y_l|,\quad |y_{n-1}| > |y_n|,\quad y_{n-1} > y_n.$$

Let us reverse this snake, changing the signs for even n. We get a D_n-snake:

$$(x_1 < x_2 > \dots x_n) = (-1)^{n+1}(y_n \dots < y_2 > y_1).$$

This is a D_n-snake, since $|x_1| < |x_2|$. The end of this D_n-snake, $x_n = (-1)^{n+1}y_1$, coincides with y_1 for odd n and with $-y_1$ for even n. Moreover, $x_1 \geqslant 0$, since the signs of x_1 and of x_2 are the same (if $x_1 \neq 0$).

Let us assume that $x_n \neq 0$, that is, $|z_1| \neq 1$.

If $x_1 > 0$, then exactly two β_n-snakes z that are not γ_n-snakes transform into this D_n-snake under the mapping. Indeed, the initial snake is reconstructed from the resulting one uniquely up to the sign of that z_k for which $y_k = 0$. This sign is arbitrary, since the signs of $z_{k-1}-z_k$ and of $z_{k+1}-z_k$ for $|z_k| = 1$ and $|z_{k\pm 1}| > 1$ are independent of the choice of the sign of z_k. The pre-images of D_n-snakes with $x_1 > 0$ are not γ_n-snakes, since from $0 < x_1 < x_2$ it follows that $0 < z_n < z_{n-1}$ for odd n and $0 > z_n > z_{n-1}$ for even n.

If $x_1 = 0$, then exactly one β_n-snake that is not a γ_n-snake transforms into this D_n-snake under our mapping. Indeed, in this case $k = n$, $z_n = \pm 1$, and the sign of z_n coincides with the sign of z_{n-1}, which is defined uniquely, since $x_2 > 0$ for any D_n-snake.

Every D_n-snake with $x_1 > 0$ remains a D_n-snake with the same end x_n when we change the sign of x_1. Therefore the number of D_n-snakes with the end $x_n = j$ is equal to the number of β_n-snakes that are not γ_n-snakes and begin with

$$z_1 = (-1)^{n+1} i, \quad i = j + \operatorname{sgn} j.$$

This proves the theorem for $x_n \neq 0$, that is, for $i \neq \pm 1$.

In the case $x_n = 0$ we have $i = \pm 1$. Then $x_1 \neq 0$, and so for the resulting D_n-snake $x_1 > 0$. Each of the sets of β_n-snakes beginning with $z_n = 1$ or -1 that are not γ_n-snakes maps in a one-to-one manner onto the set of all D_n-snakes ending with $x_n = 0$ and having $x_1 > 0$. Changing the sign of x_1 transforms a D_n-snake into a D_n-snake without changing the end (if $n > 1$). Therefore $\beta_n^1 - \gamma_n^1 = \beta_n^{-1} - \gamma_n^{-1} = D_n^0/2$, as required.

§11. Proofs of the theorems on D-snakes

Theorem 10 of §5 is proved here.

Consider a D_n-snake $(-x_2 < x_1 < x_2 > x_3 < \dots x_n)$ in which $x_i \neq \pm x_j$, $|x_i| < n$.

Definition. Those n integers with different moduli (from the $2n-1$ possible values of x_i) which occur in a snake are called its *participants*.

For fixed participants one can form other D_n-snakes. Theorem 10 is a direct consequence of the following stronger statement.

***Theorem* 19.** *For any set of participants the number of D_n-snakes x with these participants and with $|x_1| = m$ is equal to the number of D_n-snakes with these participants for which there are exactly m participants in the interval $(x_1 < x_2)$.*

Remark. Here and in what follows, the interval $(a < b)$ is regarded as the set of all integers j satisfying the inequalities $a < j < b$.

I preface a proof of Theorem 19 with three other theorems.

Definition. For a given set of participants $\{x\}$ of D_n-snakes we define $u(\{x\}, m)$ to be *the number of participants that are greater than* m.

Remark. If there exist D_n-snakes with the given participants and with $|x_1| = m$ or with m participants in the interval $(x_1 < x_2)$, then $u(\{x\}, m) > 0$. Indeed, in the first case $x_2 > m$ is a participant. In the second case the participant x_2 also exceeds m, since the number of participants in $(-x_2, x_2)$ is equal to x_2 and greater than m (among these participants there are x_1 and m participants from the interval $(x_1 < x_2)$).

Moreover, in both cases $u < n-m$. Indeed, the number of participants of absolute value less than or equal to i is equal to $m+1$. Therefore there are no more than $n-(m+1)$ participants that are greater than i.

Now we fix m and the set of participants (and hence also the number u).

Theorem 20. *The number of D_n-snakes x (with given participants) with $|x_1| = m$ is equal to the number of A-snakes of length $n-1$ $(z_1 < z_2 > \dots z_{n-1})$, $\{z_1, z_2, \dots, z_{n-1}\} = \{1, \dots, n-1\}$ with the first participant $z_1 \leqslant u(\{x\}, m)$.*

Proof. We delete x_1 from a D_n-snake whose first term satisfies $|x_1| = m$. To each of the remaining participants we assign a new value z, which is equal to the number of the remaining $n-1$ participants that are greater than or equal to the given one. We get an A-snake $\{z_1 < z_2 > \dots z_{n-1}\} = \{1, \dots, n-1\}$. Here z_1 is a new value of the participant x_2, that is, the number of participants that are greater than or equal to x_2. Therefore $z_1 \leqslant u(\{x\}, m)$.

Conversely, let an A-snake $\{z_1 < z_2 > \dots z_{n-1}\} = \{1, \dots, n-1\}$ be given and let $z_1 \leqslant u(\{x\}, m)$. We replace each z_k by the value x_{k+1} of the kth largest participant (excluding x_1). We first put $x_1 = \pm m$ (the sign is determined by the composition of the participants). We get a snake $(x_1 < x_2 > \dots)$ with the original participants. Here $-x_2 < x_1 < x_2$, since $x_2 > m$, because $z_1 \leqslant u(\{x\}, m)$ (the number of participants greater than m), and hence the participant replacing z_1 is greater than m.

It is easy to check that these mappings are mutually inverse, and so Theorem 20 is proved.

Theorem 21. *The number δ_n^m of D_n-snakes with $|x_1| = m$ is given by the formula*

$$\delta_n^m = \sum_{u=0}^{n-m-1} 2^m \binom{n-1-m}{u} a_{n-1}^u,$$

where a_{n-1}^u is the number of sequences

$$\{z_1 < z_2 > \dots z_{n-1}\} = \{1, \dots, n-1\}, \quad z_1 \leqslant u.$$

Proof. We fix the numbers m and u and consider in how many ways one can choose the structure of x in such a way that the number of participants greater than m is equal to u. The number of integer points between n and m (from which one has to choose u participants) is equal to $n-1-m$.

The number of choices of participants that are greater than m is therefore equal to $\binom{n-1-m}{u}$. It remains to choose the signs of participants with moduli $(1, \ldots, m)$. The resulting number of choices of participants is $2^m\binom{n-1-m}{u}$.

For fixed participants the number of D_n-snakes with $|x_1| = m$ is equal to a_{n-1}^n, by Theorem 20. This proves Theorem 21.

Example. Let us calculate some values of δ_n^m. It is clear that $\delta_n^{n-1} = 0$ for $n > 1$. The numbers a_n^m coincide with A_n^m if n is odd, and with A_n^{n-m} if n is even. Their values are given in the introduction (Euler–Bernoulli triangle). For example,

$$a_3^{0,1,2,3} = 0, 1, 2, 2, \quad a_4^{0,1,2,3,4} = 0, 2, 4, 5, 5.$$

If $n-m-1 = 1$, that is, $m = n-2$, we get

$$\delta_n^{n-2} = 2^{n-2}a_{n-3},$$

since $a_{n-1}^1 = a_{n-3}$.

For example, for $n = 2, 3, 4, 5, 6$

$$\delta_n^{n-2} = 1, 2, 4, 16, 80, \ldots = 1\cdot 1, 2\cdot 1, 4\cdot 1, 8\cdot 2, 16\cdot 5, \ldots$$

If $n-m-1 = 2$, that is, $m = n-3$ (and $n > 3$), then

$$\delta_n^{n-3} = 2^{n-3}(2a_{n-1}^1 + 1a_{n-1}^2).$$

But $a_{n-1}^1 = a_{n-3}$, $a_{n-1}^2 = 2a_{n-3}$. Thus,

$$\delta_n^{n-3} = 2^{n-1}a_{n-3} = 2\delta_n^{n-2}.$$

The table of the first few numbers δ_n^m is as follows (n is the number of a row, $0 \leqslant m < n$):

$$\begin{array}{ccccccccc} & & & & 1 & & & & \\ & & & 1 & & 0 & & & \\ & & 3 & & 2 & & 0 & & \\ & 11 & & 8 & & 4 & & 0 & \\ 57 & & 46 & & 32 & & 16 & & 0 \end{array}$$

According to Theorem 21, we find that

$$\delta_6^0 = \sum_{u=0}^{5}\binom{5}{u}a_5^u = 1\cdot 0 + 5\cdot 5 + 10\cdot 10 + 10\cdot 14 + 5\cdot 16 + 1\cdot 16 = 361.$$

Theorem 22. *The number of D_n-snakes x with fixed participants for which there are excactly m participants in the interval $(x_1 < x_2)$ is equal to the number of A-snakes $\{z_1 < z_2 > \ldots z_{n-1}\} = \{1, \ldots, n-1\}$ with $z_1 \leqslant u(\{x\}, m)$.*

Proof. We delete x_1 from a D_n-snake with fixed participants having exactly m participants in the interval $(x_1 < x_2)$. To each of the remaining participants x_k we assign the value z_{k-1}, equal to the number of participants that are greater than or equal to the given one. We get an A-snake $\{z_1 < z_2 > \ldots z_{n-1}\} = \{1, \ldots, n-1\}$.

***Assertion* 1.** $z_1 \leqslant u(\{x\}, m)$ = *(the number of participants greater than m).*

By definition, z_1 is the number of participants greater than or equal to x_2. But in the interval $(-x_2 < x_2)$ there are exactly x_2 participants. One of them is x_1. Also, there are m participants in the interval $(x_1 < x_2)$. Therefore $x_2 \geqslant m+1$. So the number of participants greater than or equal to x_2 does not exceed the number of participants greater than m. Assertion 1 is proved.

Consider an A-snake $\{z_1 < z_2 > \dots z_{n-1}\} = \{1, \dots, n-1\}$. Let $z_1 \leqslant u(\{x\}, m)$. Let x_2 be the z_1-th largest participant. By hypothesis, $x_2 > m$. We count another m participants in decreasing order and call the next participant x_1.

***Assertion* 2.** $|x_1| < x_2$.

The number of participants in the interval $(-x_2 < x_2)$ is equal to $x_2 > m$. Then after counting m participants downward from x_2 we do not exhaust the participants from this interval. Therefore $x_1 \in (-x_2 < x_2)$. Assertion 2 is proved.

Further points of the D_n-snake $(x_3, \dots, x_n)$ are taken from the A-snake. For this purpose we delete from the set of participants the participant x_1 already defined and then associate with z_k $(k > 1)$ that participant for which the number of other participants greater than or equal to it (besides x_i) is equal to z_k. We denote this participant by x_{k+1}.

We get a D_n-snake $(-x_2 < x_1 < x_2 > x_3 < \dots x_n)$.

It is easy to check that these mappings between the sets of D- and A-snakes concerned are mutually inverse and hence one-to-one. Theorem 22 is proved.

Proof of Theorem 19. The numbers concerned in Theorem 19 are, by Theorems 20 and 22, equal to the same third number. So they are equal to each other and Theorem 19 is proved.

Remark. The proofs of Theorems 20 and 22 also give an explicit but non-trivial one-to-one correspondence between the D_n-snakes with $|x_1| = m$ and those with m participants in the interval $(x_1 < x_2)$.

§12. Comparison of the numbers of *D*-snakes and *B*-snakes with a given beginning

Theorem 11 from §5 is proved here. In fact, more general theorems will be proved, for which we need the following notation.

Definition. A snake $(x_1 < x_2 > \dots x_n)$ of integers $|x_k| < n$, $|x_k| \neq |x_l|$, is called a λ_n-*snake*. The number of λ_n-snakes with $x_1 = i$ is denoted by λ_n^i.

Example. $\lambda_3^{-2} = 2$, $\lambda_3^{-1} = 2$, $\lambda_3^0 = 3$, $\lambda_3^1 = 1$, $\lambda_3^2 = 0$ (see Fig. 11).

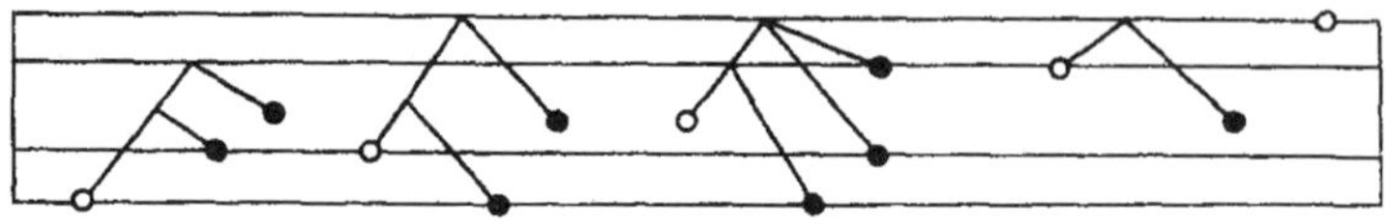

$\lambda_3^{-2}=2$ $\lambda_3^{-1}=2$ $\lambda_3^{0}=3$ $\lambda_3^{1}=1$ $\lambda_3^{2}=0$

Fig. 11. The classification of λ_3-snakes.

The table of the numbers λ_n^i begins as follows:

$$\lambda_n^i = \begin{array}{ccccccc} & & & 1 & & & \\ & & 1 & 1 & 0 & & \\ & 2 & 2 & 3 & 1 & 0 & \\ 8 & 8 & 7 & 11 & 4 & 2 & 0 \\ \cdot & \cdot & \cdot & \cdot & \cdot & \cdot & \cdot \end{array}$$

In the n-th row, i runs from $-(n-1)$ to $(n-1)$.

We recall that β_n^i is the number of snakes

$$(z_1 < z_2 > \ldots z_n), \quad 0 < |z_n| \leqslant n, \quad |z_k| \neq |z_l|, \quad z_1 = i.$$

The numbers β_n^i form a table (where the β_n^i for $i \in [-n, -1]$ and $[1, n]$ are written out separately for increasing i):

$$\beta_n^{-m} = \begin{array}{ccccccc} & & & 1 & & & \\ & & 2 & & 1 & & \\ & 4 & & 4 & & 3 & \\ 16 & & 16 & & 14 & & 11 \\ \cdot & \cdot & \cdot & \cdot & \cdot & \cdot & \cdot \end{array} \qquad \begin{array}{ccccccc} & & & 1 & & & \\ & & 1 & & 0 & & \\ & 3 & & 2 & & 0 & \\ 11 & & 8 & & 4 & & 0 \\ \cdot & \cdot & \cdot & \cdot & \cdot & \cdot & \cdot \end{array} = \beta_n^m$$

***Theorem* 23.** *The numbers β and λ are connected by the relations*

$$2\lambda_n^m = \beta_n^{m+1}, \quad 2\lambda_n^{-m} = \beta_n^{-(m+1)}$$

for $m > 0$, and for $m = 0$ by the relations

$$\lambda_n^0 = \beta_n^1 = \beta_n^{-1}.$$

Proof. We replace each element $x_k \neq 0$ of a λ_n-snake by the number z_k with the same sign and of modulus greater by 1. We replace the elements $x_k = 0$ by $z_k = \pm 1$. We get a pair of β_n-snakes. Here each β_n-snake is obtained only once. λ-snakes with $x_1 = \pm m \neq 0$ pass into pairs of β-snakes with $z_1 = \pm(m+1)$. λ-snakes with $x_1 = 0$ pass into pairs of β-snakes with $z_1 = \pm 1$. The theorem is proved.

Remark. For $m > 0$, $2\lambda_n^m = \delta_n^m$, since each λ_n-snake $(x_1 < x_2 > \ldots)$ with $x_1 = m > 0$ is a D_n-snake, being a snake obtained from it by changing the sign of x_1. For $m = 0$, $\lambda_n^0 = \delta_n^0$ by definition. Then it follows from Theorem 23 that $\delta_n^m = \beta_n^{m+1}$ for $m > 0$, and also for $m = 0$. This proves Theorem 11 of §5.

We write $\beta_n = \Sigma\beta_n^i$, $\lambda_n = \Sigma\lambda_n^i$ for the numbers of all β_n-snakes and λ_n-snakes.

Theorem 24. *The following relations hold:*

$$\beta_n = 2^n a_{n-1}, \quad \lambda_n = 2^{n-1} a_{n-1},$$

where a_{n-1} is the number of A-snakes $\{z_1 < z_2 > \dots z_n\} = \{1, 2, \dots, n\}$.

Example. $\beta_4 = 16+16+14+11+11+8+4+0 = 80 = 2^4 a_3,$

$$\lambda_4 = 8+8+7+11+4+2+0 = 40 = 2^3 a_3.$$

Proof. A β_n-snake is defined firstly by its participants, the signs of which can be chosen in 2^n ways, and secondly by their order. The number of possible orderings $(x_1 < x_2 > \dots x_n)$ of n given pairwise distinct numbers is equal to a_{n-1}, which proves the first part of the theorem. The second part follows from the first because of Theorem 23.

Theorem 25. *We have the relations*

$$B^1_{n+1} = B^{-1}_{n+1} = \lambda_n.$$

Proof. By Theorem 15 (or 5, or 2) applied to our case we get

$$B^{-1}_{n+1} - B^1_{n+1} = \beta^{n+1}_{n+1} = 0,$$
$$B^{-1}_{n+1} + B^1_{n+1} = \beta^{-(n+1)}_{n+1}.$$

According to the recurrence relation for the β (see §7),

$$\beta^{-(n+1)}_{n+1} = \beta^{-n}_n + \dots + \beta^n_n = \beta_n.$$

But $\beta_n = 2\lambda_n$ by Theorem 24, and Theorem 25 follows.

Remark. This proves Theorem 12 at the same time.

Theorem 26. *We have*

$$\lambda^{-m}_n + \lambda^m_n = B^{m-n}_n,$$

$$\lambda^{-m}_n - \lambda^m_n = B^{n-m}_n$$

for $m > 0$, and for $m = 0$ the sum is halved:

$$\lambda^0_n = B^{-n}_n.$$

Example. The sums and differences of the numbers λ^i_n symmetrically situated in the table give for $m \geqslant 0$

$$\lambda^{-m}_n + \lambda^m_n = \begin{matrix} &&&1\\ &&1&1\\ &2&3&3\\ 8&10&11&11 \end{matrix} \qquad \begin{matrix} &&&0\\ &&1&0\\ &2&1&0\\ 8&6&3&0 \end{matrix} = \lambda^{-m}_n - \lambda^m_n$$

(for $m = 0$ the sum is halved). These are rearranged b-triangles (see p. 4).

Proof. It is enough to express λ in terms of β by means of Theorem 23 and to apply Theorem 15, which in the notation of Theorem 26 has for $m \geqslant 0$ the form

$$\beta_n^{-(m+1)} = B_n^{m-n} + B_n^{n-m},$$
$$\beta_n^{+(m+1)} = B_n^{m-n} - B_n^{n-m}.$$

For $m > 0$ it follows by Theorem 23 that

$$2\lambda_n^{-m} - 2\lambda_n^{m} = \beta_n^{-(m+1)} - \beta_n^{(m+1)} = 2B_n^{n-m},$$
$$2\lambda_n^{-m} + 2\lambda_n^{m} = \beta_n^{-(m+1)} + \beta_n^{(m+1)} = 2B_n^{n-m}.$$

For $m = 0$, the same theorem gives $\lambda_n^0 = \beta_n^1 = \beta_n^{-1}$. By Theorem 15,

$$2\lambda_n^0 = \beta_n^1 + \beta_n^{-1} = 2B_n^{-n}.$$

Remark. The number of D_n-snakes

$$(-x_2 < x_1 < x_2 > x_3 < \ldots x_n), \quad |x_k| < n, \quad |x_k| \neq |x_l|,$$

with given beginning $x_1 = i$ can be expressed in terms of the quantities studied above in the following way. If $x_1 \geqslant 0$, then this number is equal to λ_n^i, as in this case the condition $-x_2 < x_1$ holds automatically. But if $x_1 < 0$, then the number of D_n-snakes beginning with $x_1 = i$ is equal to λ_n^{-i} (since changing the sign of x_1 in a D_n-snake gives a D_n-snake).

Thus, the difference $\lambda_n^{-m} - \lambda_n^m$ *is the number of* λ_n*-snakes with a given negative beginning* $-m$ *that are not* D_n*-snakes.*

This proves another combinatorial fact that is not obvious.

***Theorem* 27.** *The number of* λ_n*-snakes*

$$(x_1 < x_2 > \ldots x_n), \quad |x_k| < n, \quad |x_k| \neq |x_l|$$

with given negative beginning $x_1 = -m$ *that are not* D_n*-snakes (that is, for which* $x_1 < -x_2$*) is equal to the number* B_n^{n-m} *of* B_n*-snakes*

$$(0 < y_1 > y_2 \ldots y_n), \quad 1 \leqslant |y_k| \leqslant n, \quad |y_k| \neq |y_l|,$$

ending with $y_n = n-m$.

§13. Properties of numbers of snakes modulo p^r

These properties are generalizations of theorems of Kummer and von Staudt on the divisibility properties of Bernoulli and Euler numbers. Consider the triangle

$$c_{i,j} = a_{i,j}\,(-1)^{s(s-1)/2}, \quad s = i + j,$$

of coefficients of the exponential generating function. Let p be an odd prime number and let $r \geqslant 1$ be an integer. Consider a subtriangle of step size $p-1$, that is, for fixed (u, υ), the array

$$f_{u,\upsilon}(k, l) = c_{u+k(p-1),\, \upsilon+l(p-1)}.$$

***Theorem* 28** (see [2]). *The residue $f_{u,\upsilon}(k, l)$ mod p^r is a polynomial of combined degree less than r in k and l, provided that u and υ are large enough ($\geqslant r$).*

Remark 1. Similar congruences hold for Euler and Bernoulli numbers, that is, $f_{u,0}(k, 0)$ and $f_{0,\upsilon}(0, l)$ mod p^r are polynomials of degree less than r (the Knuth–Buckholtz theorem, [5]). This follows easily from Theorem 28. Conversely, Theorem 28 for $r = 1$ (that is, the statement that $f_{u,\upsilon}(k, l)$ mod p = const for fixed u, υ) follows from the Knuth–Buckholtz theorem, since the independence of k and l separately implies the independence of k and l together.

However, the congruences of Theorem 28 for $r > 1$ do not follow from the Knuth–Buckholtz theorem, which states only that $f_{u,\upsilon}(k, l)$ mod p is a polynomial of degree less than r in k for fixed l and in l for fixed k.

Remark 2. The polynomials of degree less than r mentioned in Theorem 28 take *integer values* at integer points. In all the cases in which they have been calculated, they even have *integer coefficients* for some reason. No general formula for these polynomials is known, and I only give some examples.

Let $p = 3$, $r = 4$, that is, we reduce the triangle modulo 81. It turns out that for $k > 1$, $l > 1$

$$c_{2k,2l} \equiv [27(kl^2 + k^2 l) - 9k^3] + (9kl - 27k^2 - 18l^2) + (30k + 9l) + 1 \mod 81,$$

$$c_{2k+1,2l} \equiv [27(kl^2 + k^2 l) - 9k^3] + (36kl + 36l^2) - 24k - 2 \mod 81.$$

Let $p = 5$, $r = 2$, that is, we reduce our triangle modulo 25. It turns out that for $k > 0$, $l > 0$

$$c_{4k,4l} \equiv 24 \mod 25,$$

$$c_{4k+1,4l} \equiv 21 + 5(k - l) \mod 25.$$

Similar congruences probably hold for the numbers in other triangles of snakes. It would be interesting to develop a complete theory of Bernoulli numbers for arbitrary Coxeter groups: ordinary Bernoulli numbers are related to the series $\zeta(n) = 1^{-n} + 2^{-n} + \ldots$; what are the ζ-functions related to the other Coxeter groups?

CHAPTER III

SPRINGER CONES AND BIFURCATION DIAGRAMS OF SINGULARITIES

§14. Morsification spaces and Maxwell strata

Definition. A *bifurcation diagram of functions* is the hypersurface in a function space, consisting of the functions with degenerate critical points or with equal critical values at different critical points.

A bifurcation diagram divides a neighbourhood of its function in the function space into connected components (of the complement of the diagram). Representatives of these components are called *morsifications* of the original function.

A morsification is called *maximal* (or an M-morsification[(1)]) if the number of non-degenerate critical points at which the given degenerate critical point splits is the maximum possible (equal to the multiplicity of the original critical point). Analogous definitions are given in the local theory for functions defined not on a compact manifold but only in a neighbourhod of a single point (see [6], [7], [8]).

Example. Consider morsifications of the function x^{n+1} of one variable. M-morsifications are equivalent to real polynomials

$$A_n\colon x^{n+1}+\lambda_1 x^{n-1}+\ldots+\lambda_{n-1}x, \quad \lambda\in\mathbb{R}^n,$$

having n different real critical values.

The number of inequivalent morsifications is equal to the number of connected components into which the bifurcation diagram divides the real n-dimensional space of polynomials of the above type.

For $n = 3$ this is a partitioning of the plane (λ_1, λ_2) by a semicubical parabola formed by the polynomials $x^4+\lambda_1 x^2+\lambda_2 x$ with degenerate critical point, and its half-tangent at zero ($\lambda_2 = 0$, $\lambda_1 \leqslant 0$) formed by polynomials with coinciding critical values (Fig. 12).

The bifurcation diagram here, as in the general case, is a union of two hypersurfaces: a *caustic* (generalizing the semicubical parabola) composed of functions with degenerate critical points and a *Maxwell stratum* composed of functions with coincident critical values at different points.

Remark. The continuation of a tangent beyond the point of contact does not belong to the Maxwell stratum of a real function. This continuation consists of the real points of the Maxwell stratum of the complexified function that are not contained in the Maxwell stratum of the real function itself (the coincident critical values of a polynomial are taken to be at complex conjugate points).

---

(1)The name M-morsification is formed by analogy with that of M-manifold, introduced by I.G. Petrovskii [9]; this denotes a real algebraic manifold for which the sum of the Betti numbers (with coefficients $\mathbb{Z}_2$) is maximal (equal to the sum of the Betti numbers of the complexified manifold).

Of the three components into which a bifurcation diagram divides the plane, the two smallest ones (shaded in Fig. 12) are filled up by M-morsifications. Let $K(f)$ denote the number of components formed by the M-morsifications of a function f in a neighbourhood of a degenerate singular point. We see that $K(A_3) = 2$.

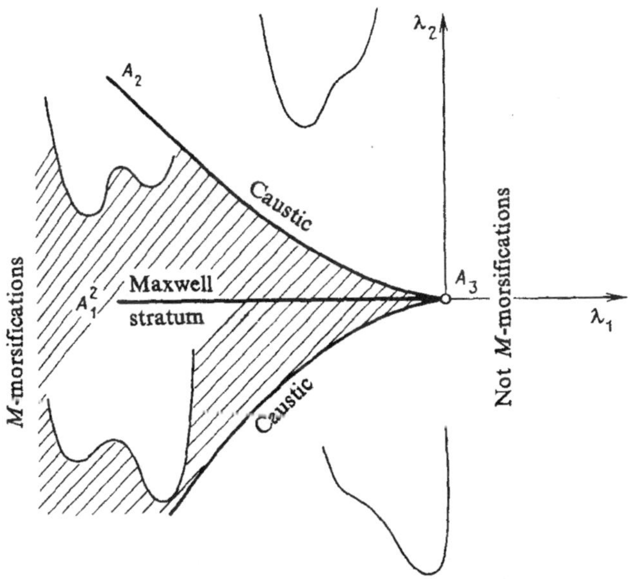

Fig. 12. Two types of M-morsifications of the singualrity A_3.

The bifurcation diagram of the function A_4 is shown in Fig. 13. The caustic is a swallow-tail and the Maxwell stratum, including complex points, is also a swallow-tail. The set of M-morsifications lies inside a pyramid bounded by the caustic. The real part of the Maxwell stratum, situated inside this pyramid, is that part of the inverted swallow-tail that lies inside the pyramid bounded by the caustic.

Thus the number of components of the manifold of M-morsifications of the function x^5 is $K(A_4) = 5$.

***Theorem* 29** (see [2], [3], [8]). *The number of components of the manifold of M-morsifications of a function A_n (with n different critical values) is equal to the number a_{n-1} of A_n-snakes with n elements*

$$(z_1 < z_2 > \ldots z_n), \quad \{z_1, \ldots, z_n\} = \{1, \ldots, n\}.$$

This number is also equal to the number of topologically inequivalent very good Morse functions of one variable with n critical values that behave like x^{n+1} at infinity.

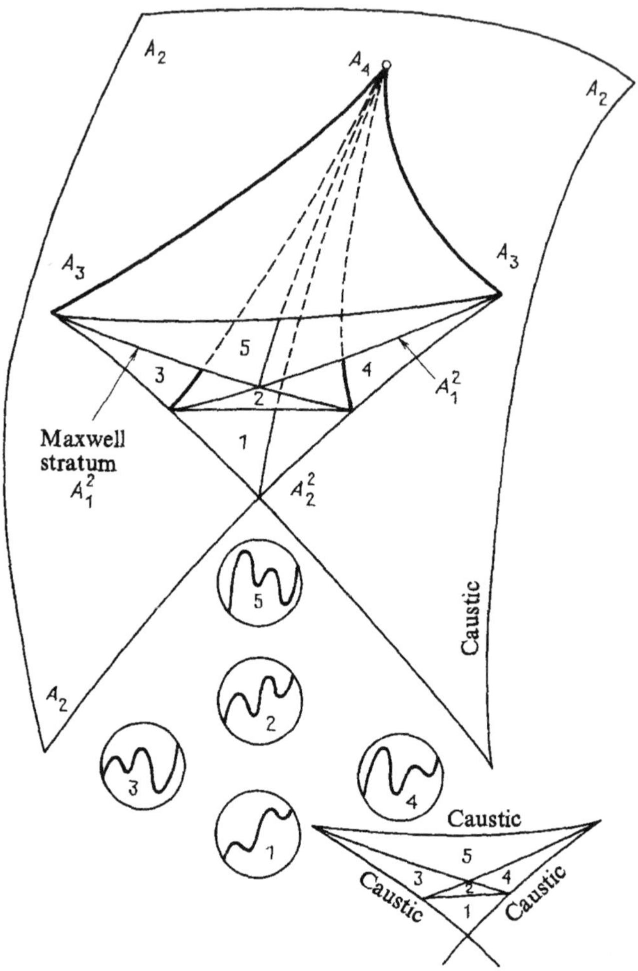

Fig. 13. Five types of M-morsifications of the singularity A_4.

A *very good Morse function* is a function whose critical points are all non-degenerate and whose critical values are all different. *Topological equivalence* is defined by orientation-preserving homeomorphisms of the axes of the independent and dependent variables.

Thus *the number of inequivalent M-morsifications of the function* x^{n+1} *is equal to the Springer number of the Coxeter group* A_{n-1}:

$$K(A_n) = a_{n-1}.$$

A comparison of Fig. 13 with Fig. 3 shows that the section by the plane $\lambda_1 = -1$ of the pyramid bounded by the swallow-tail caustic of the function A_4 and the partition of this section by the Maxwell stratum is homeomorphic (and at interior points is diffeomorphic) to the section of the principal Springer cone of the Coxeter group A_3 by a suitable plane and the partition of this section by mirrors. This is no accident.

***Theorem* 30** (see [3]). *The closure of the set of M-deformations at a singularity A_n and its subdivision into components by Maxwell strata is homeomorphic (and inside the region of M-deformations is also diffeomorphic) to the principal Springer cone of the Coxeter group A_n, and to its division by mirrors into Weyl chambers.*

An *M-deformation* is a deformation without complex (not real) critical points.

Remark. There is a unique algebraic homeomorphism of the closure of the set of M-deformations onto the principal Springer cone possessing the properties indicated in Theorem 30. In addition, the restriction of this homeomorphism to the interior of each stratum of the boundary of each component of the manifold of M-morsifications (composed of very good Morse functions) is a diffeomorphism onto the interior of the corresponding face of the Weyl chamber.

This natural algebraic homeomorphism is constructed in the following way. We enumerate the n critical points of the M-deformation of the function x^{n+1} in decreasing order ($x_1 \geqslant x_2 \geqslant \ldots \geqslant x_n$). The corresponding critical values form a (generalized) snake ($y_1 \leqslant y_2 \geqslant y_3 \leqslant \ldots y_n$). We subtract the arithmetic mean of the critical values from each of them. We have assigned to an M-morsification (a very good Morse function) a point ($z_1 < z_2 > \ldots z_n$) of the Springer cone of A_{n-1}. It is proved in [3] that this mapping possesses the properties listed above. The proof is based on the uniqueness of the holomorphic structure on the Riemann sphere and on the papers of Thom [10] and Zdravkovska [11] on the topological classification of complex polynomials.

This mapping is closely related to the Lyashko–Looijenga polynomial mapping, which maps the base of the truncated (by deleting the free term) versal deformation of an ordinary complex function with a critical point of multiplicity n into the base of the versal deformation of the singularity A_{n-1}.

This mapping assigns to a function from the versal deformation of an n-fold singularity a (non-ordered) set of critical values shifted in such a way that their sum becomes equal to zero (see [12], [15], [6], [7]).

This mapping is a covering $\mathbb{C}^{n-1} \to \mathbb{C}^{n-1}$ branched along the bifurcation diagram of functions whose images belong to the discriminant (a swallow-tail in $\mathbb{C}^{n-1}$). The complement of the bifurcation diagram of functions is mapped onto the complement of the swallow-tail as an unbranched covering. The complement of the swallow-tail in $\mathbb{C}^{n-1}$ is an Eilenberg–MacLane space $K(\pi, 1)$, where π is the braid group on n strings. Therefore the complement of the bifurcation diagram of a simple singularity is the Eilenberg–MacLane space $K(\pi', 1)$, where π' is a subgroup of the braid group on n strings.

The index of this subgroup is equal to

$$[\pi : \pi'] = n!h^n \,/\, |\, W \,|,$$

where h is the Coxeter number, and $|W|$ is the order of the reflection group corresponding to the original simple singularity. For the singularity A_n the index is equal to $(n+1)^{n-1}$.

In particular, the Lyashko–Looijenga mapping for A_3 is a polynomial mapping $\mathbb{C}^2 \to \mathbb{C}^2$ transforming the union of a semicubical parabola and its tangent at zero into a semicubical parabola and providing a 16-sheeted covering of the complement of this union over the complement of the parabola.

The coincidence of this index $(n+1)^{n-1}$ with the number of trees with $n+1$ ordered vertices (Cayley's formula) is not accidental: the trees correspond to different Dynkin diagrams.

The connection between the mapping of Theorem 30 and the Lyashko–Looijenga mapping for A_n is as follows.

Consider a pair of "*Vietà mappings*" assigning to an ordered set of complex numbers (with zero sum) an unordered set. A Vietà mapping is polynomial: it assigns to a set of polynomial roots a set of coefficients (expressed in terms of the roots as basic symmetric functions).

The first Vietà mapping $\mathbb{C}^{n-1} \xrightarrow{\mathrm{I}} \mathbb{C}^{n-1}$ assigns to an ordered set of critical points of a function from the truncated versal deformation of a singularity A_n a set of coefficients of this function (obtained by some "rescaling" of the set of coefficients of its derivatives).

The second Vietà mapping $\mathbb{C}^{n-1} \xrightarrow{\mathrm{II}} \mathbb{C}^{n-1}$ assigns to an ordered set of critical values of our function, shifted by subtracting their average, a set of coefficients of a polynomial whose zeros are these values.

Thus a commutative diagram of polynomial mappings arises (showing, incidentally, that the Lyashko–Looijenga mapping for A_n is a Tschirnhausen mapping):

$$\begin{array}{ccc} \mathbb{C}^{n-1} \ \text{(ordered critical points)} & \longrightarrow & \mathbb{C}^{n-1} \ \text{(shifted ordered critical values)} \\ {\scriptstyle \mathrm{I}}\downarrow & & \downarrow{\scriptstyle \mathrm{II}} \\ \mathbb{C}^{n-1} \ \text{(truncated versal deformation of } A_n) & \longrightarrow & \mathbb{C}^{n-1} \ \text{(versal deformation of } A_{n-1}) \end{array}$$

We replace the upper spaces by their real parts, and the lower ones by pyramids of swallow-tails (that is, by the images of $\mathbb{R}^{n-1}$ under the corresponding Vietà mappings). We get a diagram of real polynomial mappings:

$$\begin{array}{ccc} \mathbb{R}^{n-1} \ \text{(ordered critical points)} & \longrightarrow & \mathbb{R}^{n-1} \ \text{(shifted ordered critical values)} \\ {\scriptstyle \mathbb{R}\mathrm{I}}\downarrow & & \downarrow{\scriptstyle \mathbb{R}\mathrm{II}} \\ \Pi^{n-1} \ \text{(pyramid of } A_n\text{-functions with real critical points)} & \longrightarrow & \Pi^{n-1} \ \text{(pyramid of } A_{n-1}\text{-functions with real zeros)} \end{array}$$

Here the vertical arrows are homeomorphisms onto the corresponding "pyramids'.

The mapping of Theorem 30 goes from the lower left-hand corner of this diagram to the upper right-hand corner. This is the superposition of the inverse of the Vietà homeomorphism ℝI and the upper horizontal mapping.

Thus, *the mapping of Theorem* 30 *is transformed into the real part of the Lyashko–Looijenga mapping if one identifies the spaces in the right-hand column by means of the Vietà homeomorphism* ℝII.

Another way to get the sequence 1, 1, 1, 2, 5, 16, 61, ... from polynomials in a single variable was given by Catanese.

Let us consider a complex polynomial

$$z^n + \lambda_1 z^{n-1} + \ldots + \lambda_n.$$

Definition. A polynomial is called *typical* if all the $n-1$ non-zero critical values of the square of its modulus are different (of course, they are attained at the roots of its derivative).

Theorem **31** (Catanese). *The number of connected components of a manifold of typical polynomials of degree n is equal to* a_{n-2}:

degree	1	2	3	4	5	6	7...
number of components	1	1	1	2	5	16	61...

Definition. A line of positive level of the modulus of a polynomial is called a *generalized Cassini oval* and if the level is critical, a *generalized lemniscate* (ordinary lemniscates and Cassini ovals are obtained when $n = 2$).

A set of ovals and lemniscates of a typical polynomial is constructed as follows. If the modulus is large enough, then there is a single oval close to a circle. On decreasing the modulus there first appears a lemniscate in the form of a figure eight, so all the zeros of the polynomial lie in the regions bounded by the loops on the eight. Later on, decreasing the modulus inside each region, there appear new eights, and inside their loops yet more new ones, and so on, until the lowest critical value is assumed.

Example. For a polynomial of degree 2 there is only one non-zero critical value and also one lemniscate.

For a cubic polynomial there are two critical values. The two regions bounded by the first lemniscate above differ: one contains a vertex of the second lemniscate and the other does not. Therefore the topological type of the modulus of a typical cubic polynomial is uniquely defined (Fig. 14).

For a polynomial of degree 4 there are three non-zero critical values. The first lemniscate above bounds two regions. The vertices of the other two may either be both in one of these regions, or one in each (Fig. 14). There are two topologically possible configurations, both realized by typical polynomials. A topological configuration of lemniscates is a set of lemniscates ordered by heights.

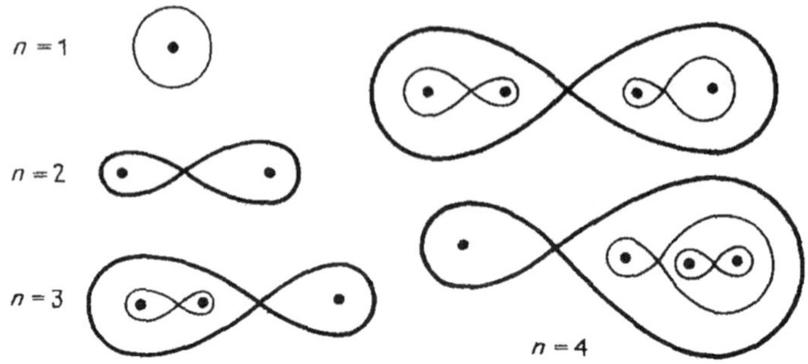

Fig. 14. The lemniscates of polynomials of degree n.

The number of topologically possible lemniscate configurations of typical nth degree polynomials is equal to a_{n-1} (this is proved by means of the recurrence relations in the first remark from §6 above).

Catanese's theorem states that, firstly, all topologically possible configurations of lemniscates are realized by typical polynomials, and secondly, that homeomorphic lemniscate configurations correspond to typical polynomials from the same connected component of a typical polynomial space.

§15. Morsifications of singularities of B-series

The construction of §14 assigns an integer $K(f)$, the number of M-morsifications, to any critical point of finite multiplicity of a real function.

Since the simplest critical points are classified by reflection groups, this enables us to assign integers to reflection groups (strictly speaking, complex functions correspond to reflection groups, and the numbers $K(f)$ for different real forms are distinct; one may take, for example, the maximum over all real forms f corresponding to a given reflection group).

We say above that the number K assigned in this manner to the reflection group A_n coincides with the Springer number, which, however, corresponds not to this reflection group but to the previous one, A_{n-1}:

$$K(A_3) = 2 = a_2, \quad K(A_4) = 5 = a_3 \text{, and so on.}$$

Now consider from this point of view the next series of Coxeter groups B_n (the symmetry groups of cubes in $\mathbb{R}^n$).

In singularity theory the group B_n corresponds to a *boundary singularity* of the function x^n on the axis $\{x\}$ with "boundary" $x = 0$ (see [14], [6], [7]).

The role of "critical values" in this case is played by the $n-1$ ordinary critical values and the value at the "boundary" point $x = 0$.

Definition. *A truncated versal deformation* of a boundary singularity B_n is a family of real polynomials

$$x^n + \lambda_1 x^{n-1} + \ldots + \lambda_{n-1}x, \quad \lambda \in \mathbb{R}^{n-1}.$$

A bifurcation diagram of functions with singularity B_n is the hypersurface formed by those values of the parameter λ for which the polynomial has a critical point or coinciding critical values.

Remark. This bifurcation diagram is thus a union of four hypersurfaces:

1) *an ordinary caustic* (there exists a degenerate critical point);
2) *a boundary caustic* (the derivative at zero is equal to zero);
3) *an ordinary Maxwell stratum* (a pair of critical points with a common value);
4) *a boundary Maxwell stratum* (a non-zero critical point with zero value).

Example. The bifurcation diagram of the function $B_3 = x^3+\lambda_1 x^2+\lambda_2 x$ is shown in Fig. 15. (In this case the ordinary Maxwell stratum is absent.)

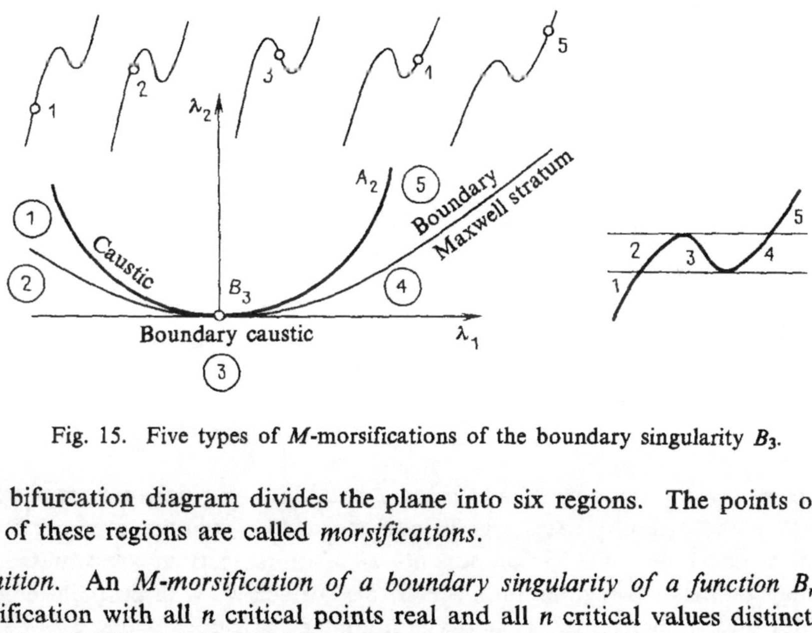

Fig. 15. Five types of M-morsifications of the boundary singularity B_3.

A bifurcation diagram divides the plane into six regions. The points of each of these regions are called *morsifications.*

Definition. An *M-morsification of a boundary singularity of a function* B_n is a morsification with all n critical points real and all n critical values distinct.

Example. Five of the six regions of morsifications of a boundary singularity B_3 in Fig. 15 are composed of M-morsifications.

Theorem 32. *The number of components of M-morsifications of a boundary singularity B_n is equal to the Springer number a_n, that is, to the number of permutations $(x_0 < x_1 > \ldots x_n)$ of the set $\{0, \ldots, n\}$.*

This unexpected theorem, proved in [2], is a motivation for the strange combinatorial Theorem 7 and the starting-point for all the theory presented in this article.

Theorem 30 carries over to the case B_n in the following way.

***Theorem* 33** (see [3]). *The closure of the set of M-morsifications of a boundary singularity B_n and its partition by the boundary caustic and the Maxwell stratum into regions of M-morsifications is homeomorphic to a gallery of Springer cones of a singularity A_{n-1} and its partition by mirrors into Weyl chambers.*

A gallery of Springer cones is defined at the end of §4.

Example. Let $n = 3$. The mirrors of the group A_2 divide the plane into six Weyl chambers, each with an angle of 60°. A principal Springer cone is an angle adjacent to the chamber. The gallery consists of this angle of 120°, an angle of 120° vertically opposite to it, and the other Weyl chamber (Fig. 16). The gallery consists of the five chambers $1-5$ and is homeomorphic to the system of regions of M-morsifications of a singularity B_3, shown in Fig. 15.

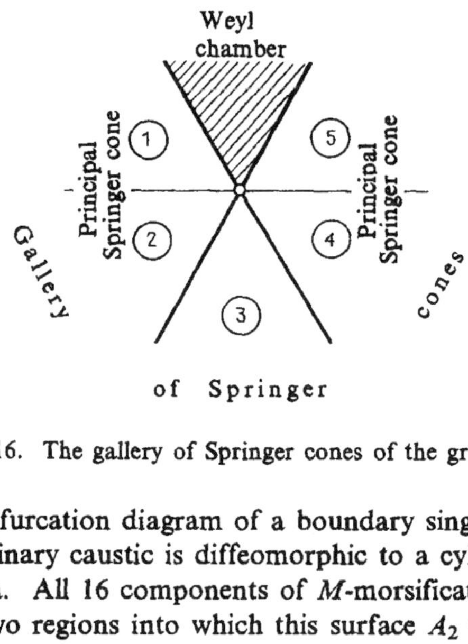

Fig. 16. The gallery of Springer cones of the group A_2.

Let $n = 4$. A bifurcation diagram of a boundary singularity B_4 is shown in Fig. 17. The ordinary caustic is diffeomorphic to a cylinder over a semicubical parabola. All 16 components of M-morsifications are situated in the smaller of the two regions into which this surface A_2 with cuspidal edge A_3 divides three-dimensional space. The boundary caustic is a plane $\lambda_3 = 0$, osculating the cuspidal edge at the point B_4. It is tangent to the caustic A_2 along the line B_3.

The ordinary Maxwell stratum A_1^2 divides the region of M-morsifications into two parts which are symmetrical to each other. Each of these parts is divided by the boundary caustic and the boundary Maxwell stratum into eight regions of morsifications. For one of the parts, these regions are enumerated by digits $1-8$. In Fig. 17 the disposition of the boundary point among the non-boundary critical points for these eight components is also shown.

In this case Theorem 33 states that this whole configuration of 16 regions of M-morsifications bounded by complex algebraic surfaces is homeomorphic to the partition of a specific polyhedral cone in $\mathbb{R}^3$ into 16 simplicial cones, namely, the gallery of Springer cones of the group A_3 in $\mathbb{R}^3$ whose simplicial cones are the Weyl chambers into which the gallery is partitioned by mirrors.

Thus, the left-hand part of Fig. 3 gives us the polyhedral topological model of the configuration of regions of M-morsifications shown in Fig. 17.

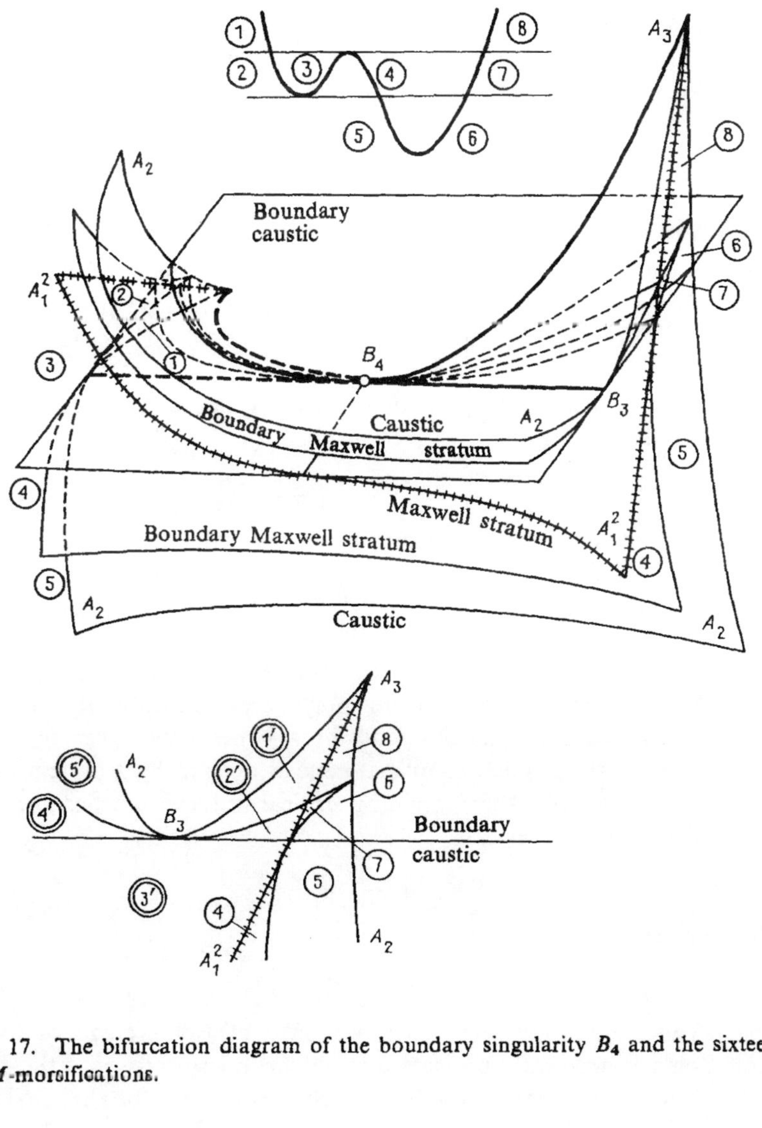

Fig. 17. The bifurcation diagram of the boundary singularity B_4 and the sixteen types of its M-morsifications.

Namely, the region above the plane of the boundary caustic corresponds in the right-hand part to one of the principal Springer cones. It contains five components of M-morsifications, which correspond to the five numbered triangles in Fig. 3. When we cross the boundary caustic downwards we arrive in three other regions of M-morsifications, which correspond to the partition of the Springer cone adjacent to the principal Springer cone in Fig. 3 into three Weyl chambers. A second crossing through the boundary caustic (in its other part, the one closer to the viewer in Fig. 17) gives three more regions of M-morsifications situated above the plane $\lambda_3 = 0$. In Fig. 3 these correspond to another triangle, next to the one representing the principal Springer cone and also divided into three parts by mirrors of the group A_3. Finally, the other five components of M-morsifications lie in the far part of Fig. 17 below the plane $\lambda_3 = 0$. They correspond to five Weyl chambers in the Springer cone vertically opposite to the initial one and are represented in Fig. 3 by the same partition of the large triangle from which we started into five small ones.

Remark. A homeomorphism from the closure of the union of the M-components of a boundary singularity B_n to a gallery of Springer cones of singularity A_{n-1} is constructed according to the scheme described above for the singularity A_n.

We arrange the critical points of a function of versal deformation of B_n, *including zero*, in decreasing order:

$$x_1 \geqslant x_2 \geqslant \ldots \geqslant x_n.$$

This ordering defines an ordered sequence of critical values, also including zero,

$$(y_1, y_2, \ldots, y_n).$$

We get a homeomorphic mapping from the closure of the union of regions of M-morsifications of a singualrity B_n into a polyhedral surface in $\mathbb{R}^n$. Namely, the faces of this surface are simplicial cones on n hyperplanes given by equations such as $y_k = 0$. Each simplicial cone is defined by a system of natural inequalities between the remaining critical values. For example, if $y_1 = 0$, then these inequalities are

$$(0 \leqslant y_2 \geqslant \ldots y_n),$$

if $y_2 = 0$, they are

$$(y_1 \leqslant 0 \leqslant y_3 \geqslant y_4 \ldots y_n),$$

and so on. The subtraction from each vector y of the average of its components homeomorphically projects this polyhedral surface onto the plane $z_1 + \ldots + z_n = 0$. The image is a gallery of Springer cones for the Coxeter group A_{n-1}.

This mapping from the closure of the union of regions of M-morsifications for B_n into $\mathbb{R}^{n-1}$ (which we may again regard as a real version of the Lyashko – Looijenga mapping) homeomorphically maps the closure onto a gallery, diffeomorphically transforming the components into the Weyl chambers of the group A_{n-1}. It is also a diffeomorphism of the interior of each stratum of the boundary onto the interior of the corresponding face of the Weyl chamber of the reflection group A_{n-1}. Moreover, our reflection is a local diffeomorphism in a neighbourhood of each point not belonging to a caustic (ordinary or boundary). Thus, onto each open Springer cone of our gallery is diffeomorphically mapped a region of M-morsifications (not necessarily "very good" ones, that is, we admit equal critical values at non-coincident points).

Definition. An equivariant singularity B_n is a singularity of an even function z^{2n} on a line $\{z\}$ with involution $z \mapsto -z$. The *truncated versal deformation of B_n* consists of all even polynomials

$$z^{2n} + \lambda_1 z^{2n-2} + \ldots + \lambda_{n-1} z^2, \quad \lambda \in \mathbb{R}^{n-1}.$$

The relation between equivariant singularities B_n and the reflection group B_n is explained in detail in [14].

Definition. An *M-morsification of an equivariant singularity B_n* is an even polynomial from this family that has n different critical values.

***Theorem* 34** (see [3]). *The closure of the set of M-morsifications of an equivariant singularity B_n and the partition into components of the region of very good M-morsifications is homeomorphic to the principal Springer cone of the reflection group A_{n-1} and its partition into Weyl chambers.*

In particular, the number of components of the set of M-morsifications of an equivariant singularity B_n is the Springer number a_{n-1} ($=$ 1, 2, 5, 16, ... for n = 2, 3, 4, 5, ...).

Example. In Fig. 17 these are the five components in the upper right-hand corner. They form the same configuration as the five Weyl chambers of the principal Springer cone for A_3 (Fig. 3).

The critical points of our polynomial are zero and the square roots of the critical points of the polynomial $x^n + \lambda_1 x^{n-1} + \ldots + \lambda_{n-1} x$. Thus, this polynomial must have $n-1$ different positive roots ($x_1 > x_2 > \ldots > 0$). The mapping constructed above sends this polynomial into the principal Springer cone of our gallery. Indeed, the values at the critical points, *together with the zero at the end,* form the alternating sequence

$$(y_1 < y_2 > \ldots 0).$$

Therefore the image of the closure of the set of M-morsifications of an equivariant singularity B_n gets into the principal Springer cone of the reflection group A_{n-1}, which also implies Theorem 34. This homeomorphism is a local diffeomorphism at all interior points of the set of M-morsifications (even at points not belonging to the Maxwell strata).

§16. Topological classification of odd functions

Thus the numbers of M-morsifications of the B-series singularities are the Springer numbers a_k of the symmetry groups of simplexes (the reflection groups of the A-series). What is the significance in singularity theory of the Springer numbers of the groups B_n of symmetries of cubes, $b_n = 1, 3, 11, 57, 361, \ldots$ $(n = 1, 2, \ldots)$? It turns out that they count the topological types of odd functions with $2n$ critical values.

Definition. Two odd functions on the real axis are called *topologically equivalent* if one can be transformed into the other by odd orientation-preserving homeomorphisms of the axes of the dependent and independent variables.

Example. There are exactly three different topological types of odd Morse functions with four critical points and four critical values that behave at infinity like x^5 (Fig. 18).

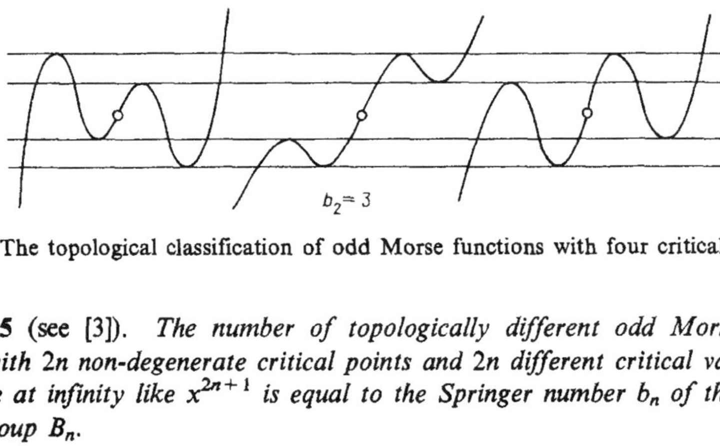

Fig. 18. The topological classification of odd Morse functions with four critical values.

Theorem 35 (see [3]). *The number of topologically different odd Morse functions with 2n non-degenerate critical points and 2n different critical values that behave at infinity like* x^{2n+1} *is equal to the Springer number* b_n *of the Coxeter group* B_n.

Proof. We enumerate the positive critical points in decreasing order $(x_1 > x_2 > \ldots > x_n)$. The corresponding critical values form a sequence $(y_1 < y_2 > \ldots y_n)$ that remains alternating after adjoining a zero on the right.

We assign to each y_k a value z_k of the same sign and equal in modulus to the number of numbers $|y_l|$ that are less than or equal to y_k. The numbers

$$(0 < w_1 > w_2 < \ldots w_n) = (-1)^{n+1} (z_n, \ldots, z_1)$$

form a B_n-snake. It is easy to check that every B_n-snake can be realized in this way, and that the topological type of the function of this snake is uniquely defined.

Definition. An *odd-versal deformation of the function* x^{2n+1} is an odd polynomial

$$x^{2n+1} + \lambda_1 x^{2n-1} + \dots + \lambda_n x, \ \lambda \in \mathbb{R}^n.$$

Such a polynomial is called an *odd M-deformation* if all its critical points are real. It is called *a very good odd M-morsification of the function* x^{2n+1} if, in addition, all its $2n$ critical values are pairwise distinct.

***Theorem* 36** (see [3]). *The number of connected components of the space of very good odd M-morsifications of the function* x^{2n+1} *is equal to the Springer number* b_n.

Let us put the non-negative critical points of an M-deformation into decreasing order ($x_1 \geqslant x_2 \geqslant \dots \geqslant x_n \geqslant 0$). The corresponding critical values form a sequence ($z_1 \leqslant z_2 \geqslant z_3 \leqslant \dots 0$). We get a continuous proper mapping of the space of odd M-deformations of x^{2n+1} into $\mathbb{R}^n$.

Whenever an odd function has a critical value z_i, it also has a critical value $-z_i$. Therefore for a very good odd M-morsification the following inequalities hold:

$$z_i \neq \pm z_j, \ z_i \neq 0.$$

Thus the vector of critical values of a very good odd M-morsification gets inside one of the Weyl chambers of the reflection group B acting on $\mathbb{R}^n$. This chamber is the same for all odd M-morsifications of the same component of the space of very good odd M-morsifications. This component is mapped diffeomorphically into the interior of the Weyl chamber (this is proved in the same way as the Lyashko–Looijenga theorem: see [7], [13]).

Our map is an algebraic homeomorphism from the space of odd M-morsifications of the function x^{2n+1} onto the principal Springer cone of the reflection group B_n. It maps components of a manifold of very good M-morsifications diffeomorphically onto the interiors of the Weyl chambers. Its restriction to the interior of each stratum of the boundary of each component is a diffeomorphism onto the interior of the corresponding face of the Weyl chamber. Moreover, our map is locally diffeomorphic at each interior point of the region of M-deformation (whether good or not).

Remark. We assign to an odd complex polynomial $x^{2n+1} + \dots$ the unordered set of the squares of its n critical values.

We obtain a polynomial mapping $\mathbb{C}^n \to \mathbb{C}^n$,

$$\{x^{2n+1} + \lambda_1 x^{2n-1} + \dots + \lambda_n x\} \to \{y^n + \varepsilon_1 y^{n-1} + \dots + \varepsilon_n\},$$

which is analogous to the Lyashko–Looijenga mapping, although its codomain is not the base of a versal deformation of an A_{n-1} singularity, but the base of a boundary singularity B_n. This polynomial mapping is a branched covering over the discriminant of the boundary singularity B_n.

The discriminant of the boundary singularity B_n (the manifold of non-regular orbits) consists of those values $\varepsilon \in \mathbb{C}^n$ for which the polynomial with coefficients ε has either a zero root or a multiple root.

The pre-image of the discriminant under this mapping is the set of odd functions having fewer than n different squares of critical values.

The complement of the discriminant of the boundary singularity B_n in $\mathbb{C}^n$ is the Eilenberg–MacLane space $K(\pi, 1)$, where π is the braid group of the B_n-series introduced by Brieskorn [15].

Therefore the manifold of odd complex morsifications of the function x^{2n+1} is also the Eilenberg–MacLane space $K(\pi', 1)$, where π' is a subgroup of finite index in the Brieskorn braid group of the B_n-series. The index of this subgroup is equal to $(2n+1)^n$ (as can be shown by counting weights).

Example. The manifold of odd morsifications of $x^5 + ux^3 + v$ forms the complement in the plane $\mathbb{C}^2 = \{(u, v)\}$ to four curves quadratically touching at 0 (this plane is the plane of symmetry of Fig. 13, and the curves are the intersections with the caustic, with its self-intersection line, with the Maxwell stratum, and with its line of self-intersection).

Our map is a branched covering $\mathbb{C}^2 \to \mathbb{C}^2$ such that the complement of the four quadratically touching smooth curves is a 25-sheeted covering of the complement to two of these curves.

Remark. The similarity between the formulae $(n+1)^{n-1}$ and $(2n+1)^n$ compels us to suppose that $(2n+1)^n$ is the answer to some problem of enumerative combinatorics, analogous to Cayley's problem on the enumeration of trees with ordered vertices.

The bifurcation diagrams of functions for a singularity D_4 contain elements that are diffeomorphic or homeomorphic to partitions of Springer cones of A_3 into Weyl chambers. It appears that real "sections" of a Lyashko–Looijenga mapping similar to that constructed above may also be defined for other simple singularities. But as far as I know, neither the number of components of M-morsification of singularities D_n, nor the sums of the Betti numbers of real or complex bifurcation diagrams for these singularities, have as yet been counted for $n > 5$.

References

[1] T.A. Springer, Remarks on a combinatorial problem, Nieuw Arch. Wisk. (3) **19** (1971), 30–36. MR **49** # 2965.

[2] V.I. Arnol'd, Bernoulli updown numbers, associated with function singularities, their combinatorics and arithmetics, Duke Math. J. **63** (1991), 537–555.

[3] ——, Springer numbers and morsification spaces, Preprint, University of Utrecht, no. 658, 1–18, Utrecht 1991. (To appear in J. Alg. Geom. 1:2 (1991).)

[4] N. Bourbaki, Eléments de mathématique, Fasc. XXXIV, Groupes et algèbres de Lie, Chs. 4–6, Hermann, Paris 1968. MR **39** # 1590.

[5] D.E. Knuth and T.J. Buckholtz, Computation of tangent, Euler and Bernoulli numbers, Math. Comp. **21** (1967), 663–688. MR **36** # 4787.

[6] V.I. Arnol'd, A.N. Varchenko, and S.M. Gusein-Zade, *Osobennosti differentsiruemykh otobrazhenii.* I, Nauka, Moscow 1982. MR **84m**:58016.
English translation: Singularities of differentiable maps. I, Monographs in mathematics 83, Birkhäuser, Boston, MA 1988.

[7] ———, V.A. Vasil'ev, V.V. Goryunov, and O.V. Lyashko, Singularities. I, Local and global theory, in: Current problems in mathematics. Fundamental directions, Vol. 6, 5-257, VINITI, Moscow 1988. MR **91h**:58010b.
English translation: Enc. Math. Sci. Vol. 6, Springer, Heidelberg 1992.

[8] ———, ———, ———, and ———, Singularities. II, Classification and applications, in: Current problems in mathematics. Fundamental directions, Vol. 39, 5-256, VINITI, Moscow 1989. MR **91h**:58010d.
English translation: Enc. Math. Sci. Vol. 39, Springer, Heidelberg 1992.

[9] I. Petrowsky [I.G. Petrovskii], On the topology of real algebraic curves, Ann. of Math. (2) **39** (1938), 189–209.

[10] R. Thom, L'équivalence d'une fonction différentiable et d'un polynôme, Topology **3** (1965), 297–307. MR **32** # 4702.

[11] S. Zdravkovska, The topological classification of polynomial mappings, Uspekhi Mat. Nauk **25**:4 (1970), 179–180. MR **46** # 2684.

[12] V.I. Arnol'd, Critical points of functions and a classification of caustics, Uspekhi Mat. Nauk **29**:3 (1974), 243–244.

[13] E. Looijenga, The complement of the bifurcation variety of a simple singularity, Invent. Math. **23** (1974), 105–116. MR **54** # 10661.

[14] V.I. Arnol'd, Critical points of functions on a manifold with boundary, the simple Lie groups B_k, C_k, F_4, and singularities of evolutes, Uspekhi Mat. Nauk **33**:5 (1978), 91–105. MR **80j**:58008.
= Russian Math. Surveys **33**:5 (1978), 99–116.

[15] E. Brieskorn, Sur les groupes de tresses (d'après V.I. Arnol'd), Séminaire Bourbaki 24ème année (1971/1972), Exp. No. 401, 21–44, in: Lecture Notes in Math. **317** (1973). MR **54** # 10660.

Steklov Institute of Mathematics

Received by the Editors 21 October 1991

Springer numbers and Morsification spaces

V.I. Arnold

In a note [4] T. A. Springer associated to any finite Euclidean reflection group an integer number. In this note we discuss the topological meaning of these Springer numbers in the theory of singularities of smooth functions. For instance, the Springer numbers, associated to the symmetry groups of cubes of dimensions $k = 1, 2, \cdots$, are

$$b_k = 1, 3, 11, 57, 361, 2763, \cdots.$$

We shall see that they count the topological types of the odd Morse functions of one variable (having $2k$ critical points and behaving like x^{2k+1} at infinity).

This approach generates an algorithm for the fast computation of the numbers b_k, similar to the classical "Euler-Bernoulli triangle" [2], which computes the Euler numbers and the tangent numbers (the numbers a_k in Springer's notation). The reader can discover the algorithm, guessing the rule, governing the following two triangules:

$$\begin{array}{cccccccccccc}
 & & & 0 & & & & & & 1 & & \\
 & & & 1\ \ 1 & & & & & & 1\ \ 0 & & \\
 & & & 0\ \ 1\ \ 2 & & & & & 2\ \ 3\ \ 3 & & & \\
 & & 11\ \ 11\ \ 10\ \ 8 & & & & & & 8\ \ 6\ \ 3\ \ 0 & & & \\
 & 0\ \ 11\ \ 22\ \ 32\ \ 40 & & & & & & 40\ \ 48\ \ 54\ \ 57\ \ 57 & & & & \\
361 & 361 & 350 & 328 & 296 & 256 & 256 & 216 & 168 & 114 & 57 & 0
\end{array}$$

The numbers a_k, which Springer associates to the symmetry group of k-simplices, coincide with the numbers of topological types of very nice Morse functions of one variable (having $k+1$ critical points and behaving at infinity like x^{k+2}).

Originally published in: *J. Algebraic Geometry* **1**(2), 197–214 (1992), © 1992 Amer. Math. Soc.

This coincidence provides an understanding of the topological structure of the complicated stratification, formed by the caustics and the Maxwell strata in the space of the truncated versal deformation of simple singularity A_{k+1} (see [3, 5]). The part of this stratification in the (closed) M-morsification cone is indeed homeomorphic to the subdivision of the (closed) Springer cone by the mirrors of group A_k.

The author thanks Professor T. Springer who has shown the author the note [4], the Mathematical Institute at Utrecht, where this paper was written, for their warm hospitality, Professor D. Siersma whose invitation has made this work possible, and Jean Arthur for fast deciphering of a difficult manuscript.

1. Definition of Springer numbers

Consider a finite group of linear transformations, generated by reflections, in a Euclidean space of dimension k. The mirrors of the group divide the space into simplicial cones, which are called chambers. Each chamber is bounded by k hyperplanes. These hyperplanes divide the space into 2^k parts.

Definition. The 2^k parts above are *Springer cones*. Two of the Springer cones are chambers. The others consist of several chambers. Springer describes an algorithm of calculation of the numbers of chambers in each of the Springer cones.

Definition. The *Springer number* of a Coxeter group is the number of chambers in the largest of the 2^k Springer cones. We shall call it the *principal* Springer cone.

Remark. Professor Springer communicated to the author that he has not considered these cones. Our notation is hence consistent with the usual notation W for a Coxeter group.

Example 1. The minimal angle between the axis of symmetry of a regular polygon in the plane is $\frac{\pi}{p}$. Hence the Springer numbers of the polygon symmetry groups $I_2(p)$ are

$$i_2(p) = p - 1, \quad a_2 = 2, \quad b_2 = 3, \quad g_2 = 5$$

(including $I_2(3) = A_2$, $I_2(4) = B_2$, $I_2(6) = G_2$).

Example 2. Any three-dimensional chamber has two orthogonal walls. These two mirrors divide the space into four congruent parts, each containing $|W|/4$ chambers, where $|W|$ is the total number of chambers (that is, the order of the reflection group). Hence the Springer number of any reflection group in 3-space is equal to

$$s_3 = |W|/4 - 1.$$

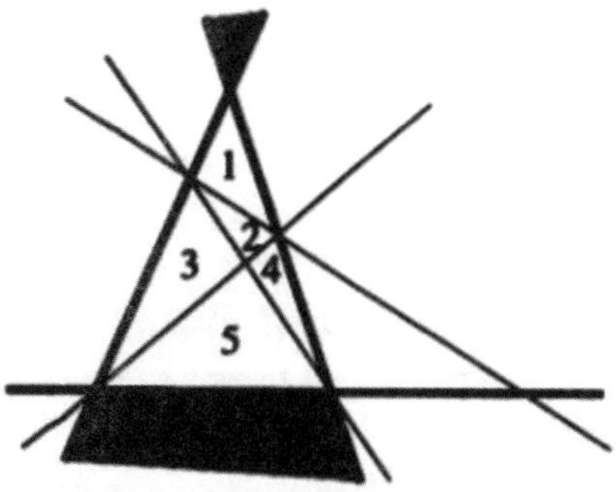

$a_3 = 5$

$24 = 2(5 + 3 + 3 + 1)$

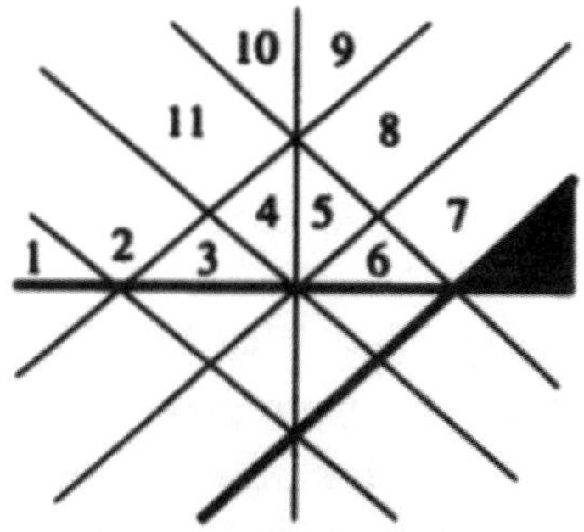

$b_3 = 11$

$48 = 2(11 + 7 + 5 + 1)$

$h_3 = 29$

$120 = 2(29 + 19 + 11 + 1)$

FIGURE 1

For instance,

$$a_3 = 4!/4 - 1 = 5, \quad b_3 = 8.3!/4 - 1 = 11, \quad h_3 = 120/4 - 1 = 29.$$

The configurations of these Springer cones and mirrors on the projective plane are shown in Figure 1. Each simplex represents two "opposite" chambers. The Springer cones are represented by larger simplices, bounded by double lines (one of these, in the b_3 case, lives at infinity). The original chamber is hatched.

Remark. The number of chambers in the Springer cells of H_3 $(1, 11, 19, 29)$ are strangely related to the degrees of the invariants of H_4 $(2, 12, 20, 30)$. In the same way, the chamber numbers of A_3 $(1, 3, 3, 5)$ generate the degrees $(2, 4, 4, 6)$ for D_4, and the chamber numbers of B_3 $(1, 5, 7, 11)$ generate the degrees $(2, 6, 8, 12)$ for F_4.

Example 3. The volume (and hence the number of chambers) of a spherical triangle is given by the well-known formula, in terms of the angles between the hyperplanes, bounding it.

This formula follows from the consideration of the areas of the biangles (which are proportional to the angles) by the standard inclusion-exclusion arguments. Iterating these arguments, one obtains the number of the chambers in the higher-dimensional Springer cones.

For instance, for H_4 we obtain

$$|W| = 14400 = 2(2281 + 1681 + 1199 + 719 + 601 + 599 + 119 + 1),$$

hence $h_4 = 2281$. This is the way Springer computed his numbers. For the exceptional groups he found

$$e_6 = 2663, \quad e_7 = 94\,121, \quad e_8 = 14\,422\,081, \quad f_4 = 169.$$

2. Springer numbers a_k and very nice Morse functions on a line

Let f be a real polynomial

$$f(x) = x^{\mu+1} + \lambda_1 x^{\mu-2} + \cdots + \lambda_{\mu-1} x,$$

where the λ_i are real parameters.

Definition. Such a polynomial is a *very nice* M-morsification of $x^{\mu+1}$ if all its μ critical values are real and different (M for *maximal*—this notion was introduced by Petrovski). Very nice M-morsifications form an open set in the space $\mathbf{R}^{\mu-1}$ with coordinates λ_i.

Theorem 1. *The pair (closure of the set of very nice M-morsifications, boundary of this set) is homeomorphic to the pair (closed principal Springer cone of $A_{\mu-1}$, its intersection with the mirrors).*

Corollary. *The set of very nice M-morsifications in the base space of a versal deformation of a function at an A_μ singularity contains $a_{\mu-1}$ connected components. Here a_k is the Springer number of A_k ($a_k = 1, 2, 5, 16, 61, 272, \cdots$ for $k = 1, 2, \cdots$).*

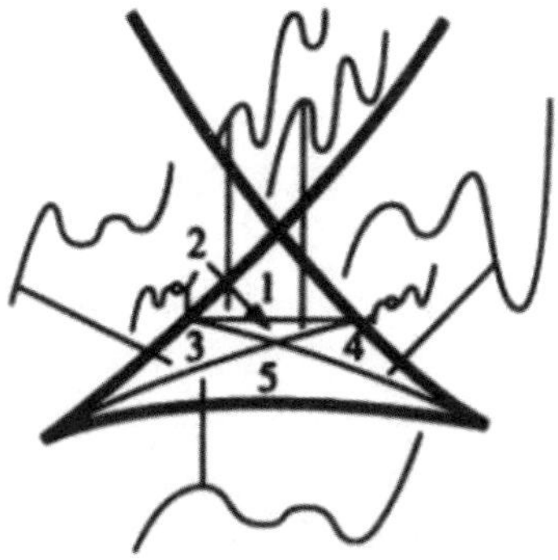

FIGURE 2

Example. The set of very nice M-morsifications in the base space of the truncated versal deformation

$$A_4 \sim x^5 + \lambda_1 x^3 + \lambda_2 x^2 + \lambda_3 x$$

is represented in Figure 2 by its intersection with the plane $\lambda_1 = -1$. It contains $a_3 = 5$ connected components. The triangle, bounded by the caustic's intersection with this plane, is divided by the Maxwell stratum into five curvilinear triangles exactly the same way as the triangle representing the principal Springer cone is divided by the mirrors in Figure 1.

Proof. Let us associate to a very nice M-morsification having μ ordered critical points $x_1 > x_2 > x_3 > \cdots > x_\mu$ the set of the corresponding critical values $z_i = f(x_i)$ ordered by the index i. Let us subtract the arithmetical mean of all these critical values from each of them and let us denote by $z_1 < z_2 > z_3 < \cdots$ the ordered set of μ real numbers with zero sum which we have obtained.

We have defined a continuous mapping from the set of very nice M-morsifications to the principal Springer cone of $A_{\mu-1}$ (since the "updown inequalities" $z_1 < z_2 > z_3 < \cdots$ define just the principal Springer cone of $A_{\mu-1}$ in the space $z_1 + z_2 + \cdots + z_\mu = 0$, according to [4]). This mapping admits a continuous extension to the boundary of the set of very nice M-morsifications. We shall prove that this extension is the required homeomorphism.

This fact is more or less equivalent to a known theorem of A. Vakulenko (see [3]). Its proof, however, has never been published. Hence, we present below the outline of this proof.

Lemma. *Any updown sequence* $z_1 < z_2 > z_3 < \cdots$, $z_1 + \cdots + z_\mu = 0$, *is generated by one and only one real polynomial of degree* $\mu + 1$ *with* *leading coefficient equal to one and vanishing next coefficient as the sequence of shifted critical values at* $x_1 > x_2 > \cdots$ *.*

Proof. The ramified covering $S^2 \to S^2$, defined by the required polynomial, is unambiguously determined topologically by the sequence of its critical values.

The curve $\operatorname{Im} f(x) = 0$ (in the complex sphere of x) consists of the real axis $\operatorname{Im} x = 0$ and of μ smooth lines, intersecting the real line orthogonally at the critical points $x_1 > \cdots > x_\mu$. These lines do not intersect each other. Each of them connects the critical point to infinity and is symmetric (invariant under the complex conjugation).

These lines divide each half-plane $\operatorname{Im} x > 0$, $\operatorname{Im} x < 0$ into $\mu+1$ discs. Each disc is sent by f to the upper (lower) half-plane $\operatorname{Im} z > 0$ ($\operatorname{Im} z < 0$) homeomorphically, including the boundaries. At the boundary the mapping f is defined by the order of the points z_i (up to a circle homeomorphism, preserving the orientation).

Hence a nice updown sequence, generated by an M-morsification, defines unambiguously the real (conjugation-equivariant) ramified covering, associated with this M-morsification.

Now let us construct the real polynomial, generating an updown sequence, given a priori. We start with $\mu + 1$ copies of the z plane. We cut the first copy along the real ray $[-\infty, z_1]$. The second copy is cut along two rays $[-\infty, z_1]$ and $[z_2, +\infty]$. We cut the ith copy along the rays $[z_i, (-1)^i\infty]$ and $[z_{i-1}, (-1)^{i-1}\infty]$. The last, $(\mu + 1)$th, copy will contain only one cut, $[z_\mu, (-1)^\mu\infty]$. We glue the points $z \pm 0i$ on the cuts with $z \mp 0i$ on the preceding and on the next copies. We obtain a symmetric (commuting with the complex conjugations) ramified covering $S^2 \to S^2$ with one simple ramification point over every point z_i and with a point of type $z = x^{\mu+1}$ at infinity. Topologically this is the required polynomial.

We equip the covering sphere with the complex structure, induced from the z-sphere. We obtain a complex structure with respect to which our ramified covering is a holomorphic mapping. The preimage of $z = \infty$ is one point ∞ on the source sphere. Hence our mapping is a polynomial. It has exactly the prescribed critical values.

The complex conjugation acts on our source sphere as an antiholomorphic involution (since the complex structure on the sphere is unique). Hence we can choose a coordinate x on the source sphere, for which the polynomial $z = f(x)$ has real coefficients and leading term $x^{\mu+1}$.

Finally, we annihilate the sum of the critical points of the polynomial by a real shift of the origin of x.

The resulting polynomial has the required critical values $z_1 < z_2 > \cdots$ at the critical points $x_1 > x_2 > \cdots$. Indeed, the derivative of f along the real line changes its sign at each critical point. Hence the sequence of critical points x_i is either decreasing or increasing. But on the real ray, corresponding to $[z_1, +\infty]$ on the first copy, the value of z is growing when the point moves from z_1 to $+\infty$. Hence this point is a minimum. If $\mu + 1$ is odd, our sequence is decreasing: $x_1 > x_2 > \cdots$. If $\mu + 1$ is even, it might be increasing, but we may change the sign of x.

Thus we have found the required polynomial. Its uniqueness follows from the above considerations (we use the uniqueness of the complex structure on S^2). The Lemma is thus proved.

Remark. A similar reasoning proves that the statement of the lemma still holds for the boundary case of nonstrictly updown sequences,

$$z_1 \le z_2 \ge z_3 \le \cdots, \qquad z_1 + \cdots + z_\mu = 0.$$

In this case the topological structure of the complex polynomial mapping is slightly different, but it is still unambiguously defined by the sequence (see, for instance, [5, 6]).

End of the proof of Theorem 1. The mapping which we have constructed is continuous, one-to-one (according to the Lemma and to the above remark), and proper. Hence it is a homeomorphism, and the theorem is proved. q.e.d.

Remark. Our mapping defines in fact a diffeomorphism from the set of nice (but not very nice) M-morsifications to the interior domain of the principal Springer cone. (Nice functions have nondegenerate critical points but may have coinciding critical values at different points.) The restrictions of this diffeomorphism to the (open) strata of the boundary caustic are also diffeomorphisms to the corresponding faces of the principal Springer cone. We obtain a homeomorphism sending the closure of the set of M-morsifications to the closed principal Springer cone.

3. Springer numbers b_k and very nice odd functions on a line

Let f be a real odd polynomial,

$$f(x) = x^{2\mu+1} + \lambda_1 x^{2\mu-1} + \cdots + \lambda_\mu x,$$

where the λ_i are real parameters. We shall call it an *M-deformation* if it has no complex (nonreal) critical points.

Definition. An odd polynomial is a *very nice* odd M-morsification of $x^{2\mu+1}$ if its 2μ critical values are all real and different.

Such polynomials form an open set in the real space $\mathbf{R}^{\mu}$ of coefficients λ.

Theorem 2. *The number of connected components of the domain of very nice odd M-morsifications of $x^{2\mu+1}$ in $\mathbf{R}^{\mu}$ is equal to the Springer number b_{μ}.*

Proof. We associate to each odd M-deformation (nice or not) the set $x_1 \geq x_2 \geq \cdots \geq x_{\mu} \geq 0$ of its nonnegative critical points and the non-strictly updown sequence $z_1 \leq z_2 \geq z_3 \leq \cdots 0$ of the corresponding critical values. This defines a continuous and proper mapping from the space of odd M-morsifications to $\mathbf{R}^{\mu}$.

Since an odd function also has the critical values $-z_i$ along with critical values z_i, a very nice odd M-morsification verifies the inequalities

$$z_i \neq \pm z_j, \qquad z_i \neq 0.$$

Hence the vector of critical values of a very nice odd M-morsification belongs to a chamber of the reflection group B_{μ}. This chamber is the same for all odd M-morsifications, belonging to the same connected component. Our mappings sends this component diffeomorphically to the chamber (this is a version of the standard Lyashko theorem, see [3]). Hence the theorem follows from the following Lemma.

Lemma. *Each nonstrictly updown sequence $z_1 \leq z_2 \geq z_3 \leq \cdots 0$ is the sequence of μ critical values (ordered by the order $x_1 \geq x_2 \geq \cdots \geq x_{\mu} \geq 0$ of critical points on the positive half line) of one and only one odd M-deformation of $x^{2\mu+1}$.*

The proof is similar to that of the Lemma in §2, and we omit the details.

Remark. Both proofs are based on the study of the mapping, sending the *ordered* sets of critical points (normalized by the convention that their sum is zero, or by the symmetry convention) to the set of critical values (normalized by the same convention), ordered by the order of critical points. This polynomial mapping sends a closed chamber homeomorphically to the closed principal Springer cone.

The closed initial chamber is also sent homeomorphically to the closure of the set of M-morsifications by the symmetrization of critical points. The subdivision of the principal Springer cone into chambers (by the mirrors) induces the subdivision of the set of nice M-morsifications into domains of very nice M-morsifications (by the Maxwell strata).

It would be interesting to study the topological properties of these poly-

nomial mappings $(x_i) \mapsto (z_i)$ in the complex domain (maybe, even for more complicated singularities, like D or E).

4. An algorithm for easy computation of Springer numbers b_n

Now let us count the topological types (up to homeomorphisms, preserving the orientations) of the very nice odd M-morsifications of $x^{2\mu+1}$. It is natural to classify them by the position of the critical value corresponding to the maximal critical point among other critical values.

Let us denote by z_1 the critical value attained at the maximal critical point.

Definition. We shall call a very nice odd M-morsification of x^{2n+1} a *right* function if $(-1)^n z_1 > 0$, and a *left* function if $(-1)^n z_1 < 0$.

Notation. Let $R_{i,j}$ denote the number of types of right functions having i positive critical values greater than $|z_1|$ and j positive critical values smaller than $|z_1|$.

Let $L_{i,j}$ denote the number of types of left functions having i positive critical values *smaller* than $|z_1|$ and j positive critical values *greater* than $|z_1|$.

Example. The initial parts of the triangles

$$
\begin{array}{ccccccccccccccc}
 & & & L_{0,0} & & & & & & & & R_{0,0} & & & \\
 & & L_{1,0} & & L_{0,1} & & & & & & R_{1,0} & & R_{0,1} & & \\
 & L_{2,0} & & L_{1,1} & & L_{0,2} & & & & R_{2,0} & & R_{1,1} & & R_{0,2} & \\
L_{3,0} & & L_{2,1} & & L_{1,2} & & L_{0,3} & & R_{3,0} & & R_{2,1} & & R_{1,2} & & R_{0,3}
\end{array}
$$

are

$$
\begin{array}{ccccccccccccccc}
 & & & 0 & & & & & & & & 1 & & & \\
 & & 1 & & 1 & & & & & & 1 & & 0 & & \\
 & 0 & & 1 & & 2 & & & & 2 & & 3 & & 3 & \\
11 & & 11 & & 10 & & 8 & & 8 & & 6 & & 3 & & 0
\end{array}
$$

The horizontal line, containing $2n$ numbers, gives the number of types of very nice odd M-morsifications of x^{2n+1}, classified according to the position of the critical value z_1 among the other critical values. The entry "zero" corresponds to the case where z_1 is the maximal critical value (which is impossible, since z_1 is a local minimum). The next entry on the same horizontal line counts the types for which z_1 is the second critical value from above, and so on. In the middle we have to jump from one triangle to the other.

Note that we move from left to right when the number of elements on the horizontal line in each triangle is odd (that is, when $i+j=n-1$ is even) and from right to the left otherwise.

Example. The three types of very nice odd M-morsifications of x^3 and the 11 types of very nice odd M-morsifications of x^5 are shown in Figure 3. Thus

$$b_2 = R_{0,1} + R_{1,0} + L_{0,1} + L_{1,0} = 0+1+1+1 = 3,$$

$$b_3 = L_{2,0} + L_{1,1} + L_{0,2} + R_{2,0} + R_{1,1} + R_{0,2}$$
$$= 0+1+2+2+3+3 = 11.$$

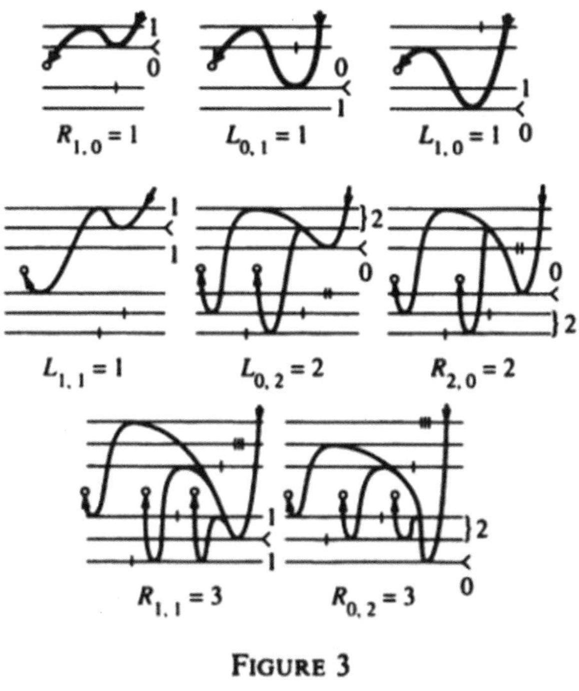

FIGURE 3

Theorem 3. *Each number in the (L, R)-table defined above is equal to the sum of the numbers in the preceding horizontal line, situated to the right of the given number if n is even, or to the left if n is odd (n being the number of elements on the horizontal line of the initial number in any of the two triangles L or R).*

The numbers on the interior sides of the triangles L and R on the same horizontal line are equal, with the exception of the vertices

$$L_{0,n} = R_{n,0} \quad \text{for } n > 1, \qquad L_{0,0} = 0, \qquad R_{0,0} = 1.$$

This theorem provides us with a fast algorithm for the calculation of numbers $L_{i,j}$ and $R_{i,j}$:

$$0+1=1,\ 1+1=2;\ 2=2;$$
$$2+1=3,\ 3+0=3.$$
$$0+3=3,\ 3+3=6,\ 6+2=8;$$
$$8=8;\quad 8+2=10,\ 10+1=11,\ 11+0=11.$$

The leftmost and rightmost numbers of our table give the values of b_k:

$$b_{2n} = R_{0,2n}, \quad b_{2n+1} = L_{2n+1,0} \qquad (n > 0).$$

The proof of Theorem 3 is bijective and very similar to that described in [2] for the A-case.

Remark. Theorem 3 also provides the number of connected components corresponding to different topological types of very nice odd M-morsifications of x^{2n+1}, since every topological type is presented by exactly one component (see Theorem 2 and its proof).

5. The exponential generating functions

To write a formula for the numbers $L_{i,j}$ and $R_{i,j}$, let us change the signs of those horizontal lines where $n \equiv 3$ or $4 \bmod 4$ (this very useful trick has been suggested by Deligne):

$$l_{i,j} = \varepsilon_{i,j} L_{i,j}, \qquad r_{i,j} = \varepsilon_{i,j} R_{i,j},$$

where $\varepsilon_{i,j} = (-1)^{(n-1)(n-2)/2}$, $n = i + j + 1$.

Let us define the exponential generating functions

$$L(x, y) = \sum l_{i,j} \frac{x^i}{i!} \frac{y^j}{j!}, \qquad R(x, y) = \sum r_{i,j} \frac{x^i}{i!} \frac{y^j}{j!}.$$

Theorem 4. *The exponential geneating functions for the numbers of topological types of very nice odd functions are given by the following formulas:*

$$L(x, y) = \frac{\sinh(x+y)}{e^y \cosh 2(x+y)}, \qquad R(x, y) = \frac{e^x \cosh(x+y)}{\cosh 2(x+y)}.$$

Proof. The numbers defined by these functions verify the properties of Theorem 3. These properties can be written in the form of linear partial differential equations,

$$L_x - L_y = L, \qquad R_x - R_y = R,$$

with boundary conditions

$$L(-x, 0) = -L(x, 0), \quad R(0 - y) = R(0, y), \quad 1 + L(0, z) = R(z, 0).$$

It is easy to check all these properties. Since they define the numbers unambiguously, the theorem is proven.

Remark. To *find* the functions L and R, observe that the numbers on the common line of the triangles L and R,

$$1, 2, 8, 40, 256, 1952, \cdots,$$

are divisible by the growing powers of two:

$$1, \ 2 \cdot 1, \ 4 \cdot 2, \ 8 \cdot 5, \ 16 \cdot 16, \ 32 \cdot 61, \cdots.$$

We recognize in 61 a member of the Bernoulli-Euler sequence (see [2, 4])

$$1, 1, 2, 5, 16, 61, \cdots,$$

counting the updown permutations. Thus we guess that perhaps

$$L_{0,n} = R_{n,0} = 2^{n-2} a_{n-2} \qquad (n \geq 2, \ a_0 = 1).$$

This fact has indeed an easy bijective proof. Since the exponential generating function for a_{n-2} is well known, we obtain the Cauchy boundary conditions $L(0, y)$ and $R(x, 0)$ for the differential equations, and we solve the equations easily.

The values

$$L(x, 0) = \frac{\sinh x}{\cosh 2x}, \qquad R(0, y) = \frac{\cosh y}{\cosh 2y}$$

computed by Springer [4] can also be used as boundary conditions.

Remark. The congruences for the Bernoulli-Euler numbers, proved in [2], imply similar congruences for the triangles L and R, which also deserve a detailed study $\bmod p^r$.

6. Galleries of Springer cones for boundary singularities

Let us consider the space $\mathbf{R}^{\mu-1}$ of real polynomials,

$$x^{\mu} + \lambda_1 x^{\mu-1} + \cdots + \lambda_{\mu-1} x,$$

on a real line with a fixed "boundary" $x = 0$.

Definition. A very nice *M-morsification* of a boundary singularity B_{μ} is a polynomial of this family, whose $\mu - 1$ critical values on the real line are all different and are different from its value at the boundary (for a motivation see [1]). The set of very nice M-morsifications is an open set in $\mathbf{R}^{\mu}$. Its closure is the set of all those polynomials which have only real critical points. It is subdivided into the connected components by the bifurcation diagram, containing three hypersurfaces:

(a) the boundary caustic consisting of the functions with a boundary critical point $(\lambda_{\mu-1} = 0)$;

(b) the ordinary Maxwell stratum, consisting of functions with equal critical values at different points;

(c) the boundary Maxwell stratum, consisting of functions whose value at the boundary is critical (such functions have a critical value, equal to 0).

The number of connected components of the spaces of very nice M-morsifications for boundary singularities B_n has been computed in [2]. In Springer's notation these numbers are equal to a_n—the Springer number of A_n. For instance, $a_4 = 16$, and the very nice quartic polynomials $x^4 + \lambda_1 x^3 + \lambda_2 x^2 + \lambda_3 x$ form in $\mathbf{R}^3$ an open set having 16 connected components. Each of these components is contractable for any n (see [2]).

Now we can obtain more information on the configuration formed by the part of the discriminant hypersurface bounding the set of polynomes with no complex critical points, and the three other bifurcation hypersurfaces, described above. Namely, this configuration (including the boundary) is homeomorphic to the configurations formed by the mirrors of the Coxeter group A_n in a polyhedral cone, which is a union of n Springer cones, forming a gallery in which two consecutive cones have a common wall. The two ends of the gallery are the two principal Springer cones.

This description provides some apparently new identities for the number of chambers in Springer cones. They have the form

$$a_n = a_{n-1} + \cdots + a_{n-1},$$

where the dots denote the number of chambers in the intermediate cones

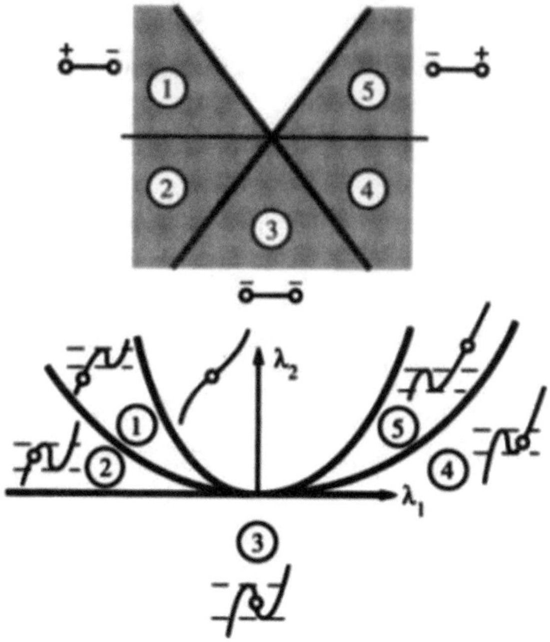

FIGURE 4

of the gallery. These identities follow from the comparison of [4] with the computation of the "Euler-Bernoulli number for B_n" in [2]

Example 1. *Case B_3 : the number of components is $a_3 = 5$.* $5 = 2+1+2$, where 2 is the Springer number of A_2 and 1 stands for the only chamber of nonprincipal Springer cones connecting two principal cones of A_2. The configurations of Springer cones and of domains of very nice M-morsifications,

$$x^3 + \lambda_1 x^2 + \lambda_2 x,$$

are shown in Figure 4.

Example 2. *Case B_4 : the number of components is $a_4 = 16$,* $16 = 5+3+3+5$. The 5's are the number of chambers in the two opposite principal Springer cones of A_3. The 3's are the number of chambers in two nonprincipal Springer cones in $\mathbf{R}^3$. See the Figure in [2].

The decompositions of a_5 and a_6 are

$$272 = 61 + 35 + 40 + 40 + 35 + 61,$$
$$1385 = 272 + 155 + 181 + 169 + 181 + 155 + 272.$$

To describe the Springer cones in A_n, forming the gallery for B_{n+1}, we define, as usual, the ith wall in

$$\mathbf{R}^n = \{(z_0, \cdots, z_n): z_0 + \cdots + z_n = 0\}$$

by the equation

$$z_i = z_{i-1} \qquad (i = 1, \cdots, n).$$

The principal Springer cone is defined by the inequalities

$$z_0 < z_1 > z_2 < \cdots.$$

To define the gallery we change the signs of these inequalities first going from the principal cone to the next one through the first wall, then to the third cone through the second wall, and so on. At the nth step we arrive at the principal cone, as opposed to the initial one:

$$z_0 > z_1 < z_2 > \cdots.$$

Now, in the base space $\mathbf{R}^n$ of the versal deformation of the boundary singularity of function B_{n+1}, let us consider those points λ for which the polynomial $x^{n+1} + \lambda_1 x^n + \cdots + \lambda_n x$ has no complex critical points.

Let us define a proper continuous mapping from this set to the space $\mathbf{R}^n$, where the group A_n acts as the group of permutations of coordinates $(z_0, \cdots, z_n)$. We first associate to a polynomial the set of its n critical values, and also 0, ordering them in the order of the critical points and 0 on the real axis:

$$x_1 \geq x_2 \geq \cdots \geq 0 \geq \cdots \geq x_n.$$

Then we normalize the resulting vector z of $\mathbf{R}^{n+1}$, projecting it orthogonally to the space $\mathbf{R}^n$ (that is, subtracting the arithmetic mean of the critical values and of 0 from each of them).

Let w be the resulting vector of $\mathbf{R}^n$, and $(w_1, w_2, \cdots, w_{n+1})$ its coordinates in $\mathbf{R}^{n+1}$ corresponding to $x_1 \geq x_2 \geq \cdots \geq 0 \geq \cdots \geq x_n$ in this order.

Theorem 5. *The mapping associating w to λ is a homeomorphism from the subset of $\mathbf{R}^n$, formed by those polynomials which have no complex critical points, onto the gallery of Springer cones of A_n in $\mathbf{R}^n$, defined above.*

Proof. Let us consider a very nice polynomial in our set.

The sequence of critical values and of the boundary value 0 consists of two updown (or downup) parts, connected by the value 0, bounded by "up-up" or "down-down" transitions. The subtraction of the arithmetical mean does not change this property. Hence the sequence w belongs to one of the Springer cones of the gallery.

Every closure of a component of the set of very nice boundary M-morsifications is sent by our mapping homeomorphically onto a chamber (according to the standard Lyashko-Looijenga arguments, see [3]).

Every sequence $(z_1, \cdots, z_{n+1})$, obtained from an updown sequence by an inclusion of a 0 at any place between two consecutive members of the updown sequence (destroying its updown character) or at one of the ends (in the last case the sequence either remains an updown sequence or is transformed into a downup sequence), may be obtained from a unique polynomial with no complex critical points as the sequence of its critical values, completed by the boundary value 0. The proof is similar to that of §2.

Hence our mapping from λ to z is a proper homeomorphism. It sends homeomorphically the set of polynomials with no complex critical points to a polyhedral hypersurface in $\mathbf{R}^{n+1}$. This hypersurface is sent homeomorphically to the hyperplane $\mathbf{R}^n$: $w_1 + \cdots w_{n+1} = 0$ by the orthogonal projection, since this projection preserves the order of the points and hence the place of the zero member of the extended sequence $(z_1, \cdots, z_{n+1})$ is defined unambiguously by the sequence w. Hence the mapping sending λ to w is a homeomorphism on each component of the space of nice boundary M-morsifications, whose image covers just once the interior parts of the Springer cones of the gallery. It follows that it is a homeomorphism of the whole set of polynomials having no complex critical points, since its restriction to every boundary stratum is a homeomorphism to the corresponding face of the corresponding cone according to the generalized Lyashko theorem.

Remark. In fact our mapping is a diffeomorphism inside each connected component of the set of boundary-nice M-polynomials (having exactly n different real critical points with nonzero critical values). This follows from the fact that the versal deformation is multiversal, and hence the critical values at different critical points move independently of each other. The image of a component covers the interior of the corresponding Springer cone.

The restrictions of our mapping to the strata of the natural stratification of the set of boundary-nice M-polynomials are also diffeomorphisms (to the interiors of the correspnding faces of Springer cones). Thus the singularities of the intersections of the union of the Maxwell and boundary Maxwell bifurcation diagrams with the starta of the natural stratification of the union of the caustic and of the boundary caustic are locally diffeomorphic to the hypersurfaces formed by the intersections of mirrors with the faces of the walls of Springer cones (at the points of the boundary of the set of boundary-nice M-polynomials).

7. Even morsifications and Springer cones

Let us consider the space $\mathbf{R}^{n-1}$ of even real polynomials

$$x^{2n}+\lambda_1 x^{2n-2}+\cdots+\lambda_{n-1}x^2$$

(the relation of this family of polynomials to the B_n Coxeter group is explained in [1]).

Definition. A (*very nice*) *M-morsification of an equivariant singularity* B_n is a polynomial of the above family, whose $2n-1$ critical points are all different and which have $n-1$ different critical values.

Theorem 6. *The pair formed by the closure of the set of very nice M-morsifications of an equivariant singularity B_n and by its boundary is homeomorphic to the pair formed by a closed principal Springer cone of A_{n-1} in $\mathbf{R}^{n-1}$ and its intersection with the union of mirrors. For instance, the number of connected components of the set of very nice M-morsifications of an equivariant singularity B_n is equal to the Springer number a_{n-1} $(=1, 2, 5, 16, \cdots$ for $n=2, 3, \cdots)$.*

Proof. The critical points of our polynomial are the origin and the square roots of the critical points of the polynomial

$$X^n+\lambda_1 X^{n-1}+\cdots+\lambda_n X, \qquad X=x^2.$$

Hence the last polynomial has n different *positive* roots. The mapping, defined in §6, sends it to the *principal* Springer cone, since the values at these positive roots $X_1>X_2>\cdots>0$ form with zero an updown sequence

$$z_1<z_2>\cdots 0.$$

Thus the proof of Theorem 5 provides the required homeomorphism. q.e.d.

Remark. This homeomorphism is in fact a diffeomorphism from the interior part of the above closure to the interior part of the principal Springer cone, sending the union of the Maxwell and the boundary Maxwell starta to the union of mirrors. Moreover, its restrictions to the strata of the caustic and of the boundary caustic, bounding the above closure, are diffeomorphisms to the interiors of the corresponding faces of the boundary of the principal Springer cone.

References

[1] V. I. Arnold, *Critical points of functions on a manifold with boundary, the simple Lie groups B_k, C_k, F_4, and singularities of evolutes*, Russian Math. Surveys **33** (1978), no. 5.

[2] ——, *Bernoulli-Euler updown numbers of functions singularities, their arithmetics and combinatorics*, Duke Math. J. **63** (1991), 537–555.

[3] V. I. Arnold, V. V. Goryunov, O. V. Lyashko and V. A. Vassiliev, *Singularities*, vols. 6 and 39, Itogi Nauki Viniti, Sovremennye Problemy Mathematiki, Fundamentaluye Napravleniia, Dynamicheskie Systemy, nos. 6 and 8, Moscow, 1989; English transl., Encyclopædia Math. Sciences, vols. 6 and 39, Dynamical Systems, nos. 6 and 8, Springer (to appear).

[4] T. A. Springer, *Remarks on a combinatorial problem*, Nieuw Arch. Wisk. (1) **19** (1971), 30–36.

[5] R. Thom, *L'equivalence d'une fonction differentiable et d'un polynome*, Topology **3** (1965), 297–307.

[6] S. Zdravkovska, *Topological classification of polynomial mappings*, Uspekhi Mat. Nauk **25** (1970), no. 4, 179–180.

STEKLOV MATHEMATICAL INSTITUTE, MOSCOW

RIJKSUNIVERSITEIT TE UTRECHT, THE NETHERLANDS

Polyintegrable flows

V.I. Arnol'd

For O. A. Ladyzhenskaya on her birthday

ABSTRACT. Dynamical systems with $m - 1$ first integrals on m-dimensional manifolds, and the topological properties of such systems, are studied, particularly in the case when the integrals are multiple-valued and are pseudoperiodic functions on covering spaces. The straightening of trajectories on a torus of dimension three or higher is proved in the case when the polyintegrable field has no singular points.

There are two important higher-dimensional generalizations of Hamiltonian vector fields on the plane. The better-known generalization is the theory of higher-dimensional Hamiltonian systems. In this theory one takes as a basis the preservation of phase area by a phase flow (the symplectic property), but the integrability property is lost.

In the generalization that will be considered below, strong integrability is preserved (for example, in the three-dimensional case the systems have two first integrals).

In this paper we describe the simplest properties of such polyintegrable systems on manifolds, particularly on tori. Here we have in mind integrals that are single-valued only on the universal covering space and multiple-valued on the tori.

The source of this work was the study of the topology of a class of flows of a three-dimensional incompressible fluid.

§1. Polyintegrable vector fields

The simplest example is the vector product of the gradients of two functions in a three-dimensional oriented Euclidean space, defining the differential equation

$$\dot{x} = [\operatorname{grad} F, \operatorname{grad} G].$$

The functions F and G are first integrals of this divergence-free vector field.

Let (M, μ) be a smooth manifold of dimension m with a fixed volume element μ, and let (N, ν) be a smooth manifold of dimension 1 less, $\dim N = m - 1$, also with a fixed volume element ν. Let $H: M \to N$ be a smooth mapping.

In this situation a vector field v on M arises, which I shall call a *polyintegrable field with poly-Hamiltonian* H.

The phase curves of this field are the inverse images of points under the mapping H, and the rate of motion over them depends on the fixed volume elements in the following way.

Definition. A *polyintegrable field* v, defined by a mapping $H: M \to N$ and the forms of the volume elements μ and ν, is defined by the condition $i_v\mu = H^*\nu$, where the $n - 1$ form on the left-hand side is given by the identity

$$(i_v\mu)\cdot(a_1, \dots, a_{m-1}) = \mu(v, a_1, \dots, a_{m-1}).$$

Originally published in: *Algebra i Analiz* **4**(6), 54–62 (1992), © 1992 Russ. Acad. Sciences
English translation: *St. Petersburg Math. J.* **4**(6), 1103–1110 (1993), © 1993 Amer. Math. Soc.

Example 1. $(M, \mu) = (\mathbf{R}^2, dp \wedge dq)$, $(N, \nu) = (\mathbf{R}, dh)$, mapping: $h = H(p, q)$. In this case v is a Hamiltonian field with Hamiltonian H.

Example 2. $(M, \mu) = (\mathbf{R}^3, dx_1 \wedge dx_2 \wedge dx_3)$, $(N, \nu) = (\mathbf{R}^2, df \wedge dg)$, mapping: $(f = F(x), g = G(x))$. In this case $\nu = [\operatorname{grad} F, \operatorname{grad} G]$.

Theorem 1. *Every component of the mapping H is a first integral of a polyintegrable field defined by H.*

Proof. $i_v(H^*\nu) = i_v i_v \mu = 0$, so that $i_{H^*\nu} v = 0$. Hence $H_* \nu = 0$.

Theorem 2. *The vector field v is divergence-free.*

Proof. In fact, $L_v \mu = d(i_v \mu) = d(H^* \nu) = H^* d\nu = 0$, since the form ν is closed.

Remark. On the curves $H = \text{const}$ there is the "Gel'fand-Leray 1-form"

$$\mu / H^* \nu = dt \quad (\text{"time differential"})$$

with singularities at the critical points of the mapping H. This differential form is related to the polyintegrable field v by the relation $i_v\, dt = 1$ (which is a generalization of the relation $dt = dp \wedge dq / dH$ familiar from the theory of Hamiltonian systems).

§2. Finite polyintegrable systems

We consider the standard fibration $H : (\mathbf{R}^m, \mu) \to (\mathbf{R}^{m-1}, \nu)$. The fibers of this fibration will be called *vertical lines.*

Definition. A *finite polyintegrable system* is a system that differs from the standard fibration into vertical lines only in a bounded domain of $\mathbf{R}^m$.

Here we discuss the simplest topological properties of generic systems of this type.

The inverse image of a generic point under this mapping H consists of a single nonsingular curve, going "from $-\infty$ to $+\infty$" along a vertical line (deviating from this line only in a bounded domain of the total space) and having a finite number of closed components.

The closed components as a whole form a finite collection of circle fibrations over open $(m-1)$-dimensional manifolds. The closures of the cells formed by these closed components are bounded by pieces of singular curves.

Example. For generic finite polyintegrable systems in a three-dimensional space a typical piece of a singular curve that enters the boundary of a cell either contains a corner point or reduces to a single point. Moreover, in generic systems pieces of exceptional singular curves that have two singular points or a semicubical cusp point can tend (in finite number) to the boundaries of cells.

The critical points of a generic mapping $H: \mathbf{R}^3 \to \mathbf{R}^2$ form a finite number of closed curves in $\mathbf{R}^3$, which enter the boundaries of cells. The generic critical points are divided into two types: *elliptic* and *hyperbolic*. The corresponding local normal forms are one-parameter trivial families of Morse functions on the plane,

$$(x, y, z) \to (x^2 + y^2, z) \quad \text{(an elliptic point)},$$

$$(x, y, z) \to (x^2 - y^2, z) \quad \text{(a hyperbolic point)}.$$

The arcs formed by elliptic and hyperbolic points can be separated further for a generic mapping by *parabolic* points with local normal form

$$(x, y, z) \to (x^3 + y^2 + zx, z) \quad \text{(a parabolic point)}.$$

Generic mappings do not have more complicated singularities.

Example. There exists a finite polyintegrable system in $\mathbf{R}^3$ without elliptic points whose set of critical values has the shape of a figure eight. One of the pieces of singular curves on the boundary of both cells formed by the closed curves is diffeomorphic to the union of the circle with the standard embedding into 3-space and its diameter. The remaining pieces are curves with one singular point.

I have not succeeded in finding a reasonable formula for this system and have not succeeded in drawing the complicated structure formed by its cells and separatrix surfaces consisting of singular curves.

The space of connected components of the inverse images of points for a typical finite system defined by a mapping $\mathbf{R}^3 \to \mathbf{R}^2$ is the union of the plane and a polyhedral surface whose singularities are easily enumerated (see, for example, the paper by Mata-Lorenzo [1]). However, I do not know what global topological types of surfaces with such singularities are realized for generic mappings.

This question is even more unresolved for mappings of more complicated manifolds, $H: M^2 \to N^2$. Burlet and de Rham [2] formulated the Poincaré conjecture as the assertion that every simply connected closed three-manifold M^3 has a generic mapping onto the plane without elliptic singular points (i.e., without boundary on the space of components).

Remark. The question for Morse functions $H: M^2 \to \mathbf{R}$ that is analogous to the question about the structure of the surface of component has a simple solution:

The space Γ of components of the level sets $H = c$ is a graph with g independent cycles if M is a surface of genus g. For example, the space of components of level curves of functions on the 2-sphere is a tree.

For functions on manifolds of higher dimension the question of the structure of the graph of the components is apparently open (and probably requires knowledge of the topology of the components in order to be solved; however, for spheres the answer is always a tree—see the theory of Kronrod et al. in [3]).

The study of the topology of the spaces of components of inverse images of points is also a necessary step of bifurcations in integrable Hamiltonian systems with several degrees of freedom. Here up to the recent work of Michèle Audin almost nothing had been done even in such classical problems as the problem of geodesics on quadrics in 3-space and the problem of Kovalevskaya tops.

In the simplest examples of finite three-dimensional systems the closed trajectories are linked with the curve of singular points (for example, this is the situation close to elliptic points).

The reasons for this phenomenon are unclear, since the linking coefficient (defined after a choice of orientations of the components of the curve of singular points, which is, in general, not canonical) can also be zero. Nevertheless, in the examples that I know the closed trajectories are linked with the curve of singular points in the sense that none of these trajectories can be moved far away by an isotopy of the ambient space that preserves the curve of singular points.

§3. Pseudoperiodic topology of polyintegrable systems

We consider a polyintegrable system with a multiple-valued multidimensional Hamiltonian on the m-dimensional torus $T^m = \mathbf{R}^m/\mathbf{Z}^m$ that is defined by a pseudoperiodic mapping

$$H: \mathbf{R}^m \to \mathbf{R}^{m-1}, \qquad H(x+k) \equiv \Omega(k) + H(x),$$

where $k \in \mathbf{Z}^m$ and $\Omega: \mathbf{Z}^m \to \mathbf{R}^{m-1}$ is a homomorphism.

The mapping H can be represented as the sum of linear and periodic terms

$$H(x) = \Omega(x) + h(x), \qquad h(x+k) \equiv h(x).$$

As a rule, I shall assume the "frequencies" $\Omega(k)$ to be maximally commensurable, so that the homomorphism $\Omega\colon \mathbf{Z}^m \to \mathbf{R}^{m-1}$ is monomorphic: $\Omega(k) = 0$ only for $k = 0$.

Example. For $m = 2$ a Hamiltonian system of the two-torus is obtained which has a multiple-valued Hamiltonian H. In this case every nonsingular level curve of a sufficiently smooth function H has exactly one unbounded component [4].

The topological and ergodic properties of such generic systems were studied in detail in [4], and then by Sinaĭ and Khanin in [5]. In particular, the phase flow of such a system on the torus is isomorphic to a special flow over a rotation of the circle and has the mixing property (although this is rather slow, logarithmic, so that the Lyapunov exponent is equal to zero).

Conjecture. *Every nonsingular level curve* $H = \mathrm{const}$ *of a generic analytic polyintegrable system on the three-dimensional torus has exactly one unbounded component.*

Remark. It is possible that this is also true for all sufficiently smooth systems on a torus of arbitrary dimension—no counterexamples have been found. Similarly to the results cited above for the 2-torus, these conjectures can be considered as generalizations of the theorems of Denjoy [6] and Kolmogorov [7] concerning systems without singular points. It is also possible to assume that *all the Lyapunov exponents of polyintegrable systems are almost everywhere zero*, as in the two-dimensional case. We do not exclude the possibility, however, that the results may depend on Diophantine properties of Ω and injectivity is not sufficient for the validity of the conjectures.

I have only succeeded in advancing a little way in the direction of these conjecture. First, it is clear that all the level curves $H = \mathrm{const}$ lie in a finite neighborhood of a totally irrational line $\Omega(x) = 0$.

Second, there is an odd number of nonsingular components going "from $-\infty$ to $+\infty$" (one, if we assume that the components going in the reverse direction have minuses).

Furthermore, there are probably no components going from $-\infty$ to $-\infty$ or from $+\infty$ to $+\infty$ (I have not proved this).

There are also no oscillating unbounded components (not going off to infinity, but returning to a fixed bounded domain after as long a time as desired) at least in analytic systems.

By the way, analyticity influences the qualitative behavior of trajectories of polyintegrable systems even in the two-dimensional case. *In an analytic system which has an analytic first integral, a bounded trajectory tends to an equilibrium position*, while in a C^∞ system it is also possible to have a winding onto a more complicated formation, for example, onto a closed curve of singular points. It is possible, however, that these distinctions do not occur for generic systems.

If there is no uniqueness for a set of values of positive measure, then there is no uniqueness for a set of values of full measure (with a complement of measure zero).

The simplest example in which the uniqueness of the unbounded curve has not yet been proved is related to the case when one of the components of the vector $H\colon \mathbf{R}^3 \to \mathbf{R}^2$ has no singular points. In this case it follows from the results of Laudenbach and Cerf [8] that by a smooth change of variables on the torus one can turn the component in question into a linear function, so that the conjecture takes the form of the following assertion concerning pseudoperiodic curves on the (x, y) plane.

Consider the analytic equation

$$y = ax + f(x, y, bx),$$

where f is an analytic function of period 1 in all three arguments and a and b are irrational arithmetically independent numbers ($pa + qb + r \neq 0$ for $(p, q, r) \neq (0, 0, 0)$).

Conjecture. *The curve defined by this equation has (if it is nonsingular) exactly one unbounded component.*

This conjecture is nontrivial even for rather simple trigonometric polynomials f. Its higher-dimensional generalizations have also not been disproved (for example, with a function f of a large number of arguments, linear in x or in x and y). Some generalization may also occur for almost periodic dependencies of f on x and y with an infinite basis of frequencies.

§4. Polyintegrable three-dimensional polyintegrable systems without singular points

We consider a system on the three-torus, defined by a mapping $H: \mathbf{R}^3 \to \mathbf{R}^2$, $H(x) = \Omega(x) + h(x)$, where $h(x+k) = h(x)$ for every $k \in \mathbf{Z}^3$ and where the linear mapping Ω gives a monomorphism $\mathbf{Z}^3 \to \mathbf{R}^3$.

This is the only higher-dimensional ($m > 2$) case in which the study has been completely carried out.

Theorem. *A polyintegrable smooth vector field v without singular points on the three-dimensional torus $T^3 = \mathbf{R}^3/\mathbf{Z}^3$ can be reduced to the form $f \cdot v_0$, where v_0 is a constant (translation-invariant) vector field on the torus and f is a positive function, by a smooth (or analytic if v is analytic) change of coordinates on the torus.*

For vector fields on the two-torus the analogous result was proved by Kolmogorov [7].

Lemma 1. *Every trajectory of polyintegrable smooth vector field without singular points on the three-dimensional torus joins $-\infty$ and $+\infty$ in a neighborhood of finite width of the line $\Omega(x) = 0$ of the universal covering space of the torus.*

Proof. As was already indicated, we can reduce one of the components of H to a standard form by a preparatory change of variables, so that the trajectories on the covering can be assumed to lie on planes:

$$H_1 = ax + by + cz, \qquad H_2 = ux + vy + wz + h(x, y, z),$$

where h is 1-periodic in all three of its arguments; also the volume element μ on T^3 has the form $A\,dx \wedge dy \wedge dz$, where A is a periodic positive function.

On T^3 and in $\mathbf{R}^3$ we fix the standard Riemannian metric $ds^2 = dx^2 + dy^2 + dz^2$, and we consider the restriction of the function H_2 to the plane $ax + by + cz = \text{const}$.

The gradients of these restrictions form a vector field ∇ in $\mathbf{R}^3$ (and on the torus), and this vector field is tangent to the planes $H_1 = \text{const}$.

We form a new vector field on the torus, $\nabla/|\nabla|^2$. This field is bounded. In fact, the lengths of the vectors ∇ are bounded above by the periodicity of h. They are also bounded from below since the torus is compact—in fact, otherwise on the torus we could find a point where $\nabla = 0$, and at this point the mapping H would have a singularity and the field v a zero.

Thus, the field $\nabla/|\nabla|^2$ on the torus is smooth, bounded, and tangent to the "planes" $dH_1 = 0$. The trajectories of this field extend unboundedly. The derivative of the function H_2 along this field is equal to 1. Therefore H_2 varies along the trajectories of the field from $-\infty$ to $+\infty$. It follows from this that every level curve of h_2 on $H_1 = \text{const}$ consists of one unbounded component.

In fact, in a domain bounded by a closed component there would be a singular point, and therefore there are no closed components. In a domain bounded by a component going "from $-\infty$ to $+\infty$", the function $H_2 = ux + vy + cz + h$ is bounded. Therefore the trajectory of the gradient cannot hit such a domain (neither entering nor leaving). Hence, there are no such domains. There cannot be several components of the curve $H_2 = d$ separating the lines $ux + vy + wz = \pm C$ on the plane $H_1 = \text{const}$, since H_2 varies monotonically along the trajectories of the field $\nabla/|\nabla|^2$, and hence it takes the value d only once on the trajectory.

Thus, the level curve $H_2 = d$ on the plane $H_1 = \text{const}$ consists of a single connected component separating the lines $ux + vy + wz = \pm C$, which proves the lemma.

Lemma 2. *A trajectory of a field v without singular points on the torus goes off to infinity in the covering space no slower than some nonzero linear function:*

$$|z(t)| > q|t| - r \quad \textit{for all } t \textit{ from } -\infty \textit{ to } +\infty$$

(*where $q > 0$ and $r > 0$ do not depend on the initial point*).

Proof. We consider a bounded domain of a fixed shape and fixed size on any of the planes $H_1 = c_1$. From the fact that a polyintegrable vector field v on a component torus does not have singular points, it follows that the length of a piece of the curve $H_2 = c_2$ hitting the indicated domain is bounded above by a constant that depends neither on the choice of c_1 and c_2 nor on the location of the domain.

Now we consider the strip between the lines $ux + vy + wz = \pm C$ on the plane $H_1 = c_1$, and we divide it into identical parallelograms by the lines $z = j \in \mathbf{Z}$. The length of the piece of the trajectory $H_1 = c_1$, $H_2 = c_2$ hitting any of the parallelograms is bounded from above, and the rate of motion is bounded from below, where the constants do not depend on the number of the parallelogram. Hence, the modulus of this number grows in time no more slowly than some linear function of time, which proves the lemma.

We assume for definiteness that $z(+\infty) = +\infty$ and $z(-\infty) = -\infty$ (cf. Lemma 1).

We form a function $Z: \mathbf{R}^3 \to \mathbf{R}$ equal to

$$Z(x) = \ln \int_0^\infty e^{-z(t)}\, dt,$$

where the integral is taken along a trajectory of the field v leaving from the point $x \in \mathbf{R}^3$ for $t = 0$, and $z: \mathbf{R}^3 \to \mathbf{R}$ is one of the coordinates.

Lemma 3. *The function Z is smooth, possesses the periodicity property*

$$Z(x + k) = Z(x) - z(k), \qquad k \in \mathbf{Z}^3,$$

and its derivative with respect to the direction of the field v is strictly negative.

Proof. The convergence of the integral follows from the linear growth of $z(t)$ to infinity, the periodicity follows from the periodicity of the field v, and the derivative is opposite to the inverse value of the integral.

End of the proof of the theorem. We consider the three functions (H_1, H_2, Z) as a mapping $T: \mathbf{R}^3 \to \mathbf{R}^3$. They define a local diffeomorphism, since $H: \mathbf{R}^3 \to \mathbf{R}^2$ does not have singular points, and the derivative of Z with respect to the direction of the level curves of H does not vanish.

We shall prove that T is a global diffeomorphism. For this it suffices to prove that $T\mathbf{Z}^3$ is an affine lattice generated by three linearly independent vectors.

Otherwise all the points of $T\mathbf{Z}^3$ would lie in a single affine plane and the whole image $T\mathbf{R}^3$ would lie in a neighborhood of this plane of finite width.

On the (H_1, H_2) plane the indicated plane is projected nondegenerately, since our system is nonresonant (the mapping Ω is a monomorphism). For fixed values $H_1 = c_1$ and $H_2 = c_2$ the function Z is not bounded on a trajectory of the field v (its derivative with respect to the direction of the field is strictly negative and periodic, and hence is bounded away from zero on the whole space; the trajectories extended unboundedly, since the field v is bounded).

Thus, the image of $T\mathbf{R}^3$ cannot lie in a finite strip around some affine plane. Therefore $T\mathbf{R}^3 = \mathbf{R}^3$ and T is a diffeomorphism, possessing the periodicity property

$$T(x+k) = T(x) + \omega(k), \qquad k \in \mathbf{Z}^3,$$

where ω isomorphically maps $\mathbf{Z}^3$ onto the lattice $\omega\mathbf{Z}^3$ generated by three $\mathbf{R}$-independent vectors.

We can therefore consider (H_1, H_2, Z) as a new coordinate system on the covering space of the torus $\mathbf{R}^3/\mathbf{Z}^3$. In this coordinate system $v = -A\partial/\partial Z$, where $A > 0$. Reducing the lattice $\omega\mathbf{Z}^3$ to the standard form $\mathbf{Z}^3$ by the inverse linear transformation ω^{-1}, we reduce v to the form required in the theorem.

Remark. The arguments at the end of the proof allows us to extend the result to polyintegrable fields without singular points on a torus of arbitrary dimension: their trajectories are also straightened by diffeomorphisms of the tori. Actually these arguments prove the special case that we needed, the theorem of Carrière [9] on Riemannian flows. Since this theorem has been published, we can replace the proof by a reference to it. In fact, on the torus we can choose a Riemannian metric so that the distances between two finitely close trajectories of a polyintegrable vector field without singular points will be the same everywhere, since the space of nonvertical hyperplanes in the tangent space of the torus is contractible and the bundle with similar fiber has a section. The straightening of the family of equidistant trajectories is guaranteed by Carrière's theorem.

After the trajectories have been straightened, we can almost always straighten the vector field itself, turning the factor A into a constant by a diffeomorphism of the torus, as Kolmogorov did in the two-dimensional case. It is also not difficult to verify that the Lyapunov exponents vanish here independently of whether the vector field can be straightened—it suffices to straighten the trajectories.

Bibliography

1. Luis E. Mata-Lorenzo, *Polyhedrons and pi-stable homotopies from 3-manifolds into the plane*, Bol. Soc. Brasil. Mat. (N.S.) **20** (1989), 61–85.
2. Oscar Burlet and Georges de Rham, *Sur certaines applications génériques d'une variété close à 3 dimensions dans le plan*, Enseignement Math. **20** (1974), 275–292.
3. A. S. Kronrod, *On functions of two variables*, Uspekhi Mat. Nauk **5** (1950), no. 1 (35), 24–134. (Russian)
4. V. I. Arnol'd, *Topological and ergodic properties of closed 1-forms with incommensurable periods*, Funktsional. Anal. i Prilozhen. **25** (1991), no. 2, 1–12; English transl. in Functional Anal. Appl. **25** (1991).

5. Ya. G. Sinaï and K. M. Khanin, *Mixing of some classes of special flows over the rotation of a circle*, Funktsional. Anal. i Prilozhen. **26** (1992), no. 3, 1–21; English transl. in Functional Anal. Appl. **26** (1992).
6. A. Denjoy, *Sur les courbes définies par les équations différentielles à la surface du tore*, J. Math. Pures Appl. (9) **11** (1932), 333–375.
7. A. N. Kolmogorov, *On dynamical systems with an integral invariant on the torus*, Dokl. Akad. Nauk SSSR **93** (1953), 763–766; English transl. in his *Selected works*. vol. I: *Mathematics and mechanics*, Kluwer, Dordrecht, 1991, pp. 344–348.
8. Jean Cerf, *1-formes fermées non singulières sur les variétés compactes de dimension* 3, Séminaire Bourbaki Lecture Notes in Math., vol. 901, Springer-Verlag, Berlin, 1981, pp. 205–219.
9. Yves Carriere, *Flots riemanniens*, Structure Transverse des Feuilletages (Toulouse, 1982), Astérisque, vol. 116, Soc. Math. France, Paris, 1984, pp. 31–52.

Received 26/AUG/92

Translated by J. S. JOEL

Bounds for Milnor numbers of intersections in holomorphic dynamical systems

V.I. Arnold

Let us move a smooth plane curve, containing the origin, by the iterations of an origin-preserving diffeomorphism of the plane to itself. It follows from the general theorems of this paper that the orders of tangency of the resulting curves with any fixed line through the origin are uniformly bounded (provided, of course, that each of these orders is finite).

We shall also prove the boundedness of the Milnor numbers, describing the tangency of a moving submanifold with a fixed submanifold of arbitrary dimension in the phase space of a higher dimensional dynamical system.

If we move a curve by a degenerate smooth plane mapping (which is not a diffeomorphism), the order of tangency of the moving curve with a fixed curve at the origin may grow with the number of iterations. We shall prove that the growth rate is at most exponential, provided that the moving curve remains smooth at the origin.

We conjecture that a similar exponential bound should hold in the general case of submanifolds of arbitrary dimension in arbitrary finite-to-one holomorphic dynamical systems.

1. Milnor numbers and the topological complexity of intersections

Let $A: (\mathbf{C}^m, 0) \to (\mathbf{C}^m, 0)$ be a local holomorphic mapping of a space to itself. Let $a: (\mathbf{C}^k, 0) \to (\mathbf{C}^m, 0)$ be an embedding onto a smooth submanifold X^k (we suppose that rank $da|_0 = k$) and let $b: (\mathbf{C}^m, 0) \to (\mathbf{C}^k, 0)$ be the projection along a smooth submanifold $Y^b = b^{-1}(0)$ of complementary dimension ($k + \ell = m$, rank $db|_0 = k$).

Originally published in: *Topological Methods in Modern Mathematics* (Stony Brook, NY, 1991). Editors: L.R. Goldberg and A.V. Phillips. Publish or Perish, Houston, TX, 379–390 (1993),

The *algebraic Milnor number*

$$\mu(n) = \mu(A^n X, Y)$$

("the number of intersection points of $A^n X$ and Y, collapsing at the origin") is defined as the multiplicity at the origin of the composed mapping $f(n) = b \circ A^n \circ a : \mathbf{C}^k \to \mathbf{C}^k$,

$$\mathbf{C}^k \xrightarrow{a} \mathbf{C}^m \xrightarrow{A^n} \mathbf{C}^m \xrightarrow{b} \mathbf{C}^k.$$

In other words, $\mu(n)$ is the dimension over $\mathbf{C}$ of the local algebra

$$Q(n) = \mathbf{C}[[x_1, ..., x_k]]/(f(n)),$$

where the denominator is the ideal generated by the components of $f(n)$.

The algebraic Milnor number

$$\mu(n) = \dim_{\mathbf{C}} Q(n)$$

is finite if and only if the origin is an isolated intersection point of $A^n X$ and Y. If it is finite, it coincides with the following "geometric Milnor numbers":

(a) it coincides with the linking number of the links of our submanifolds in a sufficiently small sphere S^{2m-1} centered at the origin:

$$\mu(n) = L((A^n X) \cap S, Y \cap S).$$

(b) it coincides with the Poincaré index of f_n at the origin, i.e., with the degree of the corresponding sphere mapping:

$$\mu(n) = \operatorname{ind}_0 f_n = \deg\left(f_n/|f_n| : S^{2k-1} \to S^{2k-1}\right).$$

The algebraic definition works also in the real domain, provided that $\mu < \infty$ (see, for instance, [1]).

Up to now we have not supposed A to be a diffeomorphism.

Example. Let $m = 2$, $A(x_1, x_2) = (x_1, x_2^2)$. Let us move the curve $X = \{x_2 = x_1\}$ by iterations of A and let us intersect the images with the x_1 axis $Y = \{x_2 = 0\}$. The curve $A^n X$ is defined by the equation $x_2 = x_1^{2^n}$ $(n \geq 0)$. Hence $\mu(n) = 2^n$ grows exponentially.

We shall see below that this growth would be impossible if A were a diffeomorphism. We shall also see that the growth cannot be faster than an exponential function, provided that the curves $A^n X$ are smooth.

These bounds are the local counterparts of the following global ones. Consider a smooth mapping of a compact manifold to itself, and two compact submanifolds. Under the iterations of the mapping, the topological complexity of the intersection of the images of the first submanifold with the second (the Betti numbers, the Morse numbers, the characteristic numbers, the total absolute curvature, ...) may grow with the number of iterations. This growth rate is generically at most exponential (at least if the mapping is a diffeomorphism, see [2]).

However, for any given function of the time there exist "exceptional" examples [3] where the complexity grows even faster (at least along some subsequence of time moments, "time" being the number of iterations). For instance, on any compact manifold of dimension higher than one and for any given function of n, there exists a C^∞ diffeomorphism for which the number of periodic points of period n grows faster than the given function of n (along some increasing subsequence of n's), all periodic points being nondegenerate [4]. Hence on every compact manifold of dimension higher than two, and for any function of T, there exists a C^∞ vector field whose singular points and periodic orbits are all nondegenerate, and yet the number of periodic orbits of period at most T grows faster than the given function on an increasing sequence of values of T [5].

In the algebraic case, superexponential growth of complexity characteristics (such as the numbers of transversal intersection points of manifolds of complimentary dimensions) is impossible (see [3], [6]). It is unknown whether it is possible in the analytic case. For instance, let us consider an analytic curve on the 2-torus. It is unknown whether the number of intersection points of the iterated images of the curve with the meridian of the torus is bounded by Ce^n, where n is the number of iterations (provided that the intersection points are isolated and counted with their multiplicities).

REMARK. Consider the diffeomorphism $B = (\mathrm{id}, A)$ of the cartesian square of the underlying manifold of the dynamical system defined by the diffeomorphism A. Consider the intersection indices of the images of the diagonal under the iterations of B with the diagonal of the cartesian square, at a fixed point of B.

The sequence of these intersection indices is bounded (even for C^1-diffeomorphisms), according to Shub and Sullivan [7]. In the analytic case our result gives slightly more in this situation: we bound the multiplicities, rather than the indices (we count the nondegenerate intersection points with + signs when we count the multiplicities, and with + or − signs according to the orientations when we count the intersection index). In fact, however, the boundedness of the multiplicities follows more or less directly from the boundedness of the intersection indices of the complexified manifolds. In the same way, the boundedness of the Milnor numbers of the intersections of submanifolds of complementary dimensions in the general case is essentially equivalent to the boundedness of the topological intersection numbers of the complexified submanifolds.

2. Milnor numbers and the Skolem theorem

Let $a\colon (\mathbf{C}^k, 0) \to (\mathbf{C}^m, 0)$ be an imbedding, $A\colon (\mathbf{C}^m, 0) \to (\mathbf{C}^m, 0)$ a local diffeomorphism of the space $\mathbf{C}^m$ to itself with the fixed point 0, and $b\colon (\mathbf{C}^m, 0) \to (\mathbf{C}^k, 0)$ the projection along a submanifold $Y = b^{-1}(0)$. Let us define the mapping $f(n)\colon (\mathbf{C}^k, 0) \to (\mathbf{C}^k, 0)$ as

$$f(n) = b \circ A^n \circ a.$$

THEOREM 1. The algebraic Milnor numbers

$$\mu(n) = \operatorname{ind}_0 f(n), \quad n \in \mathbf{Z}$$

are uniformly bounded, provided that each of them is finite.

Theorem 1 admits the following generalization to the case of pairs of submanifolds of arbitrary dimensions. Consider any strictly decreasing sequence of algebraic subvarieties of the space of jets of pairs of submanifolds of $\mathbf{C}^m$ at the origin (each subvariety is the set of zeroes of a finite set of polynomials, depending on a finite number of the Taylor coefficients of the submanifolds).

THEOREM 2. The maximal numbers $M(n)$ of the varieties of the series to which the pairs $(A^n X, Y)$ belong are uniformly bounded, provided that A is a diffeomorphism and that $M(n) < \infty$ for each n.

Example. Let $(Y, 0)$ be a hypersurface in $(\mathbf{C}^m, 0)$, defined by the equation $b(z) = 0$, $db|_0 \neq 0$. Let us consider the restrictions b_n of b to the images $A^n X$ of a submanifold X of $\mathbf{C}^m$.

THEOREM 3. The Milnor numbers of these functions b_n at the origin are uniformly bounded, provided that A is a diffeomorphism and that the critical point of each function b_n at the origin is isolated.

The proof of Theorem 1 is based on the following facts.

LEMMA 1. The set of mappings $f: (\mathbf{C}^k, 0) \to (\mathbf{C}^k, 0)$ whose Milnor numbers are at least c,

$$\{f : \mu(f) \geq c\}$$

is an algebraic variety (it is the set of zeroes of a finite set of polynomials depending on a finite number of the Taylor coefficients of f at the origin).

This fact is well known (see, for instance, [1]).

Definition. A quasipolynomial g with a spectrum $S \subset \mathbf{C} \setminus 0$ is a function of the form

$$g(n) = \sum p_i(n) z_i^n \quad z_i \in S \tag{1}$$

where the sum is finite and where the p_i arc polynomials.

Definition. A quasipolynomial g is called *admissible* (for short, g *is an AQ*), if each point of its spectrum is a product of the eigenvalues of the linear part at the origin of our diffeomorphism $A: (\mathbf{C}^m, 0) \to (\mathbf{C}^m, 0)$.

REMARK. The AQ's from a C-algebra. For instance, if we substitute AQ's for the variables of any complex polynomial we obtain an AQ.

LEMMA 2. Each coefficient of the Taylor series of the mapping $f(n) = b \circ A^n \circ A: \mathbf{C}^k \to \mathbf{C}^k$ at the origin is an AQ.

This lemma is proved in section 3 by direct calculations.

LEMMA 3. The set of integers n for which a given quasipolynoimal vanishes is the union of a finite set and of a finite union of arithmetical progressions.

This is the Skolem theorem [9].

Definition. We call an infinite set of integers a q-Skolem set if it is the union of a finite set and of a finite set of arithmetical progression with step q. We shall use the following obvious properties of q-Skolem sets.

LEMMA 4. A finite union of q-Skolem sets is a q-Skolem set. An intersection of q-Skolem sets is a q-Skolem set, provided that it is infinite. The intersection of a sequence of decreasing q-Skolem sets is a q-Skolem set. A q-Skolem set is a qs-Skolem set for any integer s.

LEMMA 5. Suppose that a quasipolynomial (1) vanishes for an infinite sequence of values of n. Then the set of its zeroes n is a q-Skolem set, where q is the smallest common multiple of the orders of the roots of unity of the form z_i/z_j (the order of a root u being the index of the subgroup $\{n \in \mathbf{Z} : u^n = 1\}$ in $\mathbf{Z}$).

This (well-known) corollary of the Skolem theorem is proven in section 4.

LEMMA 6. For any finite set of complex numbers $(w_1, \ldots, w_m)$ there exists an integer $q(w)$, such that for any quasipolynomial whose spectrum points z_i are the products of the numbers w_j, the set of zeroes $n \in \mathbf{Z}$ is a q-Skolem set (provided that it is infinite).

The point is, that q is independent of the spectrum $\{z_i\}$ of g, depending only on the generators w_j. We show in section 4 that the value of $q(w)$ is given as follows.

Let us define the *relations subgroup* $R = \{d \in \mathbf{Z}^r : w^d = 1\}$ (where $w^d = w_1^{d_1} \ldots w_r^{d_r}$). The linear subspace $\mathbf{R}^s$ generated by R in $\mathbf{R}^r$ intersects $\mathbf{Z}^r$ along a lattice $\mathbf{Z}^s = \mathbf{R}^s \cap \mathbf{Z}^r$.

Then q is the index of the relations subgroup R in this lattice, $q = [\mathbf{Z}^s : R]$.

PROOF OF THEOREM 1. Suppose that the sequence $\mu(n)$ is not bounded. Then the set $\{n \in \mathbf{Z} : \mu(n) \geq c\}$ is infinite for any c.

But the condition $\mu(n) \geq c$ is equivalent to the vanishing of a finite number of polynomials in the Taylor coefficients of the mapping $f(n)$, according to Lemma 1. Each of these coefficients is an AQ, according to Lemma 2. Hence the above polynomials evaluated at $f(n)$ are AQ's.

According to Lemma 6 the set of integers n for which $\mu(n) \geq c$ is a q-Skolem set for any c (where q is independent of c and depends only on the eigenvalues of the derivative of the diffeomorphism A at the origin).

The set $\{n : \mu(n) \geq c\}$ decreases when c increases. Hence the intersection of these sets (for $c \to \infty$) contains an arithmetical progression (Lemma 4). If n were a point of this progression, then $\mu(n)$ would be infinite, which contradicts the conditions of Theorem 1, which is thus proved.

The proof of Theorem 2 is similar (Lemma 1 being replaced by the axioms defining M). Theorem 3, like Theorem 1, is a particular case of Theorem 2, since the set of all functions, with a critical point at the origin whose Milnor number is equal to or greater than c, is an algebraic variety of the space of jets.

3. Taylor coefficients of iterations

Let us write the Taylor series of a mapping $G\colon \mathbf{C}^a \to \mathbf{C}^b$ at the origin in the form $z = G(t) = G_1(t)+G_2(t,t)+\cdots$, where G_r is an r-linear symmetric form, with values in $\mathbf{C}^b$. We shall call G_r the r-th *Taylor coefficient* of the series.

LEMMA 7. Each Taylor coefficient of the composition $w = H(G(t))$ has the form

$$(H \circ G)_r = H_1 G_r + F_r[G_1, \ldots, G_{r-1}],$$

where the r-polylinear symmetric form F_r (whose arguments are not presented in the notation) depends on $(G_1, \ldots, G_{r-1})$ as a quasihomogeneous polynomial of weight r (for the weights $\deg G_s = s$ of the variables). The coefficients of the above quasihomogeneous polynomial depend linearly on $(H_2, \ldots, H_r)$.

PROOF. Direct substitution provides at the diagonal the identity

$$\begin{aligned} H \circ G = H_1 G_1 + H_1 G_2 + H_1 G_3 + \cdots \\ + H_2(G_1, G_1) + 2H_2(G_1, G_2) + \cdots \\ + H_3(G_1, G_1, G_1) + \cdots . \end{aligned}$$

To calculate F_r, we symmetrize the results. For instance,

$$F_1 = 0,$$
$$F_2[G_1](t_1, t_2) = H_2(G_1 t_1, G_1 t_2),$$
$$F_3[G_1, G_2])(t_1, t_2, t_3) = H_3(G_1 t_1, G_1 t_2, G_1 t_3) + \frac{1}{3}\sum H_2(G_1 t_i, G_2(t_j, t_k))$$

(the summation over the 6 permutations).

The quasihomogeneity follows from the fact that the substitution of ct instead of t into $G(t)$ or $H(G(t))$ acts on G_i and $(HG)_i$ as a multiplication by c^i.

Lemma 7 implies the following:

LEMMA 8. The coefficients A_r^n of the Taylor series $A^n z = A_1^n z + A_2^n(z, z) + \cdots$ of the n-th iteration of $z \mapsto Az = A_1 z + A_2(z, z) + \cdots$ verify the equation

$$A_r^{n+1} = A_1 A_r^n + F_r[A_1^n, \ldots, A_{r-1}^n],$$

where the quasihomogeneous vector-polynomial F_r (deg $F_r = r$ for deg $A_i^n = i$) is independent of n.

PROOF. We apply Lemma 7 to $G_1 = A^n$, $H = A$ for $n = 1, 2, \ldots$.

Example. $A_1^n = (A_1)^n$, $A_2^{n+1}(t_1, t_2) = A_1 A_2^n(t_1, t_2) + A_2(A_1^n t_1, A_1^n t_2)$.

LEMMA 9. Let A_1 be nondegenerate. Then each of the coefficients A_r^n, considered as a function of n, is an AQ.

PROOF. For $r = 1$ we have $A_1^n = (A_1)^n$. Hence A_1^n is a (vector-valued) AQ. (This is *wrong* if A_1 is degenerate!) Suppose inductively that A_r^n for $r < R$ are AQ's. Then A_r verifies the linear inhomogeneous recurrent equation

$$A_R^{n+1} = A_1 A_R^n + f_R(n),\ f_R(n) = F_R[A_1^n, \ldots, A_{R-1}^n]. \tag{2}$$

The inhomogeneity f_R is an AQ by the inductive assumption (see the remark preceding Lemma 2). Now Lemma 9 follows from the standard fact of the ODE theory:

LEMMA 10. Any solution $u(n)$ of a linear inhomogeneous equation

$$u(n + 1) = A_1 u(n) + f(n),$$

where $f(n)$ is a given (vector-valued) quasipolynomial, and A_1 a given nondegenerate linear operator, is itself a quasipolynomial. The spectrum of the solution is contained in the union of the spectrum of f and of the set of eigenvalues of the operator A_1.

The proof (see, for example, [10]) also gives an estimate of the degrees of the polynomials.

A quasipolynomial solution exists also when A_1 is degenerate. However, in this case the homogeneous equation solutions ($A_1^n B$) are quasipolynomials only when n is sufficiently large, and hence the conclusion of Lemma 10 holds only for $n > n_0(A_1)$.

PROOF OF LEMMA 2. We apply Lemma 7 to $H = A^n$ and $G = a$. Each coefficient of the composition depends linearly on a finite set of coefficients of A^n, and hence is an AQ by Lemma 9. We apply Lemma 7 to $H = b$ and $G = A^n a$. Each coefficient of the composition is a polynomial in the coefficients of $A^n a$, and hence is an AQ.

REMARK. If A_1 is degenerate, our proof shows that the rth Taylor coefficient of $f(n) = bA^n a$ is an AQ for $n > n_0(r)$, but n_0 may grow with r.

4. Zeroes of quasipolynomials

PROOF OF LEMMA 5. According to the Skolem theorem, the functions $g(n) = \sum p_i(n) z_i^n$ vanish along some arithmetical progression, $n = pN + r$. Along this progression g can be considered as a quasipolynomial with a spectrum $\{Z_j = z_i^p\}$:

$$g(pN + r) = \sum P_j(N) Z_j^N, \; P_j(N) = \sum z_i^r p_i(pN + r)$$

(summation over those i for which $z_i^p = Z_j$).

If $g(pN + r) = 0$ for any integer N, then all the coefficients $P_j(N)$ vanish identically (since $Z_j \neq Z_k$ for $j \neq k$). In this case $g(pN + r + q) = 0$, provided that q is a multiple of all the orders of the roots of unity of the form z_i/z_k.

Indeed, let $z_i^p = z_k^p \neq 0$. Then z_i/z_k is a root of unity, hence $(z_i/z_k)^q = 1$. Thus $z_i^q = C_j$ is the same for all i with $z_i^p = Z_j$. Hence we can write the value of the shifted quasipolynomial as

$$g(pN + r + q) = \sum C_j P_j(N + q/p) Z_j^N, \quad C_j = z_i^q,$$

since if $z_i^p = Z_j$, then

$$\sum z_i^{r+q} p_i(pN + r + q) = C_j \sum z_i^r p_i(p(N + q/p) + r) = C_j P_j(N + q/p).$$

Since $P_j \equiv 0$, we see that $g(pN + r + q) = 0$ for all integer N. Hence, each arithmetical progression of integer zeroes of the quasipolynomial g is covered by the union of a finite number of step q arithmetical progressions, also consisting of zeroes of the quasipolynomial, which proves Lemma 5.

PROOF OF LEMMA 6. Let $(z_i/z_k)^p = 1$, where $z_j = w^{d_j}$. Then $p(d_i - d_k)$ belongs to the relations subgroup $R \subset \mathbf{Z}^r$. Hence, the order of the root z_i/z_k is the order of the cyclic subgroup of the quotient group $\mathbf{Z}^s/R$ (where $\mathbf{Z}^s = \mathbf{Z}^r \cap \mathbf{R}(R)$). The order q of this group is divisible by the orders of its subgroups. Hence q is divisible by the orders of all the roots of unity of the form z_i/z_k and Lemma 6 follows from Lemma 5.

5. Iterations of degenerate mappings

Let $A\colon (\mathbf{C}^2, 0) \to (\mathbf{C}^2, 0)$ be a holomorphic mapping of the plane to itself. Let X and Y be two holomorphic curves containing the origin 0. We suppose that all the curves $A^n X$ $(n \geq 0)$ are also smooth at the origin (in the sense that the derivative of A^n at 0 does not annihilate the tangent vector of X).

THEOREM 4. The Milnor numbers $\mu(n)$ of the intersections of the curves $A^n X$ and Y at the origin are bounded by an exponential of n, provided that all these intersections are isolated.

PROOF. Let us choose coordinates (x, y) on the plane, so as to reduce the matrix of the derivative to the diagonal form:

$$A(x, y) = (\lambda x + P(x, y), Q(x, y))$$

where $P(0) = Q(0) = 0$, $dP(0) = dQ(0) = 0$, $\lambda \neq 0$. The tangent to the curve $A^n X$ at the origin is the x-axis (if $n > 0$). If the tangent to the curve Y at 0 is different, the Milnor number is 1 for any $n > 0$. Hence we shall suppose that the curve Y is the x-axis (this is not a restriction).

From the remark at the end of section 3 we know that if the Milnor numbers $\mu(n) = \mu(A^n X, Y)$ are unbounded, then the values of n such that $\mu(n) \geq r$ form a q-Skolem set, provided that $n > n_0(r)$ is sufficiently big. Here q is the order of λ as a root of unity (or 1, if λ is not a root of unity). Replacing A by A^q and X by $A^p X, 0 \leq p < q$, we reduce the problem to the case where λ is not a root of unity different from 1. Now $q = 1$, hence the curve $A^n X$ has with Y at the origin an intersection of multiplicity at least r for any $n > n_0(r)$.

Let us consider any smooth curve $y = ax^r + \cdots$, $a \neq 0$, intersecting the axis $y = 0$ with multiplicity r.

LEMMA 11. Let $Q \neq 0$. Then there exists $r_0 = r_0(A) > 0$ such that for any $r > r_0$, any smooth curve intersecting the axis $y = 0$ with multiplicity r at the origin is transformed by A into a curve intersecting this axis with multiplicity $vr + u$, where $v(Q)$ and $u(Q)$ do not depend on the curve.

PROOF. The slopes of the segments of the Newton boundary of Q are bounded. Hence a supporting line of the boundary, having any sufficiently large slope, can't touch the boundary at more than one point. Moreover, this point (v, u) is independent of the slope of the supporting line, provided that it is sufficiently large.

To be more precise, let us consider the monomials $x^U y^V$ encountered in Q with nonzero coefficients. Let us choose the minimum v of V and then the minimum u of the U of the monomials $x^U y^v$. Call $x^u y^v$ the *principal monomial.*

Choose $r_0 = u + 1$, and let $r \geq r_0$. The principal monomial $x^u y^v$ of Q, when restricted to our curve $y = ax^r + \cdots$, is equal to

$$x^u (ax^r + \cdots)^v = a^v x^{vr+u} + \cdots .$$

All other monomials produce only terms of higher order ($Vr + U > vr + u$). Indeed, if $V > v$, then $Vr + U \geq (v+1)r + U > vr + u$, since $r \geq r_0 = u + 1$. If $V = v$, then $U > u$. Thus, $\mu(n) \geq r_0$ implies that $\mu(n+1) = v\mu(n) + u$. The Lemma is thus proved.

REMARK. The case $v = 0$ is a priori not excluded, and hence it might happen that $\mu(n+1) < r_0$. In this case $\mu(n+2)$ might be arbitrarily large, and we can't control the growth rate.

END OF THE PROOF OF THEOREM 4. We know from the beginning of the proof that if the numbers $\mu(n)$ are not bounded, then they are larger than any given number r for all $n > n_0(r)$.

Let us choose for r the number $r_0(A)$ of Lemma 11. Then $\mu(n) \geq r_0(A)$ for any $n > \mathcal{N}_0 = n_0(r_0(A))$. Hence $\mu(n+1) = v\mu(n) + u$ for any $n > \mathcal{N}_0$. Thus μ grows at most exponentially, as required.

REMARK. We have thus proved an explicit formula for the growth of the Milnor numbers:

$$\mu(n) = \begin{cases} Cv^n + u/(1-v) & \text{if } v > 1, \\ un + C & \text{if } v = 1. \end{cases}$$

Both cases appear: $A(x, y) = (x, y^2)$, $A(x, y) = (x, xy + y^2)$.

If $q \neq 1$, similar formulas hold for each arithmetic progression of step q, along which the Milnor numbers are not bounded.

The failure of attempts to construct an example of superexponential growth of Milnor numbers is an argument for the general conjecture of [2], [3], which says that the topological complexity of the intersections of analytic submanifolds in analytical dynamical systems on compact manifolds grows at most exponentially.

The author is indebted to the Mathematisch Instituut at Utrecht, where this paper was written, and to Karin Berlang for rapidly deciphering a difficult manuscript.

BIBLIOGRAPHY

1. V. I. Arnol'd, S. M. Gusein-'Zade, A. N. Varchenko, "Singularities of Differentiable Maps," Birkhäuser, Boston, 1985.
2. V. I. Arnol'd, *Dynamics of Complexity of Intersections*, Bol. Soc. Bras. Math. **21** no. 1 (1990), 1–10.
3. V. I. Arnol'd, *Dynamics of intersections*, "Analysis e.c.t." (P. Rabinowitz and E. Zehnder, eds.), Academic Press, 1990, pp. 77–84.
4. E. Gonsales Rosales, *On the growth of the number of periodic points of diffeomorphisms*, Funct. Anal. and Appl. (1991).
5. E. Gonsales Rosales, *On the growth of the number of closed orbits of dynamical systems*, Funct. Anal. and Appl. (1991).
6. M. Artin, B. Mazur, *On periodic points*, Annals of Math **81** (1965), 82–99.
7. M. Shub, D. Sullivan, *A remark on the Lefschetz fixed point formula for differentiable maps*, Topology **13** (1974), 189–191.
8. J. Milnor, "Singular Points of complex Hypersurfaces," Princeton Univ. Press, 1967.
9. Th. Skolem, *Ein Verfahren zur Behandlung gewisser exponentialer Gleichungen und Diophantischer Gleichungen*, "Proceeding of the 8th Scand. Math. Congress, Stockholm, 1934," pp. 163–188.
10. V. I. Arnol'd, "Ordinary Differential Equations," MIT Press, 1973.

Some remarks on symplectic monodromy of Milnor fibrations

V.I. Arnold

There exist interesting and unexplored relations between symplectic geometry and the theory of critical points of holomorphic functions. One of the manifestations of these relations is the Wahl-Neumann theorem, according to which the Euler characteristics of the Floer homology of the boundary of a Milnor fiber of a quasi-homogeneous function of 3 variables is proportional to the signature of the Milnor fiber, provided that the boundary is a homological 3-sphere [1].

The signature is well-defined even when the boundary is not a homological sphere of dimension three, namely for all boundaries of dimension 3, 7, 11 and so on. It is hence natural to ask whether some generalized Floer homology (or, at least, some Casson invariant) can be defined for higher dimensional boundaries. Of course, one expects that they will depend not only on the smooth structure of the boundary, but rather on some more delicate structure, related to the symplectic structure of the fiber, for instance, on the contact or on the CR structure of the boundary.

[The situation might be similar to that in Lie group theory. The natural generalization of $SL(2)$ (or of the group $SDiff(2)$ of area-preserving plane diffeomorphisms) is from many points of view not $SL(n)$ (correspondingly not the group $SDiff(n)$ of volume-preserving space diffeomorphisms) but rather the linear symplectomorphisms group $Sp(n)$ of $2n$-space (correspondingly, rather the group SpDiff of all symplectomorphisms). Thus, the natural generalization of the «jacobian determinant problem» (the classification of the plane polynomial mappings with jacobian 1) might be rather the classification of the polynomial symplectomorphisms of 4-space than that of the polynomial volume preserving mappings of 3-space.]

Floer was the only symplectic geometry expert who had expressed some optimism in discussing the preceding conjectures with me (in January 1991 at the Bombay Tata Institute). A lot of preliminary work should probably have been done before these conjectures are made precise. First of all, one has to understand the symplectic geometry of the Milnor fibers, the contact geometry of their boundaries, the Lagrange geometry of the vanishing and covanishing cycles, and the Legendre geometry of their boundaries. One has to study the higher dimensional version of the Bennequin invariants of Legendre curves, the signature of infinite dimensional quadratic forms, generalizing the Bennequin self-linking form, constructed from the symplectic or contact manifolds and from their Lagrange or Legendre submanifolds or currents. One might speculate that the Milnor fiber should have the maximal signature among all the symplectic manifolds with the same contact boundary, but even this has not been proved (or disproved).

Originally published in: *The Floer Memorial Volume*. Editors: H. Hofer, C.H. Taubes, A. Weinstein, and E. Zehnder (Progr. Math. 133). Birkhäuser, Basel, 99–103 (1995), © 1995 Birkhäuser.

In this article two simple and elementary facts on the symplectic geometry of the Milnor fibrations are presented. They might be of no use for the goals described above (but may also have other applications).

1 Symplectic geometry of regular coadjoint orbits

We start from the standard complex symplectic structure Ω on a regular orbit of the coadjoint representation of the complex Lie algebra $A_n = \mathrm{SL}(n+1, \mathbb{C})$. The theorem that follows is perhaps known for a long time, while I was unable to find it explicitly in the literature:

Theorem 1 *A regular orbit, equipped with the real symplectic structure* $\operatorname{Im}\Omega$, *is isomorphic to the cotangent bundle space of the manifold of the complete complex flags in* $\mathbb{C}^{n+1}$, *equipped with the standard Liouville symplectic structure, if and only if all the eigenvallues of some (and hence any) matrix in the orbit are real.*

Comments. A regular orbit consists of all matrices with a fixed characteristic polynomial (having no multiple roots). A point of an orbit is fixed by the choice of the invariant lines, one for each eigenvalue. Hence each orbit is complex diffeomorphic to the configuration space of the sets of $n+1$ ordered independent lines in $\mathbb{C}^{n+1}$. The union of all regular orbits is fibered holomorphically over the manifold of polynomials with no multiple roots. This fibration possesses a holomorphic locally flat connection (one may move the eigenvalues, fixing the eigendirections). The monodromy group of this connection is the standard representation of the braid group onto the Weyl group (of the permutations of the eigendirections).

As a complex manifold, the orbit is holomorphically fibered over the manifold of complete complex flags in $\mathbb{C}^{n+1}$. Namely, a matrix is sent to its first eigendirection, to the plane, spanned by the first two 1's, and so on. This fibration depends on the ordering of the eigendirection. The number of such fibrations is equal to the order of the Weyl group, $(n+1)!$, and the monodromy group permutes these fibrations.

Each of these fibrations is a holomorphic Lagrange fibration with respect to the standard symplectic structure of the coadjoint orbit. Hence its fibers have a local affine structure, as any Lagrange fibration's fibers. In fact they are holomorphic affine spaces globally. Each fibration of the orbit over the flag manifold is real diffeomorphic to the (co)tangent bundle fibration of the flag manifold as real vector bundles. However, as holomorphic fibrations, these two fibrations (with affine complex fibers) are different: the (co)tangent bundle has a zero section while the orbit fibration has no holomorphic sections.

To fix a real algebraic equivalence of the orbit fibration to the cotangent bundle space, one fixes an hermitian metric in the space $\mathbb{C}^{n+1}$. Then one obtains a real section of the orbit fibration, associating to a flag the ordered set of hermitian orthogonal lines, generating the spaces of the flag. The natural complex symplectic structure of the orbit is transformed by this identification into the sum of the standard Liouville symplectic structure of the cotangent bundle space of the flag manifold with a two-form, lifted from the flag manifold. This unitary-invariant 2-form ω on the flag manifold can be computed explicitly. The 2-dimensional homology group of the flag manifold $F(n+1)$ is generated by n spheres CP^1 (consisting of the flag , whose spaces, with one exception, coincide with those of a fixed flag). The periods of the 2-form ω on these cycles are equal to $2\pi(\lambda_j - \lambda_{j+1})$, where $j = 1, \ldots, n$ (λ_j, $j = 1, \ldots, n+1$, are the eigenvalues, $\lambda_1 + \ldots + \lambda_{n+1} = 0$).

If there exists a nonreal eigenvalue, then the symplectic structure $\operatorname{Im}\Omega$ on the orbit is not exact. If they are all real, then the calculations show that the form $\operatorname{Im}\omega$ on the flag manifold vanishes (the vanishing of the periods follows from the above formula). Thus the fibration of the orbit with the symplectic structure $\operatorname{Im}\Omega$ over the flag manifold is in this case real equivalent to the cotangent bundle of the flag manifold with its real Liouville symplectic structure. Let us fix an orbit, corresponding to real eigenvalues, and the real symplectic structure $\operatorname{Im}\Omega$ on it.

Problem *Can one choose real symplectic structures (equivalent to that one) on the other orbits in such a smooth way, that the monodromy of the resulting fibration would be symplectic?*

Comments. More precisely, the symplectic monodromy problem for a fibration with symplectic fibers consists of the following questions:

1) Let $\{p(t)\}$, $0 \le t \le 1$, be a loop in the base space. Does there exist a family of symplectomorphisms $s(t) : X(0) \to X(t)$ of the fiber over $p(0)$ onto the fibers over $p(t)$, such that $s(0) =$ identity (one may also ask whether these symplectomorphisms verify some boundary conditions, say that the boundary or «the infinity» remains pointwise fixed).
2) Is it possible to choose the symplectomorphisms $s(t)$ in such a way, that the symplectomorphisms $s(1)$ of the initial fiber onto itself, corresponding to different loops of the base space, would form a representation of the fundamental group of the base space (in our case — of the braid group) into the symplectomorphisms group of the fiber.

Remark. Weaker variants of both questions are still interesting: one may drop the boundary conditions, or one may require the symplectic condition only for the diffeomorphisms of the initial fiber into itself, $s(1)$ (forgetting the symplectic structures of other fibers).

The symplectic monodromy problem is solved below for the simplest case $A_1 = \mathrm{SL}(2, \mathbb{C})$.

2 Symplectic monodromy of Morse singularity

The regular orbits fibration for the coadjoint representation of $A_1 = \mathrm{SL}(2)$ coincides with the fibration of the regular level surfaces of the Morse function

$$f(x, y, z) = x^2 + y^2 + z^2 : \mathbb{C}^3 \to \mathbb{C}$$

where (x, y, z) are suitable linear coordinates in the dual space of the Lie algebra of zero-trace 2×2 complex matrices. Each regular level surface is real diffeomorphic to the 4-manifold T^*S^2. We choose the initial fiber $x^2+y^2+z^2 = 1$. The symplectic structure of §1 is then $\operatorname{Im}\Omega$, where $\Omega = dx \wedge dy \wedge dz/df$.

Remark. The complex symplectic structure of the Milnor fiber of complex dimension 2, induced from thc natural Poisson structure of the dual space of a simple complex Lie algebra via the Brieskorn mapping of the transversal to the subregular orbit, always has the form $dx \wedge dy \wedge dz/df$, where $f = const$ is the fiber equation (A.B. Givental, 1980). The imaginary part over the fiber corresponding to real eigenvalues is the right (exact) symplectic structure of the Milnor fiber.

The monodromy diffeomorphism lifts the loop $f = e^{2\pi i t}$. It is a four-dimensional generalization of the Dehn twist on a 2-cylinder, and we shall see that it can be made symplectic.

We start from a «finitely supported» version — from the Milnor monodromy acting on the discs bundle B^*S^2. This four-manifold with boundary is formed by the unite discs ($p^2 \leq 1$) in the fibers of the cotangent bundle. The monodromy preserves the points of the boundary, the orientation and the zero section. The monodromy action on the zero section is the antipodal involution. The image of a disc-fiber, B, is a disc B' with the same boundary, such that $B' - B$ is a cycle, homologous to the zero section.

We shall construct a symplectomorphism with these properties. Let $I : S^2 \to S^2$ be the antipodal involution, and $I^* : B^*S^2 \to B^*S^2$ — its action on the (co)tangent vectors. The mapping I^* is a symplectic involution.

Let us consider the function $H = p^2/2 : B^*S^2 \to \mathbb{R}$ (the kinetic energy of a free particle on the sphere). The transformation g^t of the phase flow with Hamiltonian function H at time t is symplectic. It preserves the points of the zero section, and it moves the (co)tangent vectors of length l along their geodesics at distance lt. For $t = \pi$ the (co)vectors of length 1 are translated by the flow to the antipodal point the same way as they are translated by the involution I^*.

Now let us consider the product I^*g^π. This symplectomorphism leaves fixed the points of the boundary and acts on the zero section as the antipodal involution. The unit cotangent disc centered at the North pole is transformed in a (co)vector field with a singularity at that pole. Hence I^*g^π acts on the fibers as required.

Theorem 2 *The mapping I^*g^π provides a symplectic monodromy of the Milnor fibration of the Morse function (in the strong sence, explained at the end of §1).*

To extend the result to the coadjoint orbits (that is to the unbounded cotangent bundle space T^*S^2 instead of B^*S^2), one replaces $H = p^2/2$ by a monotonic function of $u = p^2$ such that $\sqrt{u}dH/du$ tends monotonically to $\pi/2$ for $u \to \infty$.

Remark. A similar construction works for Morse functions in any dimension.

To study the orbits of $A_n = \mathrm{SL}(n+1, \mathbb{C})$ one may replace the involution I by the set of n involutions of the flag manifold. The involution I_j moves only the j-dimensional subspace of the flag. The new j-subspace is defined by the following condition: the line, representing it in the quotient plane of the $j+1$-subspace over the $j-1$-subspace is replaced by its hermitian orthocomplement in this plane.

The Hamiltonian of the free particle on the sphere is replaced by the set of n unitary-equivariant Hamiltonian flows on the space of the cotangent bundle of the flag manifold. These flows might be described as the left-invariant geodesic flows on the Lie group $\mathrm{SU}(n+1)$. The corresponding energy functions are then naturally defined on the space of antihermitian matrices with zero diagonal entries. Each Hamiltonian function is the squared modulus of an element immediately over the main diagonal. I have not checked whether the symplectomorphisms of the cotangent bundle space of the flag manifold, generated by this construction, form a representation of the braid group into the symplectomorphism group of the orbit, corresponding to real eigenvalues equipped with the symplectic structure $\mathrm{Im}\,\Omega$. The question whether the Milnor fibrations of the simple singularities (different from A_1) of functions of 3 or more variables have symplectic monodromies or flat symplectic connections also remains open. A locally flat connection for the A_2 singularity of functions of two variables can be constructed using the real affine structures of the elliptic curves and probably exists also in the case of three variables.

References

1 W. Neumann and J. Wahl, *Casson invariant of links of singularities*, Comment. Math. Helvetici **65** (1990), pp. 58–78.

2 V.M. Kharlamov and Ja.M. Eliashberg, *On the number of complex points of a real surface in a complex surface*, Proceedings of the Leningrad Topology Conference August 1982, Nauka, Leningrad, 1983, pp. 143–148.

3 A. Douady, *Noeux et structures de contact en dimension 3 d'après D. Bennequin*, sem. N. Bourbaki, 1982–1983, Exp. 604, pp. 60401–60420

Steklov Mathematical Institute
42, Vavilova Street
117966 Moscow GSP-1
Russia

CEREMADE
Université Paris-Dauphine
75775 Paris, Cedex 16e
France

On topological properties of Legendre projections in contact geometry of wave fronts

V.I. Arnold

Dedicated to Ludwig Dmitrievich Faddeev

The global topological problem of coexistence of singularities on wave fronts turns out to have an interesting relationship with the Sturm type theory of oscillations of linear combinations of eigenfunctions of differential operators.

This relationship can be exploited both ways. In this paper, the Sturm–Tabachnikov theorem on the oscillations of linear combinations of eigenfunctions of differential operators on a circle is used to minorize the number of cusps of a generic wave front close to a circle collapsing to a point.

The classical theorem on four vertices for curves on the Euclidean plane is shown to be a particular case of a general fact of the contact topology. This observation suggests higher-dimensional generalizations of the theorem on four vertices, as well as some nontraditional generalizations of the Sturm theory to functions on spheres.

The simplest of such generalizations corresponds to the theorem on umbilical points on a closed convex surface in the Euclidean 3-space. The general formulation in terms of the symplectic and the contact topology opens the way for various higher-dimensional and non-Euclidean generalizations of these classical results of differential geometry of the Euclidean space (to certain settings considerably more general than that of Riemannian geometry).

The present article was written during the semester spent at the MIT on an invitation from the Mathematical Department. The author is grateful to the Institute and to the Department for hospitality and support.

§1. The necessity of four cusps on a collapsing front

We start with an example. Consider the one-parameter family of circles of radius $|t|$ centered at a fixed point of the Euclidean plane. Here t is a parameter (time) indexing moving fronts (circles). As t runs from $-\infty$ to $+\infty$, the front first shrinks and collapses to a point (at $t = 0$) and then starts to grow (Figure 1) after reversion, so that the inward normals reappear as the outward normals.

This family of fronts is called the *standard Legendre collapse*.

Collapsing to a point is not a typical singularity of moving fronts. In a generic family, a shrinking front turns into an expanding one in several stages rather than instantly. The front consisting of a point never arises; instead, we obtain intermediate fronts with several singularities.

A typical singularity of a front on the plane is a (semicubical) cusp point. The main result of the present paper says that, *in any generic deformation of the standard Legendre collapse, there exist fronts having at least four cusps.*

Originally published in: *Algebra i Analiz* **6**(3), 1–16 (1994), © 1994 Russ. Acad. Sciences
English translation: *St. Petersburg Math. J.* **6**(3), 439–452 (1995), © 1995 Amer. Math. Soc.

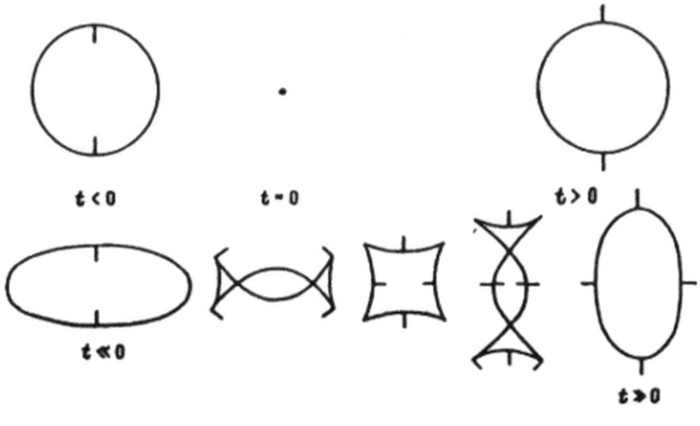

FIGURE 1. Standard Legendre collapse and its generic deformation.

Example. Starting with an initial elliptic front (for instance, an almost circular one), consider the system of the equidistant curves of this ellipse as its evolution (Figure 1). Some of these equidistant curves have four cusps. Our theorem implies that this is not by chance: for "topological reasons", it is impossible to perturb the standard collapse creating less than four cusps.

These reasons stem from certain facts pertaining to the contact topology. From the point of view of Analysis, they are topological counterparts of the Sturm type theorems on the zeros of eigenfunctions and their linear combinations.

The contact geometry is the odd-dimensional "twin brother" of the symplectic geometry. Though the latter has become one of the most basic tools of mathematical physics, the contact geometry is, unfortunately, much less known. The relationship between the symplectic and the contact geometry is similar to that between linear algebra and projective geometry. Although any calculation in the contact geometry can be reduced to a calculation in terms of the usual Hamiltonian formalism of Poisson brackets of homogeneous functions, in the topological problems we are going to study the passage to homogeneous functions is inconvenient and the language of contact geometry is practically inavoidable.

In the next two sections, the main notions of the contact geometry will presented in a suitable form and with the examples needed in the proof of the theorem on four cusps of a reversing front.

§2. CONTACT MANIFOLDS

Definition. A *contact element* on a manifold is a hyperplane (a linear subspace of codimension one) in a tangent space to the manifold.

Mostly, we shall work with *cooriented* contact elements (equipped with a transversal direction, defined, say, by a unit normal in some metric). All the cooriented contact elements on an n-dimensional manifold B^n form a manifold ST^*B^n of dimension $2n-1$, the *space of the spherized cotangent bundle*. The fiber over a point of B^n is the sphere S^{n-1} formed by all the cooriented contact elements at this point. Thus, ST^*B^n is the manifold of the directions of all momentum vectors of the phase space over the configuration manifold B^n.

Example. The manifold of cooriented contact elements of the plane $\mathbf{R}^2$ is a three-dimensional manifold diffeomorphic to $S^1 \times \mathbf{R}^2$, the interior part of a filled torus. As coordinates in this space (or in its universal covering 3-space), we shall use the functions (x, y, φ), where x and y are the usual Carthesian coordinates of the point where the contact element is tangent to the plane, and φ is the angle between the positive direction of the x-axis and the coorienting normal, measured in the direction from the positive direction of the x-axis to the positive direction of the y-axis.

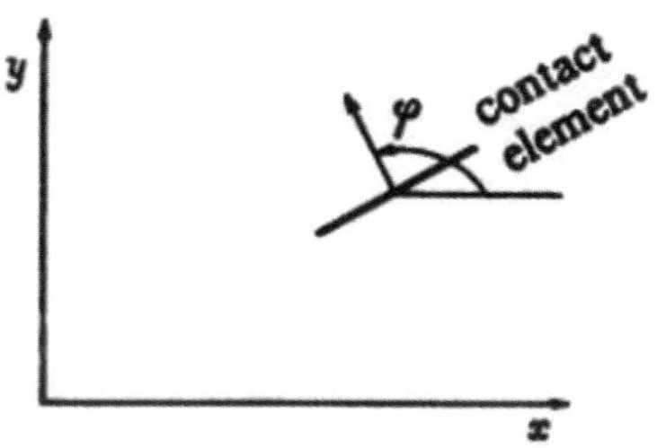

FIGURE 2. The coordinates of a cooriented contact element of the plane $\mathbf{R}^2$.

Definition. The *standard contact structure* of the manifold of contact elements of B is a certain field of tangent hyperplanes of the manifold of contact elements of B, namely, the field determined by the following "skating condition": the velocity vector of the motion of a contact element belongs to the hyperplane of the contact structure if the velocity of the point of contact belongs to the contact element in question.

Example. In the coordinates introduced above, the standard contact structure of the manifold of cooriented contact elements of the plane is given by the equation $\alpha = 0$, where α is the differential 1-form

$$\alpha = c\,dx + s\,dy$$

(from now on, $c = \cos\varphi$, $s = \sin\varphi$).

Remark 1. The form α itself is not an attribute of the definition of the contact structure: it may be multiplied by any nonvanishing function, and this will not change the structure determined by the equation $\alpha = 0$.

Remark 2. A *contact structure* on an odd-dimensional manifold is a field of hyperplanes in the tangent spaces, which is, in a sense, extremely nonintegrable: the restriction of the form $d\alpha$ to the hyperplane $\alpha = 0$ must be a nondegenerate bilinear form (a linear symplectic structure). This nondegeneracy condition is intrinsic; it is independent of a particular choice of the form defining the contact structure locally. Globally, such a form may fail to exist, since the field of hyperplanes may be noncoorientable. In our example the nondegeneracy condition is fulfilled.

Any two contact structures in a neighborhood of any point of a manifold of a fixed dimension $2m + 1$ are equivalent (the Darboux theorem). In a suitable system of local coordinates (the *Darboux coordinates*), any contact structure is determined by the equation $\alpha = 0$, where

$$\alpha = dz - p\,dq \qquad (p = (p_1, \dots, p_m),\ q = (q_1, \dots, q_m)).$$

Example. The standard contact structure of the manifold of cooriented contact elements of the plane (written above in terms of coordinates) is *globally* standard on the universal covering 3-space (i.e., it can be reduced to the Darboux normal form by a suitable choice of *global* coordinates). Indeed, the form α can be written as follows:

$$c\,dx + s\,dy = d(cx+sy) + (sx-cy)d\varphi\,;$$

hence, the triple

$$z = cx+sy\,, \quad p = -sx+cy\,, \quad \varphi$$

is a system of Darboux coordinates.

§3. Legendre submanifolds and their fronts

Definition. A *Legendre submanifold* of a contact manifold is an integral submanifold of maximal possible dimension in the contact structure (this dimension is equal to m for a contact manifold of dimension $2m+1$).

Example. Legendre submanifolds of the space of contact elements of the plane are the curves on which the form α vanishes. All the contact elements tangent to a smooth plane curve make up such a Legendre curve. A Legendre curve is also formed by all the contact elements at a given point of the plane.

Definition. The *front* of a Legendre submanifold of the manifold of contact elements of B is the projection of this Legendre submanifold to the base manifold B.

Example. As a rule, the front of a Legendre submanifold of the manifold of contact elements of the plane is a plane curve. Each (immersed) smooth curve on the plane is the front of a Legendre curve constituted by the elements of the tangents to this smooth curve at the tangent points. Each point of the plane is a (degenerate) front (constituted by the contact elements at this point).

The front of a generic Legendre curve is a plane curve the only singularities of which are semicubical cusps and transversal selfintersection points. These singularities are stable. They withstand the small deformations of the Legendre curve provided that the deformed curve remains Legendrian.

Example. The equidistant curves of an ellipse, shown in Figure 1, are the fronts of certain Legendre curves. These Legendre curves are obtained from the initial one (which is formed by the tangent elements of the ellipse) by transformations of the geodesic flow of the plane (the latter naturally acts on the manifold of the cooriented contact elements).

Remark. Since our contact elements will always be cooriented, it is always convenient to equip the fronts with an *armament* or *framing* consisting of the "unit normals" coorienting the corresponding elements. A framed front uniquely determines its Legendre submanifold in ST^*B.

Example. A point of B armed with the vectors of all directions determines a Legendrian submanifold (namely, the fiber of the spherical cotangent bundle of B over this point).

Remark. Instead of a point, we could start with an arbitrary submanifold V^k of any dimension k in B^n. The cooriented contact elements tangent to V^k constitute a Legendre submanifold L^{n-1} in ST^*B^n. The projection of this submanifold to B^n generates a mapping

$$L^{n-1} \to V^k$$

which is a *Legendre collapse over* V^k provided that k is smaller then $n-1$ (this means that V is not a hypersurface in B).

It is probable that the study of the Legendre collapse over a point (see below) can be extended to the case of the Legendre collapses over arbitrary submanifolds. The manifold corresponding to the family of fronts of the standard Legendre collapse is obtained from L^{n-1} with the help of the geodesic flow of the Rimannian metric.

Definition. A *Legendre fibration* or *foliation* is a fibration (foliation) whose fibers (leaves) are Legendrian.

Example 1. The spherized cotangent bundle is Legendrian.

Example 2. The projection

$$\mathbf{R}^{2m+1} \to \mathbf{R}^{m+1}$$

defined in the Darboux coordinates as forgetting the coordinate p,

$$(p, q, z) \mapsto (q, z),$$

is a Legendre fibration.

All the Legendre fibrations (foliations) of spaces of one and the same dimension are locally contactomorphic in a neighborhood of a point of the fibration (foliation) space: they can be transformed one into another by a diffeomorphism preserving contact structure.

The projections of the Legendre submanifolds of a Legendre fibration space to its base manifold are called *fronts*. Locally, these fronts have the same properties as those obtained under the projection to B of the space of contact elements over B.

Example. Consider the mapping

$$ST^*\mathbf{R}^n \to S^{n-1} \times \mathbf{R}$$

sending each cooriented contact element of the Euclidean space to the pair composed of the coorienting unit normal and the scalar product of this unit normal by the radius-vector of the contact point. This mapping is a Legendre fibration.

Consider a cooriented hypersurface in the Euclidean space $\mathbf{R}^n$. Its cooriented tangent elements form a Legendre submanifold in $ST^*\mathbf{R}^n$. The front of this submanifold is the graph of a function $K: S^{n-1} \to \mathbf{R}$, provided that the initial hypersurface bounds a convex body. Then K is called the *support function* of the body. Thus, the graph of the support function of a convex body is the front of a Legendre projection of a Legendre submanifold.

If we start with a nonconvex generic hypersurface in $\mathbf{R}^n$ and consider the Legendre submanifold in $ST^*\mathbf{R}^n$ formed by the tangent elements of this hypersurface, then the front of this submanifold (in $S^{n-1} \times \mathbf{R}$) is not the graph of a function. This front is a hypersurface in $S^{n-1} \times \mathbf{R}$ with singularities typical for fronts (for instance, for $n = 2$ we obtain a curve having semicubical cusps and transversal selfintersections as singularities). Thus, the (generalized) support function of a generic nonconvex hypersurface is not merely multivalued, but, moreover, has singularities typical for generic fronts of Legendre submanifolds (and for the Legendre transformations of generic nonconvex functions, whence the name).

§4. Standard Legendre collapse

Consider all the cooriented contact elements of the Euclidean space $\mathbf{R}^n$ that are orthogonal to the vector connecting the contact point with the origin.

Proposition. *These elements form a smooth submanifold diffeomorphic to the cylinder* $S^{n-1} \times \mathbf{R}$.

Proof. The mapping $S^{n-1} \times \mathbf{R} \to ST^*\mathbf{R}^n$, sending the pair (ω, t) to the contact element with the coorienting normal vector ω at the point ωt is an embedding.

Definition. The image of the embedding described above is called the *standard cylinder* over $\mathbf{R}^n$.

The standard cylinder is fibered into the spheres $t = \text{const}$. These spheres are Legendre submanifolds of the manifold of cooriented elements of $\mathbf{R}^n$.

Definition. The *standard Legendre collapse* over $\mathbf{R}^n$ is the projection of the standard Legendre cylinder from $ST^*\mathbf{R}^n$ to $\mathbf{R}^n$.

This mapping will be regarded as a family of mappings of the Legendre spheres $t = \text{const}$ onto their fronts. The sphere corresponding to $t \neq 0$ is mapped onto its front (the latter is the sphere of radius $|t|$ in $\mathbf{R}^n$) diffeomorphically. The entire Legendre sphere $t = 0$ collapses to the point 0 in $\mathbf{R}^n$.

Example. The standard Legendre collapse over the plane is contraction of the circle with reversion, as described in §1.

In what follows, we shall consider deformations of the standard Legendre collapse over $\mathbf{R}^2$. Such a deformation is a one-parameter family of Legendre embeddings of a circle into $ST^*\mathbf{R}^2$, and it is assumed that this family is not too far from the standard cylinder. An example is provided by the family of equidistant curves of a slightly deformed circle.

We shall see that, for a generic "not very large" deformation of the standard Legendre collapse over $\mathbf{R}^2$, some of the deformed fronts must have at least four cusps.

Remark. It is not clear how large admissible deformations may be. It seems essential that the fronts of the family in question correspond to nonselfintersecting Legendre curves, and, probably, even that the entire perturbed cylinder remain embedded.

The situation is similar to that having existed in the theory of fixed points of symplectomorphisms and of intersections of Lagrange submanifolds before the work of Conley and Zehnder [3], of Floer [4], etc.

In that theory, the infinitesimal perturbations of standard objects are studied with the help of the Morse theory (applied to the Hamilton function describing an infinitesimal perturbation).

The lower bounds obtained in this way for the number of fixed points of symplectomorphisms and for the number of intersection points of Lagrange manifolds were extended in [1] to certain finite (i.e., not infinitesimal, though not very large) perturbations of standard objects by using the technique of generating families of symplectomorphisms.

More recent methods of studying perturbations of quadratic forms with infinite positive and negative index have led, in many cases, to lifting all the restrictions of smallness of perturbations (e.g., for tori in [3], for Kaelerian manifolds with zero second homotopy group in [4], for torical manifolds in [9], etc.).

In the theory developed below, the infinitesimal study leads instead of the Morse theory, to the Sturm theory. The results of the present paper (as well as those of its symplectic counterpart [2], see also [6, 7]) reduce the study of finite (but not too large) perturbations to theorems of Sturm type.

However, I was able to completely renounce all the restrictions on the size of perturbations only in a single model problem concerning curves on the sphere. The result (obtained by the methods of the so-called *projective topology*, see [8]) says that *a simple closed curve splitting the sphere into two parts of equal area has at least four inflection points.*

§5. The main theorem

Consider a one-parameter family of Legendre closed curves in $ST^*\mathbf{R}^2$. If this family is not too far from the standard cylinder defining the standard Legendre collapse, then it will satisfy the conditions of smallness of perturbations formulated below (the *admissibility conditions*).

In order to formulate these conditions, we introduce the following notation. Consider the *standard projection*

$$ST^*\mathbf{R}^2 \to S^1 \times \mathbf{R},$$

sending the point $(x, y; \varphi \bmod 2\pi)$ to the point $(\varphi \bmod 2\pi, z)$, where $z = cx+sy$.

This mapping is a Legendre fibration. Indeed, $p = -sx + cy$ implies

$$\alpha = c\,dx + s\,dy = dz - p\,d\varphi.$$

The Legendre curves of the perturbed family project to the cylinder $S^1 \times \mathbf{R}$ as fronts.

Example. In the case of the standard nonperturbed Legendre collapse $x^2 + y^2 = t^2$, the fronts on the cylinder are the standard circles $z = \text{const} = t$.

Definition. A perturbation of a standard Legendre collapse over the plane is called *admissible* if the projections of the corresponding Legendre curves to the cylinder are smooth curves of the form $z = K(\varphi)$, where K is a 2π-periodic function of φ. This function K depends also on the variable t labelling the Legendre curves, and it is assumed that K converges uniformly to $+\infty$ as $t \to +\infty$ and to $-\infty$ as $t \to -\infty$.

Remark. K is the support function of the instantaneous front at the moment t. Thus, our admissibility conditions imply that the instantaneous fronts in the plane (x, y) are locally convex (they may have cusps, but not inflection points), and that the front becomes reversed inside out as t runs from $-\infty$ to $+\infty$.

Theorem 1. *For a generic admissible perturbation of the standard Legendre collapse over the plane, among the fronts of the instantaneous Legendre curves on the plane there exist fronts having at least four cusps.*

Proof. The cusps of instantaneous fronts are determined by the conditions $dx = dy = 0$. Along a Legendre curve, we have $dz = p\,d\varphi$. Hence, if $z = K(\varphi)$, $p = P(\varphi)$, then $P = K'$. From the equations

$$cx + sy = K, \qquad -sx + cy = K'$$

we find

$$x = cK - sK', \qquad y = sK + cK'.$$

Consequently, along a Legendre curve,

$$dx = (-sK + cK' - cK' - sK'')d\varphi = -s(K + K'')d\varphi,$$
$$dy = (cK + sK' - sK' + cK'')d\varphi = c(K + K'')d\varphi.$$

So, the cusps of a front on the plane (x, y) are the points where $K + K'' = 0$.

Now we use the condition $K \to \pm\infty$ as $t \to \pm\infty$. This condition implies that the integral of the function $K + K''$ along the circle vanishes for some value t_0 of the parameter t.

We prove that the front on the plane corresponding to this moment t_0 has at least four cusps.

Consider the Fourier series of the function $K+K''$. This function is orthogonal to $\sin\varphi$ and to $\cos\varphi$ for any value of t. For the value t_0 chosen above, the function is also orthogonal to the constants. Hence, its Fourier series starts with the second harmonics: for $t=t_0$ we have

$$K+K''=\sum a_n\cos n\varphi+b_n\sin n\varphi,\quad n\geq 2.$$

By Tabachnikov's Sturm-type theorem (see [5] or [2]), such a function has at least 4 zeros on the circle.

For a generic perturbation of the standard collapse, all these zeroes are simple and correspond to the usual cusps of fronts on the plane (x,y). The theorem is proved.

Remark. The existence of a front with at least 4 cusps implies the existence of at least four events of birth and death of pairs of cusps in the family of moving fronts (provided that $K+K''\neq 0$ for large $|t|$, i.e., provided that the front evolution starts and ends with nonsingular fronts). For this it is sufficient, e.g., that the standard cylinder be perturbed only in a compact domain, or that $|K''|$ be bounded.

In this way the above theorem can be used to deduce the classical theorem on four vertices (saying that any simple closed curve on the Euclidean plane has at least four extremal points of the curvature) and, also, the symplectic generalization of this theorem (see [2, 6, 7]).

However, the theorem proved above contains more information, because of the mildness of the restrictions on the evolution of Legendre curves. Our fronts are not assumed to be equidistant curves of one of them, and the corresponding Legendre manifolds are not assumed to be the images of one of them under a flow of a Hamiltonian system, as in [2]. In the classical theorem on four vertices, this Hamiltonian system describes a free particle.

§6. Minoration of the number of inflection points on a spherical curve

The main result of §5 can be reformulated in dual terms as a theorem on the inflection points of curves on a sphere.

We project the cylinder (φ, z) onto the inscribed sphere $X^2+Y^2+Z^2=1$ with the help of the family of rays issued from the origin (φ and z are the cylindrical coordinates on the surface $X^2+Y^2=1$). Each curve on the cylinder projects to a curve on the sphere. Consider the projection of a curve $z=K(\varphi)$.

Proposition. *The points where $K+K''=0$ project to inflection points (possibly, multiple) of the curve on the sphere. Conversely, assume that a curve on the sphere intersects a meridian $\varphi=\text{const}$ transversally at an inflection point of the curve (different from a pole of the sphere). Then $K+K''=0$ at the corresponding point of the projection of this curve to the surface of the cylinder.*

Proof. The graphs of the solutions of the differential equation $K+K''=0$ are sections of the surface of the cylinder by nonvertical planes containing the origin. The projections of these sections to the sphere are its nonvertical great circles.

Consider a point where $K+K''=0$, and assume that K does not satisfy this equation identically. There exists a solution of the differential equation $K+K''=0$ whose graph is tangent to the graph of K at that point. The order of tangency of the two graphs at that point is at least *two*, since both functions satisfy the same equation $K+K''=0$ at that point. Hence, the spherical curve is at least second order tangent to the great circle tangent to it at the point of the sphere corresponding to our point on the cylinder. The proposition is proved (the proof of the converse is similar).

Corollary. *Consider a smooth function on the 2-sphere whose derivative along any meridian is nonzero, except for the two poles, (which are the maximum and the minimum point), and otherwise generic. Then at least one level line of this function has at least four inflection points.*

Proof. Projecting the level lines to the cylinder, we find ourselves in the situation of the proof of the main theorem. On the surface of the cylinder, our curves are defined by the equations of the form $z = K(\varphi)$, and $K \to +\infty$ or $-\infty$ as the spherical curve shrinks to a pole.

The inflection points of the curves on the sphere are the projections of the points where $K + K'' = 0$ on the surface of the cylinder. Hence, some of the level lines on the sphere have at least four inflection points.

Now, we are going to prove that a curve splitting the sphere into two parts of equal area always has this property. Moreover, the requirement of the transversality to the meridians may be dropped.

Theorem 2. *Any closed smooth curve on the sphere splitting the sphere into two parts of equal area has at least four inflection points.*

Proof. This is a consequence of the following general result of the projective topology, which is also of independent interest.

Theorem 3. *Any closed nonselfintersecting smooth curve on the sphere having at most two points with zero geodesic curvature lies in an open hemisphere.*

Remark. As proved in [8], any closed nonselfintersecting locally convex hypersurface on a sphere of any dimension lies in an open hemisphere, and so does and any nonselfintersecting (not necessarily closed) curve (possibly, having boundary) without interior inflection points on the 2-sphere.

Theorem 3, proved in the next section, implies theorem 2. Indeed, were the number of inflection points smaller then four, the curve would lie in an open hemisphere and, therefore, would not split the sphere into two parts of equal area.

§7. Spherical curves with two inflection points

The proof of Theorem 3 is based on the results from the projective topology (see [8]). To avoid duplication of the arguments of [8], I shall omit some details. For instance, it will be assumed that the given spherical curve with two points of zero geodesic curvature is generic, so that these two points A and B are ordinary inflection points (the latter condition can be eliminated).

The inflection points A and B (Figure 3) split the curve into two locally convex segments connecting A with B.

Figure 3. The structure of a curve with two inflection points.

These curved segments are "bent likewise". This means that the orientation of the sphere determined by the following couple: (the direction from A to B; the direction to the center of curvature on the sphere) is one and the same for both segments.

In accordance with [8], each of the two curved segments AB lies in an open hemisphere. The projection $S^2 \to \mathbf{R}P^2$ sends the curved segment to an *affine* locally convex (nonclosed) curve AB (that is, to a curve not intersecting some projective line, called the line at infinity).

The main difficulty of the proof of Theorem 3 is to see that this line at infinity can be chosen to be one and the same for the two curved segments AB on the sphere. Such a choice is indeed possible, and here is the proof of that.

Consider a convex affine curve AB (Figure 4). The boundary of its convex hull consists of the convex arc CD and of the segment DC of a straight line connecting the endpoints of the arc (this result, implicitly proved in [8], will not be proved here).

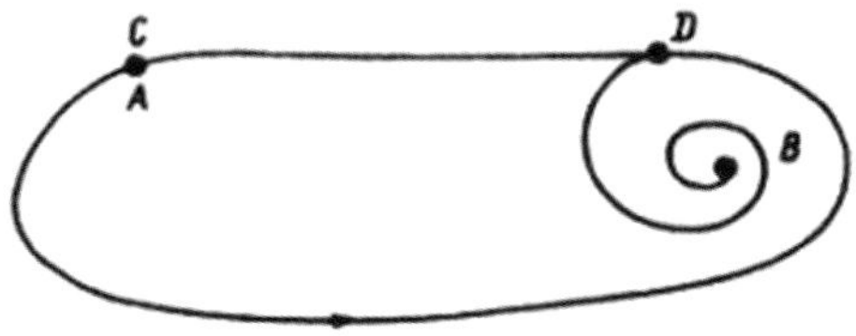

FIGURE 4. The convex hull of a convex segment AB.

The point C may or may not coincide with A, and D with B (in Figure 4 $C = A$ but $D \neq B$). The domain bounded by the curve CDC on the affine plane corresponds to a domain on the initial sphere; the latter domain contains the interval AB of the great circle, which is shorter than π. I shall call this domain the *spherical convex hull* of the convex segment AB of the initial spherical curve.

For the second convex segment AB of the initial spherical curve we also construct its spherical convex hull.

Main lemma. *One of the two spherical convex hulls constructed above contains the other.*

Theorem 3 follows from this lemma, since even the largest of the two spherical convex hulls lies in a hemisphere.

Proof of the main lemma(1). Consider the great circle through AB, call it the *equator*. Its length is 2π, and each of the two convex hulls intersects it along an interval shorter than π and containing the interval AB.

Consider an affine convex curve AB. The straight line S containing A and B intersects the curve AB. Let K and L be the endpoints of the largest interval on S with endpoints on the curve AB (Figure 5).

We denote by k and l the common points of AB and S next to K and L (it may happen that the arcs Kk and (or) Ll are trivial, then $K = k$ and (or) $L = l$).

The closed curve $KklLK$ bounds a domain containing the affine curve AB and the interval AB of the affine straight line S. This domain corresponds to a domain X on the sphere.

(1)I am grateful to M. Z. Shapiro and V. D. Sedykh, who have found and corrected a mistake in my original proof of this lemma.

In a similar way, another curved segment AB gives rise to a domain X'. It can easily be seen that one of the domains X, X' contains the other.

Indeed, each of these domains is adjacent to the affine straight line along an interval (for X this is kl) from the side where the convex arc KL lies. The corresponding hemisphere will be called *Southern*, and the curve KL will be called the *Southern curve*.

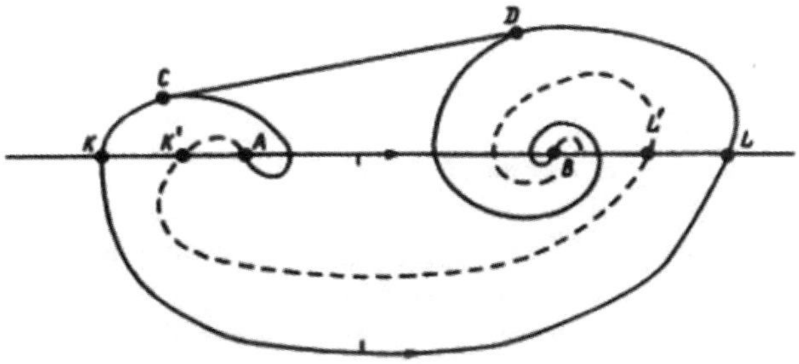

FIGURE 5. The Southern curves KL and $K'L'$.

Lemma. ***The Southern hemisphere is one and the same for the two convex segments into which the inflection points A and B subdivide the initial spherical curve.***

***Proof*. This hemisphere lies at the side of the interval AB (on the equator) for which the orientation of the sphere determined by the following couple: (the direction from A to B; the direction toward the hemisphere) is opposite to the orientation determined by the couple (the direction of the convex segment from A to B; the direction to its center of curvature on the sphere).**

But the latter orientation is one and the same for both convex segments AB (this is evident in a neighborhood of A and, so, is true everywhere). Hence, the corresponding Southern hemispheres also coincide, and the lemma is proved.

The Southern arcs KL of the first and $K'L'$ of the second convex hull both lie in the Southern hemisphere. They do not intersect each other since the initial curve was nonselfintersecting. Thus, on the equator, the couples (K, L) and (K', L') do not interlace. So, one of the short arcs of the equator (say, KL) contains the other, $K'L'$. It follows that the entire convex curved segment AB containing the points K' and L' lies inside the spherical convex hull $KLDC$ (in Figure 5 this segment is shown as a dashed line).

The proof of the statement that the dashed curve AB is confined in the convex hull $CKLD$ (which is rather evident from Figure 5) can be obtained with the help of the arguments from [8], and is omitted here.

This fact implies that the convex hull of the dashed curve lies entirely inside the convex hull of the affine curve $ACKLDB$. Thus, the Main lemma is proved, and so is Theorem 3, which implies Theorem 2.

§8. FRONTS WITH TWO INFLECTION POINTS

A Legendre curve which is far from the circles forming the standard cylinder can also be projected to the cylinder (φ, z), and then to the sphere. However, in general, the projection is neither the graph of a function $z = K(\varphi)$ nor a smooth curve on the sphere.

Such projections (of generic Legendre curves) have usual singularities of generic wave fronts: selfintersection points and cusps (of semicubical type). Moreover, the

tangents to the branches of these fronts are "nonvertical" both at the usual points, and at the points of selfintersection, as well as at the cusp points: along these tangents $d\varphi \neq 0$.

Now, consider an arbitrary family of fronts connecting a small parallel centered at the South pole with a small parallel centered at the North pole of the sphere. (If such a family is generic, then all these fronts are cooriented by the direction of the meridians from the South to the North.) How small the maximal number of inflection points of the fronts of such a family may be?

The fronts of the family depend on one parameter and, under its variation, undergo standard *perestroikas* (namely, quadratic selftangencies, triple selfintersections with distinct tangents, transversal crossings of a cusp point by a regular branch of the front and, finally, birth and death of pairs of cusps).

The example described below shows that all these conditions can be ensured in a family of fronts none of which has more than *two* inflection points.

First, we construct a front splitting the sphere into two parts of equal area. This front has no selfintersections and, by itself, provides a counterexample to an attempt of extending Theorem 2 to singular curves.

Consider two antipodal points A and A' on the equator of the sphere (Figure 6).

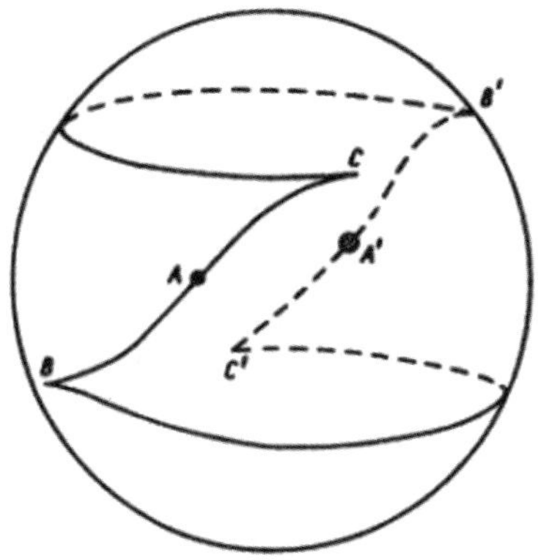

FIGURE 6. A front with two inflection points splitting the sphere into two parts of equal area.

The point A is an inflection point of the front. The front remains convex up to the cusp points B (where the tangent is horizontal, that is, touches the parallel $Z = -1/2$) and C (where the tangent touches the parallel $Z = 1/2$.) The coordinate φ is monotone increasing along the interval BAC of the front.

The segment $B'A'C'$ is central symmetric to BAC. The (mutually symmetric) parts CB' and $C'B$ are close to the parallels $Z = 1/2$ and $Z = -1/2$. Along the segment AC, the direction to the center of curvature on the sphere looks to the Southern side of the front, and along BA to the Northern side. Thus, A and A' are the only inflection points.

We are going to include the above front in a family which interpolates a parallel closed to the South pole to a parallel closed to the North pole. To this end, we describe a deformation contracting the front in question to the South pole. The deformation contracting it to the North pole is obtained by reflection.

First, we rotate the arcs BC and $B'C'$ toward each other along the sphere (as rigid bodies) about the rotation axis connecting the South and North poles of the sphere. The "parallel" BC' shrinks and the "parallel" BC' stretches. At some moment the points B and C' must pass through each other, both tangents being horizontal. (This is a codimension two nongeneric event in a one-parameter family, which can be avoided by a small deformation making the family generic.)

After this moment, two selfintersection points arise on the front. The upper self-intersection point X starts to mount along BA and $C'A'$. At some moment it will pass through the inflection points A and A' (coinciding at that moment), and then we stop the rotation (Figure 7).

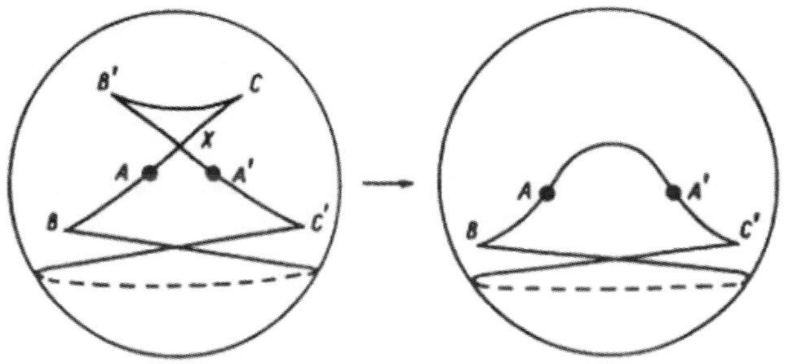

FIGURE 7. Contraction of the arc $B'C$.

The next step consists in killing the triangle $XB'C$ (this is accompanied by the standard death of the two cusps B' and C). The resulting curve has only two cusps B and C' and still has two inflection points A and A'. The latter pair of points can be killed by a smooth deformation of the arc BC' (Figure 8). The remaining two cusps die together with the lowest selfintersection point of the front.

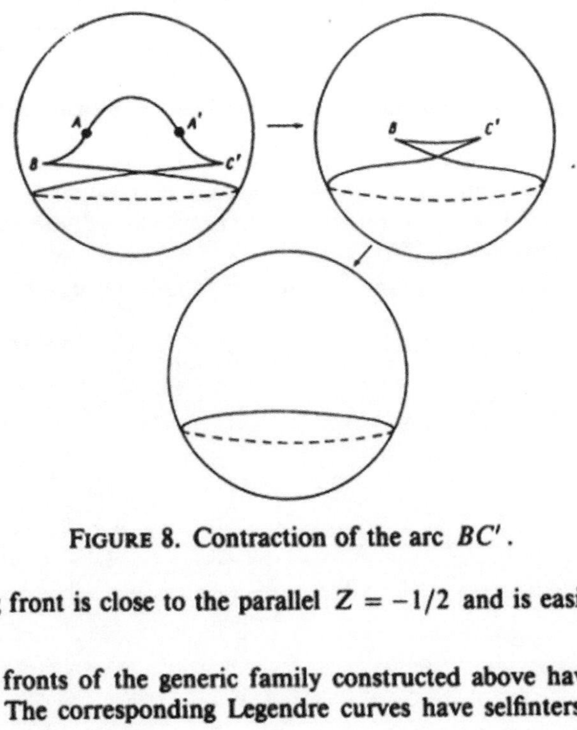

FIGURE 8. Contraction of the arc BC'.

The resulting front is close to the parallel $Z = -1/2$ and is easily shrunk to the South pole.

Remark. Some fronts of the generic family constructed above have instantaneous selftangencies. The corresponding Legendre curves have selfintersections. Unlike the embedded Legendre curves of the standard collapse, they are only immersed in $ST^*\mathbf{R}^2$.

So, maybe, on the plane, four cusps are inavoidable for an arbitrary family of nonselftangent fronts connecting a nonsingular shrinking front with an expanding one. This nontangency condition is fulfilled, for instance, if the perturbed cylinder is *embedded* into $ST^*\mathbf{R}^2$, as the standard one is.

It is interesting to study the particular case where this perturbed cylinder is connected with the standard one by a regular homotopy under which it all the time remains embedded and fibered into closed Legendre curves. This is so, for instance, if the perturbed cylinder is the image of the standard one under a contactomorphism connected with the identity by a smooth path in the group of contactomorphisms of the manifold $ST^*\mathbf{R}^2$ to itself that are different from the identity only in a compact domain.

However, even in this particular case, the inevitability of at least four cusps on one of the fronts undergoing the reversal of coorientation process has not been proved.

Bibliography

1. V. I. Arnold, *Sur une propriété topologique des applications globalement canoniques de la mécanique classique*, C. R. Acad. Sci. Paris **261** (1965), 3719–3722.

2. ——, *Sur les propriétés topologiques des projections Lagrangiennes en géométrie symplectique des caustiques*, Cahiers de Math. de la Decision no. 9320, CEREMADE, 14/6/93, pp. 1–9.

3. C. Conley and E. Zehnder, *The Birkhoff–Lewis fixed point theorem and a conjecture of V. I. Arnold*, Invent. Math. **73** (1983), 33–49.

4. A. Floer, *Proof of the Arnold conjecture and generalizations to certain Kähler manifolds*, Duke Math. J. **53** (1986), 1–32.

5. S. Tabachnikov, *Around four vertices*, Uspekhi Mat. Nauk **45** (1990), no. 1, 191–192; English transl., Russian Math. Surveys **45** (1990), no. 1, 229–230.

6. V. I. Arnold, *Plane curves, their invariants, perestroikas, and classifications*, Preprint, Forschungsinstitut für Mathematik, ETH Zürich, May 1993, 58 pp.; Singularities and Curves, Adv. Soviet Math., vol. 21, Amer. Math. Soc., Providence, RI, 1994, pp. 33–91.

7. ——, *Invariants and perestroikas of fronts on the plane*, Trudy Mat. Inst. Steklov. (to appear). Russian

8. ——, *Ramified covering $CP^2 \to S^4$, hyperbolicity, and projective topology*, Sibirsk. Mat. Zh. **29** (1988), no. 5, 36–47; English transl. in Siberian Math. J. **29** (1988), no. 5.

9. A. B. Givental, *A symplectic fixed point theorem for toric manifolds*, Progr. Math., Floer memorial volume, Birkhäuser, Boston, 1994.

Received 24/DEC/1993

Translated by THE AUTHOR

On topological properties of Lagrangian projections in symplectic geometry of caustics

V.I. Arnold

Translation of: Sur les propriétés topologiques des projections lagrangiennes en géométrie symplectique des caustiques. *Revista Matemática de la Universidad Complutense de Madrid*, **8**, no. 1 (1995), 109–119.

Translated by Emmanuel Ferrand.

This paper reproduces the text read at the CEREMADE Colloquium on May 11, 1993. The author[1] is grateful to the Newton Institute (Cambridge, UK) and to the Université Paris-Dauphine for their hospitality and support, and also to Ms Isabelle Paisant for editing a difficult manuscript.

Some misprints in the original are corrected in the Russian version of the paper in: Vladimir Igorevich Arnold. Selected-60. Technical editor: M.B. Sevryuk. PHASIS, Moscow, 1997, pp. 525–532 and in the present translation.

Abstract. The caustic of a point on a Riemannian manifold is the set of intersection points of infinitely closed geodesics starting from this point.

Using a topological reasoning, Jacobi noticed that the caustic of a point on a closed convex surface must have cusps. He also announced (without proof) that the number of such cusps for caustics on the surfaces of ellipsoids is four [1].[2]

In this note I try to place the theorems on cusps of caustics in the more general framework of symplectic topology. One immediately obtains in this way numerous generalizations (including, for example, the theorem on four vertices of a closed convex curve in the Euclidian plane) as well as conjectures on the topology of caustics and wave fronts.

Some of these theorems and conjectures can be considered as (somewhat strange) generalizations of the Sturm theory. This approach to the Sturm theory suggests its generalizations connected, for instance, with symplectization of the theorem on four umbilical points on a convex surface.

[1] Steklov Institute, Moscow, Russia and CEREMADE, Université Paris-Dauphine, France.

[2] See Lecture 6 in [1]. (Editors' remark)

Translation of: *Revista Matemática de la Universidad Complutense de Madrid* **8**(1), 109–119 (1995),

Translated by Emmanuel Ferrand

1. The Lagrangian Collapse

Definition. *The standard Lagrangian cylinder $L^n \hookrightarrow T^*\mathbb{R}^n$ is the graph of the gradient of the function $\pm|q|$:*

$$L^n = \{p, q : \exists t : p^2 = 1,\ q = pt\}.$$

This formula defines an exact Lagrangian embedding of the cylinder $S^{n-1} \times \mathbb{R}$ ($p \in S^{n-1}$, $t \in \mathbb{R}$) in the cotangent bundle $\mathbb{R}^{2n}$ of the Euclidean space $\mathbb{R}^n$.

The standard cylinder is an exact Lagrangian manifold, since $pdq = \mathrm{d}t$ on L.

Definition. *The standard Lagrangian collapse*

$$L^n \hookrightarrow T^*\mathbb{R}^n \twoheadrightarrow \mathbb{R}^n$$

is the projection of the standard Lagrangian cylinder L^n on the base of the cotangent bundle of the Euclidean space $\mathbb{R}^n$.

The projection on the base of the cotangent bundle $(p, q) \mapsto q$ induces a mapping of L^n onto $\mathbb{R}^n$ which is a two-sheeted local diffeomorphism everywhere except over the origin to which the whole sphere $S^{n-1} \times 0$ is mapped.

The set of critical values (the caustic) of the mapping $L^n \to \mathbb{R}^n$ consists of the single (infinitely degenerate) point 0. The goal of this work is a study of the caustics of Lagrangian manifolds close to L^n.

2. Examples of Lagrangian Collapses

A). Consider the family of normals to a circle in the plane. The unitary vectors of these normals form a Lagrangian cylinder (I identify tangent and cotangent vectors of the Euclidean plane).

This cylinder $S^1 \times \mathbb{R}$ is the standard Lagrangian cylinder in $\mathbb{R}^4$ and its projection on $\mathbb{R}^2$ is the standard Lagrangian collapse.

When one deforms the circle, the cylinder consisting of the unitary vectors of the normals also deforms, but it always remains an exact Lagrangian cylinder.

The critical values of the projection of the deformed cylinder are the centers of curvature of the deformed circle. The theorem on four vertices asserts that the caustic (= the focal set, = the set of the centers of curvature) of a generic convex curve has at least four cusps.

B). Consider the geodesics starting from the north pole of the sphere S^2. These geodesics meet at the south pole. The corresponding Lagrangian variety in the cotangent bundle of the sphere is the set of the unitary vectors tangent to the geodesics (as always, I identify tangent and cotangent vectors of a Riemannian manifold).

This Lagrangian manifold is smooth. Its projection on the sphere has singularities over neighborhoods of the poles. These singularities are symplectomorphic to the standard Lagrangian collapse of a cylinder in $\mathbb{R}^4$.

When one deforms the metric of the sphere (keeping the curvature positive), the manifold L formed by the unitary vectors of the geodesics starting from the north pole remains a Lagrangian variety (an immersed cylinder in T^*S^2). The critical points of its Lagrangian projection on S^2 form a sequence of closed curves whose projections are the caustics (the first one is close to the south pole, the second one is close to the north pole, etc. provided that the deformation of the metric is small).

It follows from the theorems below that each of these minuscule caustics has at least four cusps. I think that this estimate from below remains valid for long caustics of all convex surfaces. For the first caustic, this is a classical result of the variational theory. But for the next caustics, it seems that this estimate from below is not known even in the case of an ellipsoid.

It seems that the problem of the cusps of caustics on ellipsoids is resisting the efforts of algebraic geometers (see, for example, the paper [2]; this reference was given to me by H. Knörrer).

3. Singularities of the Deformed Collapse

The results below imply the following

Theorem. *The number of cusps of the caustic of the Lagrangian projection on the plane is at least four for any generic exact Lagrangian embedding of a cylinder provided that it is close enough to the standard cylinder.*

The range of the allowed perturbations is described in § 6. I think that the number of cusps of caustics is also at least four for Lagrangian[3] embeddings of cylinders far away from the standard cylinder (for instance, for embeddings that can be obtained from the standard cylinder by a Hamiltonian isotopy with compact support). There are no such estimates from below for immersions (this follows from [3]).

It is also possible that some of the restrictions used in the proof of § 6 are essential. For example, the cylinders allowed in our main theorem are "pseudo-optical" ($p \neq 0$) and satisfy the condition of "contact pseudo-convexity" (the function $S = \int p\,dq$ on L has no singular points).

The cylinders needed in the theorem on four vertices of a convex curve (or on the n-th branch of the caustic of a point on an almost spherical surface) meet the closeness conditions sufficient for the applicability of our reasoning (however, the necessary closeness of the surface to the sphere depends on n).

4. Theorem of Sturm and Tabachnikov Type

With help of generating functions, the proofs can be reduced to the following Sturm-type theorem.

Theorem. *Let a function $f(S,\varphi)$ be 2π-periodic in φ and bounded together with its second derivative with respect to φ on the whole plane. Then the restriction of the function S to the smooth curve on the surface of the cylinder $\{S,\ \varphi \bmod 2\pi\}$ given by the equation*

$$S + f(S,\varphi) + \partial^2 f/\partial\varphi^2 = 0$$

has at least four critical points.

[3] "Legendrian" in the original. (Translator's remark)

Proof. The curve consists of closed components all of which are contained in an annulus $|S| \leq c_1$. If the number of components is larger than 1, the required four critical points are provided by the maxima and the minima of S on the components.

If there is only one component, the function

$$H(S, \varphi) = S + f + \partial^2 f / \partial \varphi^2$$

is positive on one side and negative on the other side of this component, which also separates the domains $S \geq c_1$ and $S \leq -c_1$.

Suppose that the number of critical points is smaller than four. In this case the maximum (as well as the minimum) of S is attained at only one point, and any value of S between the minimum and the maximum is attained at exactly two points. The difference of the values of φ at these two points is a continuous function of S whose values at the minimum and at the maximum differ by 2π. Consequently, there exists a value S_0 attained at two opposite points: $\varphi_2 - \varphi_1 = \pi \bmod 2\pi$. The signs of $H(S_0, \cdot)$ on the two arcs separated by these two points are different.

Therefore, there exists a function

$$a \cos \varphi + b \sin \varphi$$

having the same zeros as $H(S_0, \cdot)$ and the same signs at the other points. For such a function one has

$$\oint_0^{2\pi} (a \cos \varphi + b \sin \varphi) H(S_0, \varphi) \, \mathrm{d}\varphi > 0.$$

But, for any fixed value of S, the function

$$f + \partial^2 f / \partial \varphi^2$$

is orthogonal to $\cos \varphi$ and $\sin \varphi$ (since the operator $\partial^2 / \partial \varphi^2 + 1$ kills these harmonics in the Fourier series). Thus, the positive integral is equal to zero, and this contradiction shows that there exist at least four critical points.

Remark. In the infinitesimal version of the problem of the birth of cusps on the caustic as the standard Lagrangian collapse is deformed, these cusps appear as the roots of the equation $f''' + f' = 0$, where f is the 2π-periodic function of one variable describing the infinitesimal perturbation.

Such an equation always has at least four roots on the circle, because the function $f''' + f'$ is orthogonal to 1, $\sin \varphi$, $\cos \varphi$.

S.L. Tabachnikov [4] noticed that, more generally, the function

$$\sum (a_k \cos k\varphi + b_k \sin k\varphi), \quad k \geq n$$

has at least $2n$ roots on the circle (and that a similar result holds for eigenfunctions of other Sturm–Liouville operators).

I have found applications of this theorem to estimating from below the numbers of Lagrangian singularities A_k of higher orders.[4]

Generalizations of the Sturm theory for spherical functions, which are needed to obtain lower bounds of the numbers of singularities of perturbed standard collapses in $T^*\mathbb{R}^n$, $n > 2$, have not been developed yet, even in the infinitesimal form. (These generalizations should give, for example, a symplectic version of the theorem on four umbilical points of convex surfaces, if $n = 3$.)

I should confess that I do not understand the deep sources of the theorem proven above,[5] nor the reasons why Sturm-type theorems appear in problems on deformations of the Lagrangian collapse.

5. Adapted Coordinates

To formulate the restrictions on the proximity of the perturbed cylinder to the standard cylinder, it is convenient to describe the former using special coordinates in a neighborhood of the latter. We will employ the fact that a neighborhood of a Lagrangian manifold is symplectomorphic to a neighborhood of a Lagrangian section of its cotangent bundle. We will describe such a symplectomorphism explicitly.

Consider the mapping $S^1 \times \mathbb{R}^3 \to \mathbb{R}^4$ given by the formulas

$$\begin{cases} p_x = A\cos\varphi, & p_y = A\sin\varphi, \\ x = r\cos\varphi + B\sin\varphi, & y = r\sin\varphi - B\cos\varphi. \end{cases} \tag{1}$$

Here $\varphi \bmod 2\pi$ is the coordinate in S^1, (r, A, B) are the coordinates in $\mathbb{R}^3$ and $(x, y; p_x, p_y)$ in $T^*\mathbb{R}^2 \approx \mathbb{R}^4$.

One can directly verify the following

Lemma. *There holds the identity*

$$p_x \mathrm{d}x + p_y \mathrm{d}y = P_r \mathrm{d}r + P_\varphi \mathrm{d}\varphi, \tag{2}$$

where

$$P_r = A, \quad P_\varphi = AB. \tag{3}$$

Corollary. *The formulas* (1) *describe a symplectomorphism from the domain* $P_r = A > 0$ *in* $T^*(\mathbb{R} \times S^1)$ *onto the domain* $T^*\mathbb{R}^2 \setminus \mathbb{R}^2$ $(p \neq 0)$.

Proof. The point $p = (p_x, p_y)$ determines $A > 0$ and $\varphi \bmod 2\pi$ without ambiguity. Once φ is known, the vector (r, B) is obtained from the vector (x, y) by the rotation by angle φ.

---

[4] In the Russian version of the paper (see the beginning of this text), this sentence has the opposite meaning: I have found no applications of this theorem to estimating from below the numbers of Lagrangian singularities A_k of higher orders. (Editors' remark)

[5] In the Russian version of the paper, there is the following addendum here: (for instance, I do not know its generalizations to the case of $2n > 4$ roots). (Editors' remark)

6. The Main Theorem

Let $S(r, \varphi)$ be a function 2π-periodic in φ and such that S coincides with r for $|r| \geq c_2$ and is strictly monotone with respect to r ($\partial S/\partial r > 0$).

Define an exact Lagrangian cylinder $L \hookrightarrow T^*\mathbb{R}^2$ by a generating function S, that is, via the formulas

$$P_r = \partial S/\partial r, \quad P_\varphi = \partial S/\partial \varphi \tag{4}$$

in the coordinates of § 5.

Theorem. *The caustic of the projection of a generic cylinder L to the plane (x, y) along the fibers of the cotangent bundle has at least four cusps.*

Proof. The set of critical points of the Lagrangian projection is defined by the equation

$$\det\big(\partial(x, y)/\partial(r, \varphi)\big) = 0.$$

According to (1)–(4), this equation takes the form

$$\det \begin{pmatrix} c + B_r s & s - B_r c \\ Bc + (-r + B_\varphi)s & Bs - (-r + B_\varphi)c \end{pmatrix} = 0,$$

where $c = \cos\varphi$, $s = \sin\varphi$.

Consequently, the set of critical points is the curve whose equation is

$$-r + B_\varphi - BB_r = 0.$$

Along the kernel of the derivative of the projection at a singular point, $\mathrm{d}x$ and $\mathrm{d}y$ vanish. Hence, due to (2) and (4), this kernel is defined by the equation $\mathrm{d}S = 0$.

In the setup described in the theorem, one can choose (S, φ) as independent variables instead of (r, φ). I will denote the partial derivatives by indices (u_r, u_φ) if the variables φ or r are fixed and by fractions $(\partial u/\partial S, \partial u/\partial \varphi)$ if one fixes φ or S.

I will also employ the notation of the ancients, where a function defined on a manifold, say u, is denoted by the same letter independently of the coordinate frame used.

The expression of r as a function of the coordinates S and φ can be written in the form

$$r = S + f(S, \varphi), \quad \partial f/\partial S > -1,$$

where $f = 0$ for $|S| \geq c_3$.

Passing to the differentials, one obtains

$$\mathrm{d}r = (1 + \partial f/\partial S)\,\mathrm{d}S + \partial f/\partial \varphi\,\mathrm{d}\varphi.$$

Hence, because of (4), L is described by the functions

$$P_r = (1 + \partial f/\partial S)^{-1}, \quad P_\varphi = -(\partial f/\partial \varphi)(1 + \partial f/\partial S)^{-1}.$$

Thus, the relations (4) and (3) imply a simple expression for P_φ/P_r:

$$B = -\partial f/\partial\varphi|_{S=\text{const}},$$

and, consequently, we also find that

$$B_\varphi - BB_r = \partial B/\partial\varphi|_{S=\text{const}}.$$

Finally, the equation of the curve formed by the singular points takes the form

$$S + f(S,\varphi) + \partial^2 f/\partial\varphi^2 = 0,$$

where f is the known function which defines the perturbed cylinder, is 2π-periodic in φ, and vanishes for large values of $|S|$.

The cusps of caustics are the projections of the tangency points of the kernel ($dS = 0$) with the curve of singular points. Therefore, they are the critical points of the restriction of the function S to this curve (which is a smooth curve if L and f are generic).

According to the Sturm–Tabachnikov theorem (§ 4), the number of such points is at least four, which finishes the proof of the main theorem.

Remark. The proof shows that the condition "$S = r$ for $|r| \geq c_2$" is not very important. It is enough, for example, that the function $S - r$ and its second derivatives be bounded, provided that $\partial S/\partial r > c_4 > 0$.

One can even allow a (moderate) growth of f as $|S| \to \infty$. It is sufficient that the function $H(S,\varphi) = S + f + \partial^2 f/\partial\varphi^2$ tend to $+\infty$ as $|S| \to \infty$.

This remark enables one to adapt our reasoning, for example, to proving the theorem on four vertices of a convex plane curve. This theorem is therefore a particular case of our symplectic generalization.

Other generalizations are discussed in [5–8].

References

1. Jacobi C.G.J., *Vorlesungen über Dynamik* (hrsg. von Clebsch A.). G. Reimer, Berlin, 1866. English translation: *Lectures on Dynamics.* Texts and Readings in Mathematics, **51**, Hindustan Book Agency, New Delhi, 2009.
2. v. Braunmühl A., Geodätische Linien und ihre Enveloppen auf dreiaxigen Flächen zweiten Grades. *Math. Annalen*, **20**, Nr. 4 (1882), 557–586.
3. Arnold V.I., Spaces of functions with mild singularities. *Funct. Anal. Appl.*, **23**, no. 3 (1989), 169–177.
4. Tabachnikov S.L., Around four vertices. *Russ. Math. Surveys*, **45**, no. 1 (1990), 229–230.
5. Arnold V.I., Plane curves, their invariants, perestroikas and classifications. With an appendix by Aicardi F. In: *Singularities and Bifurcations* (Arnold V.I., ed.). Advances in Soviet Mathematics, **21**, Amer. Math. Soc., Providence, RI, 1994, 33–91.
6. Arnold V.I., Invariants and perestroikas of plane fronts. *Proc. Steklov Inst. Math.*, **209** (1995), 11–56.
7. Arnold V.I., On topological properties of Legendre projections in contact geometry of wave fronts. *St. Petersburg Math. J.*, **6**, no. 3 (1995), 439–452.
8. Arnold V.I., *Topological Invariants of Plane Curves and Caustics.* University Lecture Series, **5**, Amer. Math. Soc., Providence, RI, 1994.

Plane curves, their invariants, perestroikas and classifications (with an appendix by F. Aicardi)

V.I. Arnold

A *curve* below means a smooth mapping of a circle to a plane whose derivative vanishes nowhere (an immersion).

A generic immersion has only ordinary double points of transversal self-intersections. In the space of immersions all nongeneric immersions form a hypersurface which I shall call the *discriminant hypersurface* or for short the *discriminant*.

The goal of this article is the study of the geometry and of the topology of this hypersurface. This study leads immediately to nontrivial information on generic immersions. For instance, the results of this paper imply the following facts.

Consider a generic closed path in the space of immersions.

Definition. A self-tangency point of an immersed curve is called a point of *direct tangency* if the velocity vectors point to the same direction; otherwise it is called a point of *inverse tangency*.

Theorem. *The number of values of the parameter for which the immersed curve has a direct tangency point is even.*

A moment of self-tangency of a curve on a generic path in the space of curves is *positive* (*negative*) if the number of double points increases (decreases) while the path goes through this moment.

Theorem. *The difference between the numbers of positive and negative direct self-tangency moments does not depend on a generic path connecting two generic immersions, but only on the two immersions connected by the path.*

A similar result holds for the inverse self-tangency moments.

This paper was written while visiting the Isaac Newton Institute for Mathematical Sciences, Cambridge (UK), l'Université Paris-Dauphine, and the Mathematics Research Institute, ETH Zürich. The author thanks these institutions for their hospitality and Rahel Boller for rapidly typing a difficult manuscript.

Originally published in: *Singularities and Bifurcations*. Editor: V.I. Arnold (Adv. Soviet Math. 21). Amer. Math. Soc., Providence, RI, 33–91 (1994),
Reprint by Springer Nature Switzerland AG 2022 with kind permission

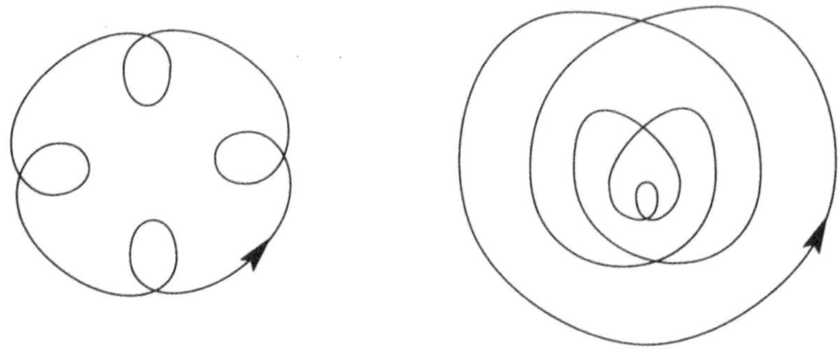

FIGURE 1. Two immersions connectable by a path in the space of immersions

The two immersions shown in Figure 1 can be connected by a path in the space of immersions.

THEOREM. *Any generic one-parameter family of immersions connecting the two immersions shown in Figure* 1, *has at least* 6 *parameter values corresponding to immersions with triple points, at least* 6 *parameter values corresponding to immersions with direct self-tangencies, and at least* 6 *parameter values corresponding to immersions with inverse self-tangencies.*

Moreover, the difference between the numbers of moments of positive and negative direct tangencies on a path from the right curve to the left one is equal to 6. For the inverse tangencies the difference is equal to -6.

For the triple points there also exists a (nontrivial) way to define the positive and negative moments. The number of moments of triple points, counted with these signs (described below) on any generic path from the left curve to the right one, is equal to 6.

The invariants of immersions responsible for these and many other results of this paper are dual to different strata of the discriminant hypersurface.

The description of all the invariants (that is, of all functions locally constant on the complement of the discriminant hypersurface) is equivalent to the classification of the immersions up to diffeomorphisms of the plane and of the circle. Such a description seems to be very complicated (more complicated than knot theory).

We shall see, however, that some simple and rather natural axioms are verified only by a small number of simplest invariants, which seem to be the most important ones. It is strange that these invariants, in particular the three most fundamental that I denote St, J^+, J^-, have not been introduced earlier.

The initial goal of my study of plane curves was an attempt to prove some symplectic topology generalizations of the 4-vertex theorem. This generaliza-

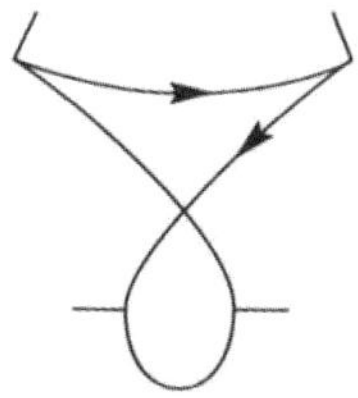

FIGURE 2. A cooriented and oriented front with two cusps

tion leads to the following conjecture.

Consider the plane cooriented curve with two semicubical cusp points shown in Figure 2. There exists a path in the space of cooriented curves with two cusps, connecting this curve to the same curve with the opposite coorientation (Figure 3).

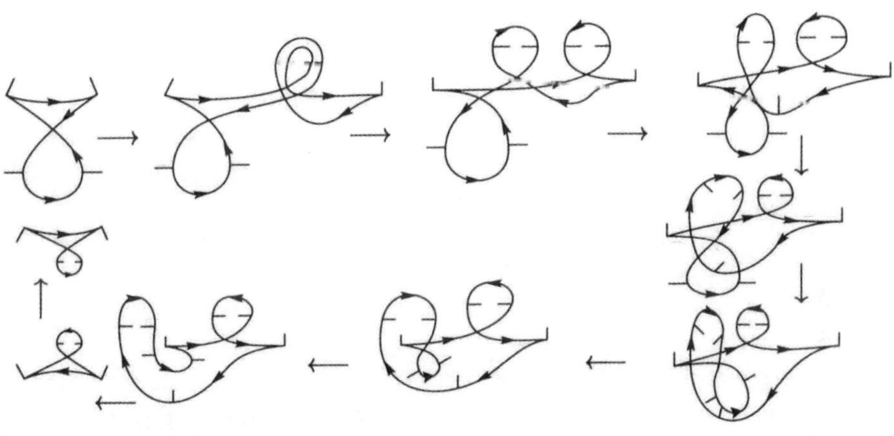

FIGURE 3. A sequence of perestroikas of a front, reversing its coorientation

CONJECTURE. *Any such generic path contains at least one moment of self-tangency with coinciding coorientations.*

The number of such moments is even (this can be proved by the methods of the present article). However, I have not (yet?) constructed an invariant of cooriented curves with two cusps and no equally cooriented self-tangencies, taking different values on the curve of Figure 2 and on the curve with opposite coorientation. It seems that such an invariant must violate the "natural" axioms verified by our invariants.

§1. The three basic invariants

These invariants are dual to the three parts of the discriminant hypersurface formed by immersions having triple points, having direct self-tangencies, and having inverse self-tangencies respectively.

LEMMA. *Each of these three parts of the discriminant hypersurface is coorientable.*

COMMENTS. A *coorientation* of a smooth hypersurface in a functional space is the choice of one of the two parts separated by this hypersurface in a neighborhood of any of its points. This part is called *positive.*

The coorientation of the smooth part of a singular hypersurface is called *consistent* if the following consistency condition holds in a neighborhood of any singular point of any stratum of codimension 1 on the hypersurface (of codimension 2 in the ambient functional space):

The intersection index of any generic small oriented closed curve with the hypersurface (defined as the difference between the numbers of positive and negative intersections) *must vanish.*

A hypersurface (a subvariety of codimension one) is called *cooriented* if a consistent coorientation of its smooth part is chosen, and *coorientable* if it exists.

The proof of the Lemma is based on the explicit construction of a coorientation. This coorientation is in fact *unique* (up to complete change of sign) in the class of *local* coorientations, defined above, and does not depend either on the orientation of the plane or of the curve.

DEFINITION. A transversal crossing of a self tangency is *positive* if the number of double points grows (by 2).

A transversal crossing of a triple point is *positive* if the new-born vanishing triangle is positive.

A *vanishing triangle* is the triangle formed by the three branches of a curve corresponding to a sub-critical or to a super-critical value of the parameter near a triple point of a critical curve.

The *sign of a vanishing triangle* is defined by the following construction. The orientation of the immersed circle defines a cyclic ordering of the sides of the vanishing triangle (it is the order of the visits of the triple point by the three branches). Hence the sides of the triangle acquire orientations induced by the ordering. But each side also has its own direction, which may coincide, or not, with the orientation defined by the ordering. For each vanishing triangle we define a quantity (which takes four values 0, 1, 2, 3)

$q =$ *number of sides equally oriented by the ordering and by their directions* (Figure 4).

DEFINITION. The *sign* of a vanishing triangle is $(-1)^q$.

EXAMPLE. A triangle is *positive* if the directions of the sides are all *opposite* to the orientation defined by the ordering.

REMARK. Reversing the orientation of the curve does *not* change the quantity q, since it reverses both the cyclic order and the directions of the sides.

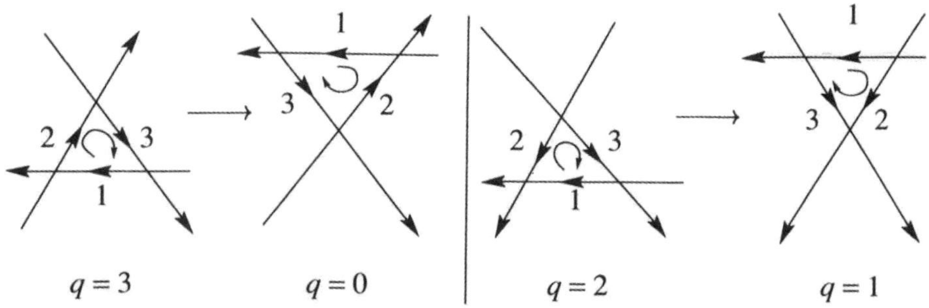

FIGURE 4. Positive and negative vanishing triangles

Hence, the coorientation of the triple point component of the discriminant chosen above does not depend on the orientation of the circle.

The orientation of the plane was not used in the construction at all.

THEOREM. *The above coorientation of the smooth part of the discriminant is consistent.*

The proof is a rather long routine check of what happens at the codimension one strata of the discriminant. In the case of self-tangencies this proof is very easy, since the number of double points is a well-defined invariant. In the case of triple points some real computations are needed; we shall present them in §5 (where the uniqueness of the coorientation is also proved).

DEFINITION. The *index* of an immersion (of an oriented circle into an oriented plane) is the number of turns of the tangent vector (the degree of the mapping sending a point of the circle to the direction of the derivative of the immersion at this point).

The index is the only invariant of immersions remaining constant along all paths in the space of immersions. In other words, the connected components of the space of immersions are counted by the indices of the immersed curves (Whitney's theorem, [1]).

Consider one of these components, that is, the space of immersions of a fixed index. In §2 we shall prove the following three theorems.

THEOREM 1. *There exists a unique (up to an additive constant) invariant of generic immersions of fixed index whose value remains unchanged while the immersed curve experiences a self-tangency perestroika, but increases by* 1 *after a positive crossing of a triple point.*

This invariant will be denoted by St (from Strangeness), when normalized by the following conditions:

$$St(K_0) = 0, \qquad St(K_{i+1}) = i \quad (i = 0, 1, \dots),$$

where K_0 is the figure eight curve (∞) and K_{i+1} is the simplest curve with i double points (Figure 5).

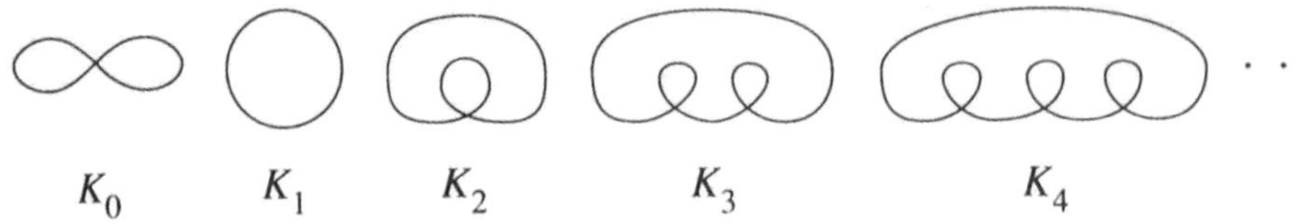

FIGURE 5. The standard curves of indices $0, \pm 1, \pm 2, \ldots$

The reason for this normalization is discussed above in §5. The curves K_0 and $K_1, K_2, \ldots$ with different orientations have indices $0, \pm 1, \pm 2, \ldots$.

THEOREM 2. *There exists a unique (up to an additive constant) invariant of generic immersions of fixed index whose value remains unchanged while the immersed curve experiences an inverse self-tangency perestroika or a triple point crossing, but increases by a constant number a_+ under a positive (increasing the number of double points) crossing of a direct tangency perestroika.*

This invariant will be denoted by J^+ when normalized by the choice $a_+ = 2$ and by the following choices of the values on the curves K_i of Figure 5 (oriented arbitrarily):

$$J^+(K_0) = 0, \qquad J^+(K_{i+1}) = -2i \quad (i = 0, 1, 2, \ldots).$$

The reasons for these choices are explained below.

THEOREM 3. *There exists a unique (up to an additive constant) invariant of generic immersions of fixed index whose value remains unchanged while the immersed curve experiences a direct tangency perestroika or a triple point crossing, but increases by a constant number a_- under a positive (increasing the number of double points) crossing of the inverse tangency perestroika.*

This invariant will be denoted by J^- when normalized by the choices $a_- = -2$ and $J^-(K_0) = -1$, $J^-(K_{i+1}) = -3i$ $(i = 0, 1, \ldots)$ for the curves K_i of Figure 5 oriented arbitrarily. The reasons for these choices are explained below.

The normalizing constants are summarized in Figure 6, where $a(\cdot)$ means the jump of an invariant under the positive crossing of the discriminant at the stratum $(\cdot)$.

To calculate the value of an invariant on a generic immersion it suffices to join it by a generic path with the standard immersion of the same index and to count the jumps at the crossings of the discriminant.

EXAMPLE. There exist exactly five classes of immersions with two double points, as shown in Figure 7 (two immersions are in the same class if one

	a()	a()	a()				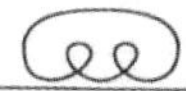		...
St	1	0	0	0	0	1	2	3	...
J+	0	2	0	0	0	−2	−4	−6	...
J−	0	0	−2	−1	0	−3	−6	−9	...

FIGURE 6. The definitions of the basic invariants St, J^+, J^-

can be transformed into the other by diffeomorphisms of the circle and of the plane; both diffeomorphisms may reverse the orientations).

The values of the basic invariants for these five immersions are shown in Figure 7.

The values of the basic invariants on the curves L (to the left) and R (to the right) in Figure 1 are equal to

$$St(L) = 4, \quad J^+(L) = -8, \quad J^-(L) = -12,$$
$$St(R) = 10, \quad J^+(R) = -20, \quad J^-(R) = -24.$$

These values imply the theorems stated in the introduction.

The values of the basic invariants on the curves with three and four double points are presented at the end of this article. These tables imply thousands of theorems similar to those in the introduction.

	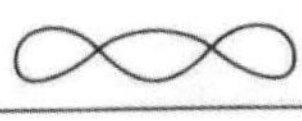	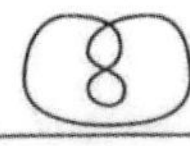	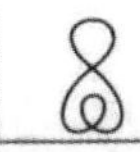		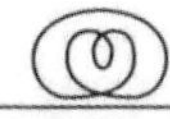
|ind|	1	1	1	3	3
St	0	0	1	2	3
J+	0	0	−2	−4	−6
J−	−2	−2	−4	−6	−8

FIGURE 7. The basic invariants of immersions with two double points

§2. Proof of the existence of the basic invariants

This proof is based on a method of S. Smale [2].

The group of Euclidean motions of the plane, $\mathbb{R}^2 \times S^1$, acts naturally on the space of immersions of the circle into the plane. This action has no fixed points in the space of immersions.

The space of immersions is fibered into the orbits of this action. This action has a section: the set of immersions sending a fixed point $(*)$ on the circle to a fixed point (0) of the plane and the orienting vector of the circle to a vector of a fixed direction tangent to the plane at 0. We shall call such immersions *normed* immersions.

The above action preserves all our discriminants, classes of immersions, and invariants. Hence it suffices to study all these objects only for normed immersions. From now on all the immersions are assumed normed. Smale theory implies the following

THEOREM. *The space of normed immersions of fixed index has trivial homotopy groups of dimension* $i > 0$.

PROOF. This follows from the exact sequence of the Serre fibration studied by Smale,

$$N \longrightarrow (\mathbb{R}^2 \setminus 0) \times \mathbb{R}^2,$$

sending a normed immersion of the segment $[0, 1]$ (with a fixed direction at the starting point 0) to the velocity vector at the end point 1.

The homotopy groups of the space N are trivial. Indeed, an immersion f of the segment $0 \leq t \leq 1$ may be contracted to its small part by the homotopy $f_s(t) = f(st)$ with parameter $s_0 \leq s \leq 1$, where s_0 is very small. The small resulting normed immersion may be then interpolated with a standard small normed immersion linearly.

We shall use the fact that the space N_i of normed immersions of a circle of index i is simply connected. This fact implies the following

THEOREM. *The intersection index of a closed curve in the space of normed immersions with any of the three branches of the discriminant, described in* §1, *is equal to zero.*

PROOF. Since the normed immersions space is simply connected, one can find a disc bounded by the curve. This disc (and the curve itself) may be made transversal to the stratification of the discriminant. As a chain, the disc is the sum of small simplices, also transversal to the discriminant.

If a transversal simplex is sufficiently small, it lies in a two- dimensional versal deformation plane of a singular point of codimension zero or 1 on the discriminant. In this case, the intersection index of its boundary with the discriminant vanishes, since the coorientation of the smooth part of the discriminant is consistent (see §1). The theorem is thus proved. It implies the existence of the invariants of Theorems 1–3 of §1.

At the same time we have proved the uniqueness of the invariants (up to the choice of additive constants depending only on the index), since the space of immersions of fixed index is connected.

§3. Properties of the invariant St

DEFINITION. The *connected sum* of two immersions of a circle (the first into the left, the second into the right halfplane) is the new immersion shown in Figure 8.

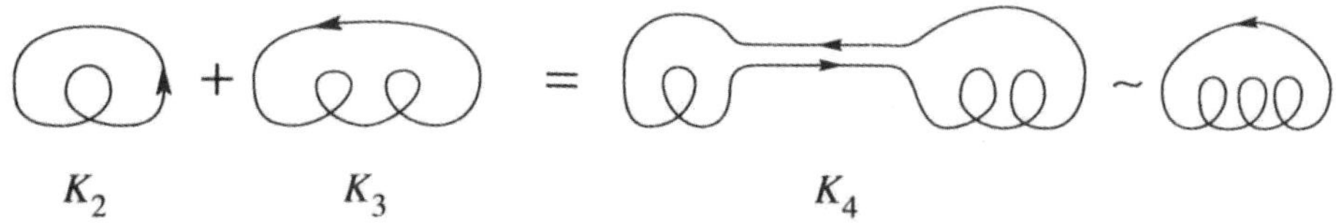

FIGURE 8. The connected sum of two immersions

REMARK. Connected summation is not an operation on immersion classes. Some pairs of immersions cannot be added (example: two standard circles with opposite orientations). When summation is possible, the class of the sum depends in general on the representatives of the classes of the summands.

THEOREM. *The invariant* St *is additive*:

$$St(a+b) = St(a) + St(b).$$

PROOF. Choose a standard representative of each class of immersions of index i, orienting the curves of Figure 5 (the figure eight curve with both orientations belongs to the same class). It follows from Figure 8 that St is additive for the standard immersions of positive indices $i > 0$, $j > 0$:

$$St(K_i) = i - 1, \qquad St(K_j) = j - 1, \qquad K_i + K_j = K_{i+j-1}.$$

One proves by a direct computation that St is additive for the standard summation with the figure eight curve (this follows from the fact that the chain of perestroikas shown in Figure 9 decreases St by 1).

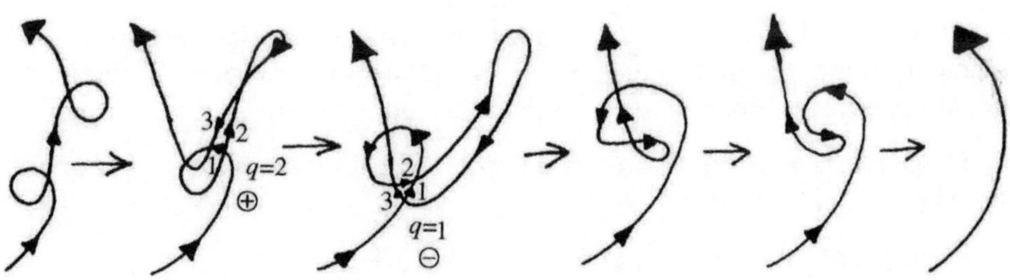

FIGURE 9. Annihilation of counteroriented loops

In any class of immersions of given index i we can choose a special immersion L_i whose exterior contour has intervals oriented in both directions. Two immersions of this kind always admit summations. Using Figure 8 and Figure 9, we see directly that the invariant St is additive in these special cases.

Now consider the sum $X + Y$ of any two generic immersions of a circle. We connect the left curve X with a standard curve L of the same index by a generic path in the space of immersions. We can assume that the immersions forming this path are all identical in the neighborhood of the point sent to the place of connection on the left curve (the other branches of the left curve may cross this point during the deformation of the immersion). We also can assume that the whole path consists of immersions into the left halfplane.

Similarly we connect the right curve Y with a standard curve R in the right halfplane. Combining both paths, we obtain a generic path connecting $X + Y$ with $L + R$.

The increment of St along this path equals the sum of its increments along the left and the right paths. Indeed, neither the orientations of the vanishing triangles nor the cyclic orderings of their sides change under the summation. Hence the signs of the triple point crossings of the left and the right curves do not change. Besides these points, the new path crosses the triple points while the double points of the mobile left or right curve cross the joining band. These crossings occur in pairs of opposite signs and do not contribute to the increment of St, since the corresponding vanishing triangles differ by the direction of one of the sides.

Thus, we find

$$St(X + Y) - St(L + R) = St(X) - St(L) + St(Y) - St(R).$$

Hence, the additivity for the summation of any two curves follows from the additivity for the summation of the special curves

$$St(L + R) = St(L) + St(R)$$

which is already proved by direct calculations.

DEFINITION. The *strange summation* of two immersions of a circle (one into the left, the other into the right halfplane) is defined by an immersion of a segment joining them such that the immersions coorient the immersed segment differently at the end points. To obtain the new circle immersion, one doubles the segment and smoothens the angles as shown in Figure 10.

THEOREM. *The invariant St is additive under strange summation of immersions.*

PROOF. The strange summation can be reduced to the ordinary one if we first push appendices from each curve toward the other (Figure 11).

This pushing does not introduce triple points, hence does not change the value of St. Thus the strange summation additivity follows from the ordinary additivity.

§4. Properties of the invariants $J^{\pm}$

The invariants $J^{\pm}$ are not additive under strange summation, but they

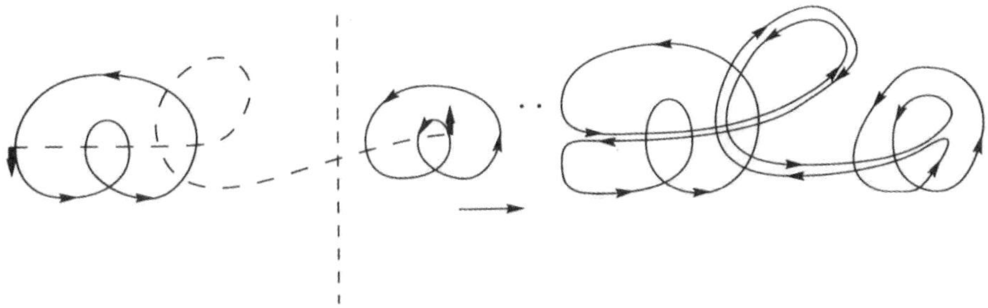

FIGURE 10. The strange summation of immersions

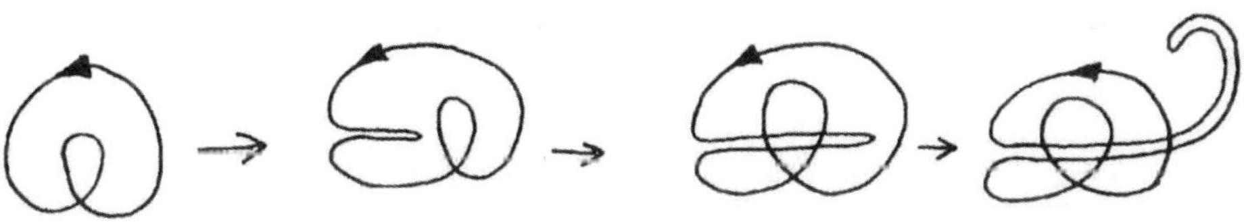

FIGURE 11. Pushing an appendix from the left curve

explicit formula for the combinations $J^{\pm} + 3St$ in terms of the rotations of the radius-vectors connecting the double points of the curve to its moving point.

Choose a representative of a generic immersion (or such a metric on the plane) such that the intersecting branches in all double points are orthogonal. Choose the ordering of the directions of the outgoing branches $(1, 2)$ at each double point for which the frame $(1, 2)$ orients the plane positively.

DEFINITION. The *halfindex* i_1 of a double point (respectively i_2) is the angle of rotation of the radius-vector connecting this double point to a point moving along the branch 1 (respectively 2) from the double point to itself divided by $\pi/2$. The *index* of a double point is the difference $i = i_1 - i_2$.

The index does not depend on the orientations of the curve and of the plane (Figure 12).

DEFINITION. The *invariants* $I^{\pm}$ are defined by summing of the indices of all the n double points, namely,

$$I^{\pm} = \frac{\sum i \pm 2n}{4}.$$

THEOREM. *The invariants $J^{\pm}$ are related to $I^{\pm}$ by the relation*

$$J^{\pm} = I^{\pm} - 3St.$$

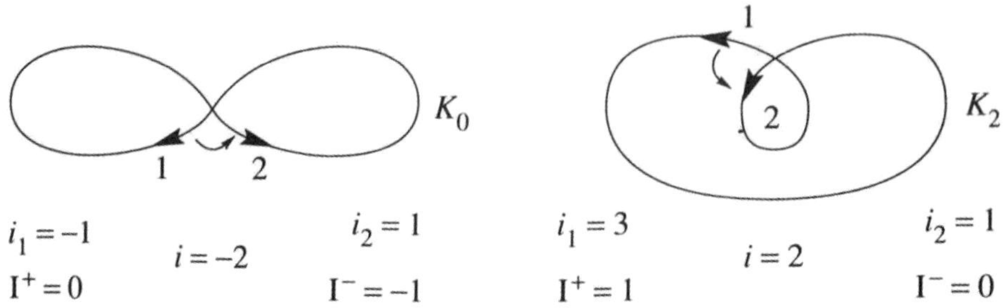

FIGURE 12. The halfindices and indices of double points

PROOF. For the special curves K_i of Figure 5, one may check this directly. For instance, $St(\infty) = 0$, $St(K_2) = 1$. Hence Figure 12 corresponds to the values

$$J^+(K_0) = 0, \quad J^-(K_0) = -1, \quad J^+(K_2) = -2, \quad J^-(K_2) = -3.$$

To prove the theorem for curves of arbitrary index, it suffices therefore to calculate the increments of $I^\pm$ under elementary perestroikas.

Case 1. Direct self-tangency. Consider the crossing of a direct self-tangency generating two double points A and B. We wish to prove that the value of I^+ increases by 2 while that of I^- does not change.

LEMMA. $i(A) + i(B) = 4$.

PROOF. First we calculate the sum for one example (Figure 13).

The lemma for the general case follows. Indeed, any other example is different from this one only outside the neighborhood of the newborn point. The replacement of any of the branches of the curve outside this neighborhood can be regarded as adding a closed curve that does not intersect the neighborhood to this branch. Such an addition changes the halfindices by the (quadrupled) number of resolutions the added curve makes around the double point A or B. Or these two numbers are equal, since the added curve does not intersect the neighborhood. However, the ordering of the branches at A and at B are opposite, and the increment of $i_1(A)$ occurs together with an equal increment of $i_2(B)$ (similarly for $i_2(A)$ and $i_1(B)$). Thus the increment of the sum

$$i(A) + i(B) = i_1(A) - i_2(B) + i_1(B) - i_2(A)$$

under the addition of the closed curve vanishes.

Hence I^+ and I^- behave correctly under the direct self-tangency perestroika.

Case 2. Inverse self-tangency.

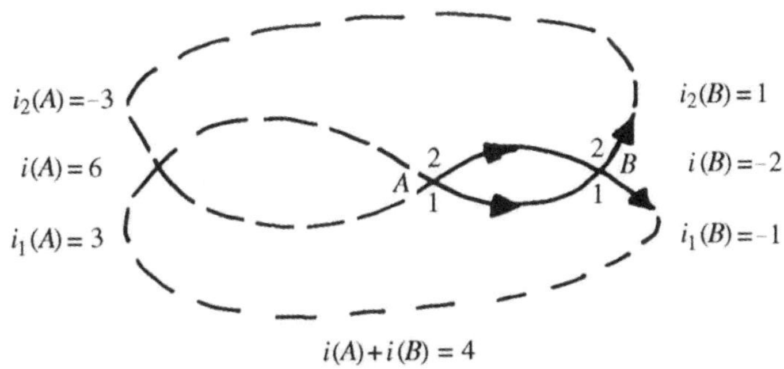

FIGURE 13. The increment of the sum of the indices of double points under the direct self-tangency perestroika

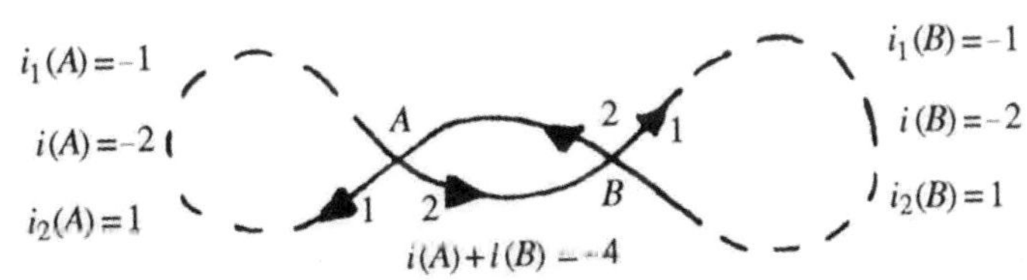

FIGURE 14. The increment of the sum of the indices of double points under the inverse self-tangency perestroika

LEMMA. $i(A)+i(B)=-4$.

PROOF. A particular example is shown in Figure 14.

The general case is reducible to that of Figure 14 by the same arguments as in the direct self-tangency case.

Hence I^+ does not change its value under the crossing of an inverse self-tangency giving birth to two double points while the value of I^- decreases by 2.

Case 3. A triple point. The number n of double points does not change under a crossing of a triple point.

LEMMA. *The index of each of the three vertices of the vanishing triangle increases by* 4 *under a positive crossing of a triple point.*

PROOF. We can fix two of the three branches of the curve and move the third one. Let A_1, A_2 be the fixed branches leaving the fixed intersection point A along the vectors (1,2) orienting the plane positively. The third visit of the triple point takes place on one of these two branches. The rotation of the radius-vector will change one of the branches as crosses the triple point, namely the one containing the third visit moment. The increment of the angle of rotation will be $\pm 2\pi$. It will be positive if, after the crossing, the third branch will define a frame (radius-vector from A to a point of the third branch, velocity vector of the third branch) positively orienting the plane.

This increment of the argument divided by $\pi/2$ is equal to the increment of the index of the point A, with the *plus* sign if the third visit point belongs to the branch A_1 (leaving A in the direction 1).

There exist four possibilities for the occurring vanishing triangles (for which the increments of the argument along the branch and of the index of the point A are both positive). They are shown in Figure 15.

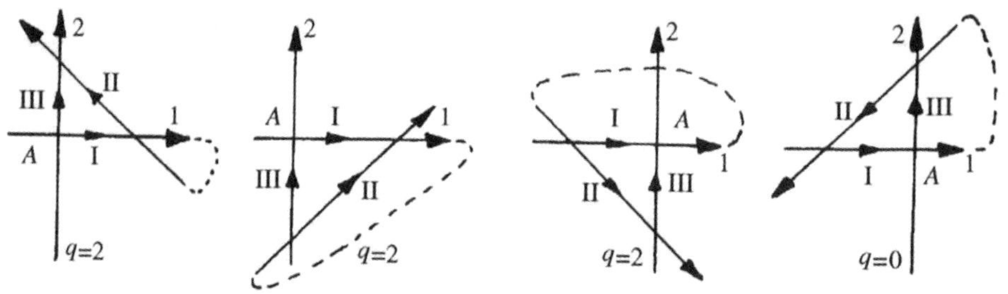

FIGURE 15. Vanishing triangles whose births imply increments of the arguments and of the index of point A

We observe that all these four vanishing triangles are *positive*.

The case of the third visit on the branch 2 can be considered similarly. But one also can avoid this, since J^+ and i do not depend on the orientations.

Thus, the quantities $I^{\pm} - 3St$ and $J^{\pm}$ behave the same way under all the three elementary perestroikas. Since their values on the standard curves K_i of Figure 5 also coincide, the theorem is proved.

COROLLARY. *The invariants $J^{\pm}$ are additive with respect to the ordinary connected sum of immersions.*

PROOF. The invariants St and n (the number of double points) are additive. Hence we only have to check the additivity of the sum of indices of the double points.

The double points of the sum are just the double points of the left and of the right curve. The addition preserves any halfindex of any double point of the left (right) curve, since the added part of the curve may be contracted in the right (left) halfplane to the end of the joining segment.

REMARK. Of course,

$$J^+ - J^- = I^+ - I^- = n, \text{ the number of double points,}$$

is also an additive invariant. However, the preceding theorems show that it is better to consider J^+ and J^- as the basic invariants. The invariant J^+ measures some kind of self-linking of the Legendre curve formed by the directions of the plane curve in the 3-dimensional space of plane contact elements.

§5. Axiomatic description of the basic invariants

DEFINITION. An invariant (of an immersion of a circle into a plane with no triple points) is *local* if its jump at the generic crossing of the hypersurface of immersions with triple points depends only on the behavior of the family in the neighborhoods of the three points sent to the triple point by the critical immersion.

This means that the jump will not change if we replace our family by any other family outside the above-mentioned neighborhoods.

THEOREM. *The jump of any local invariant of immersions with no triple points is proportional to the jump of the invariant* St (*with a coefficient independent of the immersion*).

PROOF. The only strata of codimension one on the triple point discriminant are the strata corresponding to the following singularities of immersions:

(a) a triple point with a tangency of two branches;
(b) a quadruple point;
(c) two triple points.

The jump of an invariant which is local does not change along any of the two intersecting branches of the discriminant when one crosses the other branch moving along the first in case (c).

The jumps near the strata (a) and (b) must be consistent. We shall see that the consistency condition at the (a) stratum leaves only one possibility for the jump, namely the one we have described above in the definition of St.

The versal deformation of the singularity of type (a) is a two-parameter family of curves on a plane (x, y) with parameters u and v:

$$x = 0, \qquad y = 0, \qquad y = x^2 + ux + v.$$

The discriminant is shown in Figure 16.

The stratum of triple points is hence smooth in the neighborhood of the singularity of type (a): the event happening there is the tangency to the stratum of self-tangency.

The consistency condition says that the jumps of the invariant due to the transversal crossing of the line $v = 0$, say from $v > 0$ to $v < 0$, should be the same on the left $(u < 0)$ and on the right $(u > 0)$ part of Figure 16.

The local situation in the neighborhood of a triple point crossing is described by the vanishing triangle formed by three oriented lines ordered cyclically.

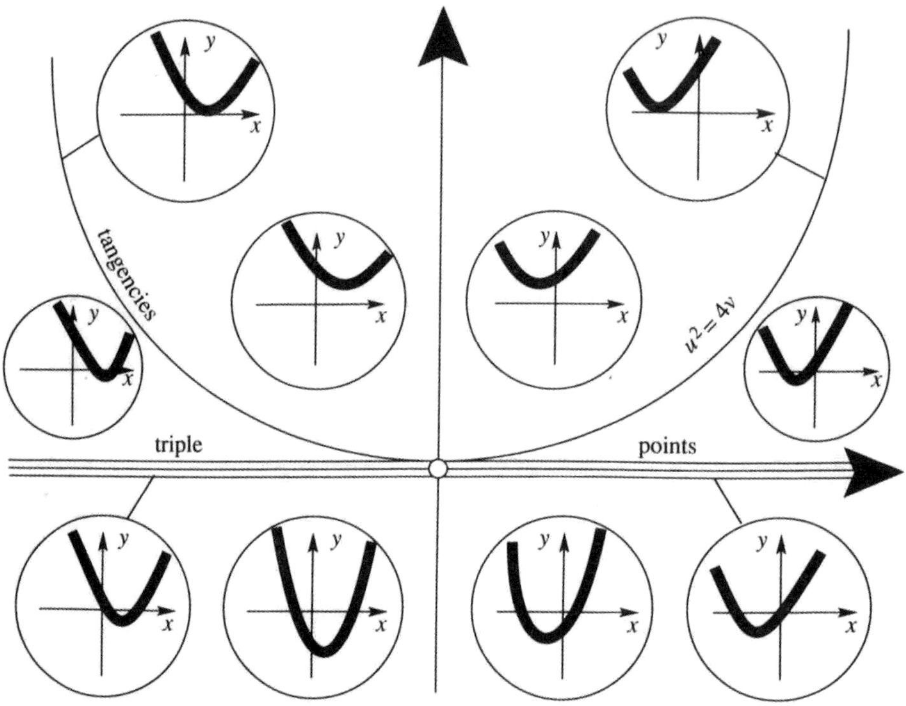

FIGURE 16. The versal deformation of a degenerate triple point

There exist eight types of such triangles in the plane (up to orientation preserving diffeomorphisms of the plane). Indeed, the angles between the three lines may be reduced to 60°. Then the angles between their directions are either all equal to 120° (case A) or are (60°, 60°, 120°) (case B).

The cyclic ordering of the three rays leaving the triple point defines an orientation of the plane which may be positive or negative. We thus define four types of triple points, A_-, A_+, B_-, B_+. In each case we have two types of vanishing triangles, one from each side of the discriminant hypersurface. We shall denote these by the types of the triple point with a superscript index q (equal to the number of sides directed conformally to the cyclical order).

All the eight resulting triangles are shown in Figure 17.

REMARK. The parity of q always changes when the discriminant of triple points is crossed. Indeed, the orientation of the plane defined by the cyclic ordering of the sides of the triangle changes while the directions do not change, hence $q(-)+q(+)=3$ for any two neighboring points $(-)$ and $(+)$ separated by the discriminant.

LEMMA 1. *The pairs of types of vanishing triangles in the domain* $v<0$

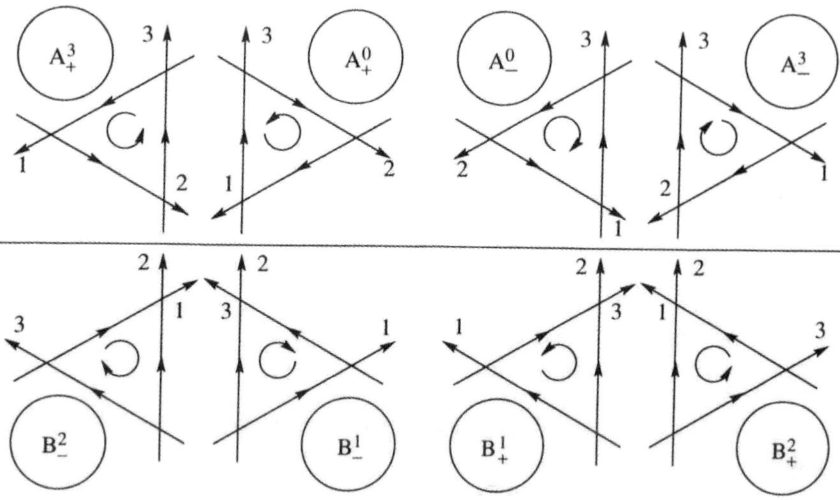

FIGURE 17. Classification of vanishing triangles

of Figure 16 *may be only*

$$A^3_+ \text{ and } B^1_+, \quad A^3_- \text{ and } B^1_-, \quad A^0_- \text{ and } B^2_-,$$
$$A^0_+ \text{ and } B^2_+, \quad B^2_- \text{ and } B^2_+, \quad B^1_- \text{ and } B^1_+,$$

depending on the orientation and ordering of the three curves of Figure 16.

To prove this, it suffices to calculate the types of both curvilinear triangles formed by the parabola and the coordinate axis for different orientations and orderings. Two crucial examples are shown in Figure 18.

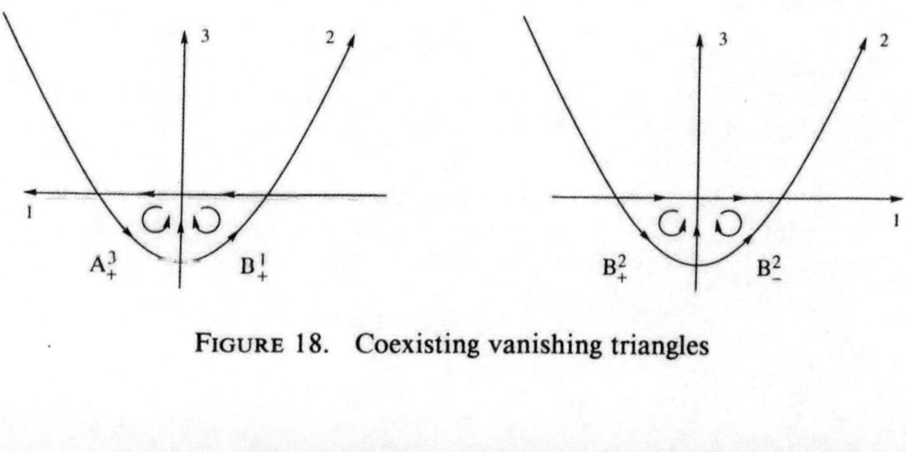

FIGURE 18. Coexisting vanishing triangles

Thus if there exists a local invariant, then, in the corresponding coorientation of the stratum of triple points, the points corresponding to the four coexistent triangles $A^0_- \sim B^2_- \sim B^2_+ \sim A^0_+$ must all be on one side of the discriminant, and those corresponding to $A^3_- \sim B^1_- \sim B^1_+ \sim A^3_+$ on the other. Thus q is even for all the triangles on one side and odd for those on the other side, in agreement with the coorientation rule defining the invariant St.

REMARK. The independence of this coorientation of the choices of orientations of S^1 and $\mathbb{R}^2$ follows thus from the localness of the jump.

LEMMA 2. *The coorientation of the triple point discriminant is consistent at points of the stratum of quadruple points.*

PROOF. The (topological) versal deformation is the two- parameter family of quadruples of lines in the plane given by the table

I	II	III	IV
$x = 0$	$y = 0$	$y = x + u$	$y = -x + v$,

where u and v are parameters. The stratum of triple points intersects the plane (u, v) along four lines corresponding to the 4 types of triple points (Figure 19)

(I II III)	(I II IV)	(I III IV)	(II III IV)
$u = 0$	$v = 0$	$u = v$	$u = -v$

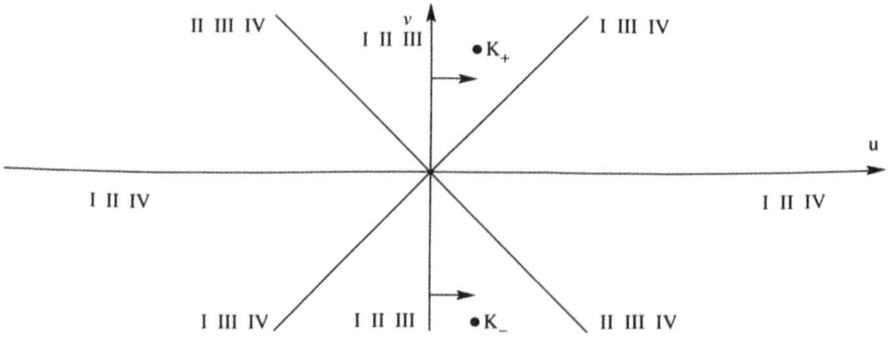

FIGURE 19. The versal deformation of a quadruple point

Proposition. *The coorientations of the triple point discriminant defines on each of the four lines the same coorientation along both rays into which this line is divided by the origin.*

Proof. Consider for example the discriminant line (I II III), $u = 0$. The configurations of the branches at the $u > 0$ side of this line for $v = 0$ and for $v < 0$ are shown in Figure 20.

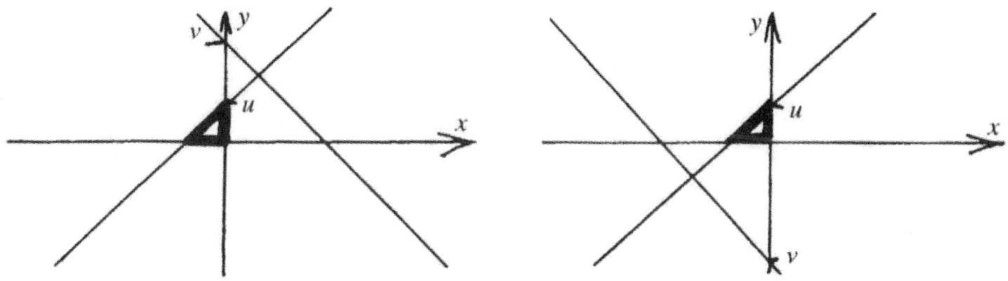

Figure 20. The consistency of coorientations at a quadruple point

The vanishing triangles for fixed and very small $u > 0$ simply coincide, they are the same for $v > 0$ and $v < 0$. Hence $q(K_+) = q(K_-)$ and both points K_+ and K_- lie on the same side of the stratum of triple points.

Theorem. *There exists exactly one invariant, St, of immersions $S^1 \to \mathbb{R}^2$ without triple points with the following properties:*

1) *it is local* (the jump only depends on the behavior near the triple point),
2) *it is additive* (under ordinary connected summation),
3) *it is orientation independent* (at least for one of the orientations: that of S^1 or that of $\mathbb{R}^2$).

Conditions 1–3 define the invariant up to a multiplicative constant. To fix the constant it is sufficient to fix the value at one curve, where the value of St is not zero. We choose the condition

4) *The invariant takes the value* 1 *at the immersion of index* 2 *with one selfintersection point* (K_2 in Figure 5).

Remark. We already know that St has all these properties (and moreover that it is additive under the more general *strange* summation and independent on *both* orientations). So we only have to prove its uniqueness.

Proof. The jump of an invariant which is local is proportional to the jump of St, according to the preceding theorem. Hence the invariant is defined by its values on the standard immersions K_i of index i of Figure 5 ($i = 0, \pm1, \pm2, \dots$), and the difference of two such invariants is constant along the set of immersions of index i. Let this difference be $f(i)$. Then we have

$$f(i_1 + i_2 - 1) = f(i_1) + f(i_2),$$

hence $f(i) = c(i-1)$. The independence of orientation implies $f(1) = f(-1)$, hence $c = 0$, which proves the theorem.

THEOREM. *There exists exactly one invariant, $J^+(J^-)$, of immersions with no direct (inverse) self-tangencies with the following properties:*

1) *it is local* (the jump only depends on the behavior near the self-tangency point),
2) *it is additive* (for ordinary connected summation),
3) *it is orientation independent* (for at least one of the two orientations of S^1 or of $\mathbb{R}^2$).

These conditions define the invariant up to a multiplicative constant, which we fix by the condition

4) $J^+(K_2) = -2 \quad (J^-(\infty) = -1)$.

REMARK. The multiplicative constant has been chosen in such a way that a crossing of the direct self-tangency discriminant increases J^+ by the *same* number as it increases the number n of double points, while the crossing of the *inverse* self-tangency discriminant increases J^- by the *opposite* of the increment of n. These choices imply $J^+ - J^- = n$. Note that the jumps of $J^\pm$ at the crossings of the corresponding discriminants are thus ± 2 and not ± 1.

§6. The "pushing away" formula

The calculation of the invariant St is greatly simplified by the following method: one can control the increment of St when a fragment of a curve is pushed away, as shown in Figure 21.

The *fragment* encircled by the dashed line in Figure 21 is an arbitrary immersed closed curve, equiped with a *tail* consisting of two parallel segments joining the segment to another *main* closed curve. We suppose that the tail does not intersect the fragment at points other than those at the joining place.

Let x be a regular point of an oriented circle immersed into the oriented plane. A double point of the immersion is called *positive* (*negative*) with respect to x, if the frame $(1, 2)$ formed by the velocity vectors of the first and of the second visit of the point by the curve orients the plane positively (negatively). (Here the ordering of the visits refers to the immersion of the oriented interval obtained from the circle by the exclusion of the point x).

DEFINITION. The *Whitney function* defined at the ordinary points x of the immersion of the circle is the difference between the numbers $w_+(x)$ and $w_-(x)$ of double points of the circle immersion that are positive and negative with respect to x:

$$w(x) = w_+(x) - w_-(x).$$

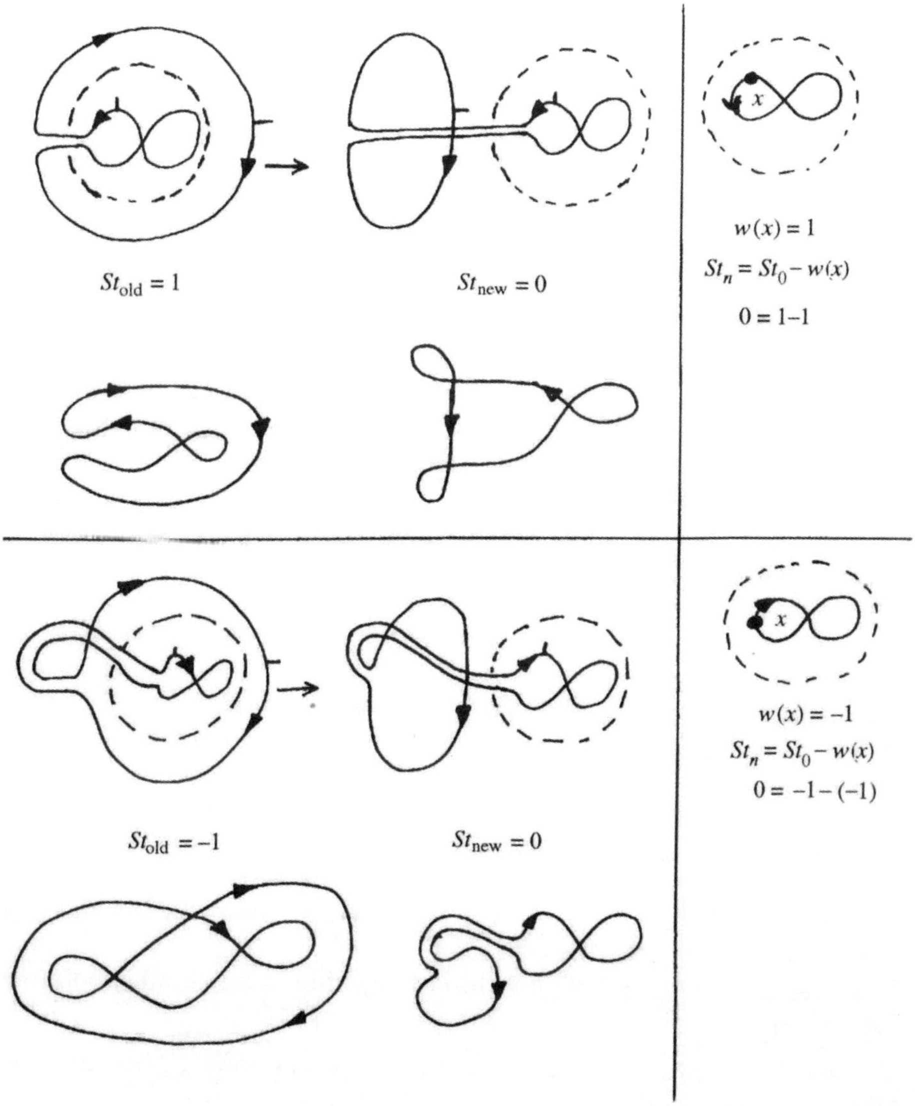

FIGURE 21. The "pushing away" perestroika

Examples are shown in Figure 22.

THEOREM. *The increment of St under the pushing of a fragment through an interval of a main curve is equal to the value of the Whitney function of the fragment at the point of the joining of the tail, provided that the frame (direction of pushing, direction of the interval of the main curve) orients the plane positively (if it orients the plane negatively, the increment is* $-w(x)$ *).*

Examples are shown in Figure 21.

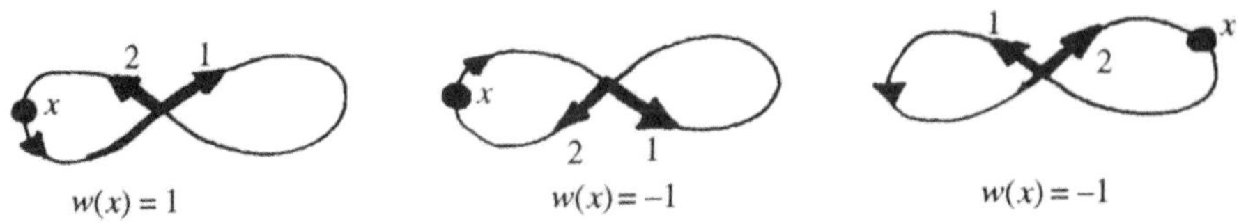

FIGURE 22. Calculation of the Whitney function

REMARK. The standard orientation of the plane may be excluded from the formulation. We must count the positive (negative) points of the curve while defining the Whitney function using the orientation of the plane defined by the direction of pushing and the direction of the main curve interval. In this way we see that the increment of St is independent of the choice of the standard orientation of the plane.

PROOF OF THE THEOREM. While we push the fragment, each of its double points will contribute ± 1 to the increment of St. The sign depends on the sign of the vanishing triangle.

The cyclic ordering of the sides of the vanishing triangle is $(1, 2, 3)$: the first visit, the second visit, the visit of the interval of the main curve.

The vanishing triangle born at the crossing of the triple point is positive if and only if the orientation $(1, 2)$ *coincides with the orientation* $(3, 4)$, *where* 3 *is the pushing direction and* 4 *direction of the the main curve interval.*

This fact is checked directly by inspection of the eight possible cases (we have done this before, cf. Figure 15). Summation over all the vanishing triangles now proves the theorem.

It is clear that the jump of the Whitney function at any double point of the curve equals ± 2.

THEOREM. *The value of the Whitney function at an ordinary point* x *of an immersed circle is*

$$w(x) = i(x) - \text{ind},$$

where $i(x)$ *is the number of half-turns of the vector connecting* x *to a point* y *moving along the curve from* x *to* x, *and where* ind *is the number of turns of the tangent vector.*

EXAMPLE. These values for a curve with two selfintersection points are shown in Figure 23.

PROOF. The jumps (± 2) of $w(x)$ and $i(x)$ at each double point of the curve are equal. Hence the difference is independent of x and is an invariant of the immersed curve.

This invariant does not change its value under the three basic perestroikas. Indeed, one may choose the point x far away from the perestroika place.

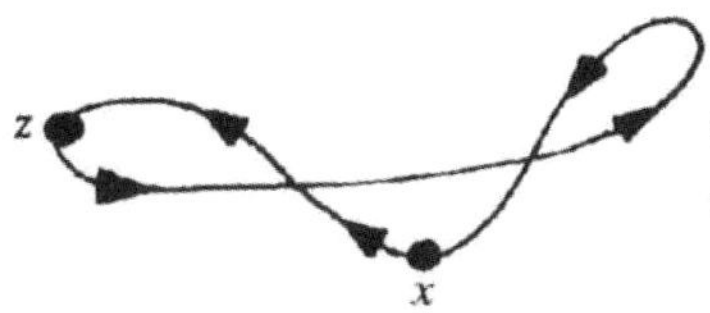

FIGURE 23. Whitney function and indices

Both $w(x)$ and $i(x)$ will then be continuous and hence constant under the perestroika.

The invariance of $w(x)$ when two double points are born follows from the fact that the orientations $(1, 2)$ at these two points are different. Thus, the difference $w(x) - i(x)$ is constant along the space of immersions of a given index. For the standard curves of Figure 5, this difference is equal to the index, whence the theorem follows.

REMARK. Applying this theorem to the points of the exterior contour, we obtain the following Whitney formula for the index. Connect a point of the exterior contour to infinity by a transversal ray which does not intersect the curve at other places.

THEOREM. *The index of an immersed circle is equal to* $a + \sum$, *where* $a = \pm 1$ *and* $\sum$ *is the sum of* ± 1 *over the double points of the curve.*

The signs are defined by the following rule: $a = i(x)$, where x is the exterior point, $\sum = -w(x)$. In other words, $a = 1$ iff the frame (ray's direction, curve's direction) is positive, and the contribution of a double point to $\sum$ is positive iff the frame $(2, 1)$ is positive (Figure 24).

COROLLARY. *The modulus of the index of a generic curve with* n *selfintersection points is at most* $n + 1$ *and takes all the values between* $-1 - n$ *and* $1 + n$ *congruent to* $1 + n \bmod 2$:

$$|\operatorname{ind}| \leq n + 1, \qquad \operatorname{ind} - n \equiv 1 \bmod 2.$$

§7. The extremal curves

DEFINITION. A generic curve with n double points is *extremal* if the absolute value of its index i takes the maximal possible value $|i| = n + 1$.

The extremal curves with four double points are shown in Figure 25.

THEOREM. *There exist natural bijections between*

1) *the set (of classes) of immersed circles of index* $n + 1$ *with* n *double points;*

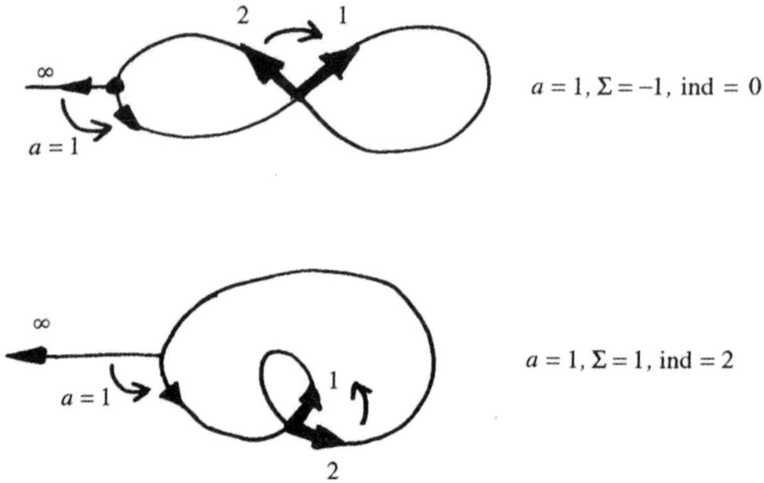

FIGURE 24. Calculation of the index by summing double points

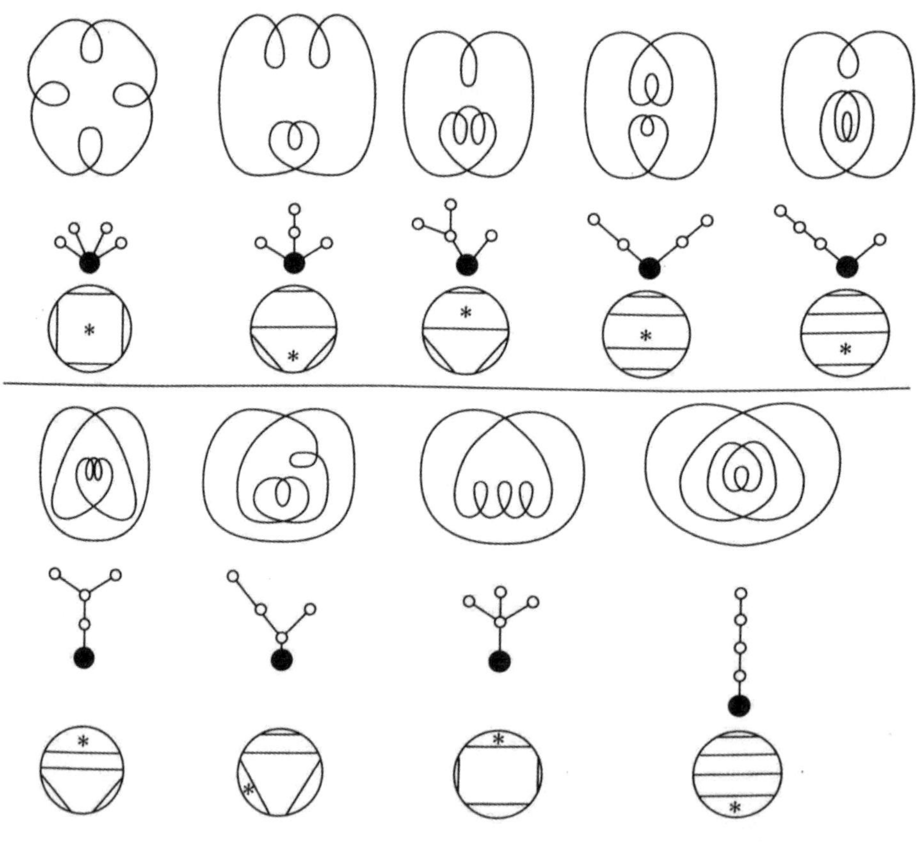

FIGURE 25. Extremal curves with $n = 4$ double points

2) *the set (of classes) of plane rooted trees with n edges*;
3) *the set (of classes) of the decompositions of a disc into parts by n chords with no common points and one part distinguished.*

REMARK. Two curves (trees, decompositions) are here *equivalent* if one can be transformed into the other by a homeomorphism of the plane (which may preserve, or not, the orientation).

A similar result holds also in the oriented case, when the equivalences preserve the orientation of the plane.

The construction of the bijections uses the following

DEFINITION. The *Gauss diagram* of a generic immersion of a circle into a plane is the system of chords of the immersing circle, connecting the points sent by the immersion to the same double point of the immersed curve. (Gauss has studied, which chord diagrams correspond to immersions).

LEMMA. *The Gauss diagram of an extremal immersion is planar (consists of nonintersecting chords).*

In other words, the two loops into which a double point breaks an extremal curve do not intersect at other points.

PROOF. The smoothening of the immersed circle at a double point transforms the curve into two immersed branches. The indices and the numbers of double points of the initial curve and of the two branches are related by the following equations

$$i = i_1 + i_2, \qquad n = 1 + n_1 + n_2 + n_{12}.$$

Since the curve is extremal, $i = n + 1 = 2 + n_1 + n_2 + n_{12}$. The Whitney inequalities (§6) $i_k \leq 1 + n_k$ imply $i = i_1 + i_2 \leq 2 + n_1 + n_2$. Hence the number n_{12} of the intersections of the branches vanishes.

Thus we have associated to each extremal curve a disc decomposition (the Gauss diagram). To each bounded component of the complement to the extremal curve there corresponds a component of the disc decomposition (it is bounded by the arcs sent to the boundary of the component of the complement by the immersion).

LEMMA. *Among the bounded components of the complement to an extremal curve in the plane exactly one component has common boundaries with the unbounded component.*

PROOF. The Whitney formula implies that: (1) The exterior contour of an extremal curve is well oriented by the immersion. (2) The loops connected to the exterior contour at the double points lie inside the domain bounded by this exterior contour (which bounds a topological disc).

These loops, as we have seen, do not intersect. Hence the complement to the domains bounded by these loops in the above disc is connected. This proves the lemma.

Among the parts of the decomposition given by the Gauss diagram, the one that we distinguish corresponds to the above distinguished bounded component (the circle arcs bounding this part are sent to the exterior contour by the immersion). We have thus associated a decomposition with a distinguished part to an extremal curve.

Choose a point in each part of the decomposition. Connect these points by edges when two parts have a common chord. We obtain a plane rooted tree. We have thus constructed the required mappings from extremal curves to plane rooted trees and to disc partitions.

The fact that these mappings are bijective is proved inductively on the number n of double points. The branches of the tree encode the places where loops must be attached to the exterior contour, and these loops are themselves standard by the inductive conjecture, since they have less double points.

REMARK. All the nine plane rooted trees with 4 edges (Figure 25) are also different as abstract rooted trees.

For trees with ≥ 5 edges the situation is different. The number of nonequivalent plane rooted trees with 5 edges is 21, while that of abstract rooted trees is only 20.

THEOREM. *The invariant St of an extremal curve is equal to the sum of the distances of the vertices of the corresponding tree from the root.*

PROOF. Induction on the number n of double points. We see directly that the result is true for $n = 1$. If the exterior contour contains several double points, we use additivity (Figure 26).

If there is only one double point, we push away the loop attached at that point, as shown in Figure 26 below. The resulting curve is the connected sum of two extremal curves (since the " $(1, 2)$ " orientation at all the double points is as it should be). Using the additivity of St and knowing its values for the summands, we can compute St of the sum.

The value of St at the initial curve is greater than this sum (accordingly to the "pushing away" theorem of §6) by the value $w(x)$ of the Whitney function at the attachement point. This value, in our case, is equal to the number of double points of the fragment since all the " $(1, 2)$ " orientations are as they should be. Thus we have an expression of the value of St in terms of the value of St at the main curve and at the fragment and of the number of double points of the fragment.

On the level of trees, our operation is the decomposition of the tree into two: we cut off a branch at distance one from the root and preserve the vertex at the cut place both on the initial tree and on the branch (where it becomes the root vertex).

The sum of the distances from the vertices to the roots decreases under this operation by the number of edges of the branch that were cut off, i.e. by $w(x)$. Thus

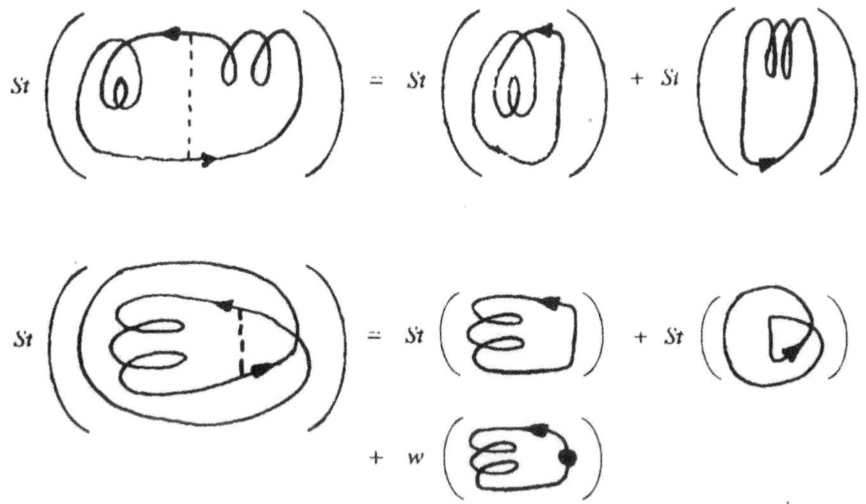

FIGURE 26. Calculation of the invariant St for the extremal curves

$$St(\text{initial curve}) = St(\text{mutilated tree}) + St(\text{branch}) + w(a)$$
$$= S(\text{mutilated tree}) + S(\text{branch}) + w(x) = S(\text{initial tree}),$$

where S means the sum of the distances of the vertices to the root. The theorem is thus proved.

COROLLARY. *The value of the invariant* St *on the extremal curves with* n *double points lies between* n *and* $n(n+1)/2$. *The equalities are reached only on the standard curves* A_{n+1} *(with the tree* $\odot - \cdot - \cdots - \cdot$*) and* K_{n+1} *(Figure* 27).

The results and conjectures on the maximal and minimal values of the invariants over the set of immersions of fixed index with fixed number n of double points are summarized in the table below.

REMARK. It is not difficult to prove that $J^+ + 2St = 0$ for any extremal curve. F. Aicardi has recently proved that this is also true for all curves having planar Gauss diagrams. Her formula for St of such curves implies that the minimum of St for curves with planar Gauss diagrams with n double points is attained on a curve having an index which is close to $n/3$.

Recall that $J^- = J^+ - n$, and hence the theorems and conjectures for J^- follow. Consider a curve with n double points whose index $n+1-2k \geq 0$ is smaller by $2k$ than the maximal possible value. Call k the *defect* and consider the maximal and the minimal values of some invariant U over the

invariant	ind	min	max	min curves	max curves
St	$n+1$	n	$n(n+1)/2$	K_{n+1}	A_{n+1}
St	$n-1$	0	$n(n-1)/2$	Figure 28	Figure 28
St	$n-3$	$2-n$	?	Figure 29	?
J^+	$n+1$	$-n^2-n$	$-2n$	A_{n+1}	K_{n+1}
J^+	$n-1$	$-n^2+n$	0	Figure 28	Figures 28, 30

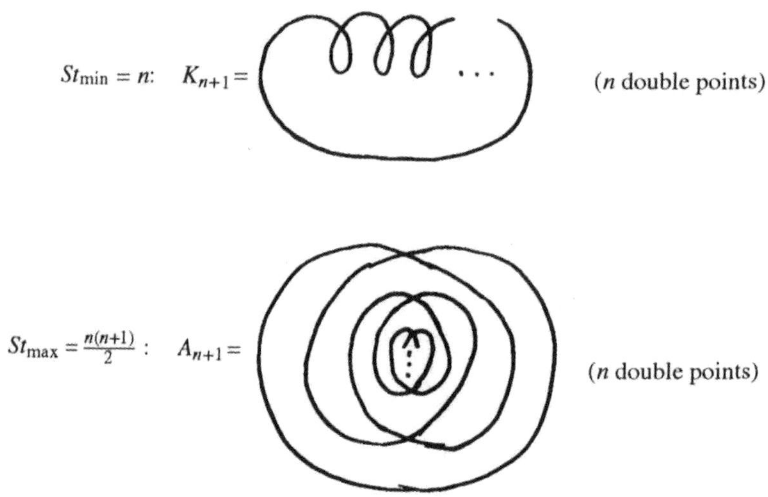

FIGURE 27. The extremal curves of index $n+1$ having maximal and minimal values of St

set of all generic immersions having a fixed number n of double points and a fixed defect k. We shall denote them by $U_{\max}(k)$ and $U_{\min}(k)$; these numbers depend on n.

CONJECTURE. *The numbers* $St_{\max}(k)$ *and* $J^{\pm}_{\min}(k)$ *are monotonic in* k *(decreasing for* St*, increasing for* $J^{\pm}$*).*

CONJECTURE. *The maximal value of* St *on all generic curves with* n *double points is attained only on the curve* A_{n+1} *of Figure 27*

$$St \leq n(n+1)/2.$$

CONJECTURE. *The minimal values of* $J^{\pm}$ *on all generic curves with* n *double points is attained only on the curve* A_{n+1} *of Figure 27:*

$$J^+ \geq -n^2 - n, \qquad J^- \geq -n^2 - 2n.$$

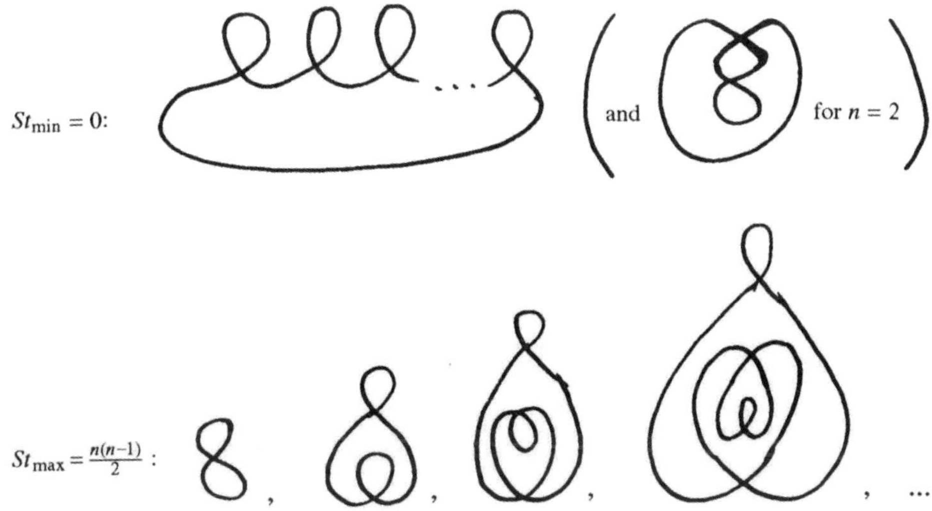

FIGURE 28. Curves of index $n - 1$ maximizing and minimizing St

FIGURE 29. Curve of index $n - 3$ minimizing St

CONJECTURE. *The minimal value of St and the maximal values of $J^{\pm}$ on all generic curves with n double points are attained on the same curve if n is sufficiently large.*

The minimum of St for $n = 2$ and the maximum of J^+ for $n = 3$ are also attained at the exceptional curves of Figure 30.

It would be interesting to study the curves with extremal values of the invariants $I^{\pm}$ §4 .

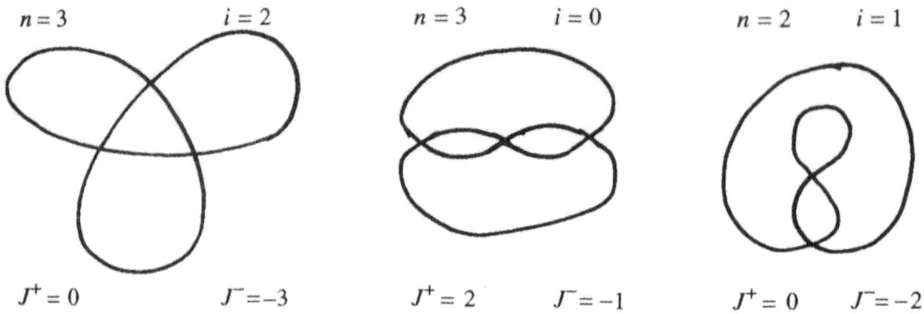

FIGURE 30. Exceptional curves

§8. The cobordisms

The final goal of our work is the study of the components of the complement of different strata of the discriminant in the space of immersions. The classification of the additive invariants constant on such components provides dual objects—the abelian groups which we may call the K-theories of the corresponding classification problems.

The classification problems I have in mind are, for instance, the classification of the components of the complement of the whole discriminant (it coincides with the classification of the immersions up to orientation preserving diffeomorphisms of the plane and of the circle), or the classification of the components of the complement to one of the three branches of the discriminant (corresponding to one of the three perestroikas—triple points $(*)$, direct self-tangency $(+)$, inverse self-tangency $(-)$), or of the components of the complement of the union of two branches. There exist also other possibilities, since each branch consists of infinitely many components (which we shall discuss later in §9).

In any case none of these K-theories is calculated. As a very rough approximation I calculate here the oriented cobordism theories corresponding to different parts of the discriminants (0, 1, 2 or 3 of the branches $(*)$, $(+)$, $(-)$). In this way we obtain eight theories, according to the lists of permitted and forbidden perestroikas.

DEFINITION. An immersion of a circle (or of a finite set of disjoint oriented circles) into the plane is called *cobordant* in the sense of one of the eight theories to an other immersion (of different system of circles) if one can join these two immersions by a chain of isotopies of immersions and of perestroikas of the types mentioned in the name of the theory together with (oriented) Morse perestroikas (shown in Figure 31).

The sum of two cobordism classes is defined as the disjoint union of two

FIGURE 31. Oriented Morse perestroikas of immersed curves

immersions: one into the left half-plane, the other into the right one.

The eight commutative semigroups formed by the cobordism classes (most of them are groups) are connected by twelve natural homomorphisms and form a commutative cube (the growth of the list of permitted perestroikas reduces the (semi-) group of cobordism classes).

Denote by X_i the *basic immersion* of total index i shown in Figure 32 ($i = 0, \pm 1, \pm 2, \dots$).

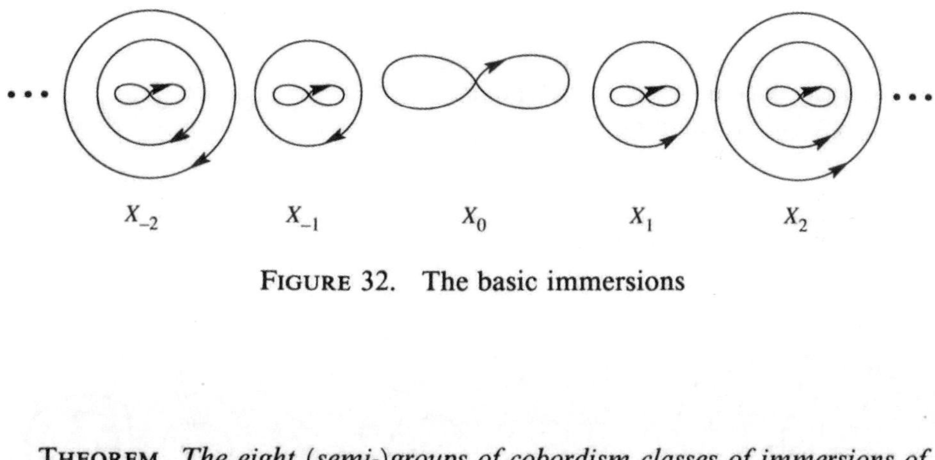

FIGURE 32. The basic immersions

THEOREM. *The eight (semi-)groups of cobordism classes of immersions of curve are given by the following table*:

permitted		$-$	$+$	$*$	$*-$	$*+$	$+-$	$*+-$
forbidden	$*+-$	$*+$	$*-$	$+-$	$+$	$-$	$*$	
answer	$\mathbb{Z}_+^\infty$	$\mathbb{Z}_2^\infty$	$\mathbb{Z}_2^\infty$	H	$\mathbb{Z}_2$	$\mathbb{Z}_2$	$\mathbb{Z}_2^\infty$	$\mathbb{Z}_2$
generators	X_i	X_i	X_i	X_i	X_0	X_0	X_i	X_0

where

$\mathbb{Z}_+$ *is the semigroup of nonnegative integers,*
$\mathbb{Z}_2$ *is the group consisting of two elements,*

H is the semigroup with generators X_i, $i = 0, \pm 1, \pm 2, \ldots$ and relations

$$3X_i = 3X_j, \quad 2X_i + X_{i+1} = X_i + 2X_{i+1}.$$

LEMMA 1. *Each curve is Morse cobordant to a sum of curves X_i (none of the perestroikas $(*)$, $(+)$, $(-)$ being permitted).*

PROOF. 1°. Each double point might be cut off of the curve by a pair of Morse perestroikas (Figure 33).

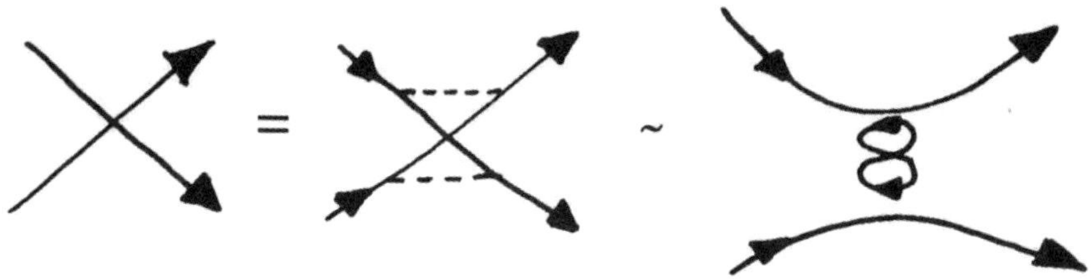

FIGURE 33. The cobordism isolating the double points

The new born figure eight curves are void, other components of the resulting curve do not intersect any other embedded circle.

2°. Each disc containing several inclusions can be transformed by a Morse perestroika into several discs, each containing only one inclusion (Figure 34).

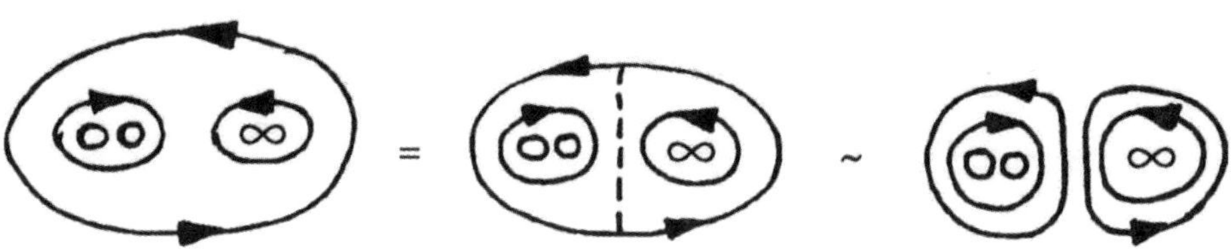

FIGURE 34. The cobordism isolating the inclusions

3°. Iterating these two procedures, we obtain a curve consisting of the nests of two types: ending by 8 or by 0 (Figure 35).

4°. The nest II is cobordant to the void curve (a series of Morse perestroikas).

5°. The nest I is cobordant to one of the nests X_i, where all the embedded circles are oriented, the same way (Figure 36). Thus the semigroup of cobordism classes is generated by the X_i.

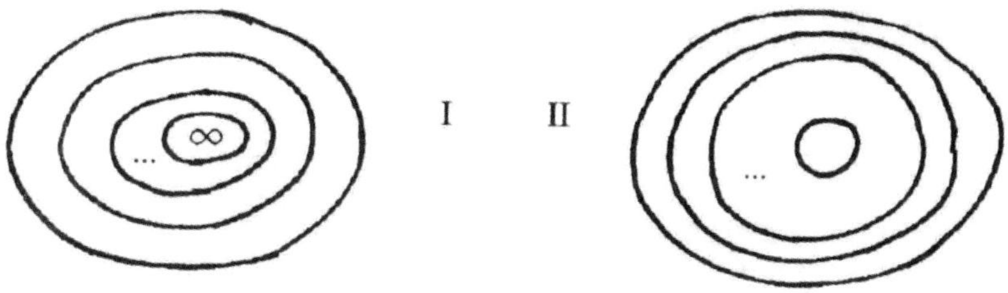

FIGURE 35. The two types of nests

FIGURE 36. The cobordism of a nest I to the standard one

DEFINITION. The *index* of a double point (of a generic immersion of a union of oriented circles into the plane) is the number of revolutions about the origin of the radius-vector from this double point to the point moving along the oriented curve obtained from the given immersed curve by isolating the double point and then eliminating the newborn figure eight curve (Figure 37).

EXAMPLE. The double point of the curve X_i has index i.

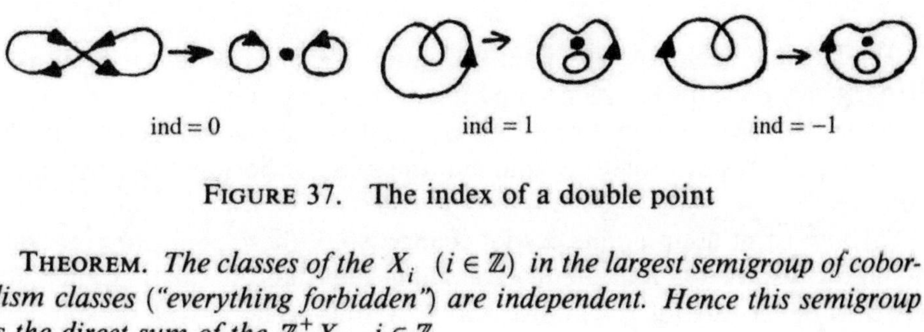

FIGURE 37. The index of a double point

THEOREM. *The classes of the* X_i $(i \in \mathbb{Z})$ *in the largest semigroup of cobordism classes ("everything forbidden") are independent. Hence this semigroup is the direct sum of the* $\mathbb{Z}^+ X_i$, $i \in \mathbb{Z}$.

Proof. The double points are neither born nor killed by Morse perestroikas and isotopies. The indices of the existing double points are the integrals of $(2\pi i)^{-1}\ d\ \ln(z - z_{\mathrm{sing}})$ along the curve obtained as the result of the isolation of z_{sing}. Hence the number of double points of index i is an invariant of our cobordisms, which proves the theorem.

Theorem. *If self-tangency perestroikas are permitted (but triple points are not), the cobordism class semigroup is the group* $\mathbb{Z}_2^\infty = \oplus \mathbb{Z}_2 X_i$.

Proof. 1°. The class of X_0 is of order 2 (Figure 38).

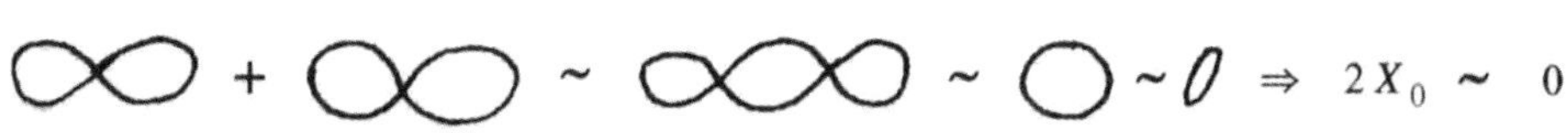

Figure 38. Calculation of $2X_0$ in the "(+), (−) permitted" case

2°. The class of any X_i is of order 2 (Figure 39).

Figure 39. Calculation of $2X_i$ in the "(+), (−) permitted" case

3°. The numbers of double points of index i change under the permitted self-tangency perestroikas by even numbers (Figure 40).

In the (+) case both newborn points, A and B, have equal indices. In the (−) case the newborn island is void and the two newborn points have equal indices.

The indices of other points do not change, since these points are far from the perestroika region and the integral along the island is small. Hence, the classes of the X_i are the generators, and $\{2X_i = 0\}$ generates all the relations of our group .

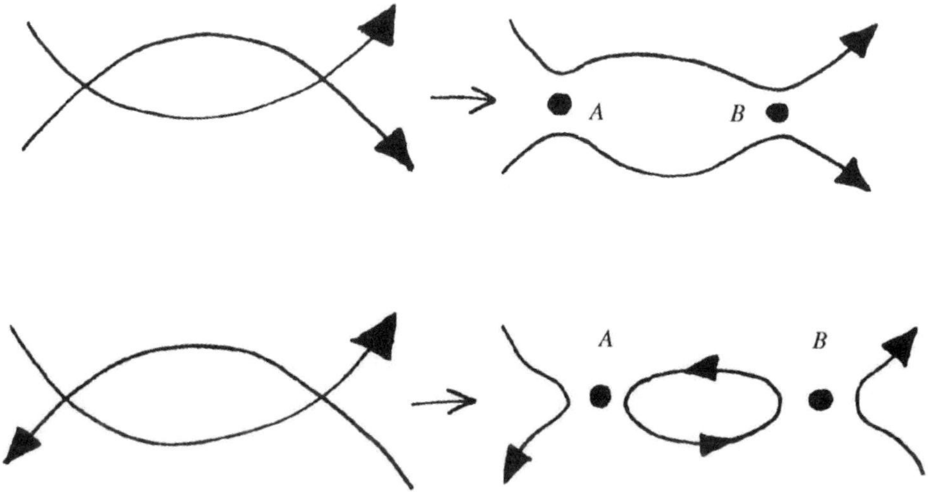

FIGURE 40. The preservation of parity of the number of double points of index i

THEOREM. *The cobordism classes "(−), (∗) permitted" form the group $\mathbb{Z}_2$ generated by X_0. The class of a curve is zero iff it has an even number of double points.*

PROOF. 1°. $2X_0 = 0$ as above.
2°. $X_1 \sim X_0$ (see Figure 41).

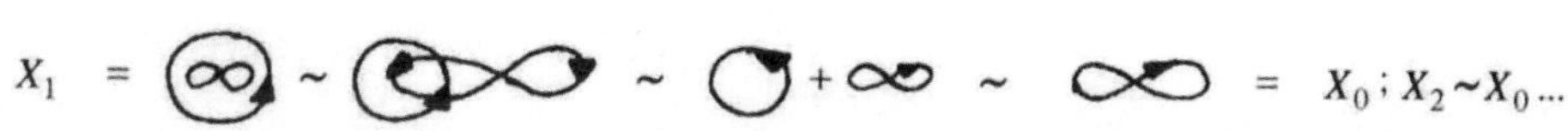

FIGURE 41. Calculation of the cobordism group "(−), (∗) permitted"

3°. The parity of the number of double points is an invariant of all permitted perestroikas.

THEOREM. *The cobordisms classes "(+), (∗) permitted" form a group $\mathbb{Z}_2$ (generator ∞, the class is zero iff the number of double points is even).*

PROOF. See Figure 42.

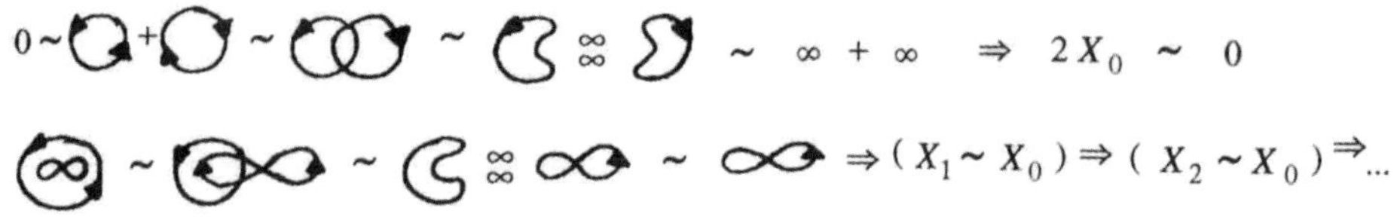

FIGURE 42. Calculation of the cobordism group "(+), (∗) permitted"

THEOREM. *The increment D of the class of a curve in the group $\mathbb{Z}_2^\infty$ of the theory "(+), (−) permitted" under the crossing of a triple point has the following form*

$$D = X_j + X_{j+1},$$

where the pair $(j, j+1)$ is

$$(j, j+1) = (c_{\text{old}} + c_{\text{new}} \pm 1)/2,$$

and where c_{old} and c_{new} are the indices of the curve with respect to the center of the vanishing triangle before and after crossing the triple point.

Examples are shown in Figure 43.

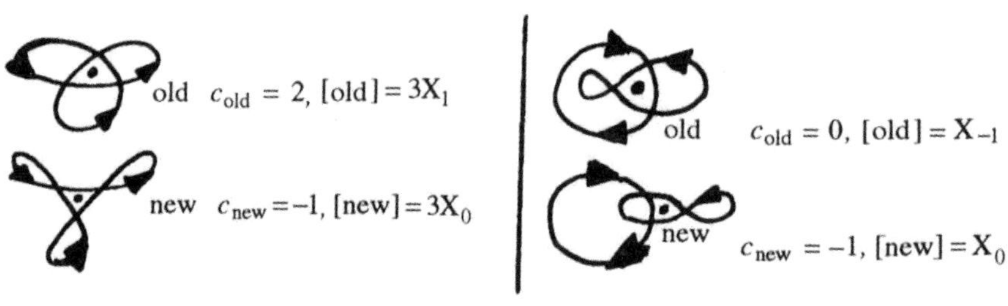

FIGURE 43. The triple point crossing

REMARK. In the semigroup of the theory "everything forbidden" the formula for the increment D has one of the following forms:

A) *case of order* 3 *symmetry* (all angles $120°$):

$$D = 3X_{i_{\text{new}}} - 3X_{i_{\text{old}}},$$

where

$$i_{\text{new}} = \frac{2c_{\text{new}} + c_{\text{old}}}{3}, \qquad i_{\text{old}} = \frac{c_{\text{new}} + 2c_{\text{old}}}{3}$$

(in this case $i_{\text{new}} - i_{\text{old}} = (c_{\text{new}} - c_{\text{old}})/3 = \pm 1$.

B) *case of order* 2 *symmetry* (angles 60°, 60°, 120°):

$$D = X_{c_{\text{new}}} - X_{c_{\text{old}}}$$

(in this case $i_{\text{new}} - i_{\text{old}} = \pm 1$, $c_{\text{new}} - c_{\text{old}} = \pm 1$).

PROOF (of the Remark and hence of the Theorem).

Case A. 1°. Consider the parts of the old and new curves near the triple crossing. The results of the isolation of the three double points are shown in Figure 44.

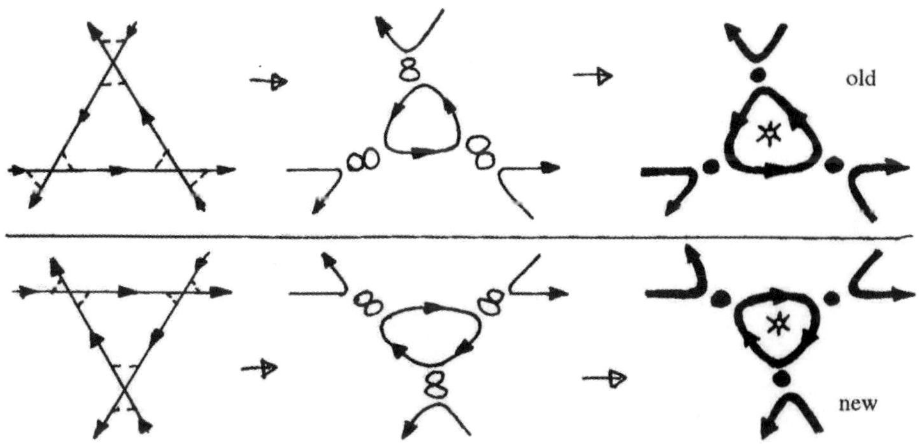

FIGURE 44. Calculation of the increment of the class of the curve under a crossing of a triple point with 3-symmetry

The indices of all the three double points of the old curve with respect to the curve obtained after isolation are equal. Denote this common index by i_c (the index of the center). Then the index of the center, with respect to the complete old curve, is equal to

$$c_{\text{old}} = i_c + a,$$

where a is the index of the vanishing triangle oriented by the directions of the sides ($a = -1$ in Figure 44).

2°. To obtain the new curve one must *subtract* the above oriented triangle from the old curve, in the sense that

$$i_{\text{new}} = i_c - a,$$

where i_{new} is the index of the curve obtained from the new curve by isolation with respect to any of its 3 double points.

3°. The orientation of the new vanishing triangle (by the direction of its sides) is opposite to that of the old one. Thus the index of the center, with respect to the complete new curve, is

$$c_{\text{new}} = i_{\text{new}} - a.$$

4°. By definition, $D = 3(X_{i_{\text{new}}} - X_{i_{\text{old}}})$, since the triple point crossing does not act on the indices of double points distant from the triple point crossing. Thus we obtain

$$c_{\text{new}} - c_{\text{old}} = i_{\text{new}} - i_{\text{old}} - 2a = -3a,$$

which gives the formula of the Remark for the case A.

Case B. We repeat the same reasoning. The results of the isolation of the double points on the old and on the new curve are shown in Figure 45.

In this case the indices of the three double points (p, q, r) of the old curve and of the three double points (u, v, w) of the new one are all different.

We deduce from Figure 45 the formulas for the indices of the center with respect to the old and to the new curves

$$\begin{aligned}
c_{\text{old}} &= i_{\text{old}}(p) = i_{\text{old}}(q) = i_{\text{old}}(r) + 1 \\
c_{\text{new}} &= i_{\text{new}}(v) = i_{\text{new}}(u) = i_{\text{new}}(w) - 1 \\
i_{\text{old}}(r) &= i_{\text{new}}(v) = i_{\text{new}}(u) \\
i_{\text{old}}(p) &= i_{\text{new}}(q) \qquad = i_{\text{new}}(w) - 1.
\end{aligned}$$

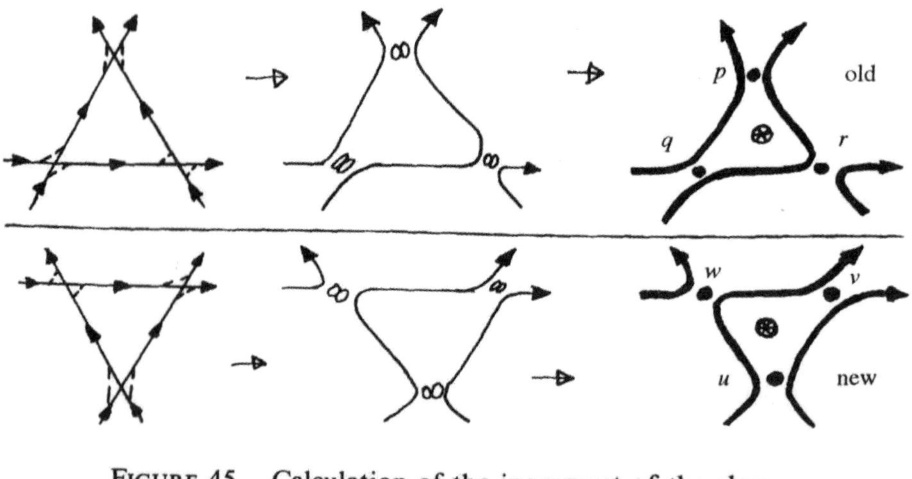

FIGURE 45. Calculation of the increment of the class of the curve under a crossing of a triple point without 3-symmetry

The indices of other points remain unchanged under the triple point crossing, since the integrals are continuous. Thus we obtain the increment of the class of the curve under the crossing

$$d = X_{i_{\text{new}}(w)} + X_{i_{\text{new}}(v)} + X_{i_{\text{new}}(u)} - X_{i_{\text{old}}(p)} - X_{i_{\text{old}}(q)} - X_{i_{\text{old}}(r)} = X_{c_{\text{new}}} - X_{c_{\text{old}}}.$$

Theorem. *The semigroup H of cobordism classes "only $(*)$ permitted" has generators X_i $(i \in \mathbb{Z})$ and relations*

$$3X_i = 3X_j\,,\ 2X_i + X_{i+1} = X_i + 2X_{i+1}.$$

Proof. To prove the relations, we consider Figure 46.

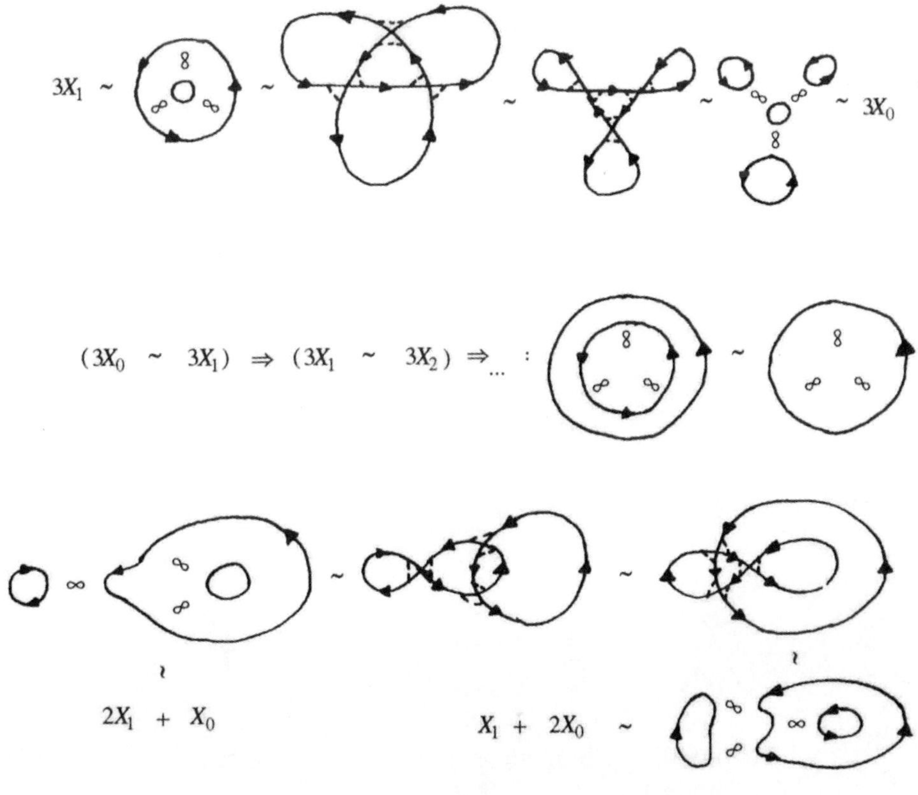

Figure 46. The relations of the semigroup H

To prove that the list of relations is complete, we consider any triple point crossing and compare the two curves K_{old} and K_{new} obtained from the old curve C_{old} and the new curve C_{new} by isolation and elimination of the three double points. The reasoning of the preceding proof gives the following cobordisms (Figure 47).

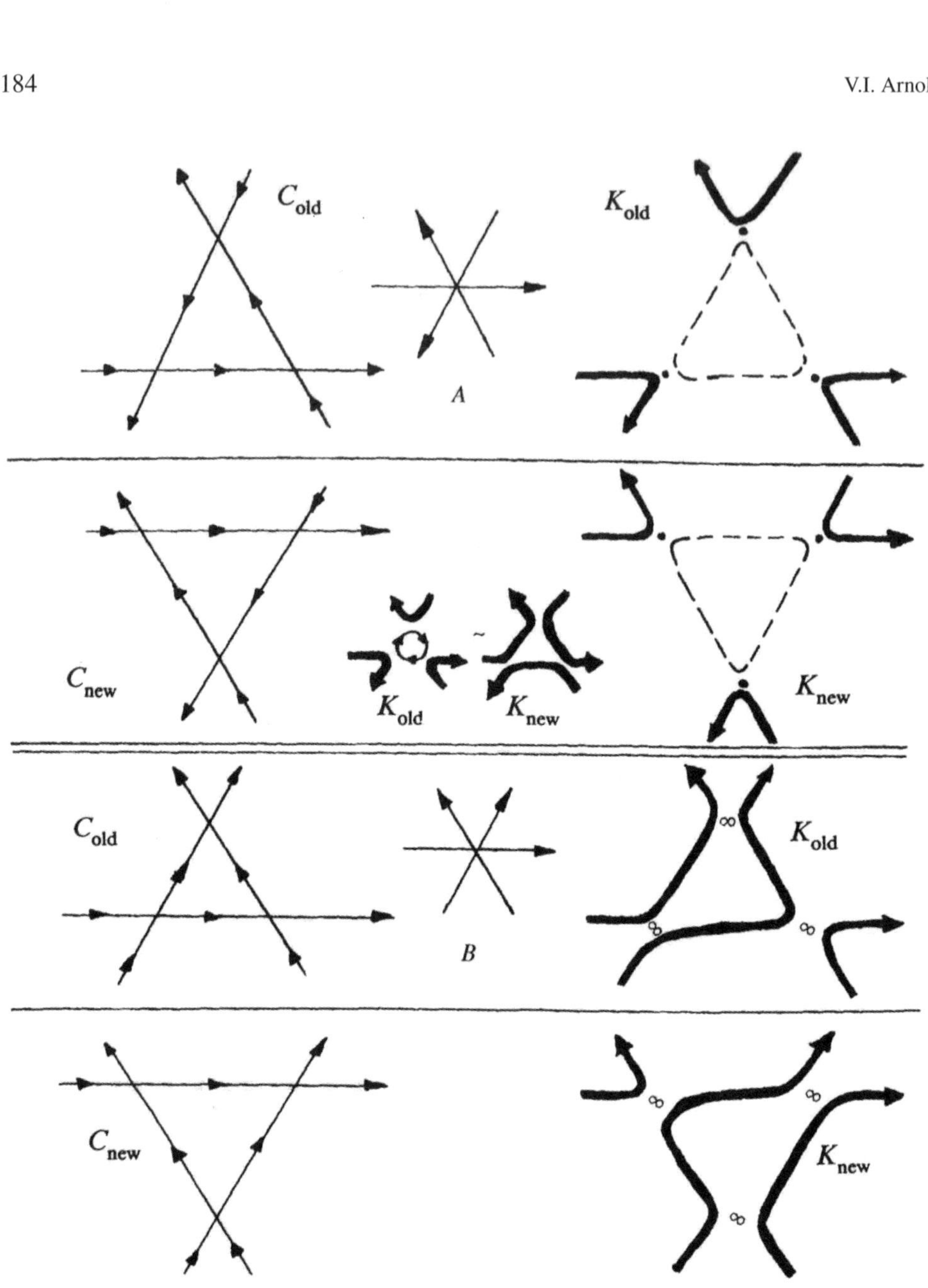

FIGURE 47. Completeness of the list of relations

Case A. $C_{old} \sim K_{old} + 3X_{i_{old}}$, $C_{new} \sim K_{new} + 3X_{i_{new}}$. Since $K_{old} \sim K_{new}$, the cobordism $C_{old} \sim C_{new}$ follows from $3X_{i_{old}} \sim 3X_{i_{new}}$.

Case B. $C_{old} \sim K_{old} + 2X_i + X_{i\pm1}$, $C_{new} \sim K_{new} + X_i + 2X_{i\pm1}$. Since $K_{old} \sim K_{new}$, the cobordism $C_{old} \sim C_{new}$ follows from $2X_i + X_{i\pm1} \sim X_i + 2X_{i\pm1}$. Hence all the relations are generated by those of the theorem, which is thus proved.

Theorem. *The cobordism classes of the theory "only* $(-)$ *permitted" form the same group* $\mathbb{Z}_2^\infty$ *as that of the theory "*$(+)$ *and* $(-)$ *permitted".*

Proof. The perestroikas of Figure 38 and 39 were either Morse or *inverse* self-tangency perestroikas, direct self-tangency perestroikas were not used.

Theorem. *The cobordism classes of the theory "only* $(+)$ *permitted" form the group* $\mathbb{Z}_2$ *(only the parity of the number of double points counts).*

Proof. We deduce $2X_0 = 0$ from Figure 42, then $2X_i = 0$ from Figure 39; the nontriviality follows from Figure 40 (the first case).

We have thus calculated the seven (semi-) groups of curve cobordisms. The eighth group ("everything is permitted") was calculated earlier [3].

§9. The long curves

Definition. A *long curve* is an immersion of a line into the plane which differs from the standard embedding of the x-axis only in a bounded domain (Figure 48).

All the preceding theory can be repeated for these curves. They are better objects than immersions of the circle for two reasons:

(1) the "addition" (which is associative but in general noncommutative) is well defined, and the index (the number of revolutions of the tangent vector) is additive.

(2) the space of long curves of a given combinatorial type is contractible.

The combinatorial type of a generic long curve is defined by its oriented Vassiliev diagram. The (nonoriented) Vassiliev diagram consists of a set of arcs in the upper halfplane, connecting the preimages of the double points belonging to the boundary. The orientations are defined in the following way.

A generic long curve may be constructed step by step, adding the segments joining two subsequent preimages of double points. At each step the new segment must join the given initial point with the given final point, which belongs either to a simply-connected domain or to its boundary (to which also the initial point belongs). This simply-connected domain is one of the components of the complement of the part of the curve constructed earlier. The last point lies inside this domain if it is the double point visited the first time or on its boundary if it is the second visit.

The possible choices of the last point and of the path form a contractible set, unless the endpoint belongs to a segment of the boundary to which the above domain is adjacent from both sides (which may happen). In this case, the space of possible choices consists of two contractible components and they are distinguished by the "orientation" supplementary structure. A similar structure exists for the strata of any codimension in the discriminant.

Thus the combinatorics of the long curves described by the oriented diagrams and by their adjacencies provide us with a variant of a cell decomposition of the complement of the discriminant and of the stratification of the

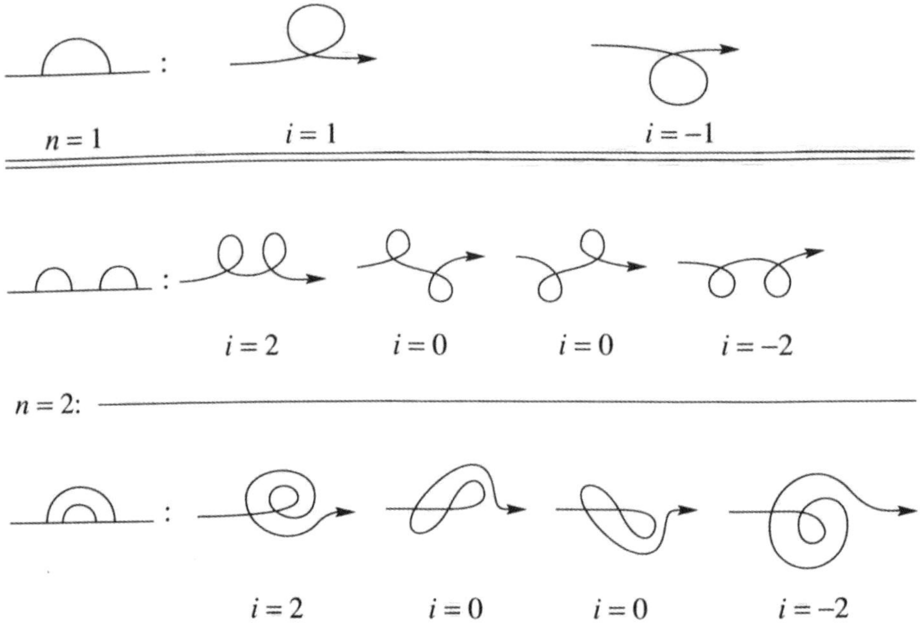

FIGURE 48. Long curves with one and with two selfintersection points

discriminant. In principle, our questions about the topology of the complement of different strata of the discriminant might be answered in terms of the algebra of oriented diagrams. However, the corresponding invariants are yet to be calculated and we present below only the first steps in this direction.

By "the discriminant" below we mean the hypersurface formed by immersions with direct self-tangencies in the space of long curves.

DEFINITION. A *Vassiliev invariant* of order one is a function of a long curve which is constant on the components of the complement of the discriminant and whose jump under the positive crossing of the discriminant does not change under the crossing of the strata of codimension one in the discriminant.

THEOREM. *The additive Vassiliev invariants of order one of long curves are uniquely defined (up to addition of a zero order invariant proportional to the index) by the collection* $\{f_i, i \in \mathbb{Z}\}$ *of their jumps at the crossings of the discriminants (in the direction of the increase of the number of double points) at the special long curves* ℓ_i *of Figure* 49.

The numbers f_i *are independent and may be chosen arbitrarily.*

REMARK. This theorem shows that the theory of plane curves is much

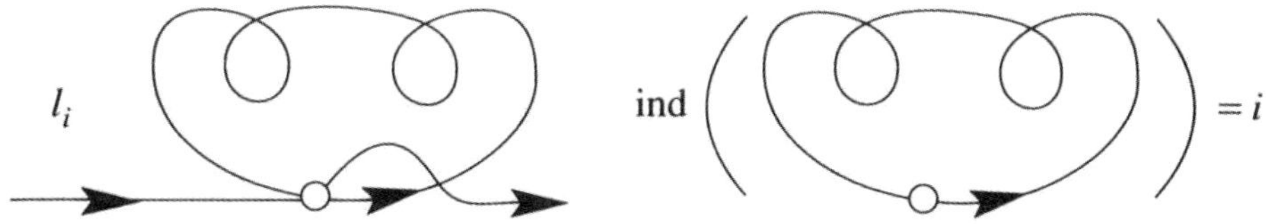

FIGURE 49. Special long curve ℓ_i having a direct self-tangency of index i

more complicated than the theory of knots, where the first order Vassiliev invariants are trivial.

PROOF. 1°. The first order condition implies that the jump might only depend on the following data:

(a) the indices j and i of the long and of the short branches of the curve with a direct self-tangency (the index j vanishes for the standard curve ℓ_i);

(b) the orientation of the plane defined by the pair (short branch, long branch) leaving the self-tangency point.

2°. Additivity implies that the jump cannot depend on j.

3°. The independence of the jump from the orientation (b) is implied by existence of the perestroika of codimension one shown in Figure 50.

This perestroika does not change the index i of the short component but changes the orientation (b).

4°. The existence of an invariant having jump 1 at the perestroikas of index i and jump 0 at the perestroikas of all other indices follows from the fact that the long curves with a direct self-tangency of index i form a codimension 1 cycle in the simply connected space of long curves of given index.

5°. To make this invariant additive, it is sufficient to add a constant depending only on the index.

Suppose that we wish to make additive the invariant F whose jump under the positive crossing of the discriminant at a direct self-tangency of index 0 is equal to a (above, a is either 1 or 0).

We choose the values of the invariant at the long curves K_i of index i shown in Figure 51 to be

$$F(K_i) = -|i|a/2.$$

Our choice implies the relation

$$F(K_i + K_j) = F(K_i) + F(K_j).$$

This relation is evident for i and j of the same sign.

The curve $K_1 + K_{-1}$ can be transformed into K_0 by one positive direct self-tangency perestroika of index 0 (Figure 52).

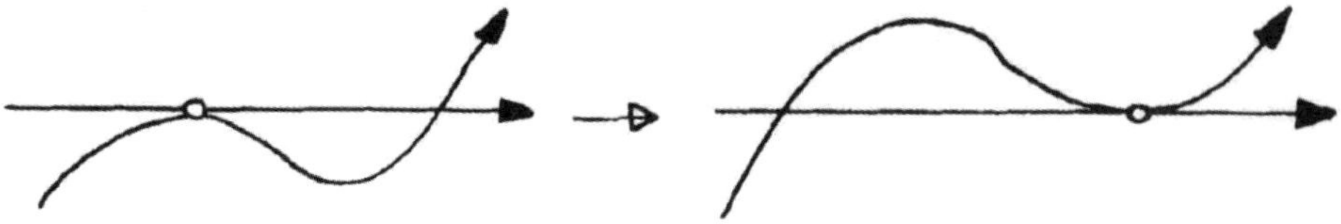

FIGURE 50. Crossing a stratum of codimension one in the discriminant

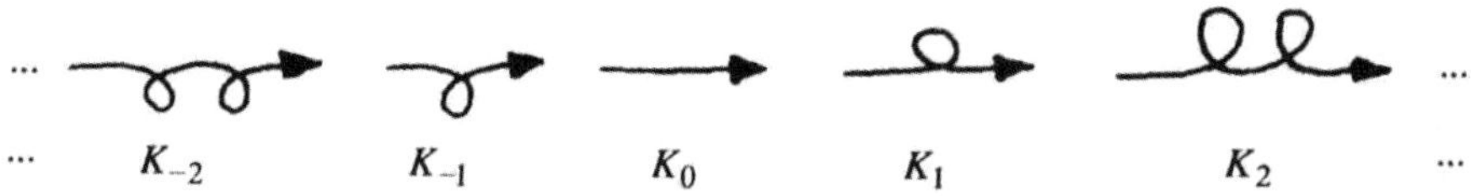

FIGURE 51. The long standard curves K_i

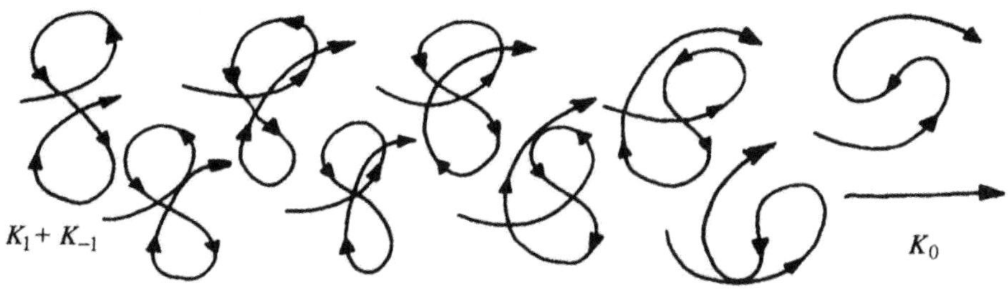

FIGURE 52. The index 0 perestroika of $K_1 + K_{-1}$ into K_0

Hence $F(K_1 + K_{-1}) = F(K_0) - a = -a$, which is also the value of $F(K_1) +$ (K_{-1}).

Iterating this argument, we see that $F(K_i + K_j) = F(K_i) + F(K_j)$ for any and j.

Any long curve $\tilde{K}_i$ of index i can be connected by a generic path with in the space of long curves. Connecting also $\tilde{K}_j$ to K_j, we construct a th connecting a long curve $\tilde{K}_i + \tilde{K}_j$ to $K_i + K_j$.

The increment of F on this path is equal to the sum of the increments of F on the paths from $\tilde{K}_i$ to K_i and from $\tilde{K}_j$ to K_j. Hence we have

$$F(\tilde{K}_i + \tilde{K}_j) - F(K_i + K_j) = F(\tilde{K}_i) - F(K_i) + F(\tilde{K}_j) - F(K_j).$$

Thus the additivity of F on the special curves K_i implies its additivity on any two curves: $F(\tilde{K}_i + \tilde{K}_j) = F(\tilde{K}_i) + F(\tilde{K}_j)$.

6°. The uniqueness of an additive invariant with given jumps: the difference of two such invariants has no jumps and hence is a function of the index. Additivity implies that this function is linear.

7°. The basic invariants that we have constructed have the nice property that their values on a fixed curve are almost all zero (only a finite number of nonzero values is possible). Indeed, any curve can be obtained from the basic curve by a finite chain of perestroikas changing a finite number of invariants, while for the basic curves all the invariants but a finite number vanish.

It follows that *any* infinite formal linear combination of the basic invariants has a well-defined value on each curve and hence represents a genuine invariant. The theorem is thus proved.

§10. Tables of curves

Here I discuss the classification of generic immersions of a circle into a plane or a sphere up to diffeomorphisms of the plane or of the sphere and of the circle. These diffeomorphisms may preserve, or not, the orientation of the plane (sphere) and that of the circle. Hence, there exist four different classification problems. The calculations (which are rather long and for which I am very grateful to F. Aicardi, SISSA Trieste) lead to the following numbers of classes.

Plane closed curves. The number of types (taking or not taking into account the orientations of the plane and of the circle) of generic immersions with n double points are given by the following table:

oriented	$n = 0$	1	2	3	4	5
$\mathbb{R}^2, S^1$	2	3	10	39	204	1262
$\mathbb{R}^2$	1	2	5	21	102	639
S^1	1	2	5	21	102	640
–	1	2	5	20	82	435

Remark 1. The standard orientation reversing involutions σ of the circle and Σ of the plane act on the space of immersions. Some immersion classes are, and some are not, invariant under the action of these involutions.

Denote by τ and by T the standard antipodal involutions of the circle and of the plane. An immersion $f: S^1 \to \mathbb{R}^2$ is *symmetric* with respect to the orientation reversals (of the circle and of the plane) if

$$f\sigma = \Sigma f \text{ (or, equivalently, } \Sigma f\sigma = f).$$

If the index of f vanishes, there are two other possibilities for the symmetry:

$$f\sigma = Tf \text{ (or, equivalently, } Tf\sigma = f),$$

$$\Sigma f = f\tau \text{ (or, equivalently, } \Sigma f\tau = f).$$

An immersion is *asymmetric, symmetric,* or *supersymmetric* if the number of different classes among the classes of the immersions

$$f, \quad \Sigma f\sigma, \quad Tf\sigma, \quad \Sigma f\tau$$

is 4, 2, or 1.

EXAMPLE. The standard "8" immersion is supersymmetric. The standard "0" immersion is symmetric ($\Sigma f\sigma = f$, $Tf\sigma = \Sigma f\tau$). If an immersion has two of the three symmetries, it has the third and is supersymmetric.

An asymmetric curve contributes the column (4, 2, 2, 1) to the table of types. A supersymmetric curve contributes (1, 1, 1, 1). A symmetric curve of type $\Sigma f\sigma = f$ contributes (2, 1, 1, 1). The symmetric curves of types $Tf\sigma = f$ and $\Sigma f\tau = f$ contribute (2, 2, 1, 1) and (2, 1, 2, 1), respectively.

REMARK 2. If a class of a curve is transformed into itself by one of the three involutions $((\Sigma, \sigma), (T, \sigma), (\Sigma, t))$, then there exists an immersion of this class which is exactly invariant under this involution.

This is also true for the equivalence classes defined by any of the three strata of the discriminant studied above. These three strata are invariant under the involutions. Hence these involutions act on the set of classes that are the components of the complement. Whenever a class is invariant, it contains an invariant point.

It seems that this observation (not too difficult to prove for generic curves with only double points) is true in the very general situation of finite (or even compact) group actions on the stratified spaces of subvarieties (or on the stratified spaces of mappings).

DISTRIBUTION OF ORIENTED CURVES ON THE ORIENTED PLANE ACCORDING TO THEIR INDICES

					1		1					
				1		1		1				
			2		3		3		2			
		4		10		11		10		4		
	10		35		57		57		35		10	
26		133		290		364		290		133		26

For instance, the number of types of curves having 5 double points and index 6 is equal to 26, of those of index 4, the number is equal to 133, and so on.

oriented	$n=0$	1	2	3	4	5
S^2, S^1	1	1	3	9	37	182
S^2	1	1	2	6	21	99
S^1	1	1	2	6	21	97
–	1	1	2	6	19	76

Curves on S^2. The number of types

DEFINITION. A circle immersion is *reducible* if some double point cuts the image into two nonintersecting loops.

The plane extremal curves are in this sense *completely reducible.*

Irreducible spherical curves with $n \leq 7$ selfintersection points. The following table gives the number N of types (no orientations taken into account).

n	0	1	2	3	4	5	6	7
N	1	0	0	1	1	2	5	9

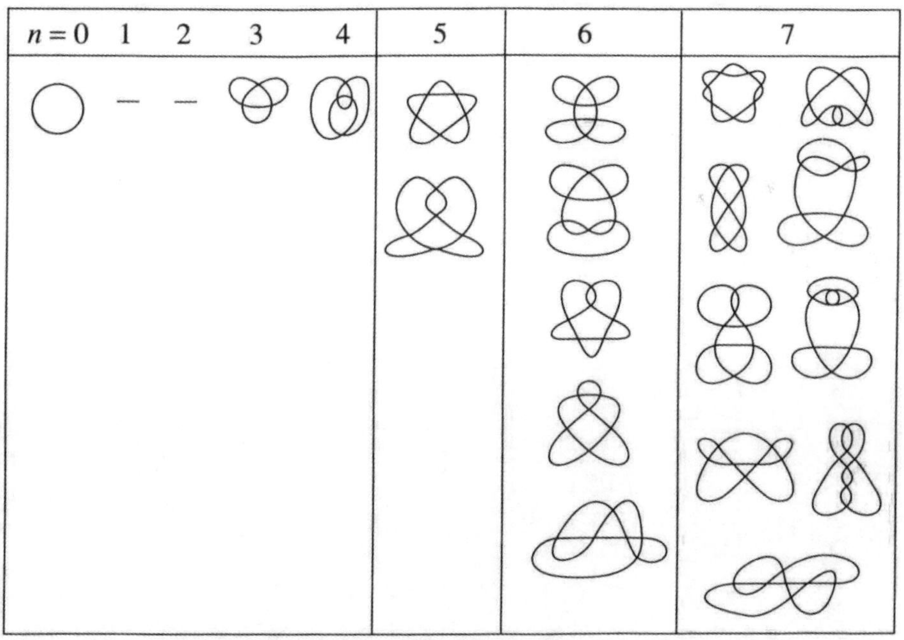

FIGURE 53. The irreducible spherical curves

The curves themselves are shown in Figure 53.

The classification of spherical curves is simpler, since the number of spherical curves with n double points is smaller than that of the plane curve (some of which are conformally equivalent). To obtain all the plane curves from the list of spherical curves it suffices to place the point ∞ inside all the components of the complement to the spherical curve.

The classifications of the spherical curves having $n \leq 4$ and $n = 5$ double points is shown in Figures 54 and 55. The classification of plane curves with $n \leq 3$ and with $n = 4$ is shown in Figures 56 and 57, where each curve is accompanied by the values of the invariants St, J^+, J^-.

As explained in the introduction, the knowledge of these values provides thousands of theorems on generic immersions, similar to the three examples described in the introduction.

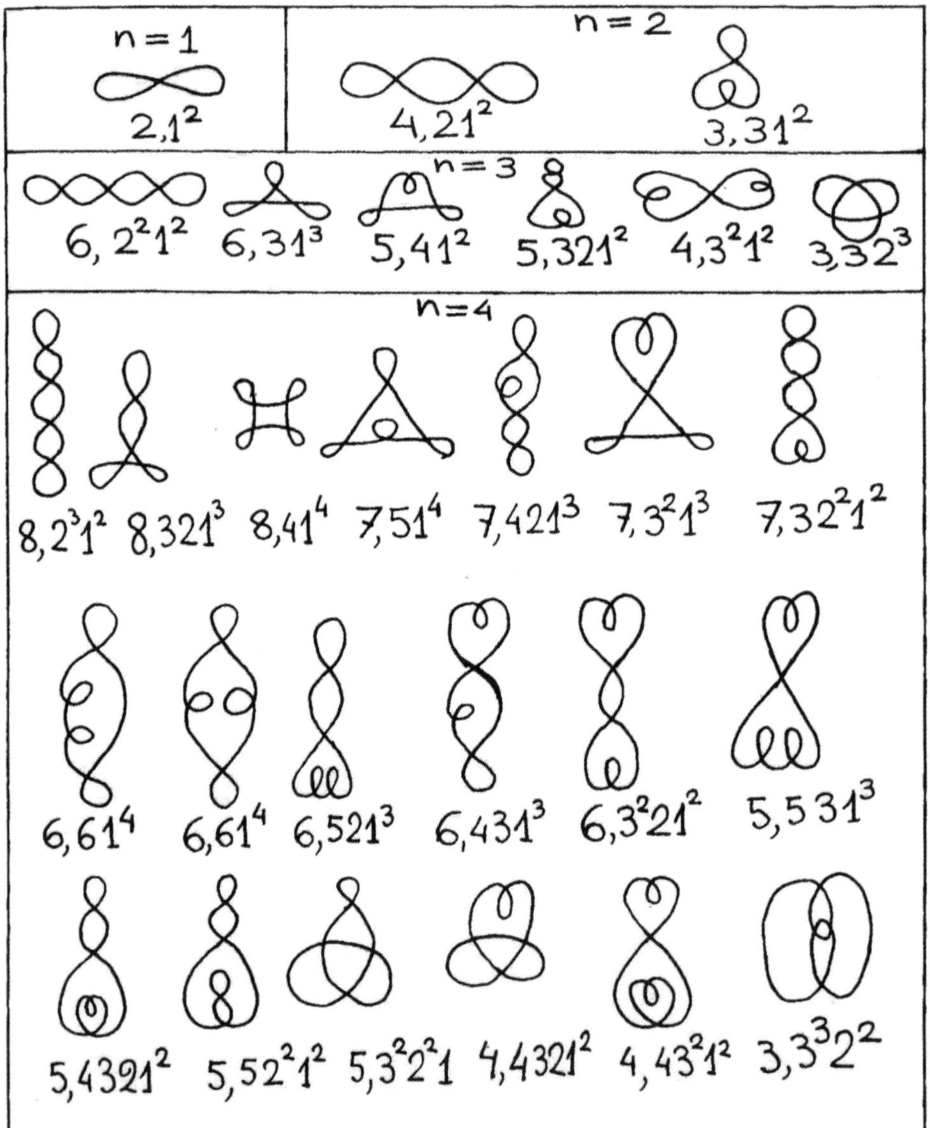

FIGURE 54. Spherical curves with $n \leq 4$ double points

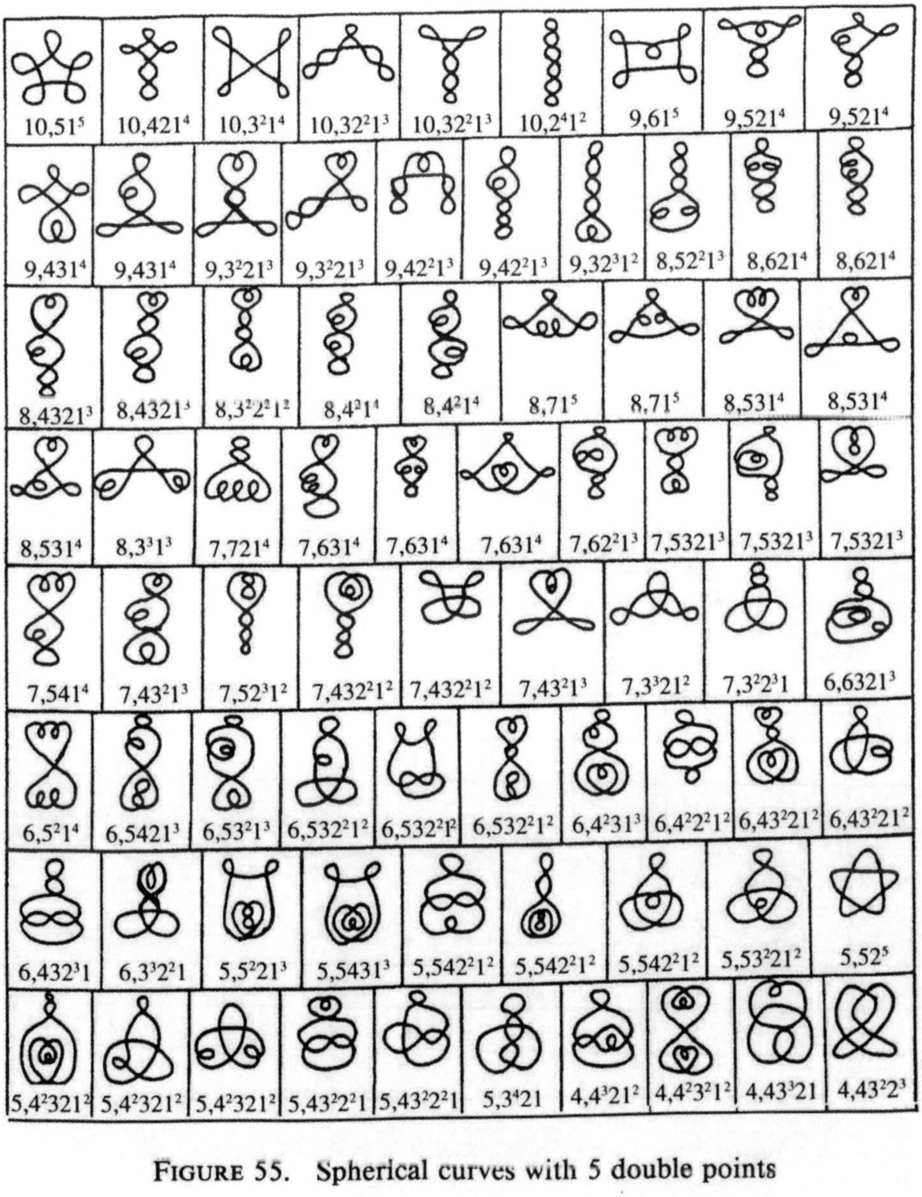

FIGURE 55. Spherical curves with 5 double points

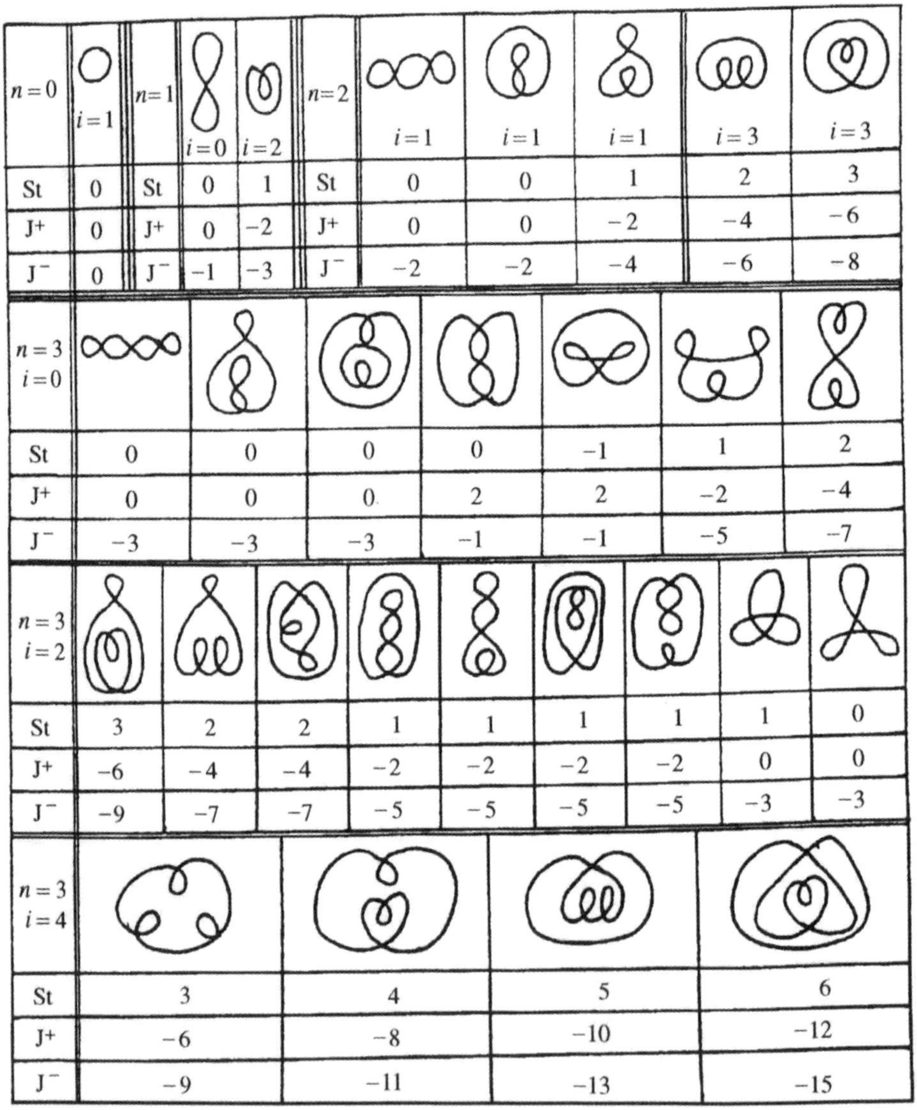

	$n=0$, $i=1$	$n=1$, $i=0$	$n=1$, $i=2$	$n=2$, $i=1$	$n=2$, $i=1$	$n=2$, $i=1$	$n=2$, $i=3$	$n=2$, $i=3$
St	0	0	1	0	0	1	2	3
J^+	0	0	−2	0	0	−2	−4	−6
J^-	0	−1	−3	−2	−2	−4	−6	−8

$n=3$, $i=0$							
St	0	0	0	0	−1	1	2
J^+	0	0	0	2	2	−2	−4
J^-	−3	−3	−3	−1	−1	−5	−7

$n=3$, $i=2$									
St	3	2	2	1	1	1	1	1	0
J^+	−6	−4	−4	−2	−2	−2	−2	0	0
J^-	−9	−7	−7	−5	−5	−5	−5	−3	−3

$n=3$, $i=4$				
St	3	4	5	6
J^+	−6	−8	−10	−12
J^-	−9	−11	−13	−15

FIGURE 56. Plane curves with $n \leq 3$ double points

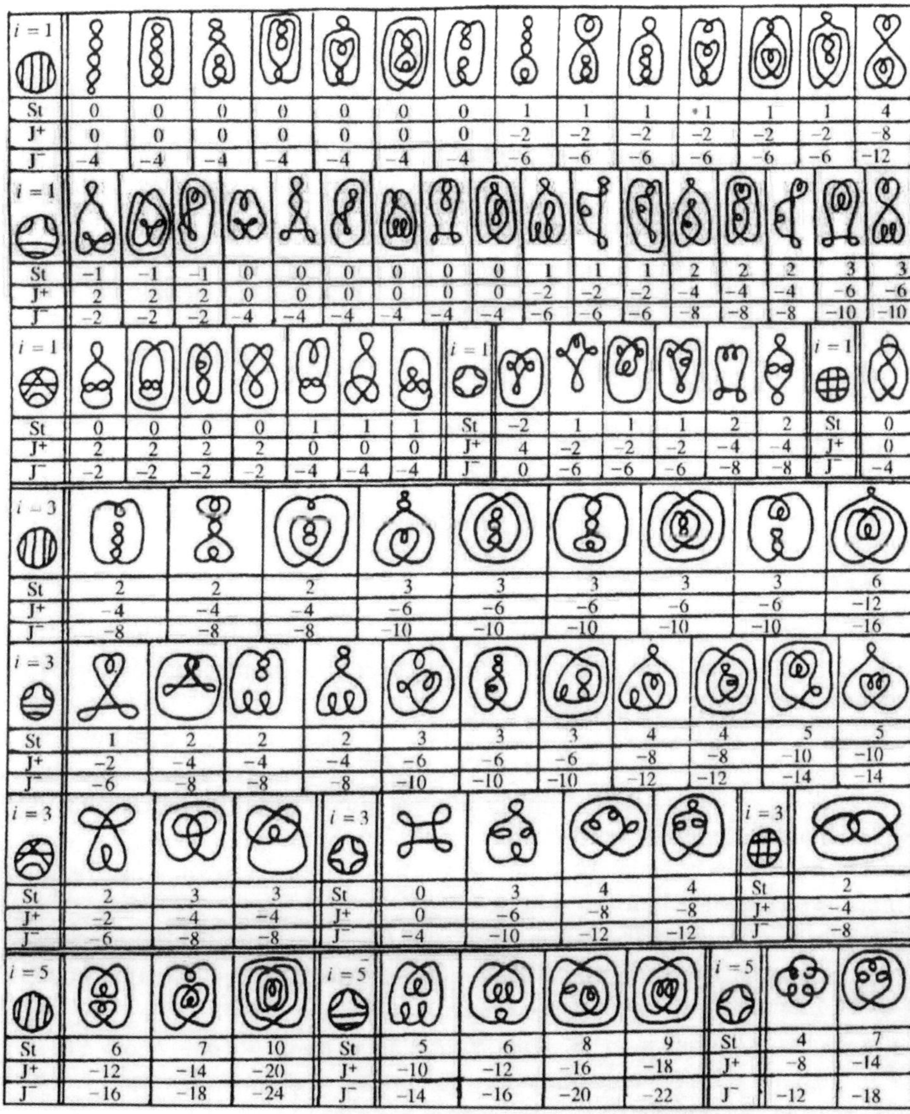

FIGURE 57. Plane curves with $n = 4$ double points

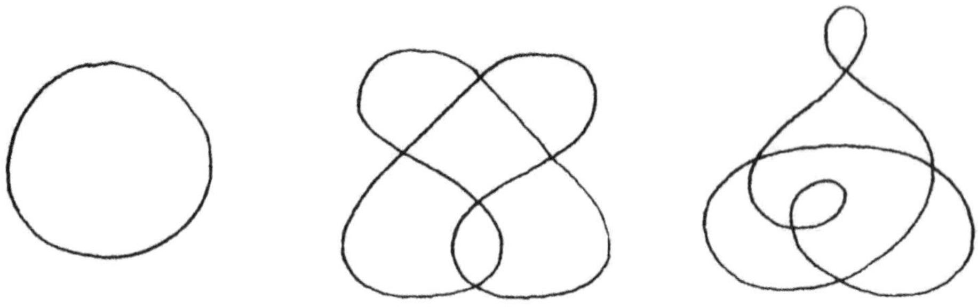

a) b) c)

FIGURE 58. Curves with different type of symmetry on the sphere

Appendix

F. AICARDI

In Figures 59a–g plane curves with $n = 5$ double points are shown. The curves are first classified by their index i and for each value of index by their Gauss diagrams.

The figures represent nonoriented curves in the nonoriented plane, and the index is a nonnegative number. The same curve may represent several classes if one or both orientations are taken into account. The number of classes depends on the type of symmetry of the nonoriented curve.

The symmetry class may be different for a curve in the plane and for the same curve considered on the sphere.

EXAMPLES. A symmetric curve of type $\Sigma f \sigma = f$ in $\mathbb{R}^2$ which is supersymmetric on S^2 is shown in Figure 58a. Figure 58b represents a curve of type $\Sigma f \tau = f$ in $\mathbb{R}^2$ which is sypersymmetric on S^2. The curve in Figure 58c is asymmetric in $\mathbb{R}^2$ and symmetric of type $T f \sigma = f$ on S^2.

For the sphere I have used the notations introduced above for the involutions defining the symmetries on the plane.

For $n \leq 4$ all the symmetric curves are of type $\Sigma f \sigma = f$. For $n = 5$ ($i = 0$) we have the first examples of symmetric curves of both types $T f \sigma = f$ and $\Sigma f \tau = f$. They are marked by σ and by τ respectively in Figure 59a,...,f. The remaining symmetric curves of type $\Sigma f \sigma = f$, which are the majority, are not marked. The supersymmetric curves are marked by x and the asymmetric ones by α.

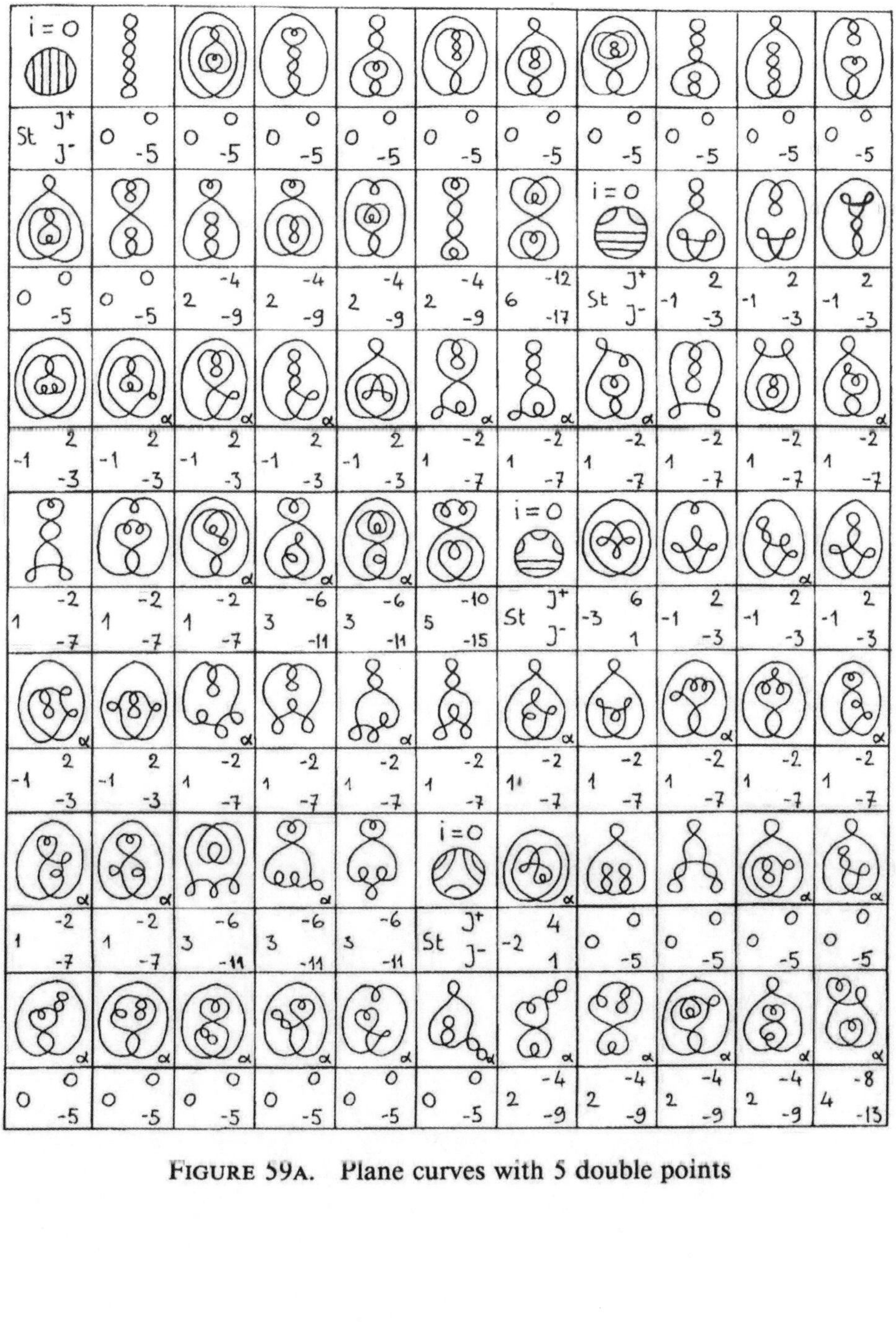

FIGURE 59A. Plane curves with 5 double points

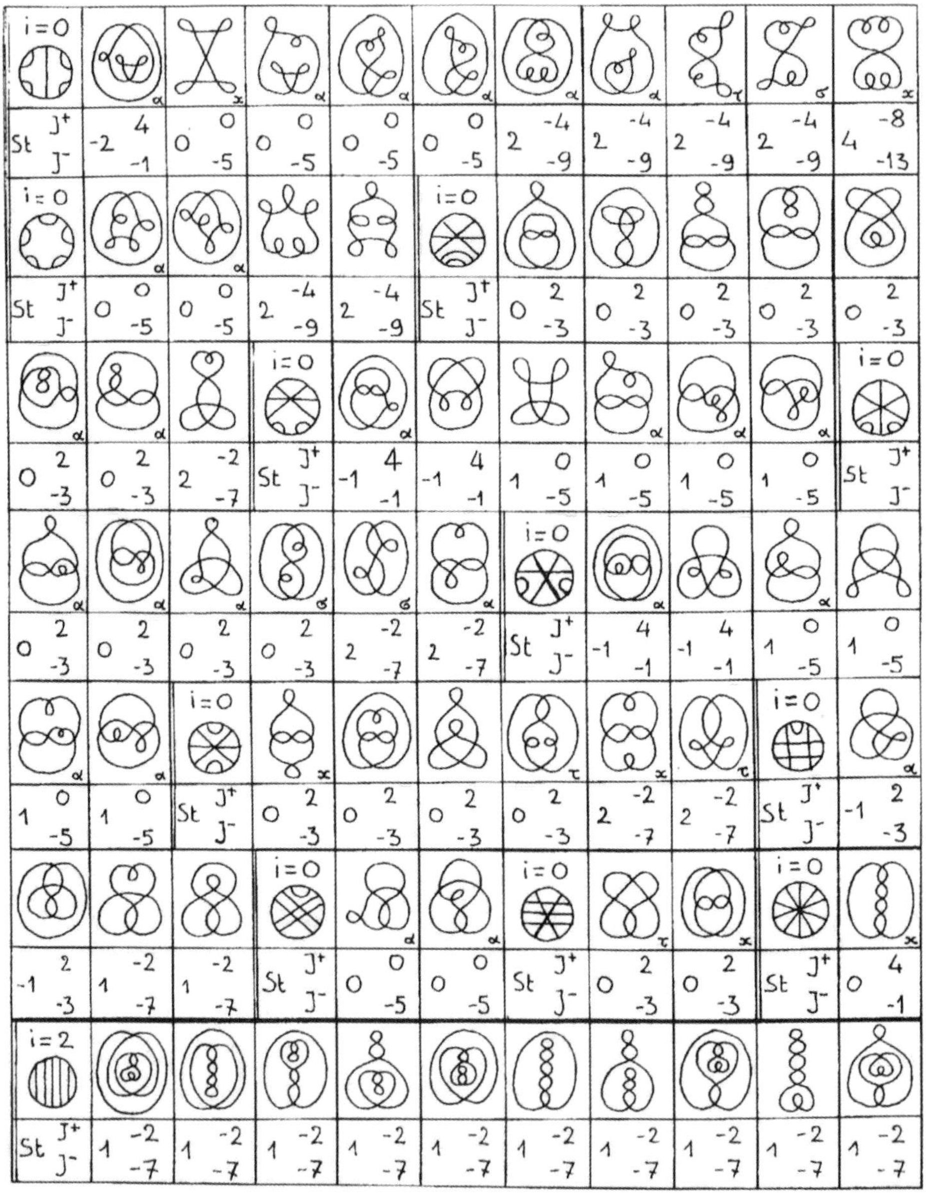

FIGURE 59B. Plane curves with 5 double points

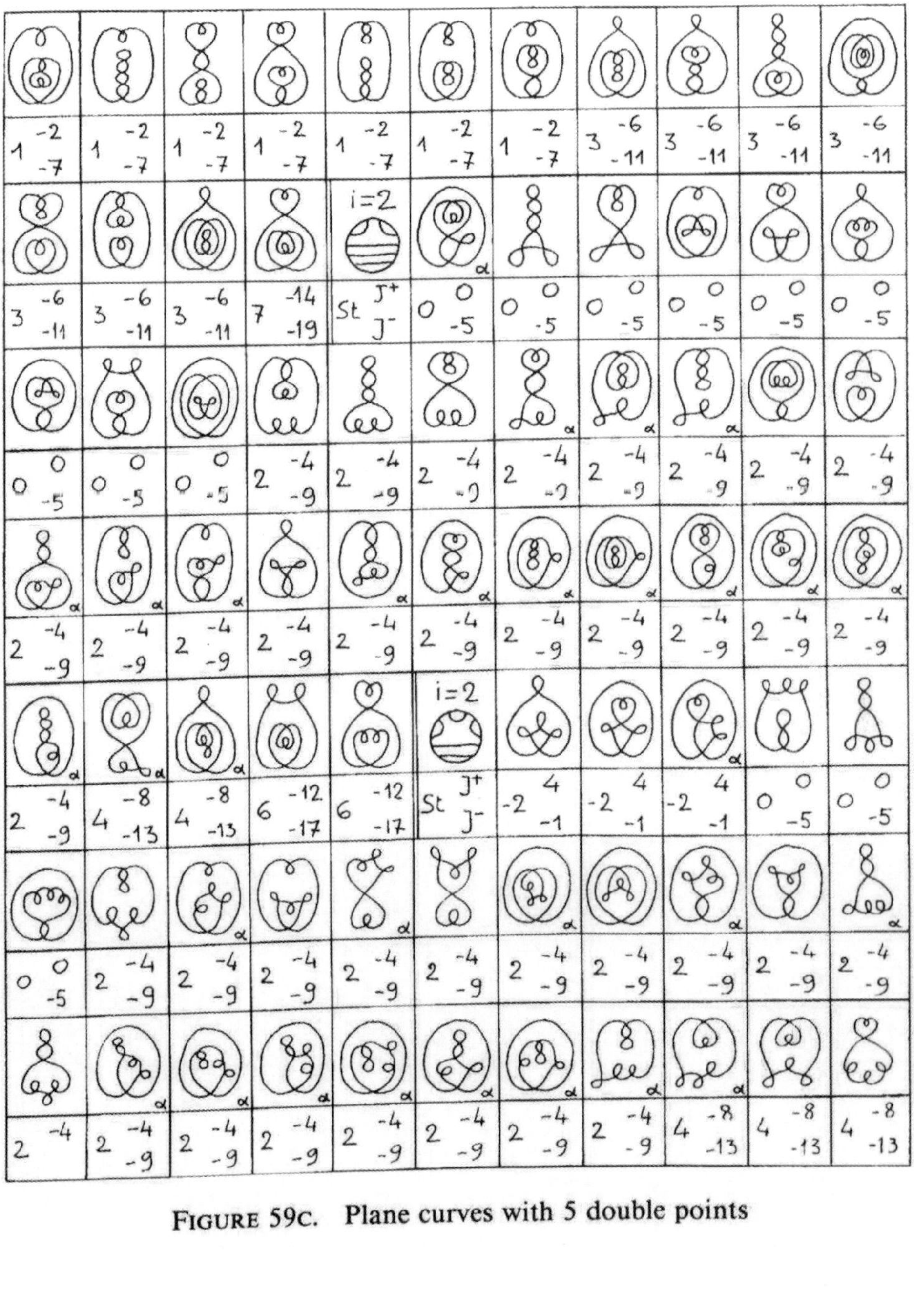

FIGURE 59C. Plane curves with 5 double points

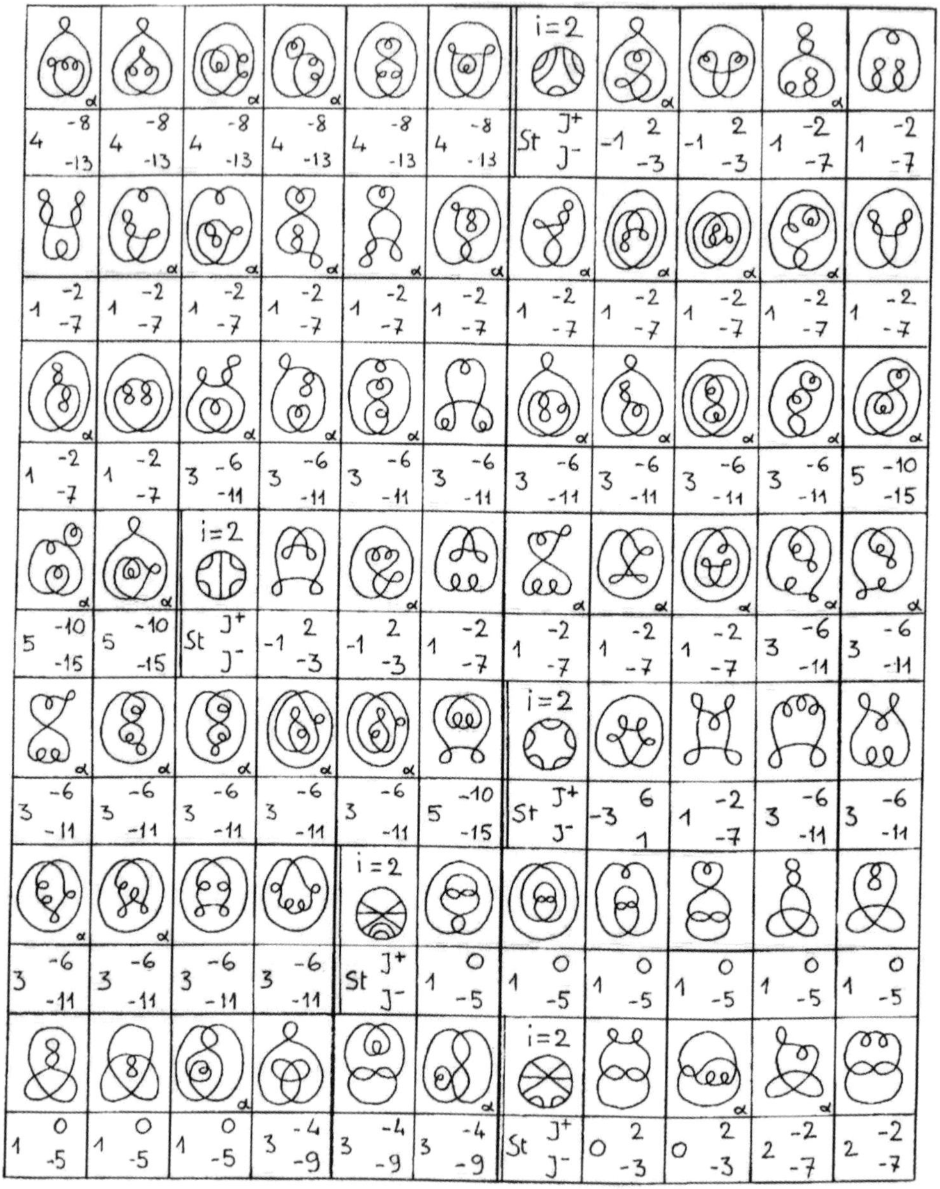

FIGURE 59D. Plane curves with 5 double points

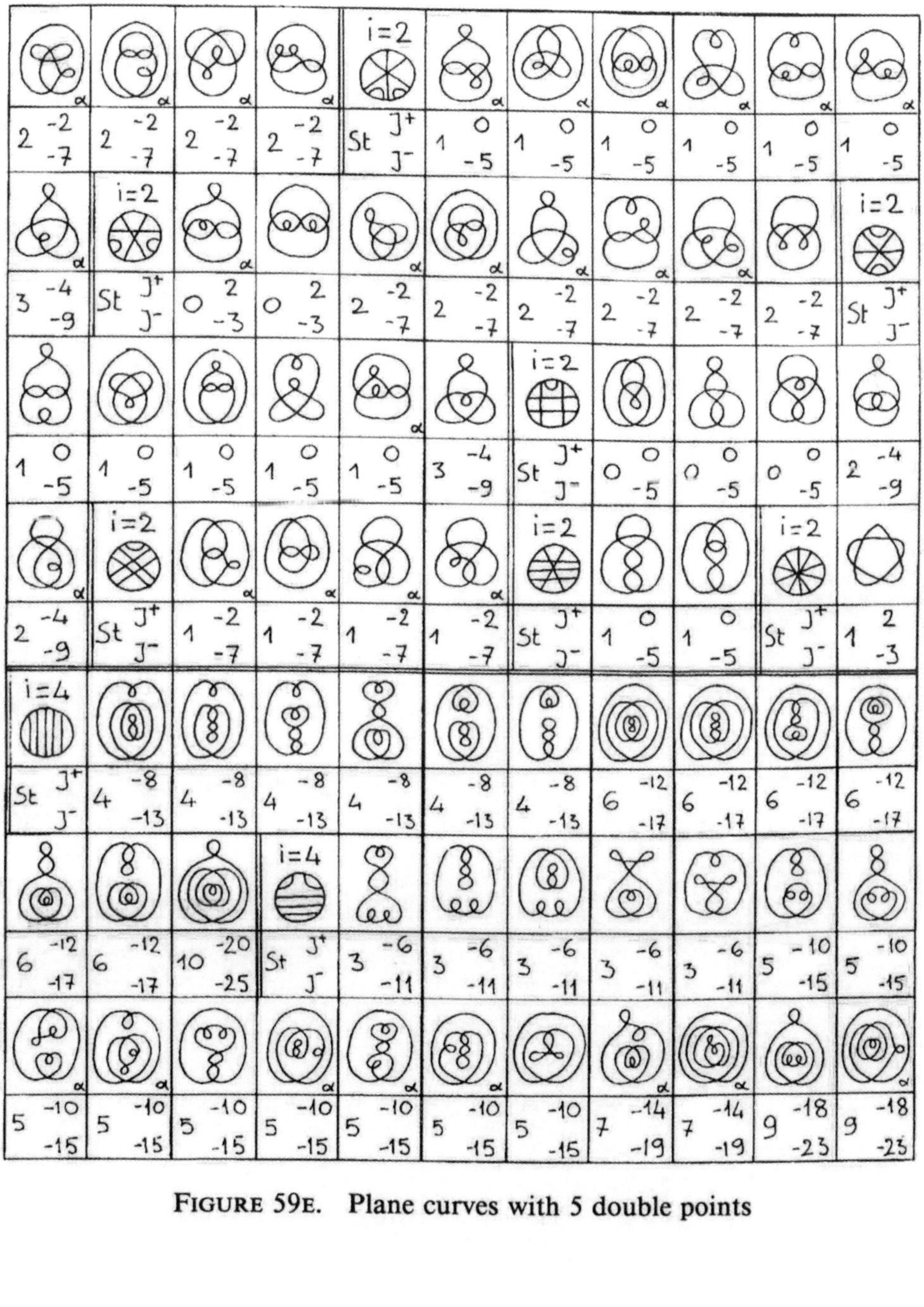

FIGURE 59E. Plane curves with 5 double points

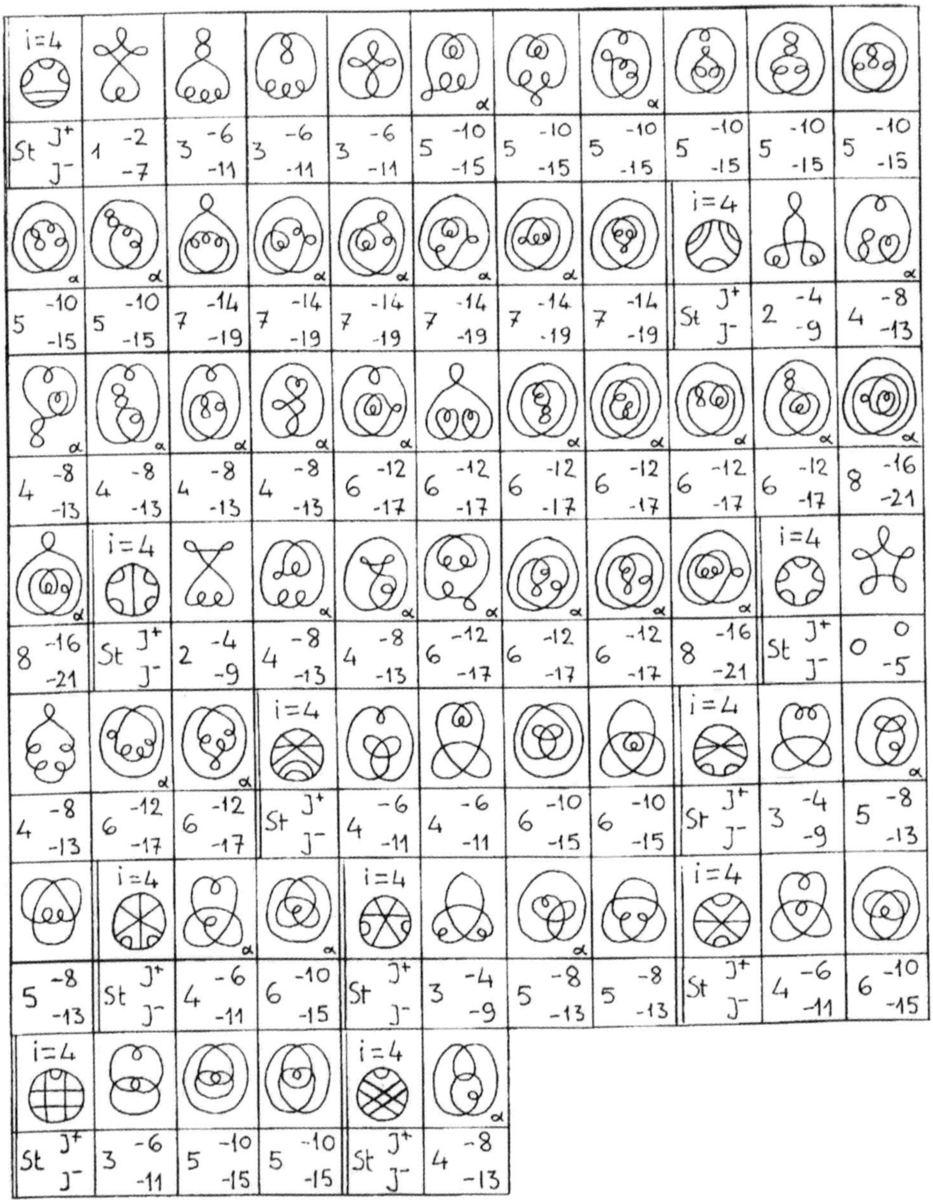

FIGURE 59F. Plane curves with 5 double points

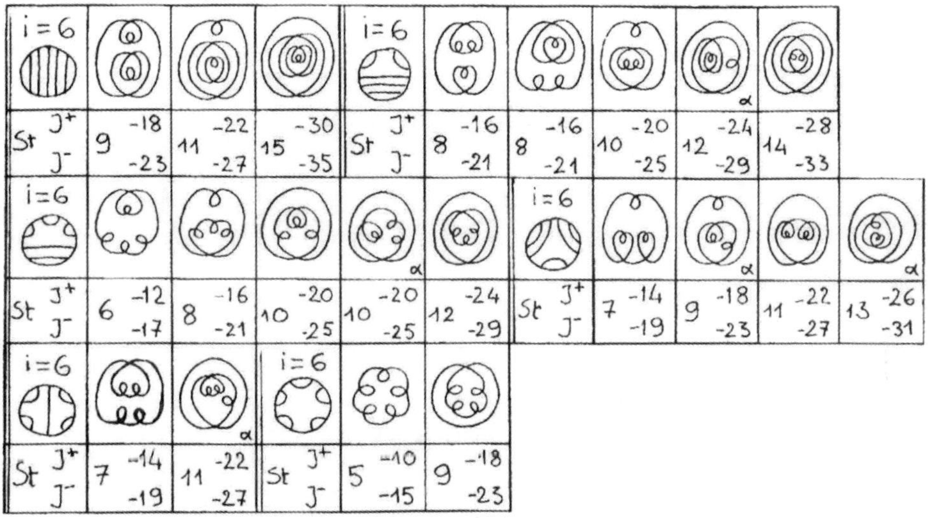

FIGURE 59G. Plane curves with 5 double points

REFERENCES

1. H. Whitney, *On regular closed curves in the plane*, Compositio Math. **4** (1937), 276–284.
2. S. Smale, *A classification of immersions of the 2-sphere*, Trans. Amer. Math. Soc. **90** (1958), 281–290.
3. V. Arnold, *Vassiliev theory of discriminants and knots*, Proceedings of the I Europ. Congress of Math., Paris, 1992.

Invariants and perestroikas of plane fronts

V.I. Arnold

INTRODUCTION

The conventional approach of the singularity theory to the problems of topological classification of objects consists in the following. We consider such a functional space of all objects of a given class—degenerate and nondegenerate—that its topological properties are as simple as possible (it is contractible, as a rule). In this space, the degenerate objects form a subspace, which is usually a hypersurface, i.e., a subset of codimension 1 which divides the complementary space into domains, each containing objects of the same topological type.

Then the classification of the types is investigated by stratifying the hypersurface of degenerate objects by the degeneracy types. In other words, we construct invariants as indices of intersection of paths from one standard object to others with the hypersurface of degenerate objects. This hypersurface should be suitably cooriented at regular points. The coorientations must be compatible near the points of greater degeneracy, and this requires a thorough study of the hierarchy of singularities and their adjacencies.

In [1], I applied this method to study the invariants of immersions of a circle into a plane, i.e., mappings of a circle in a plane with nowhere vanishing derivative.

In this paper, we shall extend certain results of [1] to the fronts of Legendre plane curves, i.e., cooriented plane curves with cusps.

One of the main invariants derived from the indices of intersection with a hypersurface of singular objects in this case occurs to be the Bennequin invariant [3] generalized by Tabachnikov [2]. The relationship of this invariant with the invariants which yield intersection index gives new formulas for the invariants of immersions of a plane circle. In particular, for the invariant, which is called *strangeness* in [1], we obtain an explicit expression as the sum of integrals of standard differential forms over curves.

Initially, these studies were keyed to proving the following hypothesis (which still remains unproven): when a circle cooriented by outer normals is metamorphosed into a circle cooriented by inner normals such that self-tangencies with coinciding coorientations are not admitted, it happen at least four metamorphoses of birth and death of a pair of cusps on the front (on the cooriented curve).

This is a generalization of the four-vertex theorem for a plane curve (which asserts that a closed convex plane curve has at least four curvature extrema points). The metamorphosing fronts in the four-vertex theorem are equidistant curves cooriented by the normals of the curve.

Although the hypothesis formulated above is still not proven, we shall state another generalization of the four-vertex theorem at the end of this paper, which converts this hypothesis into a

Originally published in: *Tr. Mat. Inst. Steklova* **209**, 14–64 (1995), © 1995 Russ. Acad. Sciences
English translation: *Proc. Steklov Inst. Math.* **209**, 11–56 (1995), © 1995 Springer.

theorem in symplectic geometry dealing with a Lagrangian collapse similar to the σ-process: an infinitely degenerate Lagrange map with a caustic consisting only of one point is transformed into a generic map; thereafter the caustic acquires at least four cusps [4].

This theorem yields information, for example, on the singularities of caustics on surfaces close enough to a sphere, for instance, ellipsoids (these singularities were first discovered by Jacobi). Let us examine all the geodesics that start from one point on a surface. The point of intersection of infinitesimally close geodesics is called the *conjugate point* of the initial point along the corresponding geodesic. All primary (secondary, ...) conjugate points form a curve on the surface—the first (second, ...) caustic of the initial point.

Jacobi proved that the first caustic on a convex surface has always a singular point. The theorem about Lagrange collapse implies that there are at least four such singular points on every caustic if the metric does not deviate much from the standard one. The last condition on the proximity of the metric to the standard metric is seemingly superfluous, but this still remains to be proved.

The theorem about Lagrange collapse has undoubtedly many fascinating generalizations: even for infinitesimal perturbations of a standard collapse (corresponding to the standard metric on a multidimensional sphere) we expect specific generalizations of the Sturm theorem to multidimensional case (the four-cusp theorem corresponds to Tabachnikov theorem [5] on the zeros of combinations of the eigenfunctions of a second-order operator on a circle). In particular, considering a two-dimensional sphere instead of a circle, we obtain the symplectic generalizations of the theorem about the umbilical points on convex surfaces; it is not even possible to speculate the answer for higher dimensions. The problem of the desintegration of singularities of a degenerate high-dimensional caustic different from a point—it is an analog of the σ-process over a submanifold seems interesting.

Although we have generalized a few results of [1] to Legendre curves in the contact manifold of cooriented plane elements, other results are yet to be generalized to the Legendre case. This is particularly true of the strangeness invariant, which is the dual of the triple points of a plane curve, Vasil'ev invariants of "long" (noncompact) curves, cobordisms, fronts on spheres, and classification of curves with a small number of singularities or extremal values of invariants [1].

I hope that appropriate theories will eventually emerge, paving the way for proving (or disproving) the hypothesis formulated above and other conjectures concerning the symplectic and contact topology of caustics and wave fronts.

1. THE SPACE OF LEGENDRE CURVES

The contact element on a manifold is a hyperplane (linear subspace of codimension 1) in the tangent space to the manifold at some point.

Consider the manifold M^3 of all transversely oriented contact elements on a plane. Its natural contact structure is the distribution of tangent planes defined by the condition: the velocity of the point of application of the contact element belongs to this element. The space $M = ST^*\mathbb{R}^2$ of spherized cotangent bundle of a plane is diffeomorphic to a solid torus, because an element at a point is defined by angle alone.

We shall largely center our attention on the space Ω of Legendre immersed curves into M^3, i.e., the immersions of a circle into M^3 for which the derivative at every point lies in the plane of the contact structure.

The manifold of contact elements is (trivially) fibered over a plane (with a circle of all directions at a given point as a fiber), so we have $M^3 \approx \mathbb{R}^2 \times S^1$. The manifold of directions in the planes of contact structure in M^3 forms over M^3 a bundle E^4 with fiber circle (which is also trivial, since we are dealing with *cooriented* elements $E^4 \approx \mathbb{R}^2 \times S^1 \times S^1$). Both these trivializations are fixed.

The Legendre immersion $L: S^1 \to M^3$ induces a Gaussian map $g_L: S^1 \to E^4$, which sends every point of the preimage to the direction of the velocity in the image (the preimage circle is assumed to be oriented).

The projections $\pi_{1,2}$ of the space E^4 on both these circles of the expansion $E^4 \approx \mathbb{R}^2 \times S^1_1 \times S^1_2$ in the composition with a Gaussian map define the maps $S^1 \to S^1_i$. We take a fixed orientation on the initial plane $\mathbb{R}^2$. Then both circles S^1_i (circle of directions in $\mathbb{R}^2$ and circle of directions in the contact plane) acquire orientations. The degrees of the mappings of the preimage circle onto these two circles are (unique) homotopic invariants of the Legendre immersion.

Definition. The degree of a map $\pi_1 \circ g_L: S^1 \to S^1_1$ is called the *index* (ind) of the Legendre curve L, and the doubled degree of the map $\pi_2 \circ g_L: S^1 \to S^1_2$ is called the *Maslov index* (μ) of the Legendre curve L.

The study of the space Ω of Legendre immersions consists in studying its component spaces $\Omega_{i,\mu}$ of Legendre immersions with a fixed index and Maslov index. Indeed, the spaces $\Omega_{i,\mu}$ are the connected components of the space of Legendre immersions. Each of these components is (at least weakly) homotopically equivalent to the torus $S^1_1 \times S^1_2$. The point is that the Gaussian map defines the (weak) homotopic equivalence of the space Ω to the space of mappings of a circle into E^4 and, consequently, to the space of mappings of a circle into $S^1_1 \times S^1_2$.

This result due to Gromov [6] is a generalization of the Smale theorem [7] on the immersions of curves and is demonstrated similarly. A key role here is played by the following consideration which is demonstrated by an explicit construction: the mapping from the space of Legendre immersions of an interval with a fixed initial velocity into the space of directions of the velocity at the end point of the interval is a Serre fibration, i.e., it has the homotopy lifting property.

This lifting is constructed explicitly from a given local perturbation of the terminal velocity of the Legendre immersion (which depends on several parameters) with the help of local Darboux coordinates in the neighborhood of the terminal element like Smale constructed the deformation of an ordinary immersion of a segment into a plane covering a given deformation of its terminal velocity.

The space of the mappings of a circle into a torus, which belongs to a fixed homotopic class (i, μ), contracts to a torus. This is clear from the fact that on the plane covering the torus all such paths with a fixed starting point have the same end point and are contracted by linear interpolation into the shortest path. Therefore, the mapping of the space $\Omega_{i,\mu}$ onto a torus, which implements the homotopic equivalence stated above, associates to every Legendre immersion of given indices a pair consisting of a contact element and the tangent direction to the Legendre curve corresponding to some chosen point of the initial circle and chosen direction along it.

To construct invariants in the form of indices of intersection with hypersurfaces in the space of Legendre immersions, it is more convenient to deal with a completely contractible (or at least simple connected) manifold of curves. This can be done in the Smale approach by fixing a point on the mapped circle and requiring that the velocity direction of the immersion at this point be always the same at the same point of the plane.

For Legendre curves, I could not succeed in completely eliminating the homologies of the space $\Omega_{i,\mu}$, but the Smale technique is nevertheless helpful in eliminating one of the two circles which form the torus while multiplying.

Definition. A Legendre mapping $S^1 \to M^3$ is said to be *distinguished* if a fixed point of S^1 is mapped into a fixed contact element of a plane (for example, an element of the x-axis at the origin which is cooriented by the positive direction of the y-axis).

The group $G^3 \approx \mathbb{R}^2 \times S^1$ of Euclidean motions of a plane (with fixed orientation and metric) acts on the space of cooriented contact elements of the plane $M^3 \approx \mathbb{R}^2 \times S^1$ without fixed points and transitively. It also acts on the space of Legendre immersions of a circle. Every such immersion is uniquely represented as an image of a distinguished immersion under the action of a unique element of the group. Thus we obtain splittings in direct products

$$\Omega \approx \Omega^0 \times (\mathbb{R}^2 \times S^1), \qquad \Omega_{i,\mu} \approx \Omega^0_{i,\mu} \times (\mathbb{R}^2 \times S^1),$$

where Ω^0 is the space of distinguished Legendre immersions (with given indices for $\Omega^0_{i,\mu}$).

It is sufficient to solve all natural classification problems for the distinguished immersions, since all natural classes are invariant under the action of the group G. Therefore, in what follows, we consider distinguished Legendre immersions and work with the space $\Omega^0_{i,\mu}$ having a (weak) homotopic type of a circle.

The mapping of a circle implementing this homotopic equivalence associates a Legendre immersion with its direction at a distinguished point of the preimage. The direction of the tangent vector of M^3 at the distinguished point of M^3 belongs to the circle of directions of the vectors on the plane of the contact structure at this point of the space M^3.

These directions, which geometrically correspond to the curvatures of fronts (the front of a Legendre curve is its projection from $M^3 = ST^*\mathbb{R}^2$ onto $\mathbb{R}^2$) are not fully equivalent—among them there are certain "vertical" directions which are tangential to the fiber S^1 and correspond to the infinite curvature of fronts. This situation complicates the construction of invariants as intersection indices, because we must verify whether the cocycle constructed with the help of the hyperplane of singular immersions is cohomologous to zero (the index of intersection with a closed curve belonging to the nontrivial homology class in the space of distinguished Legendre curves should be zero).

2. MAIN INVARIANTS St, $J^{\pm}$: THEIR DEFINITIONS

In the space of (ordinary) immersions of a circle into the plane, we take the hyperplane of singular immersions. The generic immersion has no singularities other than the simplest double points of transversal self-intersection. The simplest singularities of nongeneric immersions are triple points and simple (quadratic) self-tangency points.

In the space of all immersions the set of nongeneric immersions forms a hyperplane called the *discriminant*. The main invariants are the indices of intersection with different parts of this discriminant, namely, increment of the invariant St (called the strangeness) is equal to the index of intersection with the part of the discriminant that corresponds to the immersions with triple points and the increment of the invariants $J^{\pm}$ is equal to the (doubled) index of the intersection with the parts of the discriminant which are formed by the immersions with direct (J^+) and inverse (J^-) tangencies. The velocities of both branches at the tangency point have the same (J^+) or opposite (J^-) directions.

To reword this definition rigorously, we must state the connected components of the immersion space, indicate the coorientations and values of invariants on a curve from each component. Furthermore, we must verify whether the cohomology of the cocycle is zero.

The index (number of rotations of the tangent vector) is the only invariant that enumerates the connected components of the space of immersions of an oriented circle into an oriented plane. This Whitney theorem [8] follows, particularly, from the Serre fibration studied by Smale [7].

This fibration gives us the contractibility (weak homotopic equivalence to a point) of the space of distinguished immersions of fixed index. Hence, cocycles will be cohomologous to zero, provided we confine ourselves to distinguished immersions (and this is quite sufficient for constructing the invariants).

Coorientations are chosen as described below (this is the only choice which along with a few simple axioms stated in [1], and is mainly dictated by the conditions of compatibility of coorientations at singular points that form a hypersurface of codimension 1 in the discriminant and correspond to the passing of one branch of the immersed curve through the point of tangency of two other branches).

Case J$^\pm$. The side of the discriminant of direct tangencies that contains greater number of double points is considered positive.

Case St. The side of the discriminant of triple points where the triangle formed by three almost intersecting branches is positive is considered positive.

The sign of a triangle is determined by the cyclic order of occurrence of its three sides in passing along the curve and by the directions of the sides. The cyclic order defines the orientation of a triangle. A side is assigned "+" sign if it is directed *opposite* to the orientation and "–" sign if it is directed *along* the orientation. The sign of a triangle is the product of the signs of its sides.

Definition. The main invariants St, J^+, and J^- of immersions of a circle on a plane are determined by the following conditions:

(1) When the parts of the discriminant are intersected in positive direction, these invariants experience the following jumps:

	Triple points	Direct tangency	Inverse tangency
St	1	0	0
J^+	0	2	0
J^-	0	0	−2

(2) The values of the main invariants on simple curves K_i of index $\pm i$ (Fig. 1) with an arbitrary orientation are given in the table below

	K_0	K_1	K_2	K_3	K_4	...
St	0	0	1	2	3	...
J^+	0	0	−2	−4	−6	...
J^-	−1	0	−3	−6	−9	...

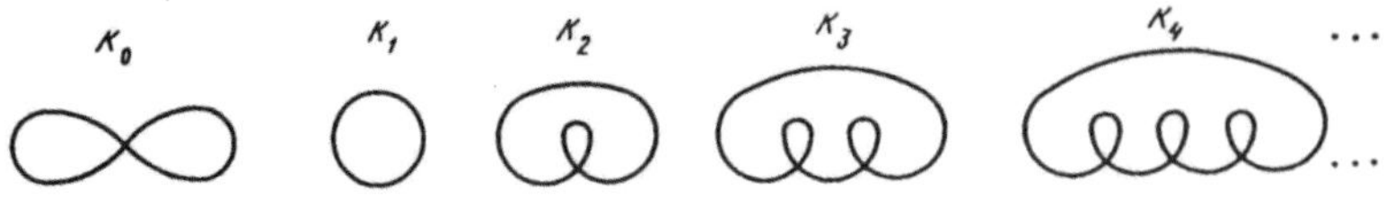

Fig. 1. Simple curves.

				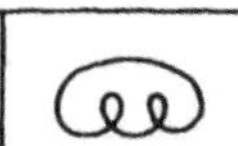	
	$i=1$	$i=1$	$i=1$	$i=3$	$i=3$
St	0	0	1	2	3
J^+	0	0	-2	-4	-6
J^-	-2	-2	-4	-6	-8

Fig. 2. Invariants of curves with two double points.

Remark. The values of the main invariants defined in this way depend on neither the orientation of curve nor the orientation of the plane. They are additive under the connected summation of curves (for example, $K_2+K_2 \sim K_3$). These conditions together with the "locality" of coorientation uniquely define invariants up to multiplicative constants given by the magnitudes of jumps at the intersection of the strata of the discriminant. The choice of the value 2 and -2 for jumps $J^\pm$ (the invariant varies like the number of double points at the intersection of direct tangency for J^+ and is opposite to the change of the number of double points at the intersection of inverse tangency for J^-) is convenient for the reasons stated below (for Legendre curves). The difference $J^+ - J^- = n$ is always equal to the number of double points on a curve.

Example. The values of the main invariants for all five curves with two self-intersection points are given in Fig. 2.

Example. A curve is said to be *tree-like* if every double point divides the curve into two loops which do not intersect each other elsewhere. All curves shown in Figs. 1 and 2 are tree-like curves. F. Aicardi [9] proved that $\mathrm{St} + 2J^+ = 0$ for all tree-like curves.

Example. A curve is said to be *extremal* if its index is greater by 1 than the number n of double points (by Whitney theorem [8], it cannot be greater than this value). Extremal curves are classified by plane root trees with n edges. The St invariant of such a curve [1] is equal to the sum of the distances of vertices to the root (Fig. 3). All extremal curves are tree-like curves.

Example. The first and the last curve in Fig. 3 lie in the same component of the immersion space. Therefore, there exists a generic path between them.

From the values of St (Fig. 3), it follows that the triple points are encountered at least six times on such a generic path.

The number of direct and inverse tangencies on any such path is also not less than six (because J^+ is equal to -8 and -20, and J^- is equal to -12 and -24, respectively).

Thus, computations of the invariants for concrete curves yield nontrivial information on the

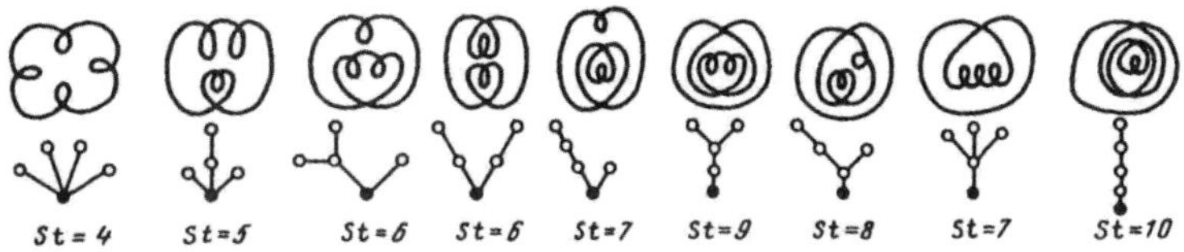

Fig. 3. Extremal curves with $n = 4$.

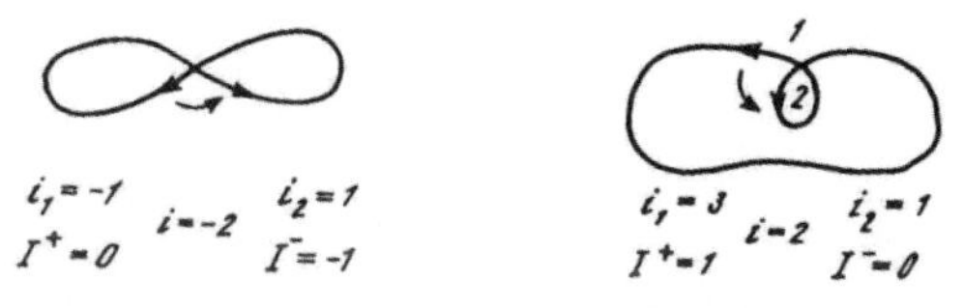

Fig. 4. Halfindices of loops and indices of double points.

metamorphoses when passing from one curve to the other. But for this purpose the values of invariants must be known.

Example. Below we give a result from [1], which we use in the sequel for computing the invariant St:

$$J^{\pm} = I^{\pm} - 3\mathrm{St},$$

where the invariants $I^{\pm}$ are determined as follows. Choose such a metric that the branches would be orthogonal at the self-intersection points.

Definition. The *halfindex* of a loop with ends at a double point is $2/\pi$ times the rotation angle of the radius-vector connecting the double point to the point which runs along the loop (recall that the orientations of the curve and the plane are fixed).

The *index* of a double point is the difference between the halfindices of the loops starting at this double point (plus is assigned to the halfindex of a loop if the frame formed by the outward directions of this and the second loop orients the plane positively (Fig. 4)).

Remark. Unlike halfindices, the indices of double points depend neither on the orientation of the curve nor the orientation of the plane.

Definition. The invariants $I^{\pm}$ are determined by summation the indices of all n double points, i.e., $I^{\pm} = \sum(i \pm 2)/4$.

3. DISCRIMINANT OF THE FRONTS OF LEGENDRE CURVES: ITS SINGULARITIES

Strictly speaking, the discriminant in the space of distinguished Legendre curves must be defined to be the hypersurface that contains the curves having nongeneric singularities, as well as the curves for which the distinguished point occurs at a singular point of the front. Here, by discriminant we mean only the hypersurface containing curves having nongeneric singularities. It is partitioned into orbits of the group of diffeomorphisms of the preimage circle which acts naturally without fixed points on the space of Legendre immersions of the circle.

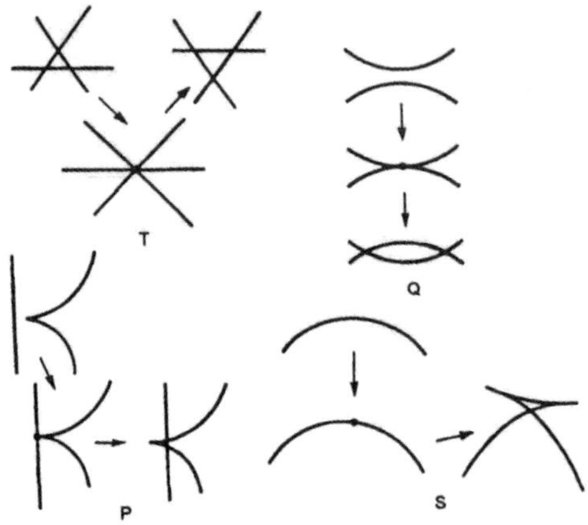

Fig. 5. Singularities and metamorphoses of fronts of codimension 1.

Remark. Unfortunately, the action of the diffeomorphism of a circle on the space of distinguished Legendre immersions (regarded as the space of orbits of the group of motions of a plane acting on the space of all Legendre immersions) has (sparse) fixed points. These fixed points, which though form a set of infinite codimension, formally hinder us from using the quotient of the action of the group of diffeomorphisms of a circle and eliminating the non-simple connectedness of the space of distinguished Legendre immersions (the multiplier S^1_2 in the notations of Section 1). Therefore, I shall construct invariants as indices of intersection of paths with hypersurfaces of the strata of the discriminant of singular fronts in a non-simply connected space of distinguished curves.

The front of a generic Legendre curve has singularities only of two standard types: transversal self-intersection points and semicubic cuspidal points. In the theory of Legendre singularities [10], singularities of codimension 1 and 2 (corresponding to the generic points of the discriminant and to the hypersurface of codimension 1 in it) are listed below.

Theorem. *The Legendre curves with nongeneric fronts are classified into the following types.*

Strata of codimension 1 (*Fig.* 5)*:*

(1) *triple point of the front with distinct tangents* (*stratum* T)*;*

(2) *quadratic self-tangency point of the front* (*stratum* Q)*;*

(3) *point of passing of a branch of the front through the cuspidal point of another branch the front* (*stratum* P); *here it is assumed that the tangents of both branches at the cusp are distinct;*

(4) *point with singularity of degree* 4/3 (*stratum* S); *this is the birth point of two cusps which corresponds to the vertex of the swallow tail on the big front.*

Strata of codimension 2 (*Fig.* 6)*:*

(1) *four-point of the front with distinct tangents* (*stratum* TT);

(2) *degenerate triple point of the front* (*stratum* QT)*; two branches touch quadratically, while the third is transversal to them;*

(3) *cubic self-tangency point of the front* (*stratum* QQ)*;*

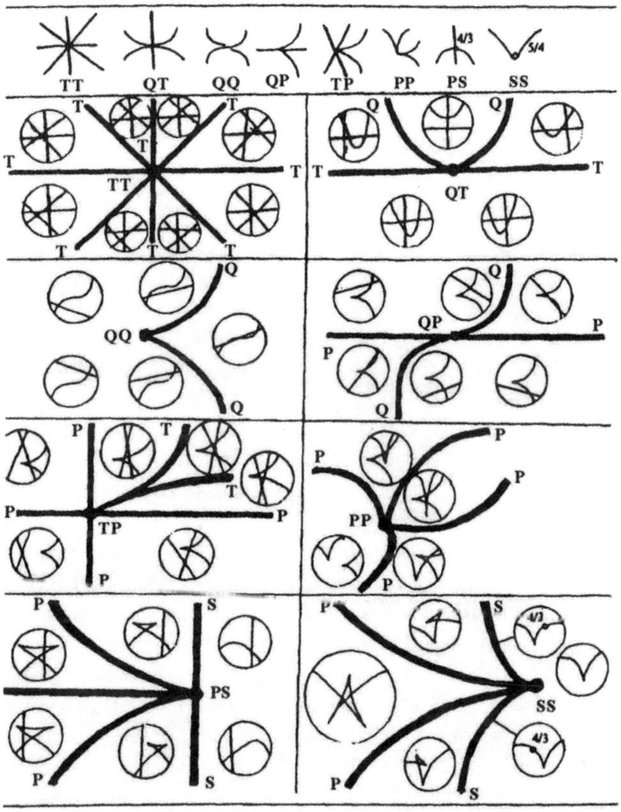

Fig. 6. Singularities and bifurcation diagrams of fronts at points of codimension 2.

(4) *point of tangential passing of a branch of the front through the cusp (stratum* QP);

(5) *the point of passing of two branches of the front through its cusp (stratum* TP); *here it is assumed that the tangents of all three branches at the point are distinct;*

(6) *the point of coincidence of two cusps (stratum* PP); *here it is assumed that the tangents of both branches at the point are distinct;*

(7) *the point of transversal passing of a branch through a point of degree* $4/3$ *(stratum* PS);

(8) *point of degree* $5/4$ *(stratum* SS).

Furthermore, the strata of transversal mutual intersection (including self-intersection) of strata of codimension 1 *corresponding to fronts with two singular nongeneric points are of codimension* 2.

The complement of the set of Legendre immersions with singularities of fronts of the type listed above has codimension 3 *in the space of Legendre immersions and codimension* 2 *in the discriminant formed by all immersions with nongeneric fronts.*

All these result hold in the space of all Legendre immersions as well as in the space of distinguished Legendre immersions.

Remark. Figure 6 shows that the closure of the strata Q, P, or S of codimension 1 defines a cocycle modulo 2, while the closure of the stratum T of triple points does not define such a cocycle, because the line T on the bifurcation diagram PS has an endpoint (ends in TS).

4. COORIENTATION OF SELF-TANGENCY HYPERSURFACE

We now study the closure of one stratum of codimension 1—the stratum Q^+ of hazardous self-tangencies. Self-tangency is referred to as *hazardous* if the coorientations of contacting branches are the same. The closure of the hypersurface of hazardous tangencies defines a cocycle modulo 2 (as well as in the case of nonhazardous tangencies).

Theorem. *The hypersurface formed by the closure of the stratum of hazardous self-tangencies in the space of Legendre curves (or in the space of distinguished Legendre curves) is coorientable. Namely, the coorientation is defined by the following construction. A self-tangency is said to be positive if the pairs (orientation of the front, coorientation of the front) for both branches define identical orientation of the plane (for hazardous tangencies this condition implies that tangency is direct, i.e., the branches at the point of contact have identical orientations).*

The side of the hypersurface of hazardous self-tangencies that contains fronts with greater (lesser) number of double points is defined to be positive if self-tangency is positive (negative).

Proof. It suffices to equip the fronts with orientations and coorientations shown in Fig. 6, because the coorientation is correctly defined over the stratum of hazardous self-tangencies Q^+, and we must only verify the compatibility of the coorientations of the branches of Q^+ of bifurcation diagrams of the strata QT, QQ, and QP in Fig. 6, where two curves of the type Q converge.

In the case QT, compatibility is obvious, because the additional branch passing through the self-tangency point changes neither the orientation nor the coorientation. It exercises identical influence on the number of double points on either side of the singularity QT.

In the case QQ, the orientations and coorientations have continuous extensions to the singular cubic tangency point and the number of double points is greater by two on that side of the bifurcation diagram which occupies lesser part of the plane near the stratum QQ. Therefore, the coorientation is consistent.

In the case QP, the number of double points is greater by two in smaller domains (top right and bottom left corners in Fig. 6). These domains lie on different sides of the curve (cubic parabola) Q–QP–Q. Both parts of this curve correspond to hazardous tangency (if the coorientations of both branches of the front at the point QP coincide) or nonhazardous tangency (if the coorientations are opposite).

The sign of tangency changes at the point QP when one passes along the curve Q–QP–Q. Indeed, the orientation vector of the front reverses its direction when passing through the cusp, whereas the coorientation vector is continuous. Therefore, the orientation of the plane (defined by the pair front orientation, front coorientation) changes when passing through the cusp of the front. The self-tangency point of the front when moving along the curve Q–QP–Q passes from one side of the cusp of the singular branch to another at the moment QP. Therefore, the orientation of the plane determined by the singular branch of the front at the point of tangency changes when passing through the point QP on the curve Q–QP–Q. But, the orientation defined by the nonsingular branch does not change, because the vectors orienting and coorienting the front are continuous at the point QP.

Thus, the coorientation of the stratum Q^+ defined in the statement of the theorem (unlike the coorientation given by the number of double points of the front) is consistent for all strata of codimension 1 on the discriminant of the stratum Q^+. This completes the proof of the theorem (the consistency on transversal intersections with other branches of the discriminant which correspond to fronts with two nongeneric singular points is evident).

Remark 1. The stratum Q^- can also be cooriented in a similar manner.

Remark 2. From the foregoing, it is clear that our coorientation is uniquely locally defined, up to a simultaneous change of sign at all points, coorientation of the stratum Q^+ of hazardous self-tangencies. (A coorientation of the discriminant at a self-tangency point is called *locally defined* if it does not change when the Legendre immersion changes outside the neighborhoods of those points which this immersion maps to a given concrete element of the plane.)

But this coorientation is not the only possible consistent one for the hypersurface of hazardous tangencies Q^+. Indeed, this hypersurface is not irreducible and the sign of coorientation on every irreducible component can be changed arbitrarily.

The reducibility of the hypersurface of hazardous self-tangencies Q^+ (and the closure of the stratum of nonhazardous self-tangencies Q^-) follows from the continuity of the rotation number of the vector which coorients the front and runs along the loop of the front with the ends at a self-tangency point. The rotation number (an integer for hazardous tangencies and a semi integer for nonhazardous tangencies) is continuous when moving along the stratum of self-tangencies (we assume that having passed through a transversal intersection with another branch of the self-tangency stratum we continue to move along the initial branch).

The derived representation of the index of Legendre immersion as the sum of two (unordered) components is an invariant characterizing the "components" of the hypersurface Q^+ (respectively, Q^-). But it is not clear whether these "components" are irreducible themselves.

In the usual (non-Legendre) immersion theory, the components defined by the partition of the index are irreducible, because all curves—loops with one point of direct self-tangency and a given index — are deformed into one another by a regular homotopy (this is implied in the Whitney theorem). For the fronts we have a Maslov index in addition to the ordinary index, and we should verify whether the Maslov index leaves any trace in the classification of loop-fronts having one self-tangency point (for instance, hazardous) with a fixed index.

Remark. The coorientation chosen above for the hazardous tangency stratum does not change when the orientations of the curve or the plane change.

5. NONCONTRACTIBLE PATHS IN THE SPACE OF DISTINGUISHED IMMERSIONS

Here we prove that the intersection index of a closed path in the space of distinguished Legendre immersions with the cooriented (see Section 4) hypersurface of hazardous collisions is always equal to zero.

To demonstrate this assertion, it suffices to compute this index for one noncontractible path in each component of the space of distinguished Legendre immersions. In order to construct such a path, we must construct a one-parameter family of fronts which start and end at the same front with a given index and a given Maslov index; furthermore, the same contact element of the plane should correspond to the chosen point of the circle in all fronts of the family.

The direction of the Legendre curve of the family in the contact plane touching M^3 at this element must commit a nonzero number of rotations when the parameter of the family changes. The following construction is convenient to compute the number of rotations. In the space of distinguished Legendre immersions, let us consider the hypersurface which is the closure of the stratum formed by the immersions for which the distinguished point is a cusp.

Theorem. *This closure defines a one-dimensional cocycle modulo 2. It becomes an integral cycle if the stratum is equipped with the following coorientation (which is consistent at the points of the hypersurface of singularities on the closure of the stratum): a distinguished curve is said to be positive if the center of curvature of the front at the distinguished point lies in the direction defined by the coorientation of the front.*

The index of intersection of a closed generic curve in the space of distinguished curves with the stratum of Legendre immersions with the distinguished cusp (equipped with this coorientation) is equal to -2 times the number of rotations executed by the immersion velocity vector at the distinguished point around the origin at the plane of contact structure (equipped with its natural orientation).

Remark. The contact structure on the space M^3 of cooriented contact elements of the plane has a natural coorientation, i.e., the corresponding differential 1-form α on M^3 is defined up to a *positive* factor. Indeed, if x and y are Cartesian coordinates on the plane (with a fixed Euclidean metric), and ξ and η are the components of the normal which coorients the contact element, then $\alpha = \xi dx + \eta dy$. Therefore, the sign of α is determined by the direction of the normal coorienting the contact element and does not depend on the orientation of the plane (x, y). The differential form $d\alpha$ when restricted to the contact plane $\alpha = 0$ defines its orientation which, consequently, does not depend on the choice of the orientation of the plane (x, y).

Having chosen this orientation, we can define the cooriented contact elements by the numbers $(x, y, \varphi \bmod 2\pi)$, where $\xi = \cos\varphi$, $\eta = \sin\varphi$ is the coorienting normal. In these coordinates, we have $\alpha = \cos\varphi dx + \sin\varphi dy$. The contact plane is spanned by the vectors $\partial/\partial\varphi$ and $\zeta = \sin\varphi\frac{\partial}{\partial x} - \cos\varphi\frac{\partial}{\partial y}$. Its natural coorientation is given by the frame $(\zeta, \partial/\partial\varphi)$ since $d\alpha(\zeta, \partial/\partial\varphi) = 1$.

This remark permits to find explicitly the direction of the velocity vector rotation of a distinguished Legendre immersion at the distinguished point.

Example. Consider the family of Legendre immersions locally defined by the formulas $x = t^2 - 2ts$ and $y = t^3 - \frac{3}{2}t^2 s$, where t is a parameter of the curve near the distinguished point $t = 0$ and s is a parameter (close to 0) of the family.

The corresponding fronts cooriented by the positive direction of the y-axis at the distinguished point are shown in Fig. 7 together with the velocities on the contact plane at the point ($x = 0$, $y = 0$, $\varphi = \pi/2$) for different values of the parameter s.

Clearly, when the distinguished point passes through the cusp of the front the velocity vector of the Legendre immersion at the distinguished point becomes vertical $(\partial/\partial\varphi)$ at the contact plane. Rotation is positive (from ζ to $\partial/\partial\varphi$) if the distinguished point passes from that part of the front where the direction toward the center of curvature coincides with the coorienting direction to the part where these directions are opposite at the distinguished point. This statement holds true even

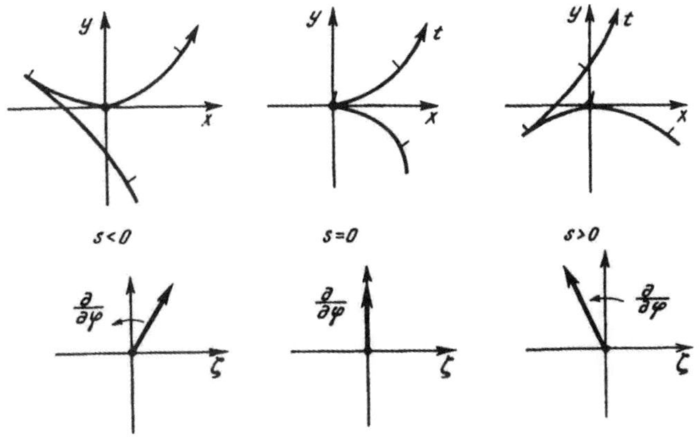

Fig. 7. Rotation of the velocity vector of the distinguished Legendre immersion when the distinguished point passes through a cusp.

if we go along the front in the opposite direction (because the velocity vector at any point rotates over 180°) and if the orientation of the plane (x, y) (which does not affect the orientation of the contact plane) is changed.

Proof. The first part of the theorem is verified by analyzing the simplest singularities of the closure of the immersion stratum with distinguished cuspidal points. In addition to transversal intersections with other branches (where the statement is evident), there is only one singularity of codimension 1 on the closure of the stratum — it corresponds to the distinguished point of the type S (of degree 4/3 on the front). In the one-parameter families of generic Legendre immersions, such a singularity is encountered for certain values of the parameter of the family as a singularity of the birth of a pair of cusps on the front.

The compatibility of coorientations near the stratum of distinguished points of the type 4/3 is directly verified by studying the versal deformation $x = t^3 - 3st$, $y = t^4 - 2st$ (Fig. 8).

Remark. Indeed, our constructions also give the coorientation of the hypersurface formed by immersions with a distinguished cusp in the space of Legendre immersions with a distinguished point for which the contact element at the distinguished point can be any element of M^3; moreover, the distinguished point on the mapped circle can, if necessary, be variable. The fact is that all our constructions are invariant with respect to the group of motions of a plane (and with respect to the rotations of the preimage circle). This situation is depicted in Fig. 8, where the compatibility is verified precisely in the space with a variable initial distinguished point t and a variable contact element in it.

The compatibility of coorientations in the initial space of distinguished immersions is implied in the compatibility of coorientations in this larger space, because we can return to the initial space with the help of parameter-dependent rotations of the circle and motions of the plane.

The second part follows from the fact that the frame $(\zeta, \partial/\partial\varphi)$ parallelizes the field of contact planes.

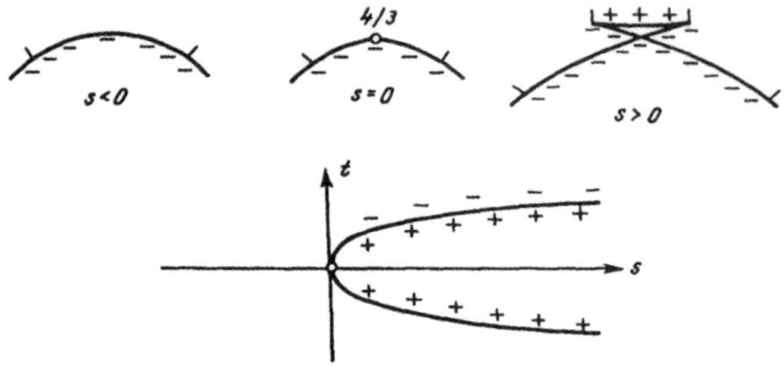

Fig. 8. Compatibility of the coorientations of the stratum of distinguished cuspidal points.

The mapping of the space E^4 of directions in the contact planes M^3 onto a plane circle with basis $(\zeta, \partial/\partial\varphi)$ defined by this parallelization is, by definition, precisely the mapping $\pi_2 : E^4 \to S^1_2$ which was used in Section 1 for defining the Maslov index and which induces the (weak) homotopic equivalence of the space of distinguished Legendre immersions $\Omega^0_{i,\mu}$ with given indices to the circle S^1_2 of directions of the plane $(\zeta, \partial/\partial\varphi)$.

The number of rotations of the tangent vector of the Legendre immersion at the distinguished point, which is equal to the degree of the mapping from the circle of the values of the parameter of the family of immersions to the circle S^1_2, is equal to half the number of times when the tangent vector passes through the vertical direction $(\pm\partial/\partial\varphi)$ with regard for the signs (passage is positive if the distinguished point passes from the positive side of the stratum to the negative side as shown in Fig. 8). This completes the proof of the second part of the theorem. The factor 2 arises because, in computing the passages through the cusp, besides the intersections of the positive vertical half-axis, we also take account of the intersections of the negative vertical half-axis. And the minus sign arise because, as is clear from Fig. 8, a positive rotation at the contact plane correspond to the passage of the distinguished point from the positive side of the stratum of distinguished cuspidal points to the negative one.

Remark. Concurrently, we obtained a method for determining the rotation number of the image of immersion with a distinguished point under mapping onto S^1_2, when this Legendre immersion depends on a parameter which traces the circle. The number of rotations is half the number of passages of the distinguished point through the cusp counted with the rule of sign from the theorem (for generic families).

Example. Fix Legendre immersion of a circle with a distinguished (generic) point. Using this mapping, let us construct a one-parameter family of distinguished immersions which is parametrized by the points of the same preimage circle S^1 (simply moving the distinguished point). From what has been proved above, we obtain the following theorem.

Theorem. *The degree of the mapping $S^1 \to S^1_2$ defined by trivialization of the fibration $E^4 \to M^3$ is equal to half the Maslov index of the initial (oriented) front.*

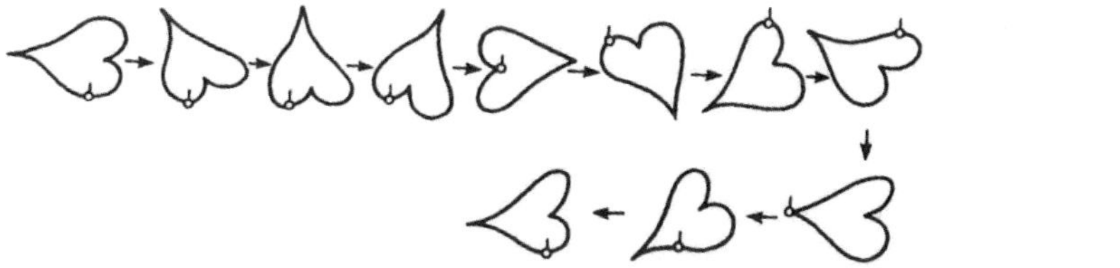

Fig. 9. Noncontractible path in the space of distinguished Legendre immersions.

Remark. This index, by definition [10], is equal to the difference between the number of direct and inverse cuspidal points of the front (a point is said to be direct if the coorienting vector is directed from the branch incoming into the cusp to the branch outgoing from the cusp; thus, the cusp in Fig. 7 is a direct point). Rotation is positive at direct points and negative at inverse points.

6. PERIOD OF A HAZARDOUS SELF-TANGENCY COCYCLE

In this section, we prove the following theorem.

Theorem. *The intersection index of a closed generic path in the space of distinguished Legendre immersions of a circle with a hazardous self-tangency stratum cooriented as described in Section* 4 *is zero.*

Proof. Since every connected component of the space of distinguished immersions is (weakly) homotopic equivalent to a circle, it suffices to verify the assertion for one noncontractible path in each component $\Omega^0_{i,\mu}$ of immersions of index i with Maslov index μ $(= 2k)$.

Case 1. The Maslov index μ is nonzero.

In this case, consider an arbitrary generic immersion with given indices and construct a family of distinguished immersions parametrized by the preimage circle marking in turn all its points (and then moving the front over the plane so that the image of a marked point should always be the same (marked) contact element of plane) (Fig. 9).

This closed path in the space of distinguished immersions is not contractible, because the mapping of the circle to the circle of directions in the contact plane is of degree $k \neq 0$ (see the example at the end of Section 5). This path does not intersect the self-tangency stratum or its closure, because the initial curve has no self-tangencies. Therefore the index of intersection with self-tangency stratum is zero. This completes the proof of the theorem for $\mu \neq 0$.

Case 2. The Maslov index is zero.

In this case, it is not so easy to construct a noncontractible path. But after finding it the verification is straight forward. The noncontractible path for a Legendre immersion of index $i = 1$ is shown in Fig. 10. It consists of three parts: the part AB joins a four-cusp curve with a circle, the part $A'B'$ joins the same curve with the circle in a different way, and the part AA' is essentially the same "roll over" as shown in Fig. 9. The closed path $ABB'A'A$ is noncontractible, because the distinguished point passes through cuspidal points only on the part $A'A$, where there are two such passages both of the same sign. Therefore, the degree of the mapping onto the circle of directions in the contact plane is ± 1.

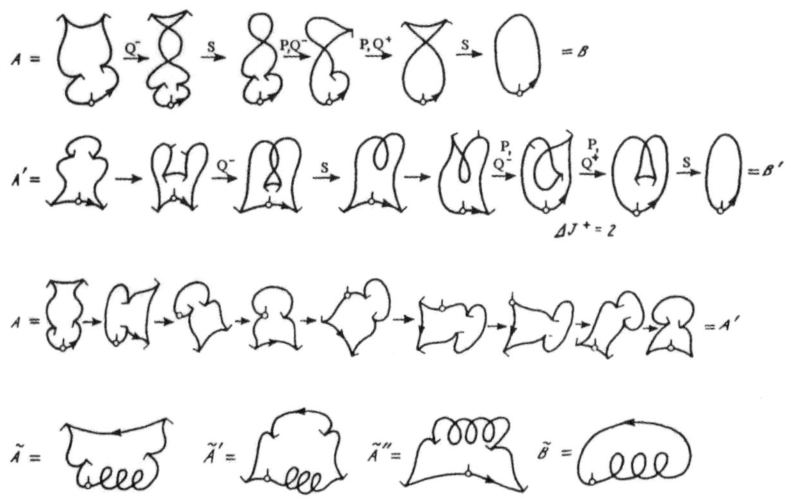

Fig. 10. Noncontractible closed path in the space of distinguished Legendre immersions with zero Maslov index.

The index of intersection of the path shown in Fig. 10 with a hazardous self-tangency stratum cooriented as described in Section 4 is zero (hazardous self-tangencies Q^+ on the parts AB and $B'A'$ give opposite contributions to the index). Therefore, Fig. 10 demonstrates the theorem for immersions of zero Maslov index and ordinary index equal to $i = 1$.

Noncontractible paths with zero Maslov index and arbitrary value of the ordinary index i can be constructed from the curve A (Fig. 10) by adding a few small loops on one or the other side of the curve near the distinguished point to get the desired value of the ordinary index i ($\tilde{A}$ in Fig. 10).

These loops have no influence on the metamorphoses along the path $\tilde{A}\tilde{B}$ (Fig. 10). Only a few somersaults occur on the path $\tilde{A}''\tilde{A}$, but no changes take place in the contribution (± 2) of this part of the path either to the degree of the mapping to the circle of the contact plane, or to the index of intersection with the hazardous self-tangency stratum (which is equal to zero), because neither the cuspidal points change, nor there is any self-tangency.

Perceptible changes are suffered only by the path segment $B'A'A$ (Fig. 10). The matter is that the additional loops constructed for the curve $\tilde{A}$ near the distinguished point must be constructed near this distinguished point on the metamorphosed circle $\tilde{B} = \tilde{B}'$ and, consequently, on the curve $\tilde{A}'$. Therefore, the curve $\tilde{A}'$ modified by the addition of loops cannot be obtained from $\tilde{A}$ by a motion of the plane: the loops on them are located differently.

In order to obtain the upturned curve $\tilde{A}$ with loops (which we denote by $\tilde{A}''$) from $\tilde{A}'$, two cuspidal points (for example, those points that are located on the right side of the curve $\tilde{A}'$,

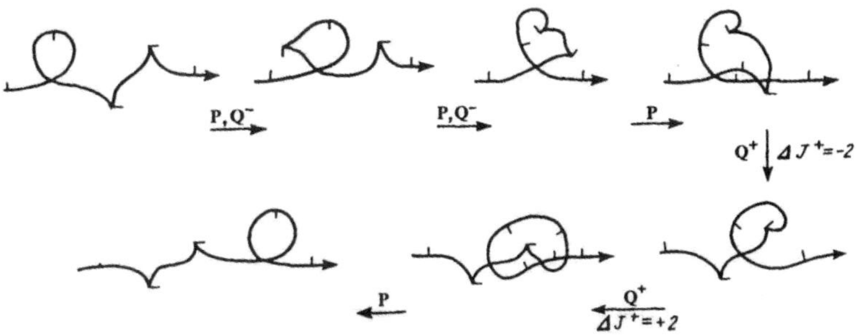

Fig. 11. Pulling of a pair of compatible cuspidal points through a loop.

Fig. 10) must be pulled through all loops. Both these points are direct ones.

Lemma. *The path created when a pair of direct (inverse) cuspidal points is pulled through a loop has zero intersection index with the stratum of hazardous self-tangencies (as well as with that of nonhazardous self-tangencies).*

Proof. The sequence of elementary metamorphoses is shown in Fig. 11. It includes four self-tangencies: two hazardous Q^+ and two nonhazardous self-tangencies. For the coorientation chosen in Section 4, the hazardous tangent points have opposite contributions to the intersection index J^+.

The closed path $\tilde{A}\tilde{B}\tilde{B}'\tilde{A}'\tilde{A}''\tilde{A}$ in the space of distinguished Legendre immersions with zero Maslov index and ordinary index i is noncontractible (the tangent vector at the distinguished point executes one rotation in the contact plane), and its index of intersection with the stratum of hazardous tangencies is zero. This completes the proof of the theorem.

7. THE INVARIANT J^+ OF LEGENDRE IMMERSIONS: ITS DEFINITION AND PROPERTIES

The invariant J^+ of Legendre immersions of a circle is the dual of the hazardous self-tangency stratum Q^+ in the sense that its increment along the generic path is equal to (double) the index of intersection with the hazardous self-tangency stratum cooriented as described in Section 4 (so J^+ varies as the number of double points if the hazardous self-tangency is positive, i.e., if the orientations of the contacting branches, like coorientations, coincide and varies in the opposite way if the hazardous self-tangency is negative).

In Section 6, we have shown that this condition defines the invariant J^+ up to an additive constant which is, in general, different for each connected component $\Omega_{i,\mu}$ of the Legendre immersion space.

In order to define the invariant J^+ uniquely, we must chose its values on one curve of every component. It happens so that this choice uniquely depends on certain natural conditions. Prior to formulate these conditions, let us state the normalization constants derived from them.

Definition. The invariant J^+ is the only invariant of the Legendre immersions without hazardous self-tangencies whose increment is +2 in passing through a positive hazardous self-tangency

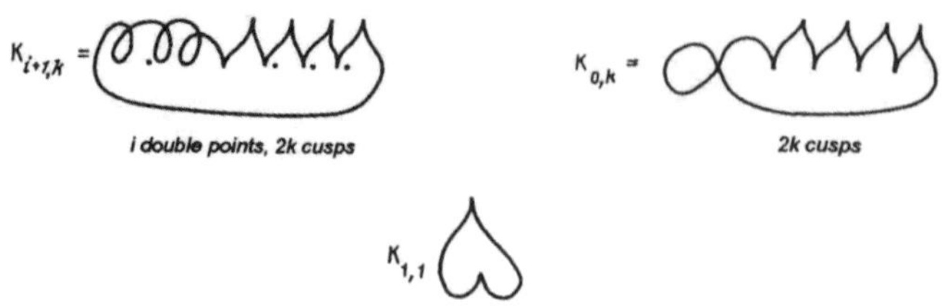

Fig. 12. Basis curves.

and −2 in passing through a negative self-tangency, and its values on basis curves shown in Fig. 12 are

$$J^+(K_{i+1,k}) = -2i - k \quad (i \geq 0,\ k \geq 0), \qquad J^+(K_{0,k}) = -k$$

for any choice of orientations and coorientations (the ordinary indices and Maslov indices of basis curves $K_{j,k}$ are equal to $\pm j$ and $\pm 2k$ respectively depending on the choice of orientations and coorientations).

Example. $J^+(K_{1,1}) = -1$; $J^+(K_{i+1,0}) = -2i$, $J^+(K_{0,0}) = 0$. Therefore, for nonsingular curves the value of J^+ are the same as stated in Section 2 (all hazardous self-tangencies on a nonsingular front are positive).

Theorem. *The invariant* J^+ *is correctly defined.*

For $j = 0$ or $k = 0$, the four immersions of curves $K_{j,k}$ with chosen orientation and coorientation belong only to two (and for $j = k = 0$ to only one) components $\Omega_{0,\pm 2k}$ or $\Omega_{\pm j,0}$ of the space of Legendre immersions of a circle. In these cases our evaluation of J^+ might be ambiguous. Fortunately, this is not true: the generic path joining two basis curves of the same component has zero index of intersection with the hazardous self-tangency stratum — this is what the theorem asserts.

Proof. For $j = 0$, this is implied in the lemma about the pulling of a pair of cuspidal points through a loop (Section 6, Fig. 11). For $k = 0$, the change in the coorientation is achieved by creating $(j-1)$ pairs of cusps and by annihilating them in another combination (for $j = 1$, i.e., for a circle it suffices to create four cusps). The paths that demonstrate the theorem for each of these cases are shown in Fig. 13.

Remark. At the same time, we have also proved that the invariant J^+ takes identical values for mirror-symmetric immersions. Indeed, Fig. 13 shows this holds for basis immersions. However, the coorientation of the hazardous tangency stratum is also invariant under the mirror-symmetry of a plane.

Theorem. *The value of the invariant* J^+ *does not depend either on the orientation or the coorientation of the front and is preserved under mirror reflection of the front.*

Proof. This has been verified only for basis immersions. For other immersions, this assertion is implied in the invariancy of the coorientation of the hazardous tangency stratum (both contacting branches suffer identical changes when orientation and coorientation change, therefore a hazardous tangency remains to be hazardous and a positive tangency remains to be positive).

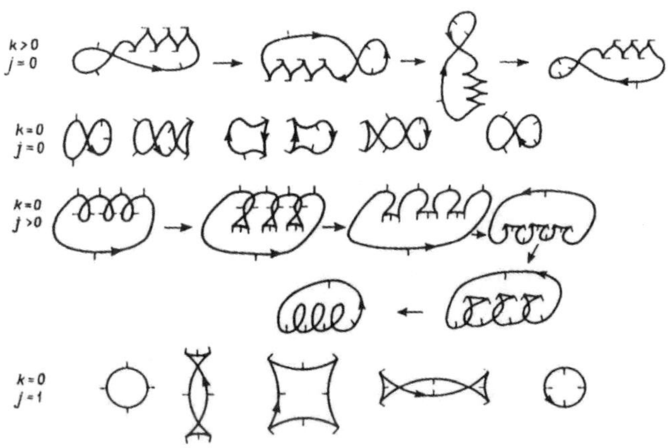

Fig. 13. The paths joining basis immersions of a component.

8. ADDITIVITY OF THE INVARIANT J^+

Theorem. *The invariant J^+ is additive under the connected summation of fronts.*

Proof. We shall demonstrate this assertion through the additivity of the invariant J^+ under the connected summation of ordinary immersions (Section 2, see [1]). First, we move the cuspidal points along the front by pulling them through other branches of cooriented and oriented fronts.

A pulling is said to be *hazardous* if the coorientation of the front at the cusp coincides with the coorientation of the front at the intersecting branch at the moment of passage. It is said to be *nonhazardous* if these coorientations are opposite (Fig. 14).

Lemma. *For a hazardous pulling of a cuspidal point, the invariant J^+ takes the increment $+2$ if the intersecting branches after pulling are alike (i.e., if the pairs — orientation and coorientation of the branches — define the same orientation of the plane.*

Remark. This increment does not change when the orientation of the front or of the plane, or the coorientation of the front changes.

Proof. The pulling operation can be replaced by a pair consisting of a passage and a hazardous tangency. A study of all four possible combinations of branch orientations leads to the assertion of the lemma (see Fig. 14).

Now we can compute the invariant J^+ for a front with cuspidal points as follows: pulling all cusps one after another to one arc of the outer contour of the front, we represent the front as the connected sum of an ordinary cuspless immersion and a front without self-intersection points. In this operation, the value of J^+ jumps up (down) by two as many times as the number of like (unlike) pullings.

Then we annihilate, one after another, the pairs of neighboring cuspidal points of the type "direct point, inverse point," thereby creating one self-intersection point for each pair (Fig. 15); the value of J^+ in such an annihilation does not change.

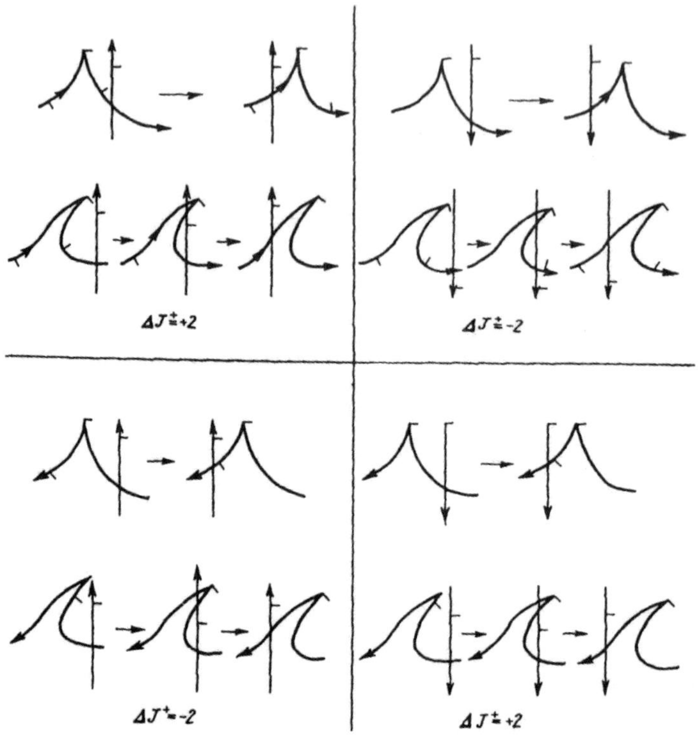

Fig. 14. Pulling of cuspidal points.

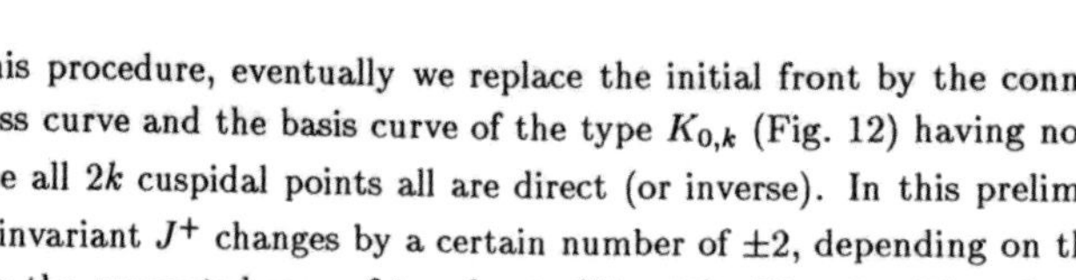

Fig. 15. Annihilation of an unlike pair of cuspidal points of a front.

Repeating this procedure, eventually we replace the initial front by the connected sum of an immersed cuspless curve and the basis curve of the type $K_{0,k}$ (Fig. 12) having no self-intersection points and whose all $2k$ cuspidal points all are direct (or inverse). In this preliminary operation, the value of the invariant J^+ changes by a certain number of ± 2, depending on the pullings.

Consider now the connected sum of two fronts (Fig. 16). The algorithm described above can be applied inside each component without disturbing the other, and thus bring all cusps to the connected points of the components. Here the change in J^+ of the sum is equal to the sum of the

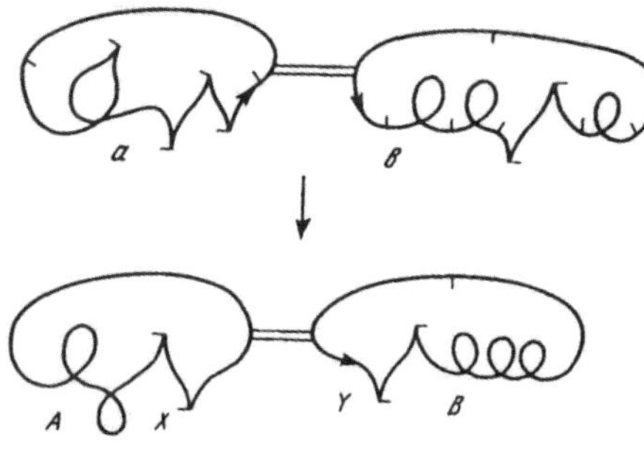

Fig. 16. Connected summation of the initial and processed fronts.

changes in J^+ of components, since pullings are alike or unlike, irrespective of whether a pulling is applied to a component or to the sum. Consequently, it suffices to prove the additivity of J^+ for the processed fronts (see Fig. 16).

The additivity of J^+ for processed fronts is implied in the additivity of cuspless fronts. Its additivity for the fronts $K_{0,k}$ is implied by the following assertion.

Lemma 1. *In connected summation of a cuspless front with a "heart" $K_{1,1}$ (Fig. 12), the invariant J^+ decreases by* 1.

Proof. A pair of alike cusps of the heart can be pulled along a cuspless front to any point without altering the value of J^+. Indeed, the arc between these points is unlike the adjacent arcs separated these cuspidal points. Therefore, by the lemma on pulling of cuspidal points, the contributions to J^+ of the pullings of both these cusps through any transversal branch are opposite.

The initial cuspless front can be metamorphosed, by the Whitney theorem, (using selftangencies and triple points) into a basis curve K_i (see Fig. 1) having (with suitable orientation) the same index. If the cuspidal points are not affected by these metamorphoses, the change in J^+ for a cuspless curve is the same as for a two-cusped curve.

If metamorphosis affects an arc containing a pair of cusps, then this pair must be pulled to another arc of the outer contour which is not affected by metamorphosis and then carry out the metamorphosis process. The value of J^+ does not change in pulling the cusps.

As a result of all these metamorphoses and pullings, the initial connected sum is converted into a connected sum of the basis curve K_i (see Fig. 1) and the heart, and the value of J^+ of the sum changes by the same amount as the value of J^+ when the smooth component is metamorphosed into K_i.

The relation $J^+(K_i + K_{1,1}) = J^+(K_i) - 1$ is verified directly (see Fig. 12) using the definition of $J^+(K_{j,k})$.

Lemma 2. *The invariant J^+ is additive under connected summation of Legendre immersions whose fronts have no self-intersection points and have only alike cusps ($K_{1,k}$, Fig. 12).*

Proof. If the cusps of both addends (cooriented and oriented) fronts are alike (on both addends all cusps are either direct or inverse), then additivity is implied in the definition, since

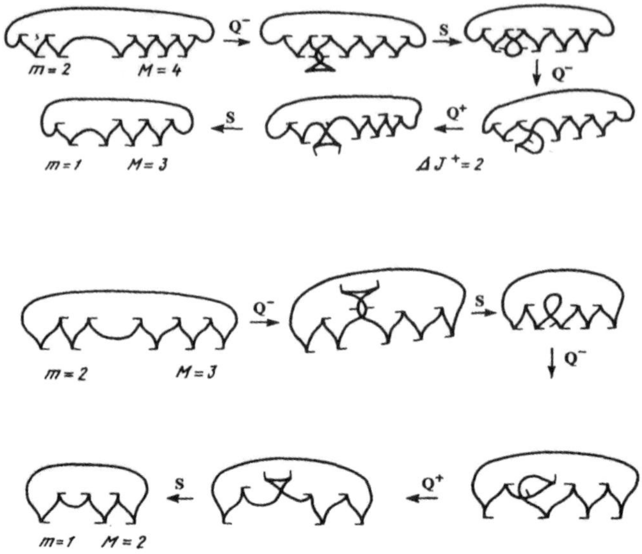

Fig. 17. Mutual annihilation of unlike cusps.

connected summation of the fronts $K_{1,k}$ and $K_{1,l}$ in this case yields $K_{1,k+l}$.

If, however, the cusps on the addends are unlike, they can be annihilated in pairs. In other words, let one of the components contain $2M$ cusps and the other $2m$ cusps, and assume that $M \geq m$. Then, by standard metamorphosis, we can annihilate $4m$ cusps as shown in Fig. 17. Thus, we obtain the front $K_{1,M-m}$. For this front, $J^+ = m - M$. But in the course of metamorphosis, the value of J^+ increases by $2m$. That means, for the initial front we have $J^+ = -m - M$. This proves that J^+ is additive under the addition of fronts of the type $K_{1,k}$, since $J^+(K_{1,k}) = -k$.

Lemma 3. *On adding k hearts $K_{1,1}$ (some of them are direct and some—inverse) to the basis curve (see Fig. 1), the value of the invariant J^+ decreases by k.*

Proof. If all added hearts are direct (or inverse), then the lemma is clear from the choice of the values $J^+(K_{j,k}) = f(j) - k$. If, however, there are both direct and inverse hearts, then we can carry out reduction as in the proof of Lemma 2 (Fig. 17) (by Lemma 2, all cusps can be gathered at one place without changing the value of J^+). Mutual annihilation of four cusps increases the value of J^+ by 2, i.e., by the number of vanishing hearts. Therefore, the validity of Lemma 3 in this case is implied in the validity of the lemma for alike hearts.

Lemma 4.. *The invariant J^+ is additive under the addition of processed curves (see Fig. 16).*

Proof. Processed curves can be represented as the sum of $A + X$ and $Y + B$, where A and B do not have cusps, whereas X and Y are the sums of hearts (all hearts in X and all hearts in Y can be assumed to be alike, but this is not necessarily true for X and Y taken together). The aggregate front is of the form $A + X + Y + B$ and is the sum of $X + Y$ and $B + A$.

By the Whitney theorem, the curves A, B, and $B+A=C$ can be converted through a chain of metamorphoses into basis curves of Fig. 1 (with appropriate orientations and coorientations). Denote these curves as A_0, B_0, and C_0. By the cusp pulling lemma, as in the proof of Lemma 1, the presence of the component $X(Y, X+Y)$ has no influence on the increment of J^+ during metamorphosis:

$$J^+(A+X)-J^+(A_0+X)=J^+(A)-J^+(A_0),$$

$$J^+(Y+B)-J^+(Y+B_0)=J^+(B)-J^+(B_0),$$

$$J^+(C+Z)-J^+(C_0+Z)=J^+(C)-J^+(C_0), \qquad Z=X+Y.$$

Consequently, the defect in the additivity in the addition $(A+X)+(Y+B)$ is

$$J^+(C+Z)-J^+(A+X)-J^+(Y+B)$$

$$=J^+(C_0+Z)-J^+(A_0+X)-J^+(Y+B_0)-(J^+(C_0)-J^+(B_0)-J^+(A_0)),$$

since $C=B+A$, and J^+ is additive for cuspless curves.

By Lemma 3, the decrease in J^+ in adding hearts to the basis curves A_0, B_0, and C_0 is equal to the number of added hearts

$$J^+(A_0+X)-J^+(A_0)=-n(X), \qquad J^+(Y+B_0)-J^+(B_0)=-n(Y),$$

$$J^+(C_0+Z)-J^+(C_0)=-n(Z).$$

But, $n(Z)=n(X)+n(Y)$, because $Z=X+Y$. Therefore, the total defect is zero. This completes the proof of Lemma 4.

Continuation of the proof of the theorem. Denote the initial fronts by a and b and the processed ones by $A+X$ and $Y+B$ (see Fig. 16). The increment in J^+ in the course of preparation does not depend on whether a and b are processed separately or whether their sum is processed:

$$J^+(A+X+Y+B)-J^+(a+b)=(J^+(A+X)-J^+(a))+(J^+(Y+B)-J^+(b)).$$

By Lemma 4, the invariant J^+ is additive for processed fronts

$$J^+(A+X+Y+B)=J^+(A+X)+J^+(Y+B).$$

Hence, it is also additive for the initial fronts

$$J^+(a+b)=J^+(a)+J^+(b).$$

9. UNIQUENESS OF THE INVARIANT J^+

Definition. The invariant of Legendre immersions without hazardous self-tangency points is called *local* if its jump for transversal passing of a hazardous self-tangency depends only on the behavior of the immersion in the neighborhood of those two points which the immersion sends to the point of tangency. The invariant is said to be *additive* if it is the sum of its values under connected summation of fronts.

Example. The invariant J^+ and the Maslov index μ are local and additive invariants.

Theorem. *Every local additive invariant is a linear combination of the invariant J^+ and the Maslov index.*

Remark. The value of the invariant J^+ does not change when the orientation or the coorientation of the front is changed. It is also not changed under mirror reflection of the plane of the front into itself.

The sign of the Maslov index changes when the orientation or the coorientation of the front changes, but its value does not change under mirror reflection of the plane.

Therefore, every additive local invariant does not change its value under mirror symmetry, and every local invariant which does not change its value when the orientation of the front changes (like every local invariant which does not change its value when the coorientation of the front changes) is proportional to the invariant J^+ (the coefficient of proportionality is the same for all components of the Legendre immersion space).

Thus, the invariant J^+ is uniquely determined by its properties of locality, additivity, orientation-independence (or coorientation-independence), and its values on one curve for which J^+ is nonzero (for example, $J^+(K_2) = -2$ for the simplest curve K_2 of index 2 (Fig. 1) or $J^+(K_{1,1}) = -1$ for the heart $K_{1,1}$ (Fig. 12)).

Proof. Some properties directly follow from additivity even without locality.

Lemma 1. *Every additive invariant J obeys the relations*

$$J(O) = 0, \qquad J(\infty) = 0, \qquad J(K_{i+1}) = iJ(K_2)$$

for any coorientation (orientations of K_2 and K_{i+1} must be compatible so that the indices are of the same sign).

Proof. When a circle O is added to any curve X, the type of the curve does not change, therefore $J(X+O) = J(X)$. Hence, $J(O) = 0$. The sum of two figure eight curves is metamorphosed into a circle without hazardous tangency. Therefore, $2J(\infty) = 0$. The sum of i copies of K_2 (with identical orientations and coorientations) is K_{i+1} (with appropriate orientation and coorientation).

The coorientation of K_j can be changed without changing the orientation through a chain of nonhazardous metamorphoses (see Fig. 13). Therefore, the value of $J(K_j)$ does not change when the coorientation of the basis curve is changed (this is also true of the figure eight curve K_0, the circle K_1, and other curves K_j).

We now consider the curves with cusps. The simplest curves with two direct or inverse cusps are four hearts $K_{1,1}$ (differing in orientations and coorientations). Denote the values of the additive invariant J on these four hearts by x, y, z, and t (for $\mu = 2$, $i = 1$; $\mu = -2$, $i = 1$; $\mu = -2$, $i = -1$; and $\mu = 2$, $i = -1$, respectively (Fig. 18)).

Lemma 2. $y = z$ *and* $x = t$.

Proof. The chain of transformations which connects the curve having the invariant $x + z$ with that of $x+y$ is shown in Fig. 19 (the sign $\sim$ denotes either a nonhazardous-type metamorphosis that does not change the value of the invariant or the rejection of the component "∞" for which, by the lemma, the invariant is zero).

Thus, we find that $z = y$. Now changing the orientation, we convert x into z, and y into t. Therefore, $x = t$.

Lemma 3. *Adding the Maslov index μ with a suitable coefficient to any additive invariant J, the invariant $J' = J + c\mu$ can be chosen to take identical values on all four hearts, moreover, the*

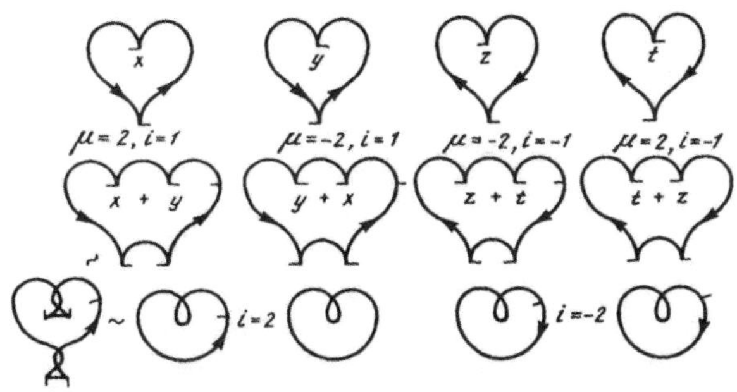

Fig. 18. Four hearts and their connected sums.

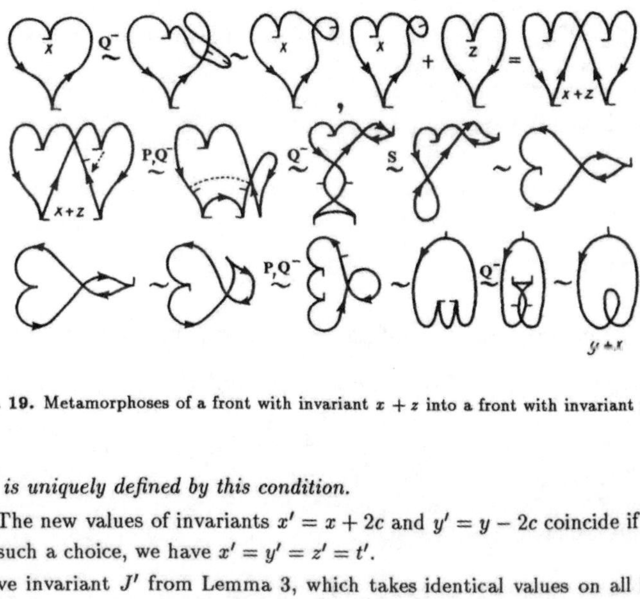

Fig. 19. Metamorphoses of a front with invariant $x + z$ into a front with invariant $y + z$.

coefficient "c" is uniquely defined by this condition.

Proof. The new values of invariants $x' = x + 2c$ and $y' = y - 2c$ coincide if $4c = y - x$. By Lemma 2, for such a choice, we have $x' = y' = z' = t'$.

The additive invariant J' from Lemma 3, which takes identical values on all hearts, is called the *normed* invariant. Let u denote the generic value of the normed invariant J on all hearts. By additivity, we find that $J(K_2) = 2u$ for any orientations and coorientations. Consequently, $J(K_{i+1}) = i(2u)$ for any orientations and coorientations (by Lemma 1). Representing the basis curves of Fig. 12 as sums ($K_{0,k} = \infty + kK_{1,1}$, and for $i \geq 0$ $K_{i+1,k} = K_i + kK_{1,1}$), we obtain the following conclusion.

Lemma 4. *The value of a normed additive invariant J on all basis curves shown in Fig. 12 depends on its value u on hearts, i.e., this invariant on basis curves is proportional to the invariant J^+ : $J(K_{0,k}) = ku$, $J(K_{i+1,k}) = (2i + k)u$.*

Corollary. *The jump of a local normed additive invariant for positive intersection of a hazardous tangency stratum is double the invariant on a heart but with minus sign.*

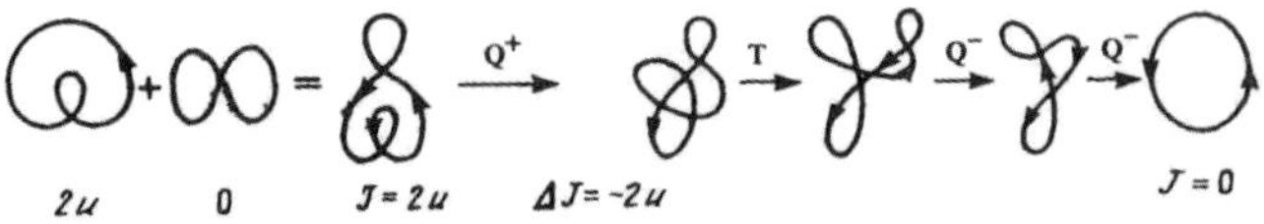

Fig. 20. Computation of the jump of the invariant for positive passing through a hazardous self-tangency.

Proof. Figure 20 shows a series of metamorphoses which converts the curve K_2 on which the invariant is $2u$ into a circle where the invariant is zero. It includes one positive intersection of the hazardous tangency stratum Q^+.

Since the invariant is local, the magnitude of the jump for any hazardous positive intersection is also equal to $-2u$ as in our example.

Remark. Hence, it immediately follows that the jump of the invariant for passing through a negative hazardous self-tangent point (where the coorientations of fronts coincide at the point of tangency and the coorientations are opposite) towards the increasing number of double points ($2u$) is opposite to the jump for passing through a positive self-tangency towards the increasing number of double points (see the diagram QP in Fig. 6).

Thus, both the jumps of a local normed additive invariant J and its values on basis immersions of each component of the immersion space are proportional (with proportionality coefficient $-u$) to those of the invariant J^+; consequently, $J = -uJ^+$. This completes the proof of the uniqueness theorem.

Remark. I am not aware of any nonlocal additive invariants.

10. INTERLACING COEFFICIENTS OF LEGENDRE CURVES

The invariant J^+ experiences jumps when the corresponding Legendre curve in $M^3 = ST^*\mathbb{R}^2$ suffers self-intersection and one of its branches passes through the other. It is, therefore, natural to associate this invariant with the "coefficient of interlacing" of the Legendre curve. Bennequin [3] defined such a coefficient for Legendre curves in $\mathbb{R}^3$. In our case, however, the Bennequin invariant is not suitable, since the contact manifold $M^3 \approx \mathbb{R}^2 \times S^1$ is not simply connected. Below we describe the generalized invariant due to Tabachnikov [2] in a slightly modified form. The following lemma is well known.

Lemma 1. *The universal covering space of a manifold M^3 of cooriented contact elements of the plane is the standard contact space $\mathbb{R}^3$ with the usual contact structure $dz = pdq$.*

Proof. In the (x, y, φ) coordinates introduced in Section 5, our contact structure takes the form $\alpha = 0$, where $\alpha = \cos\varphi dx + \sin\varphi dy$.

The diffeomorphism $(x, y, \varphi) \mapsto (p, z, \varphi)$, where $p = x\sin\varphi - y\cos\varphi$ and $z = x\cos\varphi + y\sin\varphi$, converts α to the form $dz - pd\varphi$.

Now we consider the complete preimage of a given Legendre curve in the covering space. We must distinguish two distinct cases.

Case 1. The index of the initial Legendre immersion is zero. In this case, the preimage of the immersion in the covering space $\mathbb{R}^3$ consists of a denumerable set of immersions of a circle in $\mathbb{R}^3$, which differ one from another by a displacement over an integral multiple of 2π along the φ-axis in the space with coordinates (x, y, φ). Let C_0 denote one of these immersed circles and C_j $(j \in \mathbb{Z})$ its displacement by $2\pi j$.

For a generic Legendre immersion, the front has no self-tangencies and, consequently, the curves C_j are imbedded and do not intersect pairwise.

Definition. The *generalized Bennequin invariant* L_j is defined to be the interlacing coefficient of the curves C_0 and C_j in $\mathbb{R}^3$ for $j \neq 0$ and is equal to the usual Bennequin invariant for $j = 0$.

Remark. The coefficient L_j does not change when the orientation of the curve C_0 changes (because the orientation of the C_j also changes in this case), or when the order changes ($l(C_0, C_j) = l(C_j, C_0)$), or when the curve is displaced ($l(C_i, C_{i+j}) = l(C_0, C_j)$). In particular, $L_{-j} = L_j$.

The usual Bennequin invariant, which we denote as L_0, is defined to be the coefficient of interlacing of a curve with its perturbation obtained by the following construction: we move every point of the original Legendre curve in the direction of the vector in the contact plane which is transversal to the tangent vector of the curve and which continuously changes along the curve (or—and this is more convenient—in the direction transversal to the contact plane. This leads to the same interlacing coefficient).

Such a vector field with the properties mentioned above (for example, normal to the tangent vector or to the contact plane in some Riemann metric) exists, because our contact planes are oriented (in the covering space, this obviously is a consequence of its simple connectedness; but in our case there is the orientation of contact planes on M^3 itself, since our contact elements are transversely oriented).

Remark. The coefficients L_j $(j \neq 0)$ are invariant under the orientation-preserving homeomorphisms of $\mathbb{R}^3$, particularly, under the contactomorphisms of the space $\mathbb{R}^3$ (relative to which the Bennequin invariant L_0 is also invariant). However, the Bennequin invariant is a contact invariant of the Legendre curve C_0 in $\mathbb{R}^3$, whereas all other invariants L_i are not contact-invariants. The curves C_0 and C_0' which are contact-equivalent in $\mathbb{R}^3$ can generate after displacements differently interlaced curves ($l(C_0, C_i) \neq l(C_0', C_i')$). This will not happen, if the contactomorphism, which takes C_0 into C_0', is projected into a contactomorphism of M^3: in this case, the pair (C_0, C_i) and (C_0', C_i') are contact-equivalent and have identical interlacing coefficients.

The sign of the coefficients L_j depends on the orientation of the contact manifold. But the orientation is standard, because the form $\alpha \wedge d\alpha$ defining it does not change when the sign of α changes.

Case 2. The index of the original Legendre immersion i is not 0.

For the sake of convenience, we can assume that $i > 0$ (the case $i < 0$ is reduced to this one by the change of orientation).

In this case, the curve in $\mathbb{R}^3$ which covers the original Legendre immersion in M^3 is not compact. It consists of i connected components, and φ varies from $-\infty$ to $+\infty$ over each of them. Each component is mapped into itself under displacements by $2\pi ik$. Under displacements by $2\pi(ik + j)$ $(j = 0, 1, \ldots, i-1)$, the component C_0 (which we choose arbitrarily) is transformed into all other components C_j. It is more convenient to assume that $j \in \mathbb{Z}_i$, because $C_{j+i} = C_j$ as a set (the corresponding immersions of the line differ, of course, by the $2\pi i$-displacement along the φ-axis).

Now we must define the "coefficient of interlacing" L_j between two infinite curves C_0 and C_j. For this purpose, let us consider the i-fold covering $\tilde{M} \to M$. The images $\tilde{C}_j$ of all i curves C_j

$(j = 0, \ldots, i-1)$ on $\tilde{M} \approx \mathbb{R}^2 \times S^1$ are closed and homotopic to the generator — circle, i.e., go round the solid torus $\tilde{M}$ once. Consider a (noncompact) two-dimensional chain in the solid torus $\tilde{M}$ joining $\tilde{C}_0$ with the "curve at infinity" defined in the following way: join $\tilde{C}_0$ with the circle $\{x = N, y = 0, \varphi \bmod 2\pi i$ is arbitrary$\}$ by some 2-chain (here N is quite large, so $|x| < N$ everywhere along C_0), and then add an infinite 2-chain $\{x \geq N, y = 0, \varphi \bmod 2\pi i$ is arbitrary$\}$ such that the boundary of the resultant 2-chain F is $\tilde{C}_0$.

Now we assume that L_j for $j \neq 0$ is equal to the index of intersection of the film F with the circle $\tilde{C}_j$ in $\tilde{M}$. If $j = 0$, we assume that L_j is equal to the index of intersection of F with $\tilde{C}_0'$ obtained from $\tilde{C}_0$ by a small deformation in the direction of the vector transversal to the contact structure (as in the definition of the Bennequin invariant).

Remark 1. While remaining in $\mathbb{R}^3$, we could have constructed mean asymptotic coefficients of interlacing of C_0 and C_j or C_0 and C_0'; they are equal to the coefficients L_j defined above.

Remark 2. $L_j = L_{-j} = L_{i-j}$, as in the case $i = 0$, since the interlacing coefficient is symmetric.

Remark 3. Unlike for the case $i = 0$, for $i \neq 0$ the invariants L_j are not preserved under the homeomorphisms of the space $\mathbb{R}^3$, even under those which are induced by homeomorphisms of $\tilde{M}^3$ or M^3.

However, I am not aware of any contactomorphism of $\tilde{M}_3$ changing the interlacing coefficient L_j.

Consider a contactomorphism of a solid torus (with the contact structure from M^3 or $\tilde{M}^3$) onto itself. The neighborhood of the torus bounding the solid torus can be mapped onto this torus in a standard way (in our coordinates $(x, y, \varphi) \mapsto (\theta, \varphi)$, where $x = r\cos\theta$, $y = r\sin\theta$). The image of a sufficiently thin neighborhood of the boundary torus under contactomorphism of the solid torus also falls into the neighborhood which is mapped onto the torus $\{(\theta, \varphi)\}$ in a standard way.

Consider a circle $x = \text{const}$, $y = \text{const}$ lying inside a sufficiently thin neighborhood of the boundary torus (with a sufficiently large $x^2 + y^2$). Its image falls into the neighborhood of the boundary torus projected onto the boundary torus in a standard way. After the projection we obtain a curve on the boundary torus.

If the diffeomorphism is arbitrary, then this curve may not be homologous to the projection ($\theta = \text{const}$) of the original circle of the boundary torus (it may differ from the original curve by an integral multiple of the meridian $\varphi = \text{const}$).

But, the diffeomorphism which preserves the contact structure of a solid torus seems to be incapable to implement such a twisting. This could follow from the connectedness of the group of contactomorphisms of a solid torus which preserve the coorientation of the contact planes and the nonzero class of one-dimensional homologies of a solid torus, i.e., the direction of the φ-axis.

In any case, the interlacing coefficients L_j do not depend on the coordinates (x, y) on the plane of the front in terms of which they were defined.

Definition. The *polynomial of interlacings* of a Legendre immersion of a circle without hazardous self-tangencies is a Lorant polynomial

$$L(t) = \sum L_j t^j.$$

Remark. This polynomial is regarded as an element of the group ring of the cyclic group $\mathbb{Z}/i\mathbb{Z}$, where i is the immersion index. Therefore, for $i \neq 0$ the sum has i terms for $j = 0, 1, \ldots, i-1$. For $i = 0$, the number of terms is also finite.

Definition. The *interlacing number* of a Legendre immersion of a circle without hazardous self-tangencies is defined to be the number

$$L(1) = \sum L_j.$$

Below we shall see that the interlacing number and the invariant J^+ defined in terms of the intersections with the discriminant of hazardous tangencies are related by the expression

$$J^+ + L(1) = 1.$$

11. COMPUTATION FORMULAS

First recall the formulas for the ordinary and Maslov indices of an oriented and cooriented front.

The Maslov index is equal to the difference between the numbers of direct and inverse cuspidal points. This index is always even.

Remark. I do not know whether the modulus of the Maslov index of a curve is equal to the minimal number of cusps on the curves obtained through a given chain of metamorphoses without hazardous self-tangencies (most probably, it is not). For example, can every front with zero Maslov index be converted into a cuspless front without hazardous self-tangencies? This can be done for an initial front with no self-intersection points (such a front with Maslov index equal to μ can be converted into a front with $|\mu|$ cusps).

Indeed, a front without self-intersection points is obtained from a circle by adding outward and inward cusps. Two adjacent cusps identically directed can be replaced (without hazardous self-tangent points) by a loop in the same direction. A cusp adjacent to an oppositely directed loop can be pulled through it to the other side of the loop without hazardous self-tangent points. A cusp adjacent to an identically directed loop can also be dragged to the other side of the loop without hazardous self-tangent points. For this, it suffices to replace the loop by a pair of identically directed cusps and then replace the other two cusps (out of these three identically directed adjacent cusps) by a loop.

All cusps are thus gathered at one place, then we can replace the adjacent pairs of identically directed cusps by loops. Ultimately, there remains a sequence of cusps of alternating directions. Such cusps are either all direct or all inverse, and their number is equal to $|\mu|$.

The index of a front, when a cusp is added to it, changes by 1/2, i.e., the contribution of the cusp to the index is 1/2 if the directions from the cusp to the nearby points on the incoming and outgoing branches taken in this order orient the plane positively (and the contribution is $-1/2$ if they orient the plane negatively).

This index does not depend on the coorientation of the front.

Now let us express the formulas for the coefficients of interlacing of generic Legendre immersions.

Curves of zero index. Let X denote the set of self-intersection points of the front and let X_j ($j > 0$) denote the subset of those for which the difference between the values of the argument of the contact element φ on both branches is greater in modulus than $2\pi j$. This difference is correctly

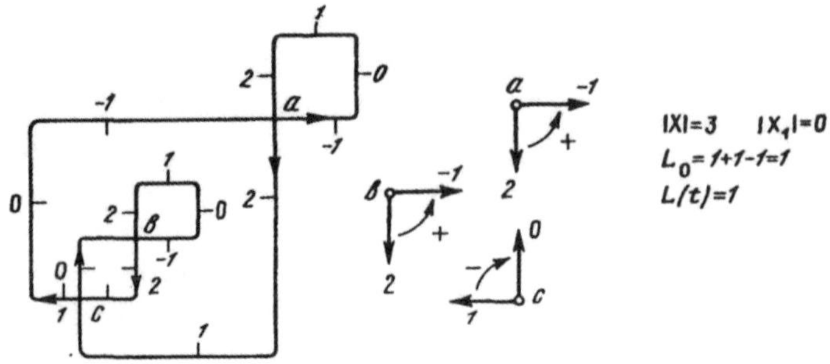

Fig. 21. Computation of the interlacing coefficient for a cuspless curve of zero index.

defined (and does not depend on which of the two loops with the ends at the self-intersection point is taken), because the complete index is equal to zero.

Theorem. *The interlacing coefficients* L_j *are given by the formula*

$$L_j = \sum_{X_j} \frac{\mathrm{op}\,(>,<)}{\mathrm{op}\,(x,y)}, \qquad j > 0,$$

where summation is taken over the set X_j, *and* op $(>,<)$ *denotes the orientation of the plane* (x,y) *defined by the vectors orienting the front at the self-intersection point. The vector with larger* φ *is taken to be the first.*

The interlacing coefficient L_0 *is given by the formula*

$$L_0 = \sum_{X} \frac{\mathrm{op}\,(>,<)}{\mathrm{op}\,(x,y)} + \frac{\nu}{2},$$

where ν *is the number of cusps (their types are unimportant).*

Example. Computation of the interlacing coefficient $L(t)$ for a curve of zero index with three self-intersection points is shown in Fig. 21. The argument of the coorienting normal is measured in units of $\pi/2$ from the x-axis to the y-axis, so the modulus of the difference of the arguments at the points X_j must be greater than four. For our example, even X_1 is empty.

Similar computations yield the other values of $L(t)$ shown in Fig. 22.

Caution. For computing op $(>,<)$, at first the branches must be ordered with respect to the argument of the coorienting normal, and then a frame must be constructed not from these normals but from tangent vectors orienting the front. For a cuspless curve, the normals and tangents are interchangeable, but this is not true if there are cusps (the orientations differ if there is an odd number of cusps on each of the loops into which a self-intersection point divides the front).

Example. A circle in $\mathbb{R}^3$ covering a figure eight curve can be transformed by a contactomorphism of $\mathbb{R}^3$ into a circle covering the fourth curve in Fig. 22 (this curve is obtained from a figure eight curve by means of one passing through a hazardous tangency). For a figure eight curve, we have $L(1) = 1$, and for the fourth curve in Fig. 22, we have $L(1) = -1$. Consequently, unlike the

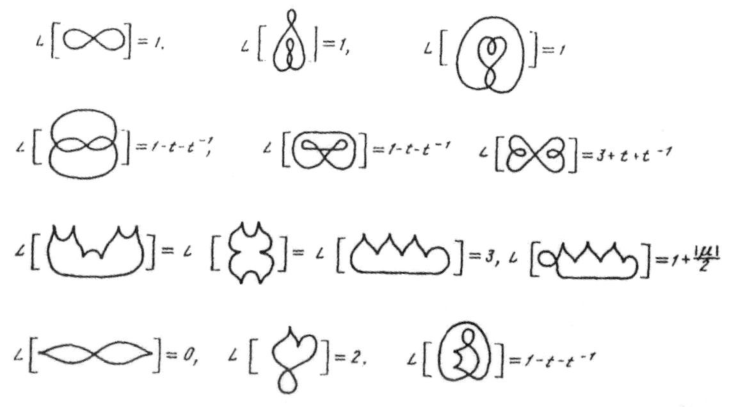

Fig. 22. Polynomials of interlacings of zero index curves.

Bennequin invariant L_0, the other numbers L_j and their sum $L(1)$ are not invariants of the contact class of a circle in $\mathbb{R}^3$ covering a zero index curve.

Proof of the theorem. Let C_0 be a circle in $\mathbb{R}^3$ which covers our curve. Construct a 2-chain with boundary C_0, which consists of a cylinder directed along φ-axis from C_0 to the plane $\varphi = N$, where N is sufficiently large (therefore the curves C_j with $j \leq 0$ do not intersect the plane $\varphi = N$). Extend the upper circle of the cylinder in the plane $\varphi = N$ by a cone.

If $j < 0$, only the points X_j make contributions to the index of intersection of the curve C_j with this 2-chain. Their orientations yields the rule of computation of L_j stated in the theorem. If $j = 0$ and there are no cusps on the front, then a shift by $-\varepsilon$ of the curve C_0 along the φ-axis deforms the curve C_0 in the direction of the contact plane. Computation of the index of intersection of the shifted curve with the cylinder yields the sum computed by the formula for L_0 over the self-intersection points of the front.

If there are cusps, then instead of shifting along the φ-axis, we move the curve C_0 transversely to the contact plane. It suffices just to move the front in the direction of coorienting normal in the neighborhood of every cuspidal point. The shifted front will intersect the original front near the cusp only once. In one half of the cusps, this intersection will not imply the intersection of the shifted curve C_0 with the cylinder, where in the other half it does imply such an intersection (these and other cusps alternate on the front). Computation of the orientations shows that the contribution to L_0 of all these $\nu/2$ cusps is positive.

Curves of index $i > 0$. Consider the projection P_0 of the curve C_0 lifted in $\mathbb{R}^3$ to the plane (y, φ) within the limits of one period (so the extension of the projection is obtained by $2\pi i$-shifting along the φ-axis). We identify the point (y, φ) with the point $(y, \varphi + 2\pi i)$. Therefore, P_0 and the projections P_j of other liftings ($j = 0, \ldots, i-1$) lie on a cylinder.

Let X_j denote the set of those intersection points of the projections P_0 and P_j for which the value of x at the point corresponding to P_0 is less than the value at the point corresponding to P_j. Let $(x_<, x_>)$ denote the respective tangent vectors of projections at these points. We assume that

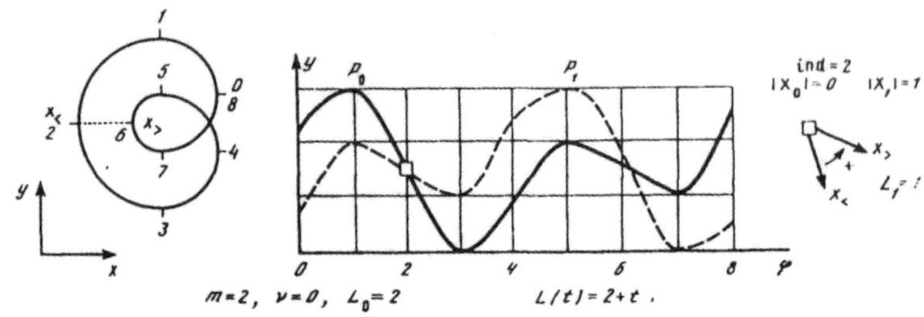

Fig. 23. Computation of the interlacing coefficients $L[K_2]$.

the coordinates (x, y) are such that the projection P_0 (and, consequently, its displacements P_j) is an immersed curve without singular points other than transversal self-intersections. Furthermore, we assume that all the intersections of P_j and P_k are transversal (all this can be attained by rotating the coordinate system).

Theorem. *The interlacing coefficients L_j for a curve of index* ind *are given by the formula*

$$L_j = \sum_{X_j} \frac{\mathrm{op}\,(x_<, x_>)}{\mathrm{op}\,(\varphi, y)}, \qquad 0 < j < \mathrm{ind}\,,$$

and the formula

$$L_0 = \sum_{X_0} \frac{\mathrm{op}\,(x_<, x_>)}{\mathrm{op}\,(\varphi, y)} + m - \frac{\nu}{2}$$

for $j = 0$, where m is the number of maxima of y on P_0, and ν is the number of cuspidal points on the original front.

Example. For cuspless basis curves (Fig. 1), we have

$$L(K_2) = 2 + t, \qquad L(K_3) = 3 + t + t^2, \qquad L(K_4) = 4 + t + t^2 + t^3, \ldots.$$

Computation for K_2 is illustrated in Fig. 23.

Example. For basis curves $K_{i+1,k}$ (Fig. 12) ($i \geq 0$ self-intersection points and $2k$ cuspidal points) and $K_{0,k}$, we have

$$L[K_{i+1,k}] = i + k + 1 + t + t^2 + \ldots + t^i, \qquad L[K_{0,k}] = k + 1.$$

Computations (for $i = 3$, $k = 2$) are illustrated in Fig. 24.

Remark. The algorithm described here is also suitable for zero-index curves. However, for these curves it is more convenient to use projection into the plane (x, y).

Proof of the theorem. The corresponding 2-chain is obtained by adding a ray $\{(\xi \geq x, y, \varphi)\}$ to the curve C_0 at the point (x, y, φ). The intersection indices are computed in the same way as for the case ind $= 0$.

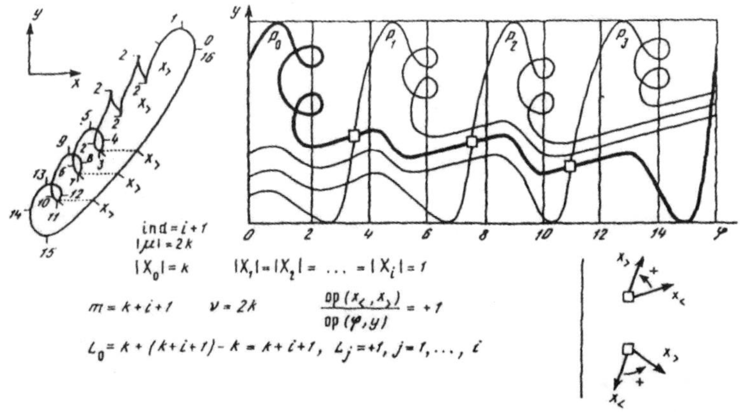

Fig. 24. Computation of interlacing polynomial $L[K_{i+1,k}]$.

12. INTERLACING NUMBER AND THE INVARIANT J^+

Theorem. *The sum of the interlacing number* $L(1)$ *and the invariant* J^+ *for any Legendre immersion without hazardous self-tangent points of the front is equal to* 1:

$$L[K](1) + J^+(K) = 1.$$

Proof. For the basis immersions of every connected component we have computed separately the interlacing number and the value of the invariant J^+ (Fig. 12). According to Section 11,

$$L[K_{i+1,k}](1) = 2i + k + 1 \quad \text{for} \quad i \geq 0,$$

$$L[K_{0,k}](1) = k + 1,$$

and, according to Section 7,

$$J^+(K_{i+1,k}) = -2i - k, \qquad J^+(K_{0,k}) = -k.$$

Therefore, it suffices to verify the following assertion.

Lemma. *When passing through a negative hazardous self-tangency, the interlacing number varies in the same way as the number of double points on the front, whereas passing through a positive hazardous self-tangent point it varies oppositely.*

Proof. For the sake of convenience, we assume that the contacting branches have oppositely directed curvatures (this can be achieved by a plane diffeomorphism which changes neither L nor J^+).

Figure 25 shows the contributions of a neighborhood of a self-tangent point to L in a coordinate system in which the common tangent to both branches of the front is transversal to the x-axis.

When passing through a tangency, the points with parallel tangents on both branches corresponding to identical values of y interchange their roles: the left branch on which before the tangency we had $x_<$ after the moment of tangency becomes the right branch with $x_>$.

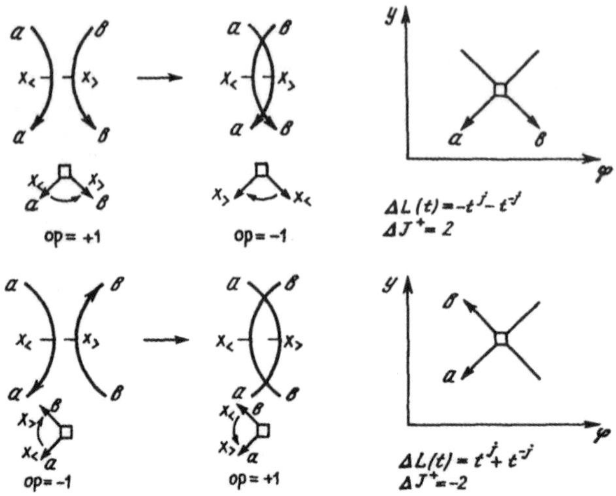

Fig. 25. Variation of $L(t)$ in passing through a hazardous tangency.

Figure 25 shows only one variant of positive and one of negative hazardous tangency; other variants can be obtained by changing the orientation of the whole front, and this affects neither $L(1)$ nor J^+.

The number j in Fig. 25 is the difference of the numbers of lifted curves. Before the moment of contact the branch a corresponds to P_0 and the branch b to P_j, whereas after it the branch b corresponds to P_0 and the branch a to the curve P_j (by assertion, we are dealing with that intersection point (out of the two) for which $x_<$ lies on the branch P_0).

This completes the proof of our lemma and, hence, of the theorem as well.

Corollary 1. *The invariant* $L(1) - 1$ *is additive under connected summation of fronts.*

Corollary 2. *The invariant* St *(strangeness) of ordinary immersions without triple points and cuspidal points can be expressed by an explicit integral formula not considering the metamorphoses of the curves.*

Indeed, we have $3\mathrm{St} = I^+ - J^+$, an explicit formula for I^+ is given in Section 2, and an explicit formula for J^+ is given by the computation algorithm of $L(t)$ in the theorem proved above.

13. TABLES OF THE SIMPLEST FRONTS

The tables below show all fronts with not more than two cusps and not more than two self-intersection points. For each front, the values of the invariant J^+ and the moduli of both indices (their signs can be alternated by a proper choice of the orientation and coorientation) are given.

The list was compiled by adding two cusps of arbitrary directions at arbitrary branches to a circle, to figure eight curve, to the simplest curve of index 2, and to the five curves with two self-intersection points shown in Fig. 2 (symmetric curves do not differ).

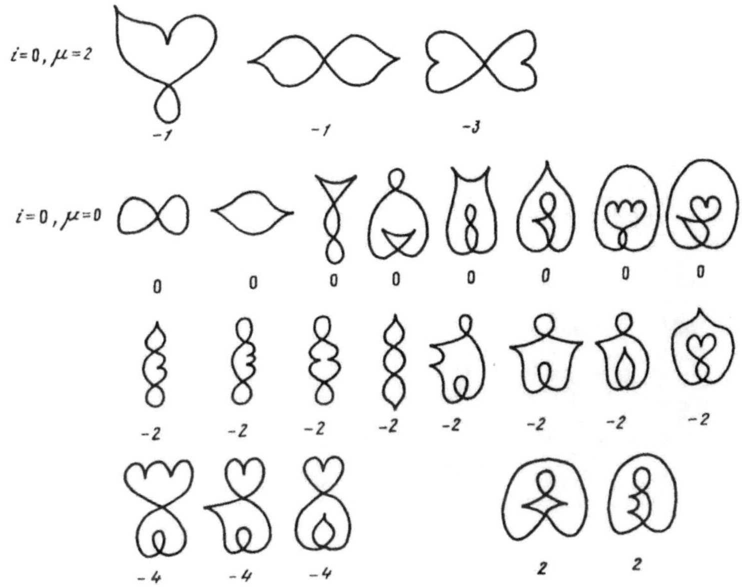

Fig. 26. Fronts of index $i = 0$, $n \leq 2$, $\nu \leq 2$.

Three fronts are obtained from a circle, six fronts from a figure eight curve, and ten curves from a curve of index 2 with one self-intersection point. The curves in Fig. 2 yield 16, 25, 25, 16, 25 fronts, respectively.

All these 134 fronts are listed in the table below by the number of fronts for different values of the index i and the Maslov index μ. They are distributed with respect to indices as follows:

μ	Number of fronts				
	$i=0$	$i=1$	$i=2$	$i=3$	$i=4$
0	21	10	33	5	12
2	3	29	4	17	0

Fronts of indices $i = 0; 1$ are shown in Figs. 26 and 27.

We readily find that there are far more front types if i and $\mu/2$ (i.e., the degrees of mappings onto both circles S_1^1 and S_2^1, Section 1) are of the same parity. This rule holds even for $i = 2$ (Fig. 28). Fronts of indices $i = 3; 4$ are shown in Fig. 29.

14. LAGRANGE COLLAPSE

The theory of singularities and perestroikas of fronts of Legendre curves in contact geometry is closely related to the theory of singularities of caustic (sets of critical values) of Lagrange mappings in symplectic geometry.

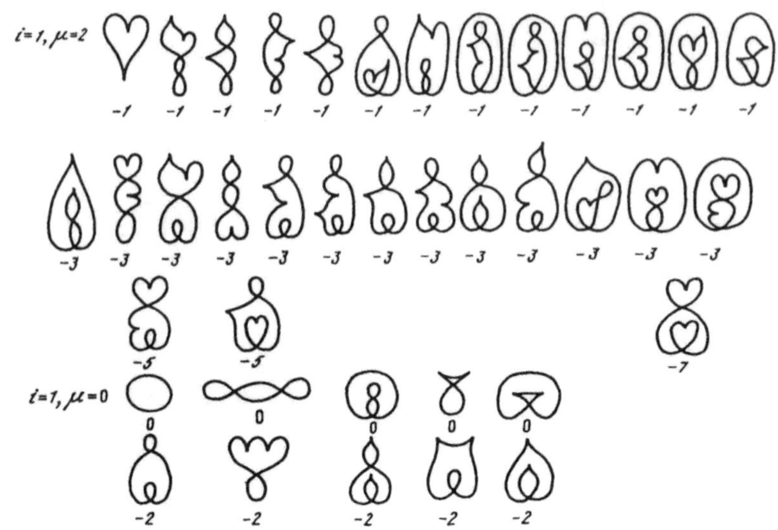

Fig. 27. Fronts of index $i = 1$, $n \leq 2$, $\nu \leq 2$.

Below we study one of the problems in this theory—singularities arising in the transformations of a point caustic to a generic curve. This problem, which was first observed in [4], is a symplectic generalization of the problem concerning caustics on Riemannian surfaces close to spheres and vertices of plane curves. The results presented below, which were preliminarily published in [4], are generalizations of the classical four-vertex theorem.

Definition. A *standard Lagrange cylinder* $L^n \hookrightarrow T^*\mathbb{R}^n$ is defined to be the graph of the gradient of the function $\pm|q|$ in the Euclidean space:

$$L^n = \{p, q : \exists t : p^2 = 1, \ q = pt\}.$$

This formula defines the exact Lagrange imbedding of a cylinder $S^{n-1} \times \mathbb{R}$ ($p \in S^{n-1}$, $t \in \mathbb{R}$) into the space of cotangent bundle $\mathbb{R}^{2n}$ of the Euclidean space $\mathbb{R}^n$.

The standard cylinder is an exact Lagrange (smooth) manifold, since $pdq = dt$ on L.

Definition. The *standard Lagrange collapse*

$$L^n \hookrightarrow T^*\mathbb{R}^n \twoheadrightarrow \mathbb{R}^n$$

is a projection of the standard Lagrange cylinder onto the base of the cotangent bundle of the Euclidean space $\mathbb{R}^n$.

The projection $(p, q) \mapsto q$ of L^n into $\mathbb{R}^n$ is a local two-sheet diffeomorphism over every point $q \neq 0$. An entire sphere $S^{n-1} \times 0$ is contracted to the origin.

The set of critical values (caustic) of the mapping $L^n \to \mathbb{R}^n$ consists of one infinitely degenerate point 0. We shall study the caustics of exact Lagrange manifolds close to L^n.

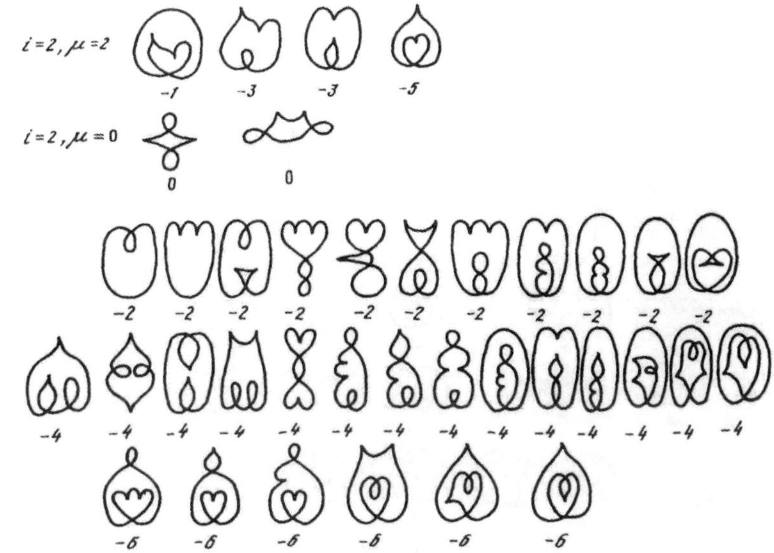

Fig. 28. Fronts of index $i = 2$, $n \leq 2$, $\nu \leq 2$.

Example 1. Consider the family of normals to a plane circle. The unit normal vectors at all points of the plane form a Lagrange cylinder (tangent vectors are identified with the cotangent ones on the Euclidean plane).

This cylinder $S^1 \times \mathbb{R}$ is a standard Lagrange cylinder in $\mathbb{R}^4$, and its projection on $\mathbb{R}^2$ is a standard Lagrange collapse.

When a circle is deformed, the cylinder of normal unit vectors suffers deformation, remaining an exact Lagrange cylinder close to the initial standard one. The critical points of the projection of the deformed cylinder onto a plane are the centers of curvature of the deformed curve. The four-vertex theorem asserts that the caustic curve (i.e., the focal set — the set of centers of curvature) of a generic convex curve has at least four cusps.

Example 2. Consider the geodesics starting from the north pole of the sphere S^2. They meet together again at the south pole. The corresponding Lagrange submanifold of the space of cotangent bundle of the sphere is formed by the unit vectors tangent to these geodesics at all points (tangent vectors are identified with the cotangents on a Riemannian manifold).

This Lagrange submanifold is smooth. Its projection onto the sphere has singularities over the poles of the sphere. In the neighborhood of the preimage of the poles, these singularities are Lagrange-equivalent to the standard Lagrange collapse of a cylinder into $\mathbb{R}^2$ (i.e., they are converted into a standard Lagrange collapse by a symplectomorphism of the phase space T^*S^2 fibered over a diffeomorphism of a neighborhood of the pole). This equivalence is implied in the Morse lemma, since the square of the distance to the pole has a nondegenerate critical point at it.

When the metric on the sphere is deformed (but the curvature remains positive), the manifold

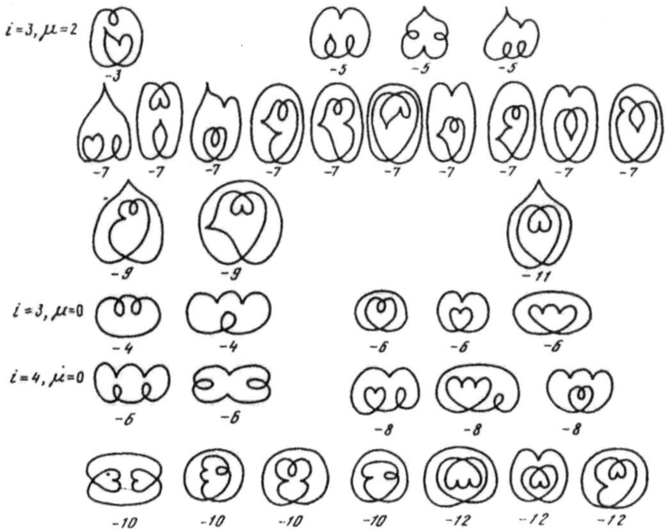

Fig. 29. Fronts of indices $i = 3$ and 4, $n \leq 2$, $\nu \leq 2$.

L of unit vectors tangent to the geodesics starting from the north pole remains to be a Lagrange manifold (this is a cylinder immersed into T^*S^2). The critical points of its Lagrange projection onto S^2 form a sequence of closed curves which are disjoint on the immersed cylinder.

Their projections into a sphere are the caustic curves of the initial pole (the first caustic curve is close to the south pole, whereas the second caustic is close to the north pole, the third is close to the south pole, and so on if the metric is close to the standard one).

The theorem formulated below asserts that each of these short generic caustics has at least four cusps. I guess that this is also true for "long" caustics of any convex surface. For the first caustic curve, this is the classical result of the variational theory of geodesics. But for the succeeding caustics, as far as I know, the lower estimate of the number of critical points is not known even for an ellipsoid.

The problem of cuspidal points on the caustics of ellipsoids was investigated by Jacobi [11], who asserted that these generic caustics had at least four cusps. But Jacobi passed away without proving the assertion. The reviewers of the last volume of his collected works made attempts to prove it but in vain. The problem of the number of singular points on the caustics of an ellipsoid so far seems to dodge the efforts of algebraic geometers, although the problem of geodesics is integrable and was resolved by Jacobi in terms of theta-functions. Particular results are reported in [12], which was pointed out to me by H. Knorrer, whom I express my indebtedness to.

Metamorphoses of fronts stated in Examples 1 and 2 are those of families of equidistant curves (the initial plane curve in Example 1 and the initial pole in Example 2) when the distance to the original object changes. Under standard collapse, equidistant curves are contracted to a point, but under perturbations they suffer near this point several transformations, which change the coorientation of the initial front which is diffeomorphic to a circle. For example, successive metamorphoses of the equidistant curves of an ellipse are shown at the bottom of Fig. 13 ($k = 0$, $j = 1$; the initial equidistant curve is on the right and metamorphosed curve—on the left).

Similar metamorphoses of fronts can also be determined for the general Lagrange collapse without any Riemannian metric. (It suffices to start with the function $t = \int p\,dq$ on the perturbed Lagrange cylinder. Its level hypersurfaces are exactly projected into the "fronts".)

It is worthwhile to study the generalization of Lagrange collapse along a submanifold of the configuration space, replacing $\pm|q|$ by $\pm$ distance to the submanifold.

15. BIFURCATION OF THE SINGULARITIES OF LAGRANGE COLLAPSE

The general results proved below as applied to the study of bifurcation of the singularity of the standard Lagrange collapse yield the following assertion.

Theorem. *There are at least four cusps on the caustic of the Lagrange projection onto a plane of a perturbed exact Lagrange cylinder if it is sufficiently close (with derivatives) to the standard Lagrange cylinder.*

Remark. The proximity constraints under which this assertion is proved are explicitly formulated in Section 18. Probably, the result holds even under less stringent constraints. For example, we may guess that it holds for imbedded exact Lagrange cylinders obtained from a standard cylinder via Hamiltonian isotopy with a compact support or at least for optical cylinders (belonging to the hypersurface of the phase space which is quadratically convex with respect to impulses); probably, the condition $p \neq 0$ on the cylinder, which is also used in the proof given below, is important.

These conditions are satisfied for cylinders which are close to the standard one. The number of cusps for immersed cylinders may be less (this is probably implied in the results of [13] concerning the uniqueness of the "index" distinguishing the families of functions in one variable without critical points of codimension 2. This result is proved in [13] only for families of functions on a line, but we need its version for families of functions on a circle).

The cylinders encountered in the four-vertex theorem of a plane curve satisfy the conditions of closeness to a standard Lagrange cylinder. Therefore the four-vertex theorem is a particular case of our general symplectic theorem.

In the problem of the caustics of the point on a convex surface the parts of the cylinder near the poles satisfy the proximity conditions only if the metric is sufficiently close to the standard one. The constraints of this proximity, however, become more and more restrictive as the number of the caustic curve increases.

The symplectic version of the four-vertex theorem is reduced with the use of generating functions of Lagrange manifolds (Section 17) to the following theorem.

16. THE STURM–TABACHNIKOV THEOREM

Consider a smooth 2π-periodic function $f(S, \varphi)$ on a plane, which is bounded together with its second derivative in φ.

Theorem. *The restriction of the coordinate function S to a curve on a cylinder $\{S, \varphi \bmod 2\pi\}$ defined by the equation*

$$S + f(S, \varphi) + \frac{\partial^2 f}{\partial \varphi^2} = 0$$

has at least four critical points (provided the curve is smooth).

Proof. The curve consists of the connected components lying in the ring $|S| \leq c_1$. If there is more than one component, the maxima and minima of S on the components give the four critical points we need.

Assume that there is only one component. The function on the cylinder

$$H(S, \varphi) = S + f + \frac{\partial^2 f}{\partial \varphi^2}$$

is positive on one side of the curve separating the domains $S > c_1$ and $S < -c_1$, and is negative on the other side.

Now assume that there are less than four critical points of the restriction of S to our curve. In this case, we have only one maximum and one minimum of S on the curve, and every intermediate value of S between the maximum and minimum occurs at two distinct points. The difference between the values of φ at these two points defines a continuous function on S. The value of this continuous function at the point of minimum differs from its value at the point of maximum by 2π. Therefore, there is a value S_0 which is attained at opposite points of the curve: $\varphi_2 - \varphi_1 = \pi \pmod{2\pi}$.

The signs of $H(S_0, \cdot)$ on two arcs separated by these points are different. Therefore, there exists a function $a \cos \varphi + b \sin \varphi$, which has the same zeros as the function $H(S_0, \cdot)$ and which has the same sign at other points of the circle. For this function, we have

$$\int_0^{2\pi} (a \cos \varphi + b \sin \varphi) H(S_0, \varphi)\, d\varphi > 0.$$

But the function $f + \frac{\partial^2 f}{\partial \varphi^2}$ for every fixed value of S is orthogonal to $\sin \varphi$ and $\cos \varphi$ (since the operator $\frac{\partial^2}{\partial \varphi^2} + 1$ annihilates these harmonics of the Fourier series). Hence, the positive integral is zero—this contradiction shows that there are at least four critical points on the restriction.

Remark. In the infinitesimal variant of the problem of birth of cusps on a caustic under infinitely small perturbations of a standard Lagrange collapse, the cuspidal points are given by the equation $f''' + f' = 0$, where f is a 2π-periodic function on a circle which describes infinitesimal perturbation.

Such an equation always has not less than four roots on a circle, since the function $f''' + f'$ is orthogonal to 1, $\sin \varphi$, and $\cos \varphi$. Tabachnikov [5] found that the function

$$\sum a_k \cos k\varphi + \sum b_k \sin k\varphi, \qquad k \geq n$$

has at least $2n$ roots on a circle (a similar result also holds for other Sturm–Liouville operators).

I did not find any application of this general theorem of Tabachnikov for estimating the number of Lagrange singularities A_n of higher order.

Perturbations of the Lagrange collapse in the spaces of higher dimensions, even in the infinitesimal formulation of the problem, yield specific analogs of the Sturm theory for functions on spheres. These analogs must yield, in particular, a symplectic generalization of the theorem of four umbilical points on a convex surface, and the lower estimate for the number of other singularities on caustics on convex hypersurfaces of higher dimensions and on focal sets of convex hypersurfaces in an Euclidean space, for example, in a three-dimensional space.

I must confess my inability to understand fully why a Sturm-like theory appears in the study of the deformations of Lagrange collapse. I believe that the transition from infinitesimal perturbations of a Lagrange collapse studied in the Sturm theory to moderate finite perturbations (studied in this paper) can be continued up to large perturbations in the same way as the passage from infinitesimal symplectic morphisms (and Lagrange intersections) studied in the Morse theory to moderate finite perturbations [14] was continued to the theory due to Conley, Sender, Floer and others.

(fixed points of infinitesimal symplectic morphisms) — (Morse theory) — (Theory of Conley, Sendler, Floer)

| | |

(Infinitesimal deformations of Lagrange collapse) — (Sturm theory) — (?).

17. ADAPTED SYMPLECTIC COORDINATES

To formulate the conditions of proximity of a perturbed cylinder to an unperturbed one, it is more convenient to describe a perturbed cylinder in a coordinate system constructed for the initial unperturbed cylinder. The matter is that the neighborhood of a Lagrange submanifold of any symplectic manifold is symplectomorphic to the neighborhood of the zero section in the space of its cotangent bundle (A. Weinstein theorem).

For a standard Lagrange cylinder, this neighborhood consists of the whole complement to the zero section $p = 0$ of the space of the cotangent bundle $T^*\mathbb{R}^n$. If also $n = 2$, the cotangent bundle of the cylinder is naturally trivialized, and we obtain not only a symplectomorphism onto the neighborhood of the section in the space of the cotangent bundle of the cylinder, but also a system of coordinates.

The resulting formulas look as follows. Consider the manifold $S^1 \times \mathbb{R}^3$ with coordinates $(\varphi \bmod 2\pi, r, A, B)$ and the space $\mathbb{R}^4 = T^*\mathbb{R}^2$ of the cotangent bundle of the plane $\{(x, y)\}$ with the usual symplectic coordinates (x, y, p_x, p_y) and action form $p_x dx + p_y dy$.

Definition. An *adapted chart* of the standard Lagrange collapse is defined to be the mapping $S^1 \times \mathbb{R}^3 \to \mathbb{R}^4$ given by the following formulas:

$$\begin{cases} p_x = A\cos\varphi, & p_y = A\sin\varphi, \\ x = r\cos\varphi + B\sin\varphi, & y = r\sin\varphi - B\cos\varphi. \end{cases} \tag{1}$$

Lemma 1. *Mapping* (1) *is a diffeomorphism of the domain* $A > 0$ *in* $S^1 \times \mathbb{R}^3$ *onto the domain* $p \neq 0$ *in* $\mathbb{R}^4$.

Proof. From formula (1), for the point $p \neq 0$ we get unique values of $A > 0$ and $\varphi \bmod 2\pi$. The vector (r, B) is obtained by rotating the vector (x, y) by angle φ.

Lemma 2. *The following identity holds:*

$$p_x dx + p_y dy = P_r dr + P_\varphi d\varphi, \tag{2}$$

where

$$P_r = A, \qquad P_\varphi = AB. \tag{3}$$

Proof. The formulas (1) imply that, in the expansion $p_x dx + p_y dy$, the coefficients of dA and dB are zero. Computing the coefficients of dr and $d\varphi$, we find that

$$A\cos^2\varphi + A\sin^2\varphi = A, \quad Ar(\sin\varphi\cos\varphi - \cos\varphi\sin\varphi) + AB(\cos^2\varphi + \sin^2\varphi) = AB.$$

Corollary. *Mapping* (1) *is a symplectomorphism of the domain $P_r > 0$ in the space of the cotangent bundle of the cylinder $\{\varphi \bmod 2\pi, r\}$ (with standard symplectic coordinates $(r, \varphi, P_r, P_\varphi)$ and action form $P_r dr + P_\varphi d\varphi$) to the complement of the zero section $p = 0$ of the space of the cotangent bundle of a plane (with standard symplectic coordinates (x, y, p_x, p_y) and action form $p_x dx + p_y dy$). The action forms are transformed one into another, and the section $(P_r = 1, P_\varphi = 0)$ of the space of the cotangent bundle of the cylinder $\{\varphi \bmod 2\pi, r\}$ is mapped onto the standard Lagrange cylinder in $\mathbb{R}^4$.*

An exact Lagrange cylinder close to the standard cylinder is given in adapted coordinates $(r, \varphi, P_r, P_\varphi)$ by the generating function $S(r, \varphi)$, which is 2π-periodic in φ, according to the formulas

$$P_r = \partial S/\partial r, \qquad P_\varphi = \partial S/\partial \varphi. \tag{4}$$

The function S is close to r if the cylinder is close to the standard cylinder.

Our main result concerning the perturbations of Lagrange collapse is formulated in these notations in the following section.

18. THEOREM ON MODERATE FINITE PERTURBATIONS OF A STANDARD LAGRANGE COLLAPSE

Let $S(r, \varphi)$ be a smooth function which is φ 2π-periodic in φ, which coincides with r everywhere except the domain $|r| < C_2$, and which is strictly monotone everywhere in r (so we have $\partial S/\partial r > 0$).

Theorem. *The caustic curve of the projection on the plane (x, y) of a generic perturbed Lagrange cylinder defined by the generating function S (by formulas* (4) *of Section* 17) *has at least four cusps.*

Proof. The critical points of the projection are given by the equation

$$\det\left(\frac{\partial(x, y)}{\partial(r, \varphi)}\right) = 0.$$

This equation, by virtue of (1)–(4), for our perturbed cylinder takes the form

$$\det\begin{vmatrix} c + B_r s & s - B_r c \\ Bc + (-r + B_\varphi)s & Bs - (-r + B_\varphi)c \end{vmatrix} = 0,$$

where $c = \cos\varphi$ and $s = \sin\varphi$ (from now on, the signs in formulas of [4] are corrected).

Thus, the curve of the critical points of the projection of the cylinder to the plane (x, y) is given by the equation

$$-r + B_\varphi - BB_r = 0.$$

The differentials dx and dy vanish on the kernel of the derivative of the projection of the cylinder at the critical point. Consequently, according to (2) and (4), this kernel is given by the condition $dS = 0$.

The cusps on the caustic curve are the images of the points where the kernel touches the curve of the critical points. Hence, we must prove that the restriction of the function S to the curve of the critical points of the projection of the cylinder has at least four critical points.

Under the conditions of the theorem, we can use S and φ instead of r and φ as the coordinates on the cylinder. In these coordinates, the study of the critical points of the restriction of S to the curve of critical points leads us to a theorem similar to the Sturm–Tabachnikov theorem of Section 16.

From now onwards I will denote partial derivatives in the coordinate system (r, φ) by subscripts and in the coordinate system (S, φ)—by fractions, for example, u_φ is the derivative of the function u with respect to φ for constant r, and $\partial u/\partial\varphi$ is the derivative of the function u with respect to φ for constant S. Furthermore, I will use the age-old notations in which the same letter (for example, u,) with arguments (say, $u(r, \varphi)$ or $u(S, \varphi)$) is used to denote the same function of a point on our cylinder, whose expressions in different coordinates are quite different.

We can express r in terms of the coordinates S and φ as

$$r = S + f(S, \varphi), \qquad \partial f/\partial S > -1,$$

where $f = 0$ for $|S| \geq C_3$.

For differentials, we obtain

$$dr = \Big(1 + \frac{\partial f}{\partial S}\Big)dS + \Big(\frac{\partial f}{\partial\varphi}\Big)d\varphi.$$

According to (4), the cylinder L is given by the functions

$$P_r = \Big(1 + \frac{\partial f}{\partial S}\Big)^{-1}, \qquad P_\varphi = -\Big(\frac{\partial f}{\partial\varphi}\Big)\Big(1 + \frac{\partial f}{\partial S}\Big)^{-1}.$$

Therefore, formulas (4) and (3) yield a simple expression for the ratio P_φ/P_r

$$B = -\frac{\partial f}{\partial\varphi}\Big|_{S=\mathrm{const}}.$$

Consequently, the combination of the derivatives of B in the equation for the singular point curve has also a simple meaning

$$B_\varphi - BB_r = \frac{\partial B}{\partial\varphi}\Big|_{S=\mathrm{const}}.$$

The final equation of the singular point curve of the projection of a Lagrange cylinder onto the plane (x, y) is written in our coordinates as

$$S + f(S, \varphi) + \frac{\partial^2 f}{\partial \varphi^2} = 0.$$

By the theorem proved in Section 16, the restriction of S to this curve has at least four critical points.

Hence, the caustic curve of the projection of the perturbed Lagrange cylinder onto a plane has at least four cusps. This completes the proof of our theorem.

Remark. The compactness of the support of perturbation ($S = r$ for $|r| \geq C_2$) is unimportant in proving the theorem. It is clear from the proof—it suffices, for example, that the difference $S - r$ would be bounded together with the second derivatives (provided $\frac{\partial S}{\partial r} > C_4 > 0$).

We can also assume that the function f increases (moderately) as $|S| \to \infty$. It is sufficient that the function $H(S, \varphi) = S + f + (\frac{\partial^2 f}{\partial \varphi^2})$ tends to $\pm\infty$ as S tends to $\pm\infty$.

This remark permits to modify our proof to the usual four-vertex theorem for a convex plane curve. Thus, our symplectic version is indeed a generalization of the usual four-vertex theorem.

19. THE FOUR-VERTEX THEOREM AND METAMORPHOSES OF FRONTS WITHOUT HAZARDOUS TANGENCIES

Consider an optical Hamiltonian in the cotangent bundle space, i.e., a hypersurface V which is quadratically convex with respect to impulses, compact in every fiber, and separates the zero section $p = 0$ from infinity.

Example. The eikonal equation $p^2 = 1$ in $T^*\mathbb{R}^n$ yields such a hypersurface. Any close (with derivatives) hypersurface is also an optical hypersurface, for example, the eikonal equation of the Riemann metric.

Examining the possible generalizations of the four-vertex theorem to the case $n = 2$, we obtain the hypotheses concerning the metamorphoses of fronts, which we stated in the Introduction.

Let us begin with a Lagrange cylinder. Consider a fiber of the bundle V over B^n, i.e., the set V_0 of impulses belonging to the hypersurface V at some point $0 \in B^n$. This is a convex surface, a "sphere" in $T_0^*B^n$.

The characteristics of the hypersurface V starting from the points on this "sphere" V_0 (locally) form a cylinder, since they cannot be tangent to the fiber of the cotangent bundle.

Indeed, if $H = 0$ is the equation of V, then for the characteristic (Hamiltonian) field X and every vector v the equality $\omega(X, v) = dH(v)$ holds, here ω is the symplectic structure. If the vector X is tangent to a fiber of the cotangent bundle, then for the vector v which is transversal to V and touches this fiber, the left side would vanish, while the right side would not.

Hence, the continuations of the characteristics starting from the points of our "sphere" V_0 define an immersion of the cylinder $S^{n-1} \times \mathbb{R}$ into T^*B^n (for the sake of simplicity, we assume that characteristics can be continued to infinity), because the phase flow consists of diffeomorphisms.

This cylinder is a Lagrange cylinder and is exact as well, because the phase flow consists of symplectomorphisms, and all characteristics are skew-orthogonal to the initial isotropic "sphere."

The exactness of the cylinder implies the exactness of the form of action pdq lifted to the cylinder. This exactness follows from the fact that $dq = 0$ on V_0.

Define a function t on a cylinder as the integral of this exact form (along any path from any fixed point of the initial "sphere" to the current point). We obtain an imbedding of the cylinder into the space $J^1(B^n, \mathbb{R})$ of one-jet functions on B^n. This space is the product $J^1(B^n, \mathbb{R}) \approx T^*B^n \times \mathbb{R}$ of the cotangent bundle and the t-axis. The $(2n+1)$-dimensional space J^1 is a contact space with contact structure $dt = pdq$. The lifted cylinder in this contact space is a Legendre (n-dimensional integral) imbedded submanifold.

Lemma. *The function t has no critical points on the lifted cylinder.*

Proof. By the opticality condition, the "spheres" V_q which are the intersection of a hypersurface V with the fibers of a cotangent bundle are convex and contain the point $p = 0$ in the interior. Therefore they are star shaped. Hence, the equation of V can be defined by a Hamiltonian function H such that the value of dH on the vector p belonging to $T_b^*B \approx T_pT_b^*B$ at every point of V is 1. This condition uniquely defines the characteristic vector field X on the surface of V touching the Legendre cylinder. On a vector of this field, we have $dt(X) = pdq(X) = dH(p) = 1 \neq 0$.

Therefore, the sections $t = \text{const}$ of the Legendre cylinder in J^1 are smooth imbedded integral surfaces of the contact structure. Their projections into V are also imbeddings, since they are obtained from the initial imbedded "sphere" by diffeomorphisms of the phase flow of the field X.

Consider now the natural projection of a hypersurface V in T^*B^n onto the space of cooriented contact elements ST^*B^n (i.e., the mapping which assigns to every cotangent vector $p \in V \subset T^*B$ its direction). Since the hypersurface V is a star shaped, in every fiber over a point belonging to B, this mapping is a diffeomorphism. Therefore, the projections to ST^*B^n of the sections $t = \text{const}$ of a Legendre cylinder lying in J^1 are imbedded Legendre submanifolds with respect to the natural contact structure of the space of cooriented contact elements of the base. For $n = 2$ this means that the fronts of these Legendre submanifolds have no hazardous self-tangent points.

Thus, we associated to the optical hypersurface $V \subset T^*B^n$ and the initial point in the base — a Legendre cylinder lying in $J^1(B^n, \mathbb{R})$, fibered into the isochrones $t = \text{const}$ diffeomorphic to spheres and a family of t-dependent Legendre imbeddings of these isochrones into a $(2n-1)$-dimensional contact manifold ST^*B^n of cooriented contact elements of the base. When t is varied, the caustic of the Lagrange projection of the initial Lagrange cylinder is swept by the singularities of the moving fronts of these Legendre imbeddings.

For $n = 2$ the caustic is a curve and its cusps correspond to metamorphoses of birth and death of a pairs of cusps on the moving front. Therefore, the hypothesis, which asserts the inevitability of four metamorphoses of birth or death of cusps for the cooriented front moving on a plane without hazardous self-tangencies and passing from a circle cooriented by internal normals to a circle cooriented by outer normals, is a direct generalization of the four-vertex theorem.

Precisely this generalization stimulated the study of invariants of fronts without hazardous self-tangencies and thereby the results of [1] and this work.

REFERENCES

1. Arnold, V.I., Plane Curves, Their Invariants, Perestroikas, and Classifications, *Prepr. Forschungsinst. für Math. ETH*, Zürich, 1993.
2. Tabachnikov, S.L., Computation of the Generalized Bennequin Invariant with the Help of Fronts, *Funktsion. Analiz Pril.*, 1988, vol. 22, no. 3, pp. 89–90.
3. Bennequin, D., Entrelacements et équation de Pfaff, *Astérisque*, 1983, vol. 107/109, pp. 87–162.
4. Arnold, V.I., Sur les propriétés topologiques des projections Lagrangiennes en géométrie symplectique des caustiques, *Cah. Math. Decision*, 1993, no. 9320, pp. 1–9.

5. Tabachnikov, S.L., Around four Vertices, *Uspekhi Mat. Nauk*, 1990, vol. 45, no. 1, pp. 229–230.
6. Gromov, M.L., *Inequalities with Partial Derivatives* [in Russian], Moscow: Mir, 1991.
7. Smale, S., A Classification of Immersions of the 2-sphere, *Trans. Amer. Math. Soc.*, 1958, vol. 90, pp. 281–290.
8. Whitney, H., On Regular Closed Curves in the Plane, *Compos. Math.*, 1937, vol. 4, pp. 276–284.
9. Aicardi, F., Classification and Invariants of Tree-like Plane Curves, *Prepr. ICTP*, Trieste, 1993.
10. Arnol'd, V.I., Lagrangian and Legendre cobordism, Part II, *Funktsion. Analiz Pril.*, 1980, vol. 14, no. 4, pp. 8–17.
11. Jacobi, K., *Lectures on Dynamics* [Russian translation], Moscow: ONTI, 1936 [B., Reimer, 1884].
12. Braunmuhl, A.V., Geodatische Linien und ihre Enveloppen auf Dreiaxigen Flachen Zweiten Grades, *Math. Ann.*, 1982, vol. 20, pp. 557–586.
13. Arnol'd, V.I., Space of Functions with Mild Critical Points, *Funktsion. Analiz Pril.*, 1989, vol. 23, no. 3, pp. 1–10.
14. Arnold, V.I., Sur une propriété topologique des applications globalement canoniques de la mécanique classique, *C.r. Acad. Sci.*, 1965, vol. 261, pp. 3719–3722.

The Vassiliev theory of discriminants and knots

V.I. Arnold

In these times, the angel of topology and the devil of abstract algebra fight for the soul of each individual mathematical domain

H. Weyl*

The study of the discriminant variety in a functional space of smooth mappings is a traditional and fundamental part of the theory of singularities. The discriminant variety is the set of those points of the functional space which represent the mappings having nongeneric singularities. The topological, homotopical and even homological invariants of the complement to the discriminant variety (that is, of the space of generic mappings) are important for many applications. However, progress in these difficult global problems of singularity theory was rather slow until Vassiliev [1] over the last few years has demonstrated the new perspectives opened up by the singularity theory approach in knot theory.

A *knot* is a connected component in the space of smooth embeddings of a circle into 3-space. Hence, we start with the functional space $\mathcal{F}$ of all smooth mappings of S^1 into $\mathbb{R}^3$ and we define the *discriminant variety* Σ as the set of mappings, which are not embeddings (that is, those that have either self-intersections or singularities (see Figure 1).

The discriminant variety is a *hypersurface* in the space $\mathcal{F}$ of all mappings since the self-intersections occur in generic one-parameter families of mappings of a curve in 3-space. This hypersurface subdivides the complement into connected domains which are the knots.

We wish to study the topological properties of the knot space $\mathcal{F} - \Sigma$. For instance, the elements of its 0-dimensional cohomology group are locally constant functions, that is knot invariants. Such functions can be multiplied, forming a ring:

$$H^\circ(\mathcal{F} - \Sigma) = \text{knot invariants ring.}$$

* *Invariants*, Duke Math. J., **5** (1939). This description seems to be an allusion to a painting by Uccello (at the Urbino castle) "l'hostie profannée," representing an event that happened in Paris in 1290. The event is also represented in a series of pictures in the church Saint-Jean-Saint François in Paris.

Originally published in: *First European Congress of Mathematics (Paris, 1992)*.
Vol. I. Invited Lectures. Part 1. Editors: A. Joseph, F. Mignot, F. Murat, B. Prum, and R. Rentschler (Progr. Math. 119). Birkhäuser, Basel, 3–29 (1994), © 1994 Birkhäuser.

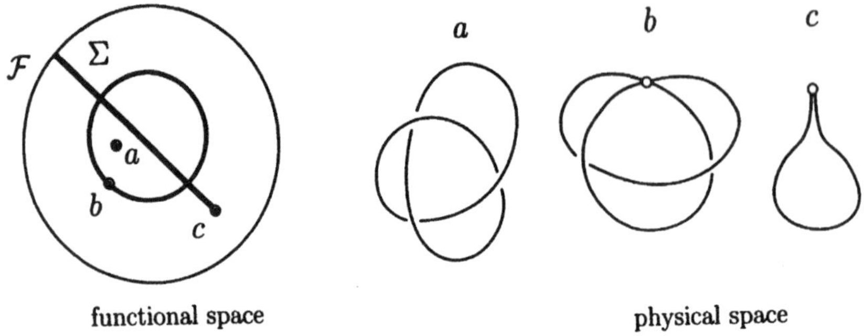

Figure 1. Generic mappings and the discriminant variety in the space $\mathcal{F}$ of mappings $S^1 \to \mathbb{R}^3$

The space $\mathcal{F}$ is linear and hence contractible. The study of the cohomology of the space of knots is therefore reducible to that of the discriminant (modulo ∞), by the Alexander duality. The difficulties of the infinite dimensionality of $\mathcal{F}$ can be overcome by the standard finite dimensional approximation technique of singularity theory (see e.g., [2]–[10]). For instance, one can replace $\mathcal{F}$ by the space $\mathcal{F}_N$ of trigonometric polynomials of degree at most N. Then each homology group $H^i(\mathcal{F}_N - \Sigma)$ stabilizes for $N \to \infty$:

$$H^i(\mathcal{F}_N - \Sigma) \approx H^i(\mathcal{F} - \Sigma) \text{ for } N \gg i.$$

Thus the Alexander duality is essentially used only in finite dimensional cases.

The advantage of the discriminant variety (over its complementary knot space which is our main object of study) is that this variety is naturally stratified according to the hierarchy of the singularities (while the knot space is smooth). Thus, to study homology, we need to cut the knot spaces into pieces, while for the discriminant variety, the pieces are provided by the strata of the stratification. This stratification induces an additional structure in the homology of the discriminant which survives also in the cohomology of the knot space, for instance, in the ring of its zero-dimensional cohomology. This talk is an introduction to the study of the Vassiliev structure in the ring of knot invariants.

The works of J. Birman, X.S. Lin, D. Bar-Natan, and M. Kontsevich ([11]–[15]) have shown that this Vassiliev structure is a fundamental general combinatorial mathematical object, related to the Jacobi identity, Yang-Baxter and Knizhnik-Zamolodchikov equations, the hierarchy of Feynman integrals of perturbative theory in the Chern-Simons action, the D. Zagier ζ-functions of several variables, and to the cohomology of the Lie algebra of Hamiltonian vector fields on infinite dimensional spaces.

1. Vassiliev invariants

These invariants form an increasing sequence of finite dimensional subspaces in the ring of knot invariants, similar to the sequence of spaces of polynomials of increasing degree in the ring of power series. Together these finite dimensional subspaces form the subring V of the Vassiliev invariants:

$$H^0(\mathcal{F}\backslash\Sigma) \supset V \supset \ldots \supset V_n \supset \ldots \supset V_1 \supset V_0.$$

The subspace V_n (or the subgroup, if we consider cohomology with integer coefficients) is called *the space (group) of Vassiliev's invariants of order* n. The product of invariants of orders m and n will be an invariant of order $m+n$.

The polynomials of degree at most n are defined by the condition $d^{n+1}p = 0$. The Vassiliev invariants of order n are defined by a similar condition

$$\nabla^{n+1}i = 0, \quad i \in H^0(\mathcal{F} - \Sigma),$$

with the *jump operator* ∇ replacing the derivative (and which we shall see is also similar to the residue) is defined by the following construction.

Lemma. *The discriminant hypersurface in* $\mathcal{F}$ *has a natural coorientation* (Figure 2).

Indeed, fix the orientations on the circle and in 3-space. A generic (nonsingular) point of the discriminant hypersurface is represented in the physical space by an immersed curve γ with one point of transversal self-intersection. A small displacement of the point from the discriminant hypersurface in a direction, transversal to it, transforms the curve γ into an embedded curve γ^+ or γ^-. The self-intersection point is represented on each of these embedded curves by two points 1 and 2. The velocity vectors of the embedding at points 1 and 2, together with vector 12, form a frame in 3-space. Its orientation is positive in one case (γ^+) and negative in the other (the result does not depend on the choice of points 1 and 2, for instance, not on their ordering). □

Definition. The *jump* of an invariant i at a point of the discriminant hypersurface is the difference of the values of the invariant, evaluated at

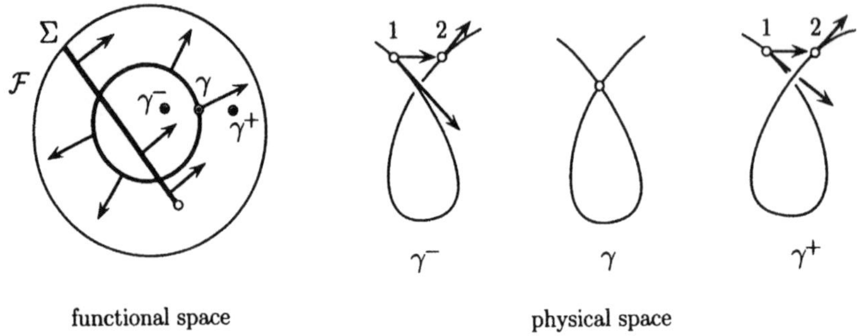

Figure 2. Coorientation of the discriminant variety

both sides of the hypersurface:

$$(\nabla i)\gamma = i(\gamma^+) - i(\gamma^-), \quad i \in H^0(\mathcal{F} - \Sigma).$$

Thus, ∇i is a locally constant function on the set of nonsingular points of the discriminant.

Iterating this construction, one defines the n-th jump, $\nabla^n i$, which is a locally constant function on the set of immersions whose images have n double points.

Example. The second jump of an invariant is defined at the self-intersection points of the discriminant hypersurface as the jump of the first jump of the invariant at the first branch of the discriminant hypersurface (Figure 3). Its value does *not* depend on the choice of the branch of the discriminant hypersurface which was called above the first one. Similarly, the higher jumps are well defined.

Definition. A *Vassiliev invariant of order* n is a knot invariant whose $n + 1$-th jump vanishes identically.

Theorem. *The Vassiliev invariants form a subring of the ring of all knot invariants. Indeed, the following version of the Leibniz formula holds:*

$$\nabla(ij) = i^+j^+ - i^-j^- = i^+j^+ - i^+j^- + i^+j^- - i^-j^- = (i^+)\nabla j + (j^-)\nabla i.$$

Hence the product of Vassiliev invariants of orders m and n is a Vassiliev invariant of order at most $m + n$.

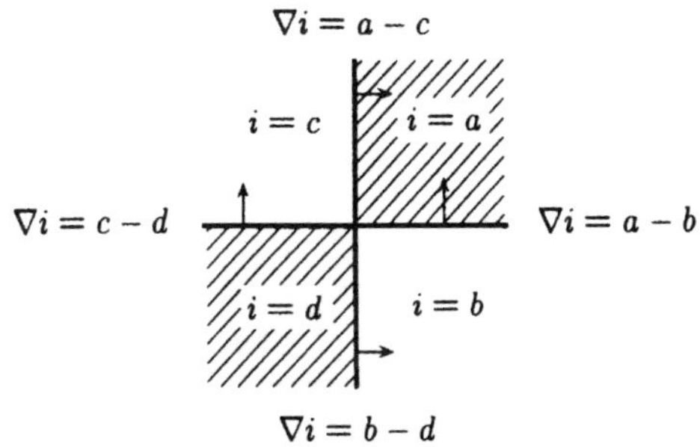

$$(\nabla^2 i) = (a-b) - (c-d)$$
$$= (a-c) - (b-d)$$

Figure 3. The independence of the second jump of an invariant on the ordering of the branches of the discriminant variety

Before we start to calculate the ring of Vassiliev invariants, let us discuss the motivations behind its definition.

The standard technique of topological work with discriminant varieties is the following *resolvent* construction. Replace each self-intersection point by two copies of it (one at each branch) and add a segment so that these are joined points. Then replace all the triple points by triads of points and glue a closed 2-simplex to each such triad. Glue 3-simplices to the resolved quadruple points, and so on (Figure 4).

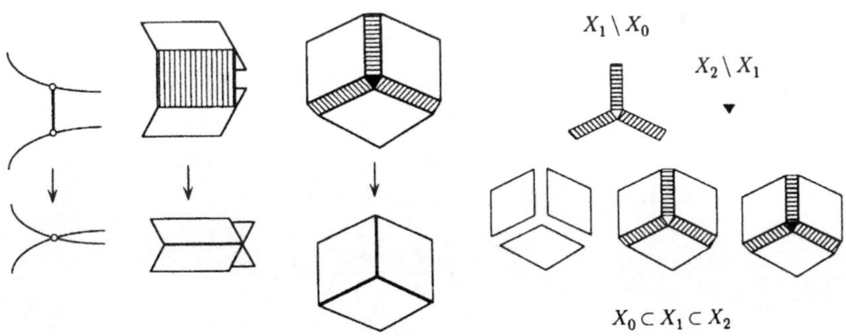

Figure 4. The resolution of self-intersections

The resulting topological space is homotopy equivalent to the initial one. It has an increasing filtration $X_0 \subset X_1 \subset X_2 \subset \ldots$, where $X - X_i$ replaces the self-intersections of multiplicity greater than i. The difference $X_i - X_{i-1}$ is the closure of the space of the fibration into open i-simplices over the set of self-intersection points of multiplicity i. The space X_0 is the closure of the set of those points of the initial (discriminant) variety, which are not self-intersection points.

Now one considers the spectral sequence associated to this filtration. Vassiliev's iterated jumps occur naturally in the study of the first differential of this spectral sequence (see [1]). If this sequence converges to the cohomology of the knots space, then the Vassiliev invariants distinguish all knots. This way of thinking, so natural from the singularity theory point of view, was rather unusual for the knot theorists. Vassiliev theory had not been noticed by the knot theory community until I explained it to Joan Birman, and posed the problem of whether Vassiliev invariants distinguish more knots than do the one variable Jones polynomials (a question which she and X. S. Lin ssubsequently settled affirmatively).

Kontsevich stated in his Bonn lectures in March 1992 that the Vassiliev spectral sequence degenerates at the first term (at least when tensored with $\mathbb{C}$).

Remark 1. Vassiliev has conjectured that his invariants distinguish any two knots. This conjecture has been neither proved nor disproved. In any case, the Vassiliev invariants distinguish at least as many knots as all other known invariants. For instance, if one substitutes e^t for the variable in the Jones polynomial and develops the resulting function in a Taylor series, then the coefficient of the term containing t^n will be a Vassiliev invariant of order n (Birman and Lin [11]). Hence all knots, distinguished by the Jones polynomials, are distinguished also by Vassiliev invariants. Similar results hold for all other known polynomial invariants.

Remark 2. The Vassiliev ring has not yet been computed explicitly. However Bar-Natan and Kontsevich announced that the corresponding graded ring (tensored with $\mathbb{C}$) is isomorphic to the graded ring of polynomials in an infinite set of indeterminates whose degrees are such that the number $\#(n)$ of indeterminates of any fixed degree n is finite:

n	1	2	3	4	5	6	7	8	9
$\# n$	0	1	1	2	3	5	8	12	?

One thus finds the dimensions of the spaces of Vassiliev invariants of small order n to be:

n	0	1	2	3	4	5	6	7	8	9
$\dim V_n$	1	1	2	3	6	10	19	33	60	?

For $n < 5$, these dimensions and spaces had been calculated by Vassiliev [1] and for higher n they have been calculated by Bar-Natan (using many hours of Cray computations).

2. Calculation of the Vassiliev invariants

The dual of the free finitely generated abelian group V_n/V_{n-1} admits an explicit combinatorial description: it is generated by the Feynman diagrams of a special form (Vassiliev diagrams), and their relations are described below. These relations, while rather complicated, are as fundamental as the relations in braid groups, the Jacobi identity, the Yang-Baxter and Knizhnik-Zamolodchikov equations mentioned above (which are closely related to the combinatorics of the relations between the Vassiliev diagrams).

To understand the nature of these relations, we shall start to calculate the Vassiliev invariants of small order n. It is technically convenient to represent the knots by embeddings $\mathbb{R} \to \mathbb{R}^3$ with boundary conditions at infinity (the corresponding functional space of mappings $\mathcal{F}$ is an affine space).

2.1 Invariants of order 0

The defining relation $\nabla i = 0$ means that the invariant i is constant globally. Hence *the space of zero order Vassiliev invariants is the space of constants.*

$$V_0 = \mathbb{Z}$$

(similar to the space of polynomials of degree zero)

2.2 Invariants of order 1

The defining equation $\nabla^2 i = 0$ means that *the first jump of the invariant i is constant on all the immersions with just one point of transversal self-intersection* (Figure 5).

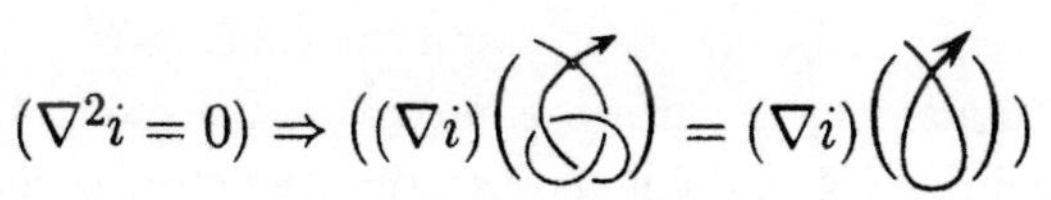

Figure 5. The constancy of the jump of an invariant of order 1

Indeed, each pair of immersions of this class is joined by a finite chain of surgeries ("perestroikas") during which one branch of the curve moves through the other (introducing at that moment one new double point of transversal self-intersection of the immersed curve). The jump of the first jump at any such surgery vanishes, since $\nabla^2 i = 0$. Hence the value of the jump is the same as for the standard plane curve γ (Figure 6)

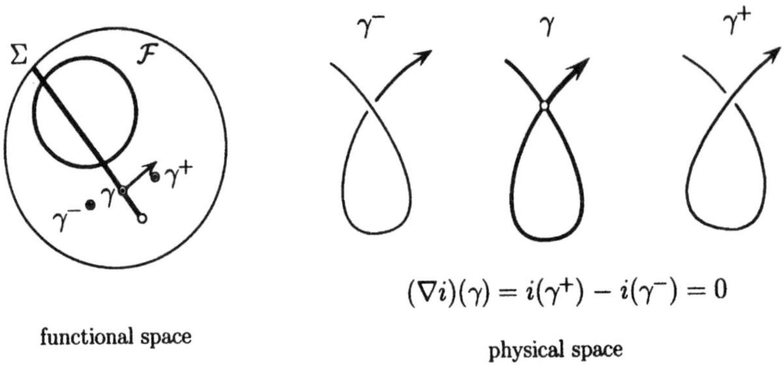

Figure 6. The calculation of the jump of an invariant of order 1

For the standard plane curve γ with one self-intersection point the discriminant is surrounded by the same component of the complement from both sides (the curves γ^- and γ^+ are regularly isotopic). Hence $(\nabla i)\gamma = 0$, and thus *any first order invariant is a zero order invariant*:

$$V_1 = V_0 = \mathbb{Z}.$$

Remark. In terms of the functional space $\mathcal{F}$, the preceding result expresses the following information on the discriminant hypersurface:

(1) the strata, corresponding to more complicated singularities of the discriminant hypersurface, as well as the transversal self-intersection of two branches, *do not divide* the discriminant hypersurface.

(2) the stratum (of codimension 2) in $\mathcal{F}$, formed by the simplest (cusped) singular curves in $\mathbb{R}^3$, is *the boundary* of the discriminant hypersurface.

The mini-versal deformation of a semi cubical cusp is two-parametrical, and the discriminant hypersurface intersects the plane of the parameters along a ray, ending at the point representing the cusped curve. (Figure 6 is thus rather realistic).

It is clear that the points of the plane at both sides of a ray belong to the same component of the complement to that ray. That explains the existence of a regular isotopy between the embeddings γ^+ and γ^-.

The calculations of the higher order invariants are similar to what we have done; only the simplest information on the stratification of the discriminant hypersurface, corresponding to the hierarchy of singularities, is used. This information is provided by the versal deformations of some few very simple singularities. The next step, where the relevant singularity is the triple point, is crucial for the whole theory.

2.3 Invariants of order 2

The defining equation $\nabla^3 i = 0$ means that the second jump of the invariant i does not change under surgery of an immersion whose image has two points of transversal self-intersection, which introduces for a moment a third self-intersection point

Unlike the immersions whose image has one self-intersection point, *the immersions with two such points cannot in general be connected by a finite chain of surgeries each of which introduces momentarily one more self-intersection point.*

Indeed consider the preimages of the double points on the oriented line by examining their mappings in 3-space. There are 4 preimages and they form two pairs (the two points of a pair have the same images in 3-space).

It is convenient to describe a decomposition of the set $\{1, 2, \ldots, 2n\}$ into n pairs by a system of arcs in the upper halfplane (connecting the i-th point with the j-th one iff (i, j) is a pair). I shall call any such system of n arcs *a Vassiliev diagram* of order n (Figure 7).

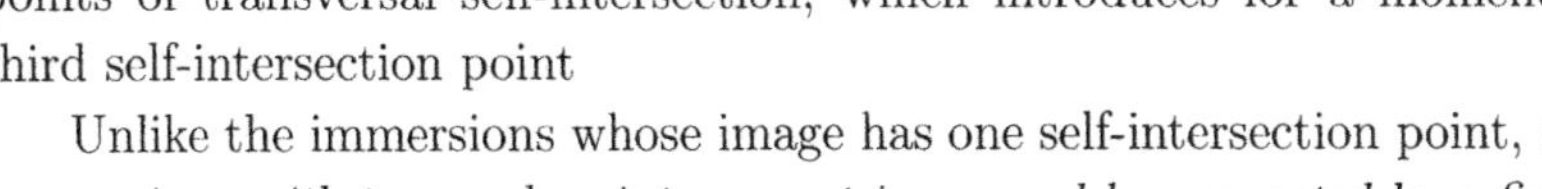

Figure 7. The Vassiliev diagrams of order 2

Of course, many people have previously studied these diagrams, which, for instance, describe the classes of complete flags in a linear symplectic space of dimension $2n$. The components of the knot space are the orbits of the coadjoint representation of $S\,\mathrm{Diff}\,\mathbb{R}^3$, which may be more than just a coincidence.

There exist exactly 3 Vassiliev diagrams of order 2 (Figure 7).

The Vassiliev diagram of an immersion with n double points does not change under a surgery, which introduces momentarily one more double point of the immersed curve. Hence *there exist at least three immersions of a line with 2 double points which cannot be reduced to one another by a chain of such surgeries* (Figure 8).

Any immersion with two self-intersection points on the immersed curve can be reduced to one of these three standard curves by a finite chain of standard surgeries, which introduces a third self-intersection point.

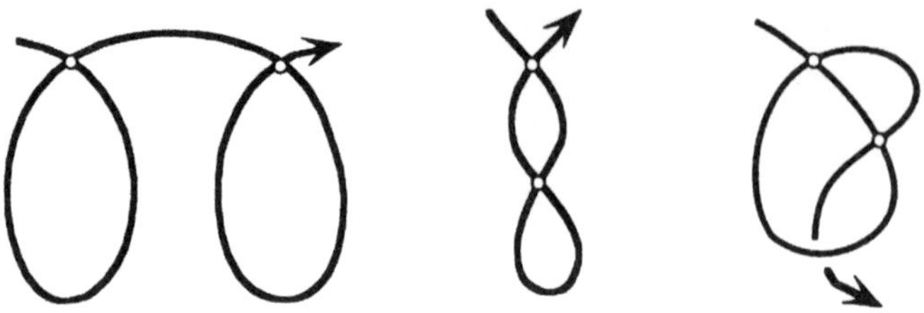

Figure 8. The three standard immersed curves with 2 double points

Therefore, any Vassiliev invariant of order 2 is determined (up to an additive constant) by the values of its second jump on the three standard curves of Figure 8.

The values of the second jump of any invariant on the first two standard curves of Figure 8 vanish. This follows from the fact that both resolutions of one of the self-intersections produce equivalent (smoothly isotopical) immersed curves with one transversal self-intersection. (Figure 9).

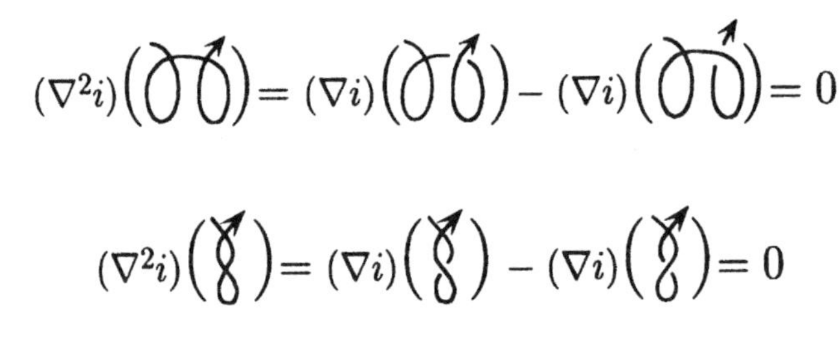

Figure 9. Evaluation of the second jump

Thus, the second jump of an invariant of the second order is unambiguously defined by its value on the third curve. If this value does not vanish (i.e., if the invariant is genuinely of second and not of the first order) then we can multiply it by a constant in such a way that the value on the third curve of Figure 8 will be equal to 1.

A second order Vassiliev invariant with these properties exists and is unique (up to an additive constant). Thus, $V_2 \approx \mathbb{Z}^2$. We can eliminate the constant, choosing the value of the invariant on an unknot to be zero. The calculation of this invariant for the trefoil knot is presented in Figure 10 (where, for simplicity, the signs are neglected):

$$1 = (\nabla^2 i)(K) = (\nabla i)(K^+) - (\nabla i)(K^-), \quad (\nabla i)(K^-) = 0;$$
$$1 = (\nabla i)(K^+) = i(K^{++}) - i(K^{+-}), i(K^{+-}) = 0;$$
$$1 = i(K^{++}) = \text{ the value of the invariant evaluated at a trefoil knot.}$$

The existence of this invariant is proved in [1].

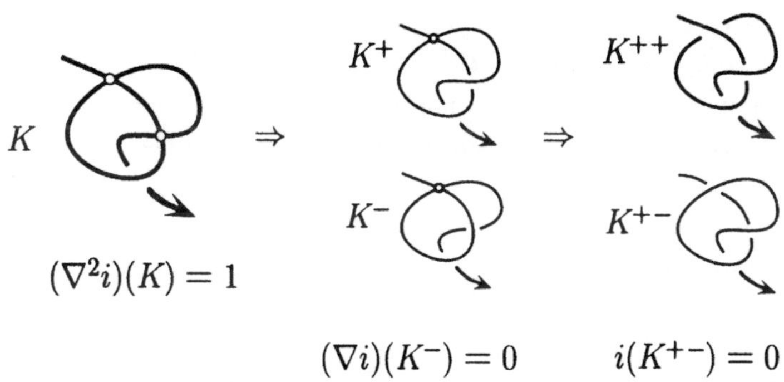

Figure 10. Calculation of the Vassiliev invariant of order 2

This invariant is nontrivial. But it can be reduced to the known ones (it is equal to the x^2 coefficient in the Conway version of the Alexander polynomial). In this case Vassiliev's approach gives an algorithm for calculation of an old invariant. In more complicated cases it generates invariants automatically, by standard combinatorial calculations similar to the preceding ones.

3. The group of diagrams

The calculation of the Vassiliev invariants of order n is similar to the calculations of those of order 2. The defining equation $\nabla^{n+1} i = 0$ means that the n-th jump of the invariant i is a locally constant function on the space of immersions, whose images have n self-intersection points. which does not change under the surgeries, which introduces momentarily one more double point. It follows that *the n-th jump, $\nabla^n i$, depends only on Vassiliev diagram of the immersion with n double points.* □

The number of Vassiliev diagrams formed by n arcs is equal to

$$(2n-1)!! = 1 \cdot 3 \cdot 5 \cdot \ldots \cdot (2n-1).$$

An invariant of order n is defined by the values of its n-th jump on those diagrams up to the addition of an invariant of a smaller order. Hence we obtain the inequality

$$\dim V_n/V_{n-1} \leq (2n-1)!!$$

showing that *the space of Vassiliev invariants of any given order is finite-dimensional.*

To describe explicitly the space V_n/V_{n-1}, it is convenient to start from the free abelian group $\mathbb{Z}^{(2n-1)!!}$, whose generators are the diagrams of order n. The n-th jump of an invariant of order n is a linear function on the additive group, generated by the diagrams. However, as we have seen above for $n = 2$, some of these linear functions are not equal to the n-th jump of any n-th order Vassiliev invariant. For example, for $n = 2$ the values of this function on the first two diagrams of Figure 7 must vanish.

In the general case of arbitrary n, the admissible linear functions are those which vanish on some special diagrams or linear combinations of diagrams. We shall describe below these diagrams and combinations. It is convenient to introduce the following.

Definition. The group of diagrams of order n is the abelian group A_n whose generators are the Vassiliev diagrams consisting of n arcs and whose relations subgroup (in the free abelian group generated by the diagrams) is generated by the two types of relations, as described below:

$$A_n = \frac{\mathbb{Z}^{2n-1!!}}{(\text{relations 1 and 2})}.$$

Relation 1. (The *easy relations*). Each diagram, containing an arc joining two neighboring points belongs to the relations subgroup (Figure 11).

Relation 2. (The *4-term relations)*. The combination of four diagrams

$$S_1 - S_2 + S_3 - S_4$$

belongs to the relations subgroup. Here S_i are the diagrams consisting of n arcs, which are described in (Figure 13) below.

The 4-term relation is a fundamental combinatorial relation whose role in Vassiliev invariants theory is similar to that of the Jacobi identity in Lie algebra theory.

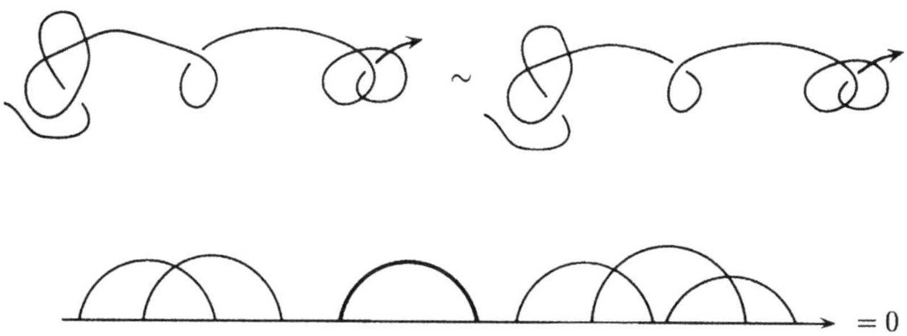

Figure 11. An easy relation and its motivation

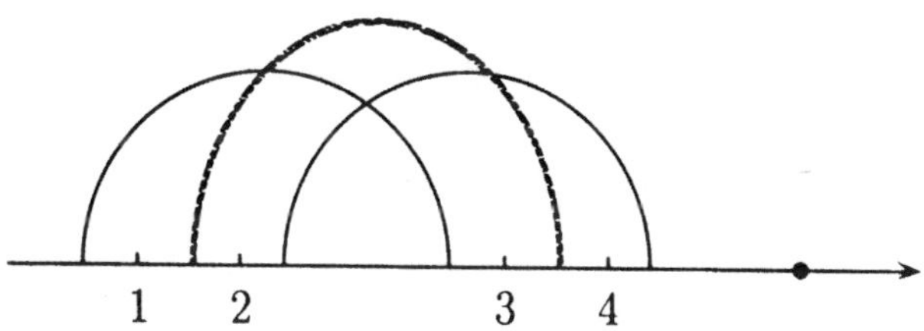

Figure 12. The construction of a 4-term relation

To write the Jacobi identity as a system of relations between the structure constants we have to fix the value of four indices (i, j, k, l) and then add the corresponding products of the structure constants with those indices. Thus the Jacobi identity is in fact a family of numerical equations, parameterized by the choice of indices.

The parameter of the 4-term relations of the group A_n consists of the following data:

(1) a Vassiliev diagram of order $n-2$ (shown in Figure 12 by the ordinary lines);
(2) one more distinguished arc in the upper halfplane (shown in Figure 12 by a wavy line);
(3) one more distinguished point on the border line (0 in Figure 12).

Thus the total number of points at the border line is $2n-1$. These points divide the line into parts. Let us consider the 4 parts adjacent to the endpoints of the distinguished arc (some of these parts may coincide). We denote them by the numbers $(1, 2, 3, 4)$ in the order defined by the orientation of the border line.

The diagram S_i is the union of the $n-1$ arcs defined by the above data and of one more arc joining the distinguished point to a point of the part i. A 4-term relation, corresponding to the data in Figure 12, is represented in Figure 13.

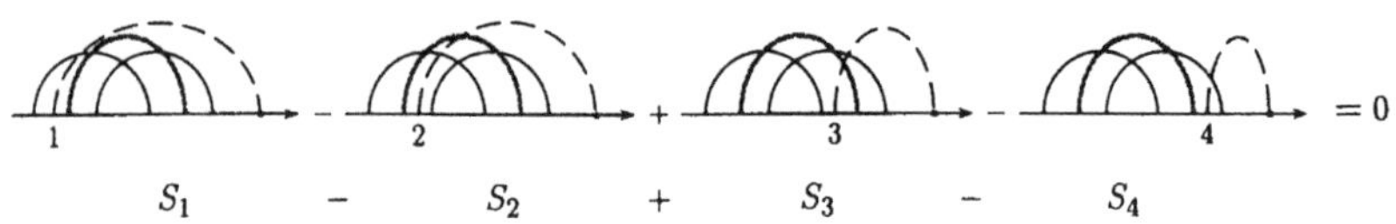

Figure 13. A 4-term relation

Remark. The 4-term relations, which were implicit in Vassiliev's initial work [1], have been written in the form described above by Birman and Lin [11].

The number of independent relations among the relations 1 and 2 is at present known only for small n. According to the computations of Vassiliev ($n < 5$) and Bar-Natan, the ranks of the free abelian groups A_n are given by the following table:

n	0	1	2	3	4	5	6	7	8	9
$\dim A_n$	0	0	1	1	3	4	9	14	27	?

Any function on the set of diagrams with n arcs defines a linear function on the free abelian group generated by the diagrams.

Theorem. *The value of the n-th jump of any Vassiliev invariant of order n on each relation of the diagram group A_n vanishes.*

Proof. Fix an easy relation, that is, a diagram containing a short arc. Consider an immersed curve with n double points whose diagram has a short arc. Introducing one more double point at the moment of the surgeries, we can transform this immersion into an immersion for which the short arc is represented by a standard short simple loop in a ball of 3-space containing no other parts of the curve (Figure 14).

The value of the n-th jump of any invariant on such a curve is equal to the difference of the values of the preceding jump on two regularly isotopical immersed curves with $n-1$ double points; hence it vanishes. □

Figure 14. Evaluation of the n-th jump on an easy relation

The 4-term relation appears naturally in the study of the generic triple points of immersions (where the tangents of the three branches are 3 linearly independent lines). Such points occur unavoidably in generic 3-parameter families of mappings of a curve in 3-space. The mappings with a triple point form a variety of codimension 3 in the space of mappings. Its transversal 3-space intersects the discriminant hypersurface along three surfaces, intersecting each other transversally (Figure 15). The first surface corresponds to the first return to a point visited by the immersion. The second and third surfaces correspond to the subsequent return to one of the two intersecting branches of the immersed curve (visited at the first and at the second instances).

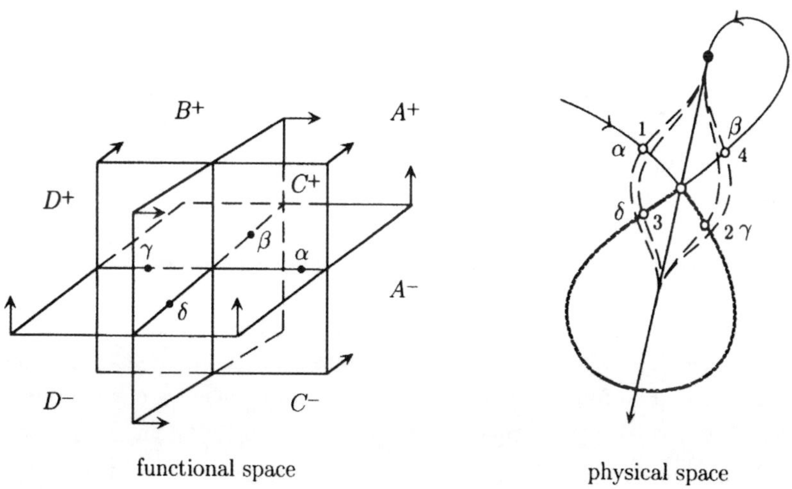

Figure 15. The origin of the 4-term relations: deformations of a triple point

Deform slightly the immersion near the third visit in such a way that the intersection with the initial part of the immersed curve at the triple point is replaced by the intersection with one of the 4 rays of the cross formed at the initial self-intersection. The four deformed immersions are shown in Figure 15 by the broken lines.

These four deformed immersed curves are represented in the functional space (and in the versal deformation 3-space, shown in Figure 15) by 4 points $(\alpha, \beta, \gamma, \delta)$ belonging to the codimension 2 strata of the discriminant hypersurface (namely, to its simple self-intersection strata). All these four points belong to one of the branches of the discriminant hypersurface (represented in Figure 15 by a horizontal plane).

Calculate the second jumps of an invariant at these points and denote them as the points themselves.

Lemma. $\alpha - \beta - \gamma + \delta = 0$

Proof. By definition

$$\begin{aligned}\alpha &= A^+ + C^- - A^- - C^+,\\ \beta &= A^+ + B^- - A^- - B^+,\\ \gamma &= B^+ + D^- - B^- - D^+,\\ \delta &= C^+ + D^- - C^- - D^+. \qquad \square\end{aligned}$$

The relation between the values of the n-th jump of a Vassiliev invariant of order n on the 4 diagrams S_i (Figure 13) follows from the same arguments, applied to the four deformations of an immersion having one triple point, and $n-2$ double points. The value of the n-th jump of an invariant on the deformed immersion with n double points can be considered as the second jump of the $n-2$-th jump (as of a locally constant function on the space of mappings with $n-2$ double points.)

The parameters of the corresponding 4-term relation have the following meaning. The $n-2$ arcs form the diagram of an immersion in which the triple point disappears completely. The distinguished arc corresponds to the first return to the triple point (preserved under all the four deformations). The distinguished point describes the place of the last return among the moments of the other visits of the double points. $\square$

The theorem that we have proved implies that *any (rational) Vassiliev invariant of order n defines a homomorphism $A_n \to \mathbb{Q}$ and is defined by this homomorphism up to an addition of an invariant of a smaller order.*

Kontsevich has stated that any homomorphism $A_n \to \mathbb{Q}$ is the n-th jump of some (rational ?) Vassiliev invariant of order n. In other words, all the relations between the values of the n-jumps follow from relations 1 and 2 above:

$$(V_n/V_{n-1}) \otimes \mathbb{Q} \approx \text{Hom}\,(A_n,\ \mathbb{Q}).$$

Kontsevich's proof based on complete integration is sketched in Section 4 below. In the original approach of Vassiliev [1], the existence of his invariants was proved by purely combinatorial methods.

Remark 1. (cyclic invariance) The element of the group of diagrams, corresponding to an immersed closed curve with n double points, is well defined: *it does not depend on the place where we cut the circle to obtain a line* (which we have used in the construction of the diagram).

Indeed, consider any diagram and replace the leftmost point by a new point at the extreme right (connecting it by a new arc to the right end of the destroyed leftmost arc, see Figure 16).

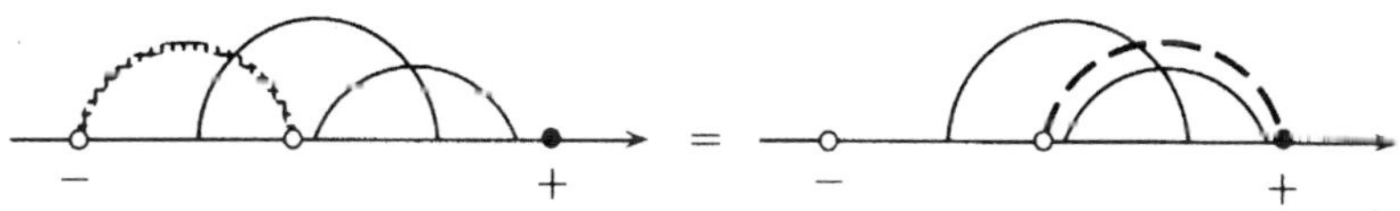

Figure 16. The cyclical invariance of a diagram's class in A_n

Lemma. *The new diagram defines the same element of the group of diagrams as the old one.*

Proof. Destroy the leftmost arc and sum the four-term relations, corresponding to all the choices of the distinguished arc among the remaining $n-1$ arcs, the distinguished point being at the right end of the left arc. □

Remark 2. The same reasoning proves that *any diagram containing an arc which does not intersect any other arc is equal to zero in the group of diagrams.* Indeed, one can transport the left end of this arc towards its right end, jumping over the intermediate arcs using the same operation as in the above proof (Figure 17).

Remark 3. One can combine the diagram groups into the *diagram ring* $A = \oplus A_n$, defining the product $A_n \otimes A_n \to A_{m+n}$ as the concatenation of corresponding diagrams (Figure 18).

Figure 17. A corollary of relations 1 and 2

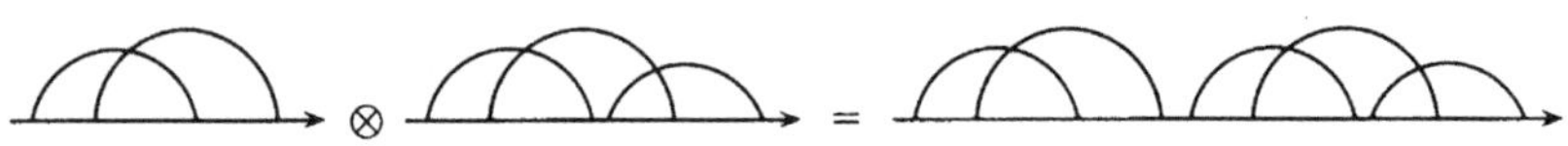

Figure 18. Multiplication of diagrams

This ring is commutative (by cyclical invariance (proved above in Remark 1).

In fact A has also a structure of a commutative and cocommutative Hopf algebra (the comultiplication is dual to the multiplication of the Vassiliev invariants). The elements of the ring A (or rather of its completion) may be viewed as models of knots—the Vassiliev invariants defining linear functions on it.

According to a general algebra theorem, the graded algebra A is isomorphic to the algebra of polynomials. It would be interesting to represent the multiplicative generators by linear combinations of special knots. The arithmetical properties of the coefficients of these combinations are also interesting.

4. Kontsevich integrals for Vassiliev invariants

Recently M. Kontsevich has presented some explicit formulas for the Vassiliev invariants of order n in a form of n-dimensional integrals, similar to the Gauss integral for the linking number.

Represent $\mathbb{R}^3$ as the product of the *horizontal plane* of a complex coordinate z and of the *vertical axis* of a real coordinate t. Represent a knot K as a "*very nice Morse embedding*" $S^1 \to \mathbb{R}^3$, for which all the critical points of the restriction of t to the knot curve are Morse nondegenerate and all the critical values are different.

The construction starts from the iterated integrals defining *Morse knot invariants*, which are constant along the components of the set of embeddings having only Morse critical points.

Choose n noncritical values $t_1 < \ldots < t_n$. Choose two different points (z_i, z_i') among the points of intersection of the knot with the horizontal plane $t = t_i$ (Figure 19).

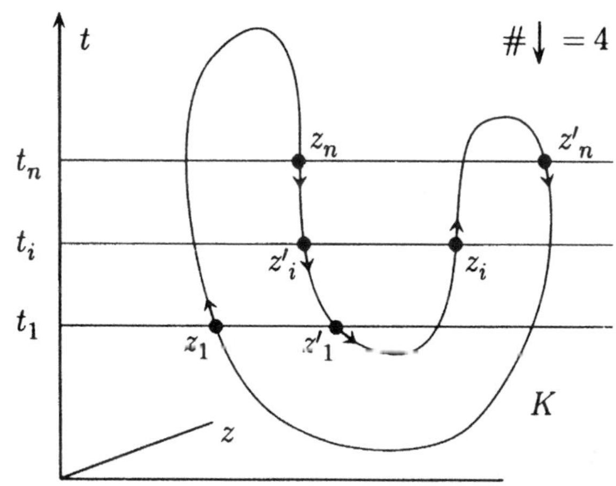

Figure 19. The construction of the Kontsevich integral

The knot branches define locally the smooth functions $z_i(t), z_i'(t)$. The n-times iterated *Kontsevich integral* is the integral with values in $A_n \otimes \mathbb{C}$,

$$\tilde{I}(K) = \int \ldots \int_{t_1 < \ldots < t_n} \sum_{\{z_i, z_i'\}} [w \bigwedge_{i=1}^{n} \frac{dz_i - dz_i'}{z_i - z_i'}(-1)^{\#\downarrow}],$$

where the *weight* $w \in A_n$ represents the Vassiliev diagram, formed by n arcs connecting (z_i, z_i') on the oriented circle K, and where $\#\downarrow$ is the number of *descending points* among $\{(z_i, z_i')\}$ (points where the orientation of K is opposite to that defined by dt). The summation is over all choices of the points z_i and z_i' for all i.

Remark. The integral is absolutely convergent. Indeed, the weight w vanishes at a neighborhood of a zero of the denominator (according to the easy relation 1 in A_n, see Section 3)

Thus *the integral depends on the Morse embedding K continuously.*

The crucial property of the Kontsevich integral is its *constancy along the deformations of the embedding K in the class of the Morse knots.* This property depends on the following, elementary but strange

Lemma. $\dfrac{dz_1 - dz_2}{z - z_2} \wedge \dfrac{dz_2 - dz_3}{z - z_2} +$ *cyclic permutations* $\equiv 0.$

Proof. Compute. □

Remark . This identity first appeared in [16] as the generator of the identities in the exterior algebra of the differential forms in the configuration space $\mathbb{C}^n-$diag, generated (over $\mathbb{C}$) by the standard forms $\omega_{i,j} = d\ell n(z_i - z_j)$. This subalgebra is isomorphic to the cohomology algebra of $\mathbb{C}^n - \cup(\text{diag})$.

The above identity is closely related to the Knizhnik-Zamolodchikov equation [17].

The Kontsevich construction depends on the choice of a closed complex $n-1$-form ω on $\mathbb{R}_1^n \times \mathbb{R}_2^n-$diag, verifying 3 conditions:

(1) the cohomology class $[\omega]$ is nonzero;
(2) ω is antisymmetric, that is $\sigma^*\omega = (-1)^n\omega$, where σ is the involution exchanging the factors;
(3) let $\omega_{i,j}$ be the form in $\mathbb{R}_1^n \times \mathbb{R}_2^n \times \mathbb{R}_3^n - \cup(\text{diag})$, induced from ω on $\mathbb{R}_i^n \times \mathbb{R}_j^n-$diag under the natural projection (where $i, j \in \{1, 2, 3\}$); then

$$\omega_{1,2} \wedge \omega_{2,3} + \text{ cyclical permutations } \equiv 0.$$

For $n = 2$ such a form is given by the above lemma:

$$\omega = d\ell n(z - z').$$

For $n > 2$ no smooth form verifying the conditions 1–3 is known. The Kontsevich integrals correspond to a generalized solution in the class of currents. One represents $\mathbb{R}^n$ in the form $\mathbb{C} \times \mathbb{R}^{n-2}$ with coordinates $(z, t_1, \ldots, t_{n-2})$. The solution used by Kontsevich is the current

$$\omega = d\ell n(z - z') \wedge d\theta(t_1 - t_1') \wedge \ldots \wedge d\theta(t_{n-2} - t_{n-2}'),$$

where $(\theta)(t)$ is equal to 1 for positive t and to zero for negative t.

To prove the deformation invariance of a Kontsevich integral, one writes its variation as an integral of some differential form along K. *This form vanishes identically according to the preceding lemma.* (We leave the details to the interested reader. It is here that the four-term relations will be needed).

The deformation invariance implies the second crucial property of the Kontsevich integral; it can be considered as a "Vassiliev invariant of Morse knots with values in $A_n \otimes \mathbb{C}$". Kontsevich has stated that the n-th *jump of the integral* $\tilde{I}_n$*, evaluated at a Morse immersion* K with n double points, is equal to *the product of* $(2\pi i)^n$ *with the diagram of this immersion* (considered as an element of the diagram group A_n). The idea is to deform K near the singular points, as shown in Figure 20 for $n = 1$. It could follow from the Kontsevich iterated jump formula that *the* $n+1$-th *jump vanishes identically.* Thus $\tilde{I}_n$ may be considered as a generalized vector-valued Vassiliev invariant of Morse knots.

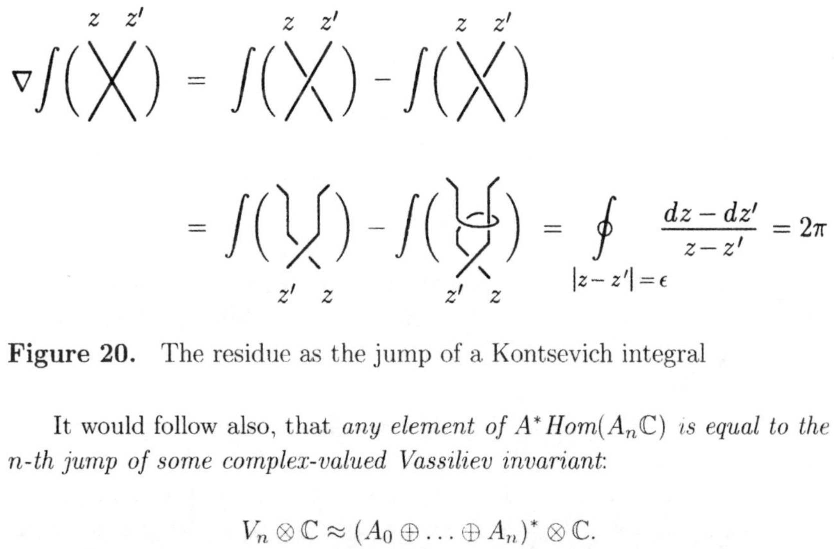

Figure 20. The residue as the jump of a Kontsevich integral

It would follow also, that *any element of* $A^* Hom(A_n \mathbb{C})$ *is equal to the* n*-th jump of some complex-valued Vassiliev invariant*:

$$V_n \otimes \mathbb{C} \approx (A_0 \oplus \ldots \oplus A_n)^* \otimes \mathbb{C}.$$

To write Kontsevich's formulas for the ordinary knots (providing invariants which do not depend on the choice of the Morse knot, representing a given knot class), introduce the *total integral* with values in the completion of the algebra $A \otimes \mathbb{C}$,

$$\tilde{I}(K) = \oplus \tilde{I}_n(K).$$

Consider an unknotted closed curve K_0 with two Morse maxima and two minima of t (to calculate the integrals, we may replace K_0 by the nonclosed plane curve $t = x^3 - x + i0$, since the integrals are deformation-invariant).

To make the Kontsevich integral invariant under the deformations which change the number m of maxima of the function t, Kontsevich had suggested to twist it in the following way:

$$I(K) := \tilde{I}(K)/\tilde{I}(K_0)^m.$$

The division here is understood in the sense of the completion of the algebra $A \otimes \mathbb{C}: \ \tilde{I}_0(K_0) = 1$ and $(1-a)^{-1} = 1 + a + a^2 + \ldots$, if the order of a in A is positive.

Example. The only number-valued Vassiliev invariant of order 2 (normed by the conditions that it vanishes on the unknot and takes value 1 on the trefoil knot) is equal to

$$\Phi(K) = \frac{1}{4\pi^2} \iint_{t_1 < t_2} \sum_{\{z,z'\}} \frac{dz_1 - dz_1'}{z_1 - z_1'} \wedge \frac{dz_2 - dz_2'}{z_2 - z_2'} (-1)^{\#\downarrow} + \frac{m-1}{6}$$

where K is a Morse embedding of a circle with $2m$ critical points of t on it and where the summation is over all the choices of the four points $(t_i, z_i), (t_i, z_i')(i = 1, 2)$, such that the points of the first pair ($i = 1$) alternate with those of the second along the closed curve K.

The Kontsevich integrals would equip the ring $V \otimes \mathbb{C}$ with a $\mathbb{Z}^+-$ grading (generated by that of A). However the arithmetical properties of this transcendental grading are not clear. Conjecturally the values of $I_n(K)$ belong to $(2\pi i\mathbb{Q})^n \otimes A_n$.

This arithmetic reflects the arithmetical nature of the constants involved in the formulas for the integer-valued invariants (like $4\pi^2$ and $1/6$ in the preceding formula). These constants depend on the values of the D. Zagier *ζ-functions of several variables* at the positive integer points,

$$\zeta(a_1, \ldots, a_n) := \sum k_1^{-a_1} \ldots k_n^{-a_n}$$

(the summation over the integer points in the Weyl chamber $0 < k_1 < k_2 < \ldots < k_n$).

The integer linear combinations of the numbers $\zeta(a)$ form a ring Z. Kontesevich has stated that his integrals values on any knot, $\tilde{I}_n(K)$ and $\tilde{I}(K)$, belong to $A_n \otimes Z$.

To understand how the ζ-function enters in the formulas, it suffices to consider the simplest case of the double Kontsevich integral

$$i_2(K_0) = 1 + \frac{1}{4} + \frac{1}{9} + \ldots = \frac{\pi^2}{6}$$

for the plane curve $K_0 : t = x^3 - x,\ z = x + i0$ (Figure 22).

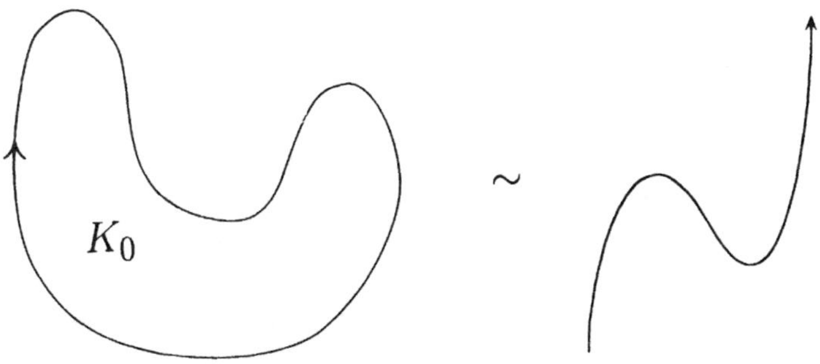

Figure 21. The standard curve K_0

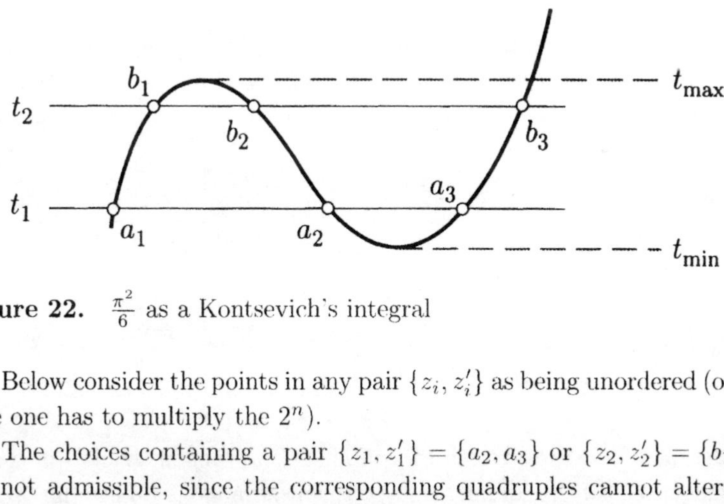

Figure 22. $\frac{\pi^2}{6}$ as a Kontsevich's integral

Below consider the points in any pair $\{z_i, z_i'\}$ as being unordered (otherwise one has to multiply the 2^n).

The choices containing a pair $\{z_1, z_1'\} = \{a_2, a_3\}$ or $\{z_2, z_2'\} = \{b_1, b_2\}$ are not admissible, since the corresponding quadruples cannot alternate. The remaining 4 possibilities of the choices provide 4 terms in the integrand, of the form

$$d\ell n a_{1,2} \wedge d\ell n b_{1,3}(-1) + \ldots$$

where $a_{1,2} = a_1 - a_2$, and so on. Taking the signs into account, one can reduce the integrand to the form

$$d\ell n(a_{1,2}/a_{1,3}) \wedge d\ell n(b_{2,3}/b_{1,3}).$$

This expression explains the rather mysterious *invariance of the integral under the deformations of K_0 in the class of Morse embeddings.* Indeed, the integration domain is the triangle

$$t_{\min} < t_1 < t_2 < t_{\max}.$$

The ratios $u = a_{1,2}/a_{1,3}$, $v = b_{2,3}/b_{1,3}$ send the boundary of this triangle onto the boundary of the standard triangle $u+v \geq 1$, $u \leq 1, v \leq 1$ which is invariant under the deformations in the class of Morse embeddings. Thus the integral is reduced to the standard integral along the standard triangle,

$$\iint d\ell n u \wedge d\ell n v = \int_0^1 \ell n(1-u)\frac{du}{u}.$$

Taylorizing the logarithm, one obtains the value

$$i_2(K_0) = \sum_{k=1}^{\infty} \int_0^1 \frac{u^k}{k}\frac{du}{u} = \sum_{k=1}^{\infty} \frac{1}{k^2} = \frac{\pi^2}{6} = \zeta(2).$$

Similar wonderful cancellations are responsible for the independence of other Kontsevich integrals on the choice of the Morse representative in a knot class. The *standard integrals* occurring in these calculations always have the form

$$\iint_{0<t_1<\ldots<t_N<1} \frac{dt_1}{1-t_1}\frac{dt_2}{t_2}\cdots\frac{dt_{a_1}}{t_{a_1}}\frac{dt_{a_1+1}}{1-t_{a_1+1}}\cdots\frac{dt_N}{t_N}$$

(n groups similar to the first product of a_1 forms.) *This number is the value of $\zeta(a_1,\ldots,a_n)$.*

It is clear that the theories described above will be soon developed in many directions.

Vassiliev has started from the stabilization problem ([2]–[4]) of the cohomology rings of the complements to the discriminants and to the caustics in the complex versal deformation spaces of critical points of holomorphic functions of n complex variables ([5]–[8]). These stable rings are isomorphic respectively to the rings

$$H^*(\Omega^{2n}S^{2n+1}),\ H^*(\Omega^{2n}\Sigma^{2n}U(n)/O(n)).$$

generalizing the May-Segal ([18],[19]) result for the braid groups cohomology ($n = 1$) and providing a homological version of the Gromov h-principle while its homotopical version fails there.

Then, applying his methods to the real functions of one variable with restricted singularities (see [9],[10]) Vassiliev has realized that these also work for the vector functions, for instance, for the knots.

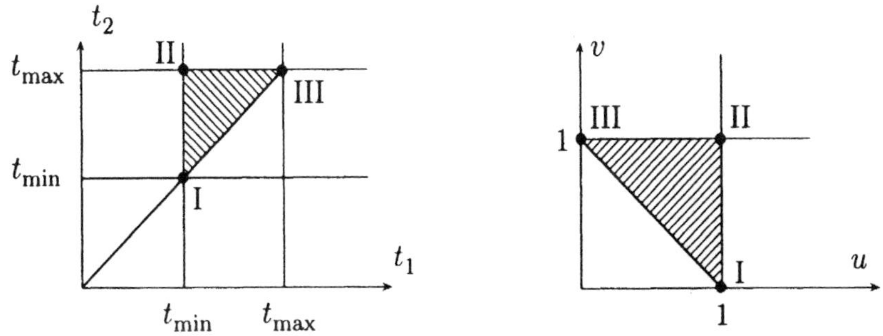

Figure 23. Integration domains for the Kontsevich integral equal to $\zeta^{(2)}$

Vassiliev has also discussed the applications of his theory to the higher dimensional embeddings. Bar-Natan, Lin and Kontsevich have defined Feynman diagram groups, starting from more general Feynman diagrams than those of Vassiliev, and using more relations (inspired by the Jacobi identity in Lie algebras).

The resulting diagram groups are isomorphic to those of Vassiliev. Birman, Lin, Bar-Nathan and Kontsevich have used these constructions to associate a Vassiliev invariant to any representation of a simple Lie algebra; Kontsevich has promised applications to the topology of 3- and 4-manifolds, to the cohomology of infinite-dimensional Lie algebras and to associative algebras.

The success of the singularity technique in knot theory should not obscure the fact that many fundamental problems of the topology of the functional spaces of mappings with restricted singularities are still open both in the real and in the complex domain, even for the functions of one variable (see [3], [4], [9], [10], [20], [21], [22]). Other probable domains of application include symplectic and contact geometry and the theories of immersed plane curves and of evolution of wave fronts.

References

[1] V.A. Vassiliev, Cohomology of knot spaces, in: *Theory of Singularities and its Applications*, V.I. Arnold (ed.), Advances in Soviet Mathematics, AMS **1** (1990), 23–70.

[2] V.I. Arnold, On some topological invariants of algebraic functions, *Trans. Moscow Math. Soc.* **21** (1970), 27–46.

[3] V.I. Arnold, Some open problems in the theory of singularities, *Proc. S. L. Sobolev Seminar* **1** (1976), Novosibirsk, 5–15. Translated in *Singularities, Proc. Symp. Pure Math.* **40** AMS, 1983, 57–69.

[4] V.I. Arnold, "On some problems in singularity theory", in: *Geometry and Analysis, papers dedicated to the memory of V. K. Patodi*, M. F. Atiyah (ed)., Bangalore (1980). Reproduced in the *Proceedings of the Indian Acad. Sci. Math Sci.* **90** (1981), 1–9.

[5] V.A. Vassiliev, Stable cohomology of the complements to the discriminants of the singularities of smooth functions, *Itogi Nauki and Techn. VINITI, Prob. Math., Noveishie Dostijenia* **33** (1988), 3–29, Moscow. English translation: *J. of Sov. Math.* (1990).

[6] V.I. Arnold, V.A. Vassiliev, V.V. Gorjumov, O.V. Ljashko, Singularities–1, *Itogi Nauki and Techn. VINITI, Sovzem. Probl. Math. Fund Napr.* **6** (1988), English translation: *Encycl. of Math. Sci.* **6**, 1992, Springer.

[7] V.I. Arnold, V.A. Vassiliev, V.V. Gorjunov, O.V. Ljashko, Singularities–2, *Itogi Nauki and Techn. VINITI, Sovzem, Probl. Math. Fund. Napr.* **39** (1989), English translation: *Encycl. of Math. Sci.* **39**, 1993, Springer.

[8] V.A. Vassiliev, "Topology of complements to discriminants and loop spaces", in: *Theory of Singularities and its Applications*, V.I. Arnold (ed)., Advances in Soviet Mathematics 1, AMS, 1990, 9–21.

[9] V.I. Arnold, Spaces of functions with mild singularities, *Funct. Anal. Appl.* **23**(3) (1989), 1–10.

[10] V.A. Vassiliev, Topology of the spaces of functions without complicated singularities, *Funct. Anal. Appl.* **23** (1989), 24–36.

[11] J.S. Birman, X.-S. Lin, *Knots polynomials and Vassiliev invariants*, preprint, Columbia University, **19** (1991), 39–56.

[12] X.-S. Lin, *Vertex Models. Quantum Groups and Vassiliev knot invariants*, preprint, Columbia University **19** (1991).

[13] D. Bar-Natan, *Perturbative aspects of the Chern-Simons topological quantum-field theory,* Ph.D. Thesis, Princeton University, 1991.

[14] D. Bar-Natan, *Coefficients of Feynman diagrams and Vassiliev knot invariants*, preprint, Princeton University, 1991.

[15] M. Kontsevich, *Integrals representing Vassiliev's knot invariants*, Lectures at Bonn MPI, February–March 1991.

[16] V.I. Arnold, The cohomology ring of dyed braids, *Mat. Zametki (Math. Notes)* **5** (1969), 227–231.

[17] V.G. Knizhnik, A.B. Zamolodchikov, Current algebra and Wess-Zumino models in two dimensions, *Nucl. Phys.* **B247** (1984), 83–103.

[18] J.P. May, The geometry of iterated loop spaces, *Lecture Notes in Math.* **268** (1972), Springer-Verlag, New York.

[19] G.B. Segal, Configuration spaces and iterated loop spaces, *Invent. Math.* **21** (1973), 213–221.

[20] V.I. Arnold, Bernoulli-Euler updown numbers associated with functions singularities, their combinatorics and arithmetics, *Duke Math. J.* **63** (1991), 537–555.

[21] V.I. Arnold, Springer numbers and morsifications spaces, *J. Alg. Geom.* **1** (1991).

[22] V.I. Arnold, Calculus of snakes and combinatorics of Bernoulli. Euler and Springer numbers of Coxeter groups, *Russ. Math. Surveys* **47** (1992), 1–45.

Added December 1992: meanwhile new expositions of Vassiliev's theory and of its generalizations have appeared, namely,

[23] V.A. Vassiliev, *Lagrange and Legendre Characteristic Classes*, Gordon and Breach, 1988, 268 pp.

[24] D. Bar-Natan, *On the Vassiliev knot invariants*, preprint, Harvard University, Oct. 1992, 60 pp.

[25] D. Bar-Natan, *Vassiliev homotopy string link invariants*, preprint, Harvard University, Dec. 1992, 20pp.

[26] J. Birman, New Points of View in Knot Theory, preprint, Columbia University, 1992, 40 pp, to appear in *Bull. Amer. Math. Soc.*

[27] S. Piunikhin, *Weights of Feynman diagrams, link polynomials, and Vassiliev knot invariants*, preprint, Moscow State University, 1992, 23 pp.

[28] T. Stanford, *Finite type invariants of knots, links and graphs*, preprint, Columbia University, 1992, 26 pp.

[29] V. A. Vassiliev, *Complements of discriminants of smooth maps.* Topology and Applications. AMS, Providence, Translations of Mathematical Monographs, **18** (1992), 210 pp.

Steklov Mathematical Institute
Moscow, Russia

Received July 8, 1992
Revised December 18, 1992

The geometry of spherical curves and the algebra of quaternions

V.I. Arnol'd

Contents

Originally published in: *Uspekhi Mat. Nauk* **50**(1), 3–68 (1995), © 1995 Russ. Acad. Sciences
English translation: *Russ. Math. Surveys* **50**(1), 1–68 (1995), © 1995 London Math. Soc.

Introduction

The spreading of a wave over the surface of a two-dimensional sphere may be described by means of various curves—wave fronts, rays, caustics, and so on.

A system of wave fronts is a system of equidistants of a single curve. Even if this curve is smooth the wave fronts in general have singularities. The singularities of all the wave fronts of the system themselves form a curve, called the caustic. The caustic in general has cusps, just as the wave fronts do.

Another description of the spreading of a wave is given by a Lagrangian curve. Such a curve divides the area of the sphere in half but in general has no singularities apart from self-intersections.

The connections between all these curves, their singularities and their topological invariants are clarified by means of three natural complex structures of quaternion space: namely the connections between three projections of the same Legendrian knot along three Hopf bundles $S^3 \to S^2$.

The following are studied below as consequences of these connections:

1) several generalizations of the classical four-vertex theorem for a plane curve;
2) an unusual version of the Gauss–Bonnet formula for an immersed circle in the sphere and for wave fronts;
3) an explicit formula for the Maslov index of a Lagrangian curve on the 2-sphere;
4) unexpected connections between the indices of points with respect to hypersurfaces on even-dimensional spheres;
5) conformal invariants of an immersed circle in the plane that generalize the Bennequin invariant;
6) new—partly proved and partly conjectured—generalizations of the Morse inequalities and theorems of Chekanov and Givental' on Lagrangian intersections and Legendrian links in symplectic and contact topology;
7) theorems on a duality between area and length in spherical geometry.

In a natural way all these facts arise directly from consideration of the influence exerted by the symmetry of the quaternion units on the symplectic geometry of the 2-sphere and the contact geometry of the 3-sphere. Without these connections the facts stated might have remained unnoticed for a long time.

Since, however, the quaternions can be banished from the formulation of the theorems in question (and, if it were desired, from their proof), I begin straight away with their formulation (the motivation of which, for the most part, lies in the symmetry of the quaternions).

CHAPTER I

THE GEOMETRY OF CURVES ON S^2

We present here geometrical constructions which relate caustics and wave fronts on the 2-sphere. The meaning of these constructions would be clearer if considered from the point of view of contact and symplectic geometry. However I begin with very elementary formulations, which do not require any prior knowledge for their understanding.

§ 1. The elementary geometry of smooth curves and wave fronts

With every curve on the 2-sphere of radius 1 we associate two other curves. The first of these is obtained by moving each point of the curve a distance $\pi/2$ along normals to the original curve. This curve consists of the centres of curvature of great circles of the sphere tangent to the original curve. For this reason it is said to be dual to the original curve (Fig. 1).

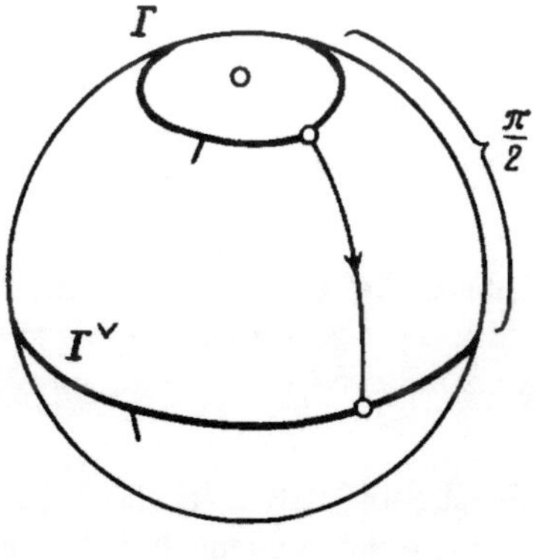

Figure 1

Definition. The curve *dual* to a given co-oriented curve on the sphere is the curve obtained from the original curve by moving a distance $\pi/2$ along the normals on the side determined by the co-orientation.

This definition applies not only to smoothly immersed curves, but also to wave fronts, having cusps (of semicubical type or, in general, of type $x^a = y^{a+1}$).

The dual curve itself is naturally co-oriented and is a wave front equidistant from the original one (lying at a distance $\pi/2$ from it).

The cusps on the original front correspond to points of inflection on the dual front, while points of inflection on the original one correspond to cusps on the dual (a point of inflection of a spherical curve is a point having an order of contact greater than one with the tangent great circle).

The second dual of a front is antipodal to the original one, while the fourth coincides with the original one.

Definition. The *derivative* of a co-oriented curve on the oriented standard sphere S^2 is the curve obtained by moving each point a distance $\pi/2$ along the great circle tangent to the original curve at that point (Fig. 2). The direction of motion along the tangent is chosen so that the orientation of the sphere, given by the direction of the co-orienting normal and the direction of the tangent, is positive.

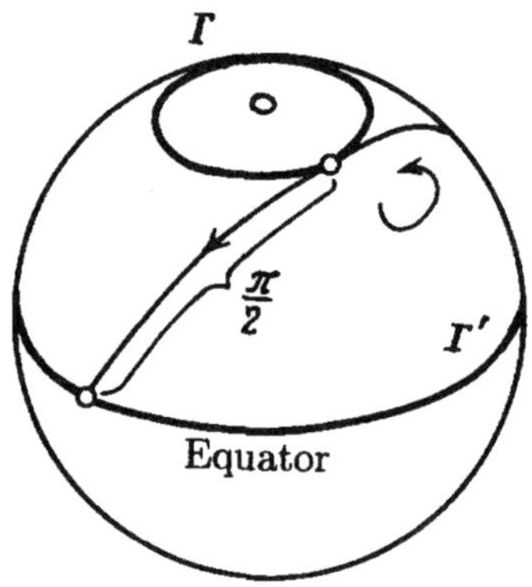

Figure 2

This definiton applies also not only to smoothly immersed curves, but also to wave fronts.[1]

Example. On the unit sphere the parallel of latitude of Euclidean radius $\cos\theta$ is dual to the parallel of lattitude of Euclidean radius $\sin\theta$. The length of the first parallel is equal to the area between the second parallel and the equator of the sphere.

The derivative of any parallel of latitude is the equator parallel to it.

It turns out that the coincidence of length and area and the division by the equator of the sphere into halves is not accidental.

Theorem 1. *The derivative of a wave front coincides with the derivative of any of its equidistants and is a smoothly immersed curve on S^2 even if the original wave front has generic singularities.*

[1] Derivatives of smoothly immersed curves are studied in a recent work by B. Solomon (B. Solomon, Tantrices of spherical curves, Preprint, University of Indiana, 1993) that refers to earlier results of Jacobi and Kagan. The author is grateful to S.L. Tabachnikov for communicating this preprint.

Remark. In the definition of the derivative the movement along tangents by a distance $\pi/2$ may be changed to moving by an arbitrary constant distance s. The curve so obtained is also smoothly immersed in S^2, provided that s is not a multiple of π (as B.A. Khesin has shown to me).

For a plane wave front one can move all the points along tangents by an arbitrary non-zero constant distance: one always obtains a non-singular (smoothly immersed) curve.

The derivative curve of a closed wave front is not an arbitrary curve immersed in the sphere: it satisfies a topological condition of quantization.

If the derivative curve has no points of self-intersection, then this condition consists simply in that, just like the equator, it divides the area of the sphere into two equal parts.

On the other hand if the derivative curve of a closed co-oriented front has points of self-intersection, then the condition of quantization is given by the formula

$$\text{the area bounded by the derivative curve} = -\pi\mu,$$

where μ is the Maslov index of the original wave front. The Maslov index of a wave front, defined in [1], is equal to the algebraic number of cusps of the front. In calculating μ a semi-cubical cusp is considered to be positive if the chords that join points of the branch approaching the cusp to points of the branch leaving it co-orient the curve positively in a neighbourhood of the cusp. The number μ so defined is the Maslov index of the closed curve that corresponds to the front on the two-dimensional Lagrangian manifold of the cotangent bundle of the sphere, that is, on the symplectization of the front (see [1]).

To clarify the quantization condition it remains to define 'the area bounded by an immersed curve on S^2'. Of course, this area is the integral of the standard area form over a 2-chain, whose boundary is the given immersed curve in S^2.

However the 2-chain defined by this condition is not unique, but only modulo the 2-cycle S^2. So 'the area bounded by an immersed curve in S^2' is defined only modulo 4π.

It turns out, however, that among all the 2-chains bounded by a given curve immersed in the 2-sphere there is a distinguished 2-chain (called below the *characteristic chain*).

The characteristic chain appears in an analogue of the Gauss–Bonnet formula for curves immersed in S^2. There is an analogous construction also for hypersurfaces on all even-dimensional spheres.

The geodesic curvature of a co-oriented curve is considered to be positive if the curve turns away from its geodesic tangent on the side of the co-orienting normal.

Theorem 2. *The integral of the geodesic curvature of a closed co-oriented curve immersed in S^2 is equal to the area of its characteristic 2-chain.*

The integral of the area of this characteristic 2-chain for an arbitrary closed co-oriented front is equal to $-\pi\mu$, where μ is the Maslov index of the front.

The co-orientation of the derivative curve is chosen here so that the co-orienting vector together with the orientation form a positively oriented frame of S^2. Such a co-orientation is said to be *correct*.

I now introduce a formula for calculating the characteristic chain of a co-oriented hypersurface on an oriented even-dimensional sphere.

This formula has the form of the product of the volume element of the sphere by a locally-constant function on the complement of the hypersurface. This locally-constant function depends only on the co-orientation of the hypersurface and does not depend either on the orientation of the hypersurface or the orientation of the sphere. On change of sign of the co-orientation of the hypersurface this function changes sign.

We select one of the points of the complement of the hypersurface on the sphere. We call this the 'point at infinity' and denote it by B.

Stereographic projection maps the complement to the point B to an oriented Euclidean space. The hypersurface projects to a compact co-oriented hypersurface lying in this space. This hypersurface of the Euclidean space has a certain index i_B (the degree of the Gaussian map to a sphere). Moreover, for any point A of the complement to this hypersurface there is defined the winding number i_A^B of the hypersurface around A (the degree of the map of the hypersurface to a sphere, sending a point C to the direction of the vector AC).

Theorem 3. *The difference $i_A^B-(i_B/2)$ is conformally invariant (that is, it depends only on the point A and does not depend on B, the chosen point at infinity).*

This difference, considered as a multiplicity function of the point A (integrally or semi-integrally valued), turns the complement to the original hypersurface into a chain whose boundary is the original hypersurface. This is the chain called the *characteristic chain* of the original hypersurface.

The conformal invariant of the winding number of the hypersurface A (corrected by subtracting $i_B/2$) is explained by the following elementary (but apparently not previously observed) identity.

Theorem 4. *There is a formula connecting the indices of points with respect to a hypersurface and the indices of the hypersurface in even-dimensional space, namely*

$$2i_A^B = i_B - i_A.$$

Consider now a closed co-oriented wave front Γ on S^2 with Maslov index μ, its correctly co-oriented derivative curve λ with length element dl, and its characteristic 2-chain c (so that $\partial c = \lambda$).

Theorem 5. *The integral of the area form ω over the characteristic chain, the Maslov index of the original wave front and the geodesic curvature $\varkappa$ of its derivative are connected by the relations*

$$\iint_c \omega = -\pi\mu = \int_\lambda \varkappa\, dl.$$

Remark. The equality of the left and right parts of these equalities is a general formula of Gauss–Bonnet type, satisfied by any closed immersed curve λ (not necessarily occurring as the derivative of a closed front).

A similar formula of Gauss–Bonnet type holds also for hypersurfaces immersed in a sphere of arbitrary even dimension.

The reason for the connection between the derivative curve and the Maslov index is the following, also elementary, fact concerning the geometry of curves on the standard two-dimensional Riemannian sphere.

Let us recall that the derivative curve is obtained from the original wave front on the sphere of radius 1 by moving its points a distance $\pi/2$ along the tangent great circles.

Theorem 6. *The directions of these great circles at points of the derivative curve are parallel in the sense of the standard Riemannian geometry of the sphere S^2.*

The preceding theorems allow one to determine necessary and sufficient conditions on a curve immersed in S^2 for it to be the derivative of a closed co-oriented wave front: the area of its characteristic chain must be a multiple of 2π.

In particular, a curve embedded in S^2 is the derivative of a closed co-oriented front if and only if it divides the sphere into two parts of equal area (like an equator of the sphere).[2]

The cusps of a system of fronts equidistant from one another form the caustic (Fig. 3).

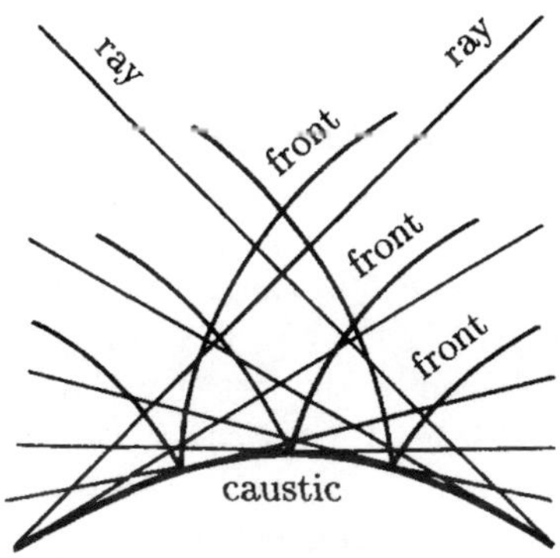

Figure 3

Theorem 7. *The caustic of a system of equidistants from a closed co-oriented wave front on the standard two-dimensional Riemannian sphere is dual to the common derivative curve of any of the fronts equidistant to one another that form the family.*

Remark. In particular, the caustic does not have any points of inflection, since its dual is the derivative curve of fronts smoothly immersed in the sphere and so has no cusps.

[2]The derivative of a smoothly immersed closed curve has full geodesic curvature equal to 0 and does not have an arc of full geodesic curvature π; each closed smoothly immersed curve in the sphere with these properties is the derivative of a smoothly immersed closed curve, as B. Solomon has shown in the work cited earlier.

Theorem 8. *The points of inflection of a family of fronts equidistant from one another lie on two smoothly immersed curves in S^2, each of which bounds an area which is a multiple of 2π, these being the derivative curves of the caustic.*

In fact the points of inflection of a front lie at a distance $\pi/2$ from the cusps of the dual front along the great circle orthogonal to both fronts.

At a cusp of a front this great circle of the sphere is tangent to the caustic. Therefore the points of inflection of all the fronts of the family lie on the pair of derivative curves of the caustic (furnished, for each of the two curves of the pair, with one of the two possible co-orientations).

The topological restriction on the area bounded by the derivative curve is transformed on transfer to the dual curves into a quantization condition on the length of the caustic.

Definition. The *oriented length* of a generic caustic is the alternating sum of the lengths of its segments between successive cusps.

Theorem 9. *The oriented length of the caustic of a system of closed equidistant wave fronts on the standard Riemannian two-dimensional sphere is equal to an integral multiple of 2π (in fact equal to $\pi\mu$, where μ is the Maslov index of any front of the family).*

Which of these segments of the caustic to be counted here as positive will be settled in § 24.

Theorem 9 generalizes a known property of the caustic of a closed plane curve (the oriented length of such a caustic, for example the astroid, arising as the caustic of an ellipse, is equal to zero).

The stated property is explained by the fact that the front is obtained from the caustic 'by unwinding a thread from it' (just as an evolvent from its evolute). The increase in the length of the free part of the thread is equal to the length of the thread unwound from the caustic, so long as the caustic is convex (Fig. 4).

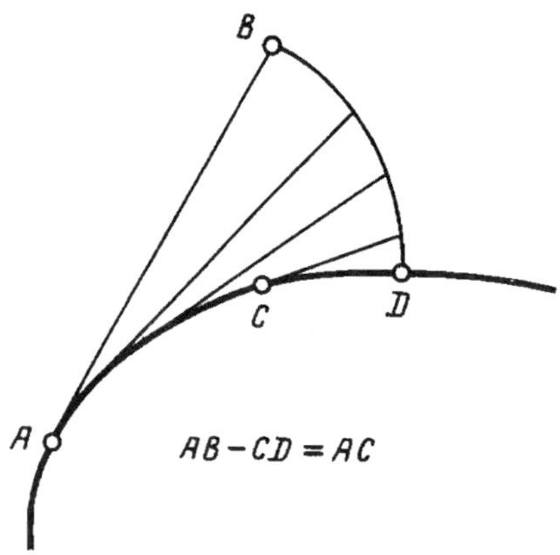

Figure 4

On passing through a cusp of the caustic the direction of convexity changes. It follows that the length of the unwound thread after such a point should be counted with opposite sign.

If the evolvent of a plane curve is closed, then, for a complete circuit of it, the increase in length of the free end of the thread must be equal to zero. Therefore the algebraic sum of the lengths of the segments of the plane caustic between successive cusps (with alternating signs) must be equal to zero.

On the sphere this sum is not zero, but is a multiple of 2π, since the return to the original point of the evolvent no longer requires the return of the length of the free part of the thread to its original value. The increase of this length by 2π corresponds to adding to the free part of the thread a complete circuit of the sphere along the great circle containing the free part of the thread. In that case the increase of the length of the free part of the thread by an integral multiple of 2π does not change the end point of the thread.

The preceding theorem shows that this increase of the length of the free part of the thread after a full circuit of the caustic is equal to $\pi\mu$, where μ is the Maslov index of the front (which for a co-oriented front is always even).

Remark. The equality of the area of the characteristic chain of a closed curve immersed in the sphere and the oriented length of its dual curve is independent of the fact that the immersed curve is a derivative: this is a general 'area-length duality' in the geometry of S^2 (the sign of the length changes on passage through each cusp of a generic front).[3]

To formulate all these formulae with their boring signs with a reasonable degree of generality I must first of all recall some definitions from the general theory of wave fronts and caustics in contact and symplectic geometry. After this, general theorems will be proved (with the aid of quaternions). From these the theorems of this section will follow. Theorem 1 is proved in §§ 6 and 9 of Chapter II. Theorems 2, 5 and 9 are proved in §§ 22, 23 and 24 of Chapter V. Theorems 3 and 4 are proved in § 18 of Chapter IV. Theorems 7 and 8 are proved in § 4. Theorem 6 is proved in § 12 of Chapter III.

§ 2. Contact manifolds, their Legendrian submanifolds and their fronts

A linear hyperplane $\mathbb{R}^{n-1}$ of a tangent space to an n-dimensional smooth manifold M^n is said to be a *contact element* of M^n. A co-orientation of the contact element is the choice of one of the two halves into which it divides the tangent space. A co-orientation of a contact element of a Riemannian manifold is determined by a choice of direction on the line normal to it.

All the (co-oriented) contact elements on M^n form the bundle space E^{2n-1} of a smooth fibration

$$p\colon ST^*M^n \to M^n$$

with fibre S^{n-1} (the cotangent sphere bundle of M^n). This manifold E^{2n-1} is equipped with the natural 'tautological' field of tangent hyperplanes. The hyperplane of the tautological field at a point s of E^{2n-1} is the inverse image by p_* of

[3]The area–length duality was encountered earlier by Santalo in works on integral geometry; it has recently been employed by S.L. Tabachnikov in Russian Math. Surveys 48:6 (1993), 81–109 and by J.C. Alvares (Rutgers University, 1994).

the hyperplane in the space tangent to M^n, which is represented by the point s of the manifold E^{2n-1}.

The tautological field of hyperplanes on E^{2n-1} is called the *natural contact structure* of the manifold of contact elements on M^n.

Example. The manifold of co-oriented contact elements of S^2 is the projective space $\mathbb{R}P^3 = S^3/\pm 1$. Its natural contact structure is obtained from the field of planes orthogonal to the fibres of the Hopf bundle $S^3 \to S^2$, under the projection $S^3 \to S^3/\pm 1$.

The identification $ST^*S^2 \approx S^3/\pm 1$ will be described in detail below, in § 6.

Definition 1. A *Legendrian curve* in a three-dimensional contact manifold E^3 is an integral curve of the contact structure, that is, an immersed curve in E^3, whose tangent at each point lies in the plane of the contact field of planes.[4]

Example. Each curve Γ immersed in a surface M^2 determines a Legendrian curve in the contact manifold $E^3 = ST^*M^2$. This curve consists of the contact elements on M^2, tangent to Γ.

A point on M^2 also determines a Legendrian curve in E^3. This curve consists of the contact elements of M^2, applied at the point (that is, it is a fibre of the cotangent sphere bundle).

Definition 2. The *front* of a Legendrian curve $L : S^1 \to E^3$ on the manifold E^3 of contact elements of a surface M^2 is the projection of this Legendrian curve onto the surface $pL(S^1) \subset M^2$.

The front of a generic Legendrian curve on the surface has semicubical cusps and points of transversal self-intersection as its only singularities.

The contact elements forming the original Legendrian curve determine a *co-orientation* of that front, namely a concordant choice of normal to the front (at double points of the front there are two normals).

Such co-oriented fronts Γ of generic Legendrian curves on the 2-sphere will be the starting material of our constructions.

§ 3. Dual curves and derivative curves of fronts

We consider a co-oriented front Γ in an oriented sphere S^2 of radius 1.

The great circles of the sphere, orthogonal to the front, will be called its *rays*. The rays are oriented (by the co-orientation of the front).

On moving each point of the front along the ray by a distance t we obtain a new curve Γ_t, called the *t-equidistant* of the front. For example, $\Gamma_{2\pi} = \Gamma$.

According to the Huygens' principle, an equidistant is orthogonal to the rays emanating from the original front. An equidistant co-oriented by the directions of these rays is the front of a Legendrian curve, diffeomorophic (and contactomorphic) to the original one.

[4] A Legendrian submanifold of the contact manifold E^{2n-1} is an integral submanifold of maximal possible dimension (equal to $n-1$).

Definition 1. The front $\Gamma^\vee$ equidistant by $\pi/2$ from the original one is said to be *dual* to Γ.[5]

Example. The curve dual to a parallel of Euclidean radius $\sin\theta$ on S^2 is a parallel of Euclidean radius $\cos\theta$.

The following lemma (clarifying the connection with projective duality) is obvious.

Lemma 1. *The dual front is formed from the centres of the great circles tangent to the original curve. The second dual of a front is antipodal to the original one:* $\Gamma^{\vee\vee} = -\Gamma$.

In particular, points of inflection of the original front correspond to cusps of the dual (and conversely).

The co-orientation of the original front on the oriented sphere determines at every point of the front (including singularities) a positive tangent direction (such that the frame determined by the co-orientation and orientation directions orients the sphere positively).

Definition 2. The *derivative curve* Γ' of a front Γ is formed from those points of the great circles tangent to Γ which lie at a distance $\pi/2$ from the point of tangency on the positive side.

Example. The derivative of a parallel of latitude of the sphere is its equator.

Remark. The derivative curve may be considered to be the curve of normals in the same way that the dual curve may be considered to be the curve of tangents.

The following lemma (clarifying the term 'derivative') is almost obvious.

Lemma 2. *The derivative of a front coincides with the derivative of an arbitrary front equidistant from it.*

Example. $(\Gamma^\vee)' = \Gamma'$: the derivatives of dual fronts coincide. The derivatives of anitpodal fronts also coincide: $(-\Gamma)' = \Gamma'$. Change of the orientation of a front changes the sign of the derivative.

Theorem 1. *The derivative of the front of a generic Legendrian curve is a smooth curve (immersed in the sphere), bounding the area* $-\pi\mu$*, a multiple of* 2π.

The topological meaning of the even number μ appearing here (in fact the Maslov index) will be considered in Chapter V.

Theorem 2. *The great circles tangent to a given front form at the points of the derivative curve of this front a framing, parallel in the sense of the Riemannian geometry of the sphere.*

The connection between the integral of the geodesic curvature of the derivative curve and the Maslov index of the front will be discussed below in Chapter V.

[5]The differential geometry of a pair of dual hypersurfaces on spheres has been a classical object of study (see, in particular, Chapter 17 of Santalo's book); it is remarkable that here it is still possible to find new results, even for curves on the two-dimensional sphere.

Theorem 3. *The angle between the great circles tangent to two equidistant fronts at the point of their intersection on the derivative curve is equal to the distance between the equidistants (and, in particular, is constant along the derivative curve).*

Theorem 3 is obvious.

The proof of Theorems 1 and 2 will be given below in § 18 and § 12.

§ 4. The caustic and the derivatives of fronts

Consider the family of all the equidistants of a given co-oriented front on the standard sphere. Even if the original front is non-singular, some of its equidistants will have singularities.

A *singular point* of a front is a critical value of the projection of the corresponding Legendrian curve.

The singular points of generic fronts are semicubical cusps. The double points of a front in general are not singular.

Definition 1. The *caustic* of a family of fronts equidistant from one another is the curve formed from their singular points.

The caustic of a generic front has as its only singularities semicubical cusps and points of transversal self-intersection.

Theorem 1. *The derivative of a front is dual to its caustic.*

Proof. The front at a cusp is orthogonal to the caustic. Therefore to move such a point a distance $\pi/2$ along the tangent to the front is the same as moving it a distance $\pi/2$ along the normal to the caustic (Fig. 5).

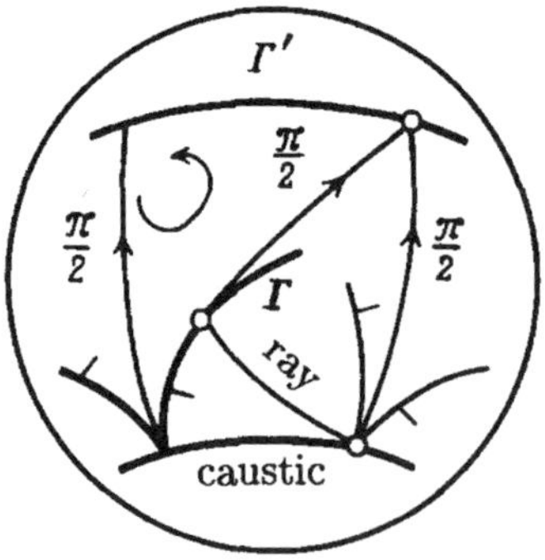

Figure 5

Remark 1. The caustic, in general, is not co-oriented. In this theorem the co-orientation of the front is not taken into account. Therefore the derivative of the front in Theorem 1 is understood to be the union of both (mutually antipodal) components, obtained by moving along the tangent a distance $\pi/2$ in either direction.

Remark 2. The caustic of a co-oriented front on the sphere consists of two different (mutually antipodal) components. This follows easily from the evenness of the number of cusps of a co-oriented curve on the sphere.

Corollary. *The caustic of a family of fronts on the sphere equidistant from one another has no points of inflection.*

In fact, its dual, being the derivative of any of the fronts equidistant from one another, has no cusps (by Theorem 1 of § 3).

Theorem 2. *The points of inflection of all the fronts equidistant from a given one form an immersed curve on the sphere, bounding an area that is a multiple of* 2π *(namely the derivative of the caustic, which is consequently the second derivative of the front).*

Proof. The points of inflection of the front are obtained from the cusps of the dual front (lying on the caustic) by moving a distance $\pi/2$ along the ray orthogonal to the front at the points of inflection and the cusps. At a cusp it is tangent to the caustic, and this proves Theorem 2 (in view of Theorem 1 of § 3).

CHAPTER II

QUATERNIONS AND THE TRIALITY THEOREM

Passing from the wave front to the dual front and to the derivative curve of them both corresponds in algebra to passing from the imaginary unit i to the imaginary units j and k in the space of quaternions (see § 9).

In order to explain this we need some elementary facts from the geometry of quaternions, set out in §§ 5–8. Basic to these facts is the easily proved identity

$$e^{ks}e^{it}e^{-ks} = e^{\varkappa t}, \qquad \varkappa = i\cos 2s + j\sin 2s$$

(where s and t are real). The other computations of this chapter are just as simple, but it is necessary to carry them out in order not to get lost among numerous signs and orientations.

§ 5. Quaternions and the standard contact structures on the sphere S^3

Let $\varkappa$ be an imaginary quaternion of length 1 (for example, i, j or k). Multiplication of all the quaternions by $\varkappa$ on the *right* imparts a complex structure (depending on $\varkappa$) to the space of quaternions $\mathbb{R}^4$.

We associate with each quaternion q the operator of multiplication on the *left* by q in the space of quaternions $\mathbb{R}^4$.

Proposition 1. *The operator of multiplication by the quaternion* q *on the left is* $\varkappa$*-complex.*

Proof. In fact, denoting quaternion multiplication by a dot we have $q \cdot (z \cdot \varkappa) = (q \cdot z) \cdot \varkappa$ for every quaternion z.

Proposition 2. *The determinant of the operator of multiplication on the left by a quaternion of length* 1 *is equal to* 1.

Proof. Both length and determinant are multiplicative. Therefore the image of the map det : $S^3 \to \mathbb{C}\backslash 0$ is a compact connected subgroup. The connected compact subgroups of $\mathbb{C}\backslash 0$ are two in number: the point 1 and S^1. But S^3 does not fibre over S^1. Accordingly, $\det(S^3) = 1$.

Corollary. *The determinant of the $\varkappa$-complex operator of multiplication on the left by an arbitrary quaternion is equal to the square of its length.*

Proof.

$$\det(q) = \det(|q|) \cdot \det(q/|q|) = |q|^2 \cdot 1,$$

since the determinant of the operator of multiplication of a vector in the complex plane by a real number is equal to the square of that number.

Example. The vectors 1 and j form an i-complex basis for $\mathbb{R}^4$. This means that the quaternion $x+yi+uj+vk$ may be considered as the complex vector $(x+iy, u-iv)$ (pay attention to the sign!).

With respect to this basis the operator of multiplication on the left by the quaternion $q = a + bi + cj + dk$ has the matrix

$$\begin{pmatrix} a+bi & -c-di \\ c-di & a-bi \end{pmatrix} = \begin{pmatrix} A & -\overline{C} \\ C & \overline{A} \end{pmatrix}, \qquad \det(q) = a^2 + b^2 + c^2 + d^2.$$

Accordingly, the group S^3 of quaternions of length 1 is isomorphic to the group $SU(2)$ of unitary matrices of degree 2 with determinant 1.

If we were to start from another complex structure involving $\varkappa$ instead of i we should obtain another isomorphism $S^3 \to SU(2)$.

For each quaternion $\varkappa$ of length 1 consider now its corresponding Hopf bundle

$$\pi_\varkappa \colon S^3 \to S^2_\varkappa \approx \mathbb{C}\mathrm{P}^1,$$

associating with each non-null point of the $\varkappa$-complex line passing through 0 the direction of that line.

The fibre of this principal S^1-bundle is the circle

$$\{z \cdot e^{\varkappa t} : t \in \mathbb{R}/(2\pi\mathbb{Z})\},$$

where $z \in S^3$ is a point of the fibre. The action by S^1 on the fibre is determined by shifts of the real axis t.

Warning. Our notations are not complex, but quaternionic. If $I_\varkappa$ is the operator of the complex structure on $\mathbb{C}^2$ such that

$$I_\varkappa z = z \cdot \varkappa,$$

then $z \cdot e^{\varkappa t} = e^{I_\varkappa t} z$.

Consider on the sphere S^3 of quaternions of unit length the vector field $\varkappa$, tangent to the fibres of the bundle $\pi_\varkappa$,

$$\varkappa(z) = z \cdot \varkappa = (d/dt)\big|_{(t=0)}(z \cdot e^{\varkappa t}).$$

Proposition 3. *The vector field $\varkappa$ is left-invariant (under the action of the group S^3 of quaternions of length 1 on itself, by multiplication by elements of S^3 on the left).*

Proof.

$$q \cdot (z \cdot e^{\varkappa t}) = (q \cdot z) \cdot e^{\varkappa t}.$$

Consider the field of planes on S^3 orthogonal to the vectors of the field $\varkappa$. This field provides S^3 with a contact structure, which we call the $\varkappa$-structure.

Proposition 4. *The contact $\varkappa$-structure is left-invariant (with respect to the action of the group S^3 of quaternions of length 1 on itself by multiplying elements of S^3 on the left) and gives an S^3-left-invariant connection to the principal S^1-bundle $\pi_\varkappa$.*

Proof. Left shifts preserve the metric on S^3 and the vector field $\varkappa$, and therefore also the field of planes orthogonal to it. The S^1-structure of the fibres is also preserved, since it is given by the parameter t.

We have constructed on S^3, in particular, three S^3-left-invariant vector fields $\mathbf{i}, \mathbf{j}, \mathbf{k}$ and the contact structures that correspond to them and determine S^3-left-invariant connections on the three principal Hopf bundles $\pi_i : S^3 \to S^2_i$, $\pi_j : S^3 \to S^2_j$, and $\pi_k : S^3 \to S^2_k$.

§ 6. Quaternions and contact elements of the sphere S^2_j

Suppose that the i-contact structure is assigned to the space of the j-bundle π_j constructed in § 5.

Theorem 1. *The bundle $\pi_j : S^3 \to S^2_j$ is naturally isomorphic to the double covering over the bundle of co-oriented contact elements of the 2-sphere, $\tilde{\pi}_j : (ST^*S^2 \approx \mathbb{R}P^3) \to S^2_j$. Under this isomorphism the contact structure on S^3 projects to the natural contact structure on the space of contact elements.*

Proof of Theorem 1. We begin with the following obvious remark.

Lemma 1. *The fibres of the bundle π_j are Legendrian (that is, simply integral) curves of the i-structure.*

In fact, the fields $\mathbf{i}$ and $\mathbf{j}$ are orthogonal at every point. Therefore the direction of the fibre of $\mathbf{j}$ is tangent to the orthogonal $\mathbf{i}$-plane of the contact i-structure.

Thus the two-dimensional planes of the contact i-structure project along the fibres of the j-bundle to one-dimensional contact elements of the sphere S^2_j. These elements are co-oriented (by the projection of the direction of the field $\mathbf{i}$).

Definition. The co-oriented contact element of the sphere S^2_j *corresponding to a given plane of the contact i-structure on* S^3 is the projection of the plane co-oriented by the vector $\mathbf{i}$ along the fibres of the j-bundle.

We prove that the constructed map $S^3 \to ST^*S^2_j$ is a double covering and has the properties stated in the theorem.

Lemma 2. *On moving a point on S^2 with velocity 1 along a fibre of the j-bundle the contact i-plane changes in such a way that its projection on the sphere S_j^2 turns with angular velocity 2.*

Proof. We compare the directions of the vectors of the field $\mathbf{i}$ at points z and $z \cdot e^{jt}$ of a fibre of the j-bundle. We bring the vector $z \cdot e^{jt} \cdot i$ of this field from the point $z \cdot e^{jt}$ back to the point z by acting on the right with the quaternion e^{-jt} (which does not change the projection onto S_j^2). We obtain at the point z the vector $z \cdot e^{jt} \cdot i \cdot e^{-jt}$.

On varying t this vector turns in the plane of the vectors at z of the fields $\mathbf{i}$ and $\mathbf{k}$, orthogonal to the direction $\mathbf{j}$ of the fibre. The rotation is from the direction $\mathbf{i}$ to the direction $-\mathbf{k}$, and has (constant) angular velocity 2. For example, for small t

$$e^{jt} \cdot i \cdot e^{-jt} \approx i + (ji - ij)t + \dots .$$

From Lemma 2 it follows that after the time $t = 2\pi$ of a complete turn along the fibre of the j-bundle on S^3 the corresponding contact element on S_j^2 has turned through an angle 4π. Consequently, our natural map $S^3 \to ST^*S_j^2$ is fiberwise a double covering of principal S^1-bundles.

The projection of the contact i-structure by this map is the field of planes generated by the projections of the vector fields $\mathbf{j}$ and $\mathbf{k}$. Such a plane contains the velocities of motion of the contact elements for which the velocity of the point of application belongs to the element (since the projection of the vector $\mathbf{k}$ onto the sphere belongs to the element, while the direction $\mathbf{j}$, by Lemma 2, corresponds to the rotation of the element).

The constructed double covering $S^3 \to ST^*S_j^2$ allows one to identify the latter space in a natural way with the factor-space $\mathbb{R}\mathrm{P}^3 = S^3/\pm 1$. In fact, $e^{tj} = -1$ for $t = \pi$. Therefore both the antipodal points of S^3 project to one and the same co-oriented contact element on the sphere S_j^2. The theorem is proved.

Note that on the manifold

$$\mathrm{SO}(3) \approx \mathbb{R}\mathrm{P}^3 = S^3/\pm 1$$

there are naturally induced from the sphere S^3 a metric, vector fields $\varkappa$ (in particular $\mathbf{i}, \mathbf{j}, \mathbf{k}$), the contact structures orthogonal to them, the principal S^1-bundles

$$\tilde{\pi}_\varkappa \colon S^3/\pm 1 \to S_\varkappa^2$$

and the S^3-invariant connections given by the corresponding contact structures.

Consider an i-Legendrian curve L on S^3.

Project it to the three spheres S_i^2, S_j^2, S_k^2 by means of the three natural projections π_i, π_j, π_k. The projection $\pi_j L$ is a front Γ of the i-Legendrian curve L on the sphere S_j^2. Let us clarify the meaning of the other two projections.

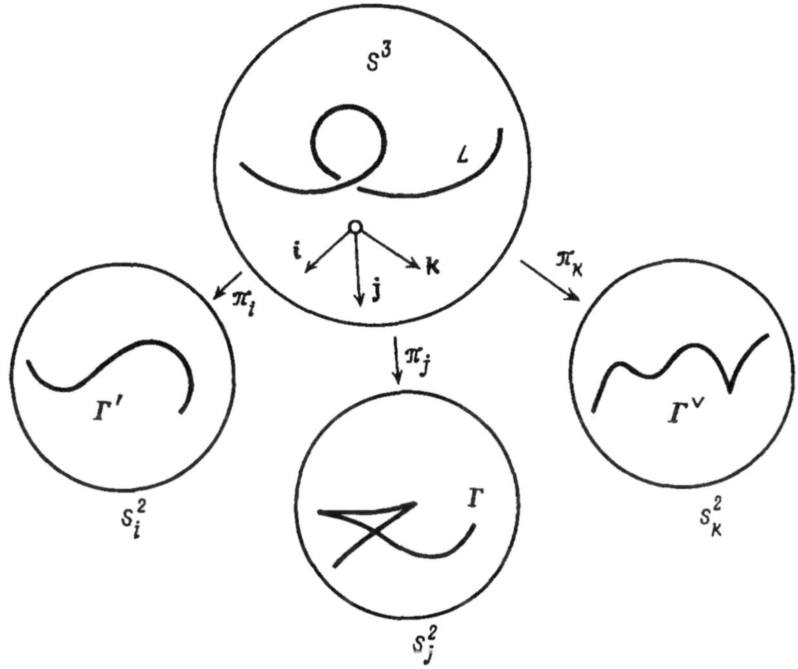

Figure 6

The triality theorem. *The projection $\pi_k L$ is congruent to the front $\Gamma^\vee$ dual to Γ, while the projection $\pi_i L$ is congruent to the derivative Γ' of Γ.*

In this way a front, the dual front, and the derivative of them both, are three personifications of one and the same Legendrian curve (Fig. 6).

The next few sections are dedicated to the proof of the triality theorem, completed below in § 9. The proof establishes also explicit forms of the isometries $S^2_k \to S^2_j$ and $S^2_i \to S^2_j$ that realize the congruences stated in the theorem.

To be precise, the first is induced by multiplication of S^3 on the right by $e^{i\pi/4}$, and the second by multiplication by $e^{k\pi/4}$.

From the triality theorem it follows that the derivative curve is smoothly immersed. In fact the i-Legendrian curve L, smoothly immersed by definition, is transversal to the direction of the fibres of the i-bundle. Therefore its projection $\pi_i L$ on the sphere S^2_i also is smoothly immersed. So Theorems 1 of § 1 and § 3 follow from the triality theorem.

Remark. The proof of the smoothness of the curves obtained from a front by moving along tangents by a distance $2s$ that is not a multiple of π is similar to the proof of the smoothness of the dual curve given below, except that multiplication by $e^{k\pi/4}$ has to be replaced by multiplication by e^{ks}.

§ 7. The action of quaternions on the contact elements of the sphere S^2_j

Let $S^3/\pm 1$ be identified with the space $ST^*S^2_j$ of co-oriented contact elements of the sphere S^2_j by means of the complex structure j and the contact structure i, as has been described in § 6. Let t be a real parameter.

The action of multiplication on the right by the quaternions e^{it}, e^{jt} and e^{kt} on S^3 induces on $S^3/\pm 1 \approx ST^*S^2_j$ three dynamical systems, described in the following lemmas (Fig. 7).

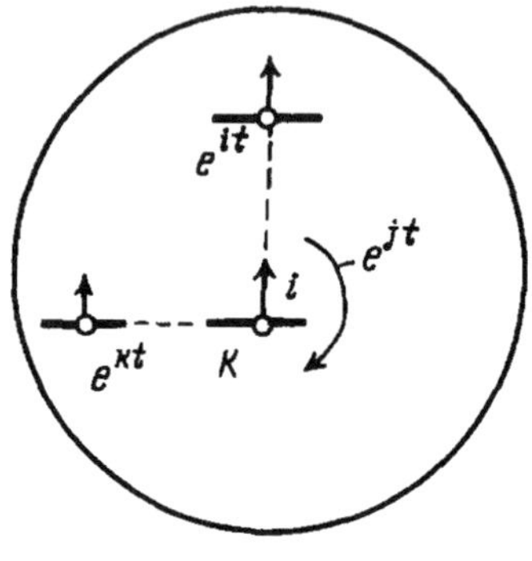

Figure 7

Lemma 1. *Under the action of e^{it} on $ST^*S^2_j$ a contact element moves along a geodesic at all times orthogonal to it in the direction of its co-orientation with speed* 2.

Lemma 2. *Under the action of e^{jt} on $ST^*S^2_j$ a contact element turns around its fixed point of application with angular velocity -2 in the direction given by the j-complex structure on the sphere S^2_j.*

Lemma 3. *Under the action of e^{kt} on $ST^*S^2_j$ a contact element moves with speed* 2 *along a geodesic at all times tangential to it in the direction obtained from the co-orientation direction by turning through an angle $\pi/2$ in the direction given by the j-complex structure on the sphere S^2_j.*

Proof. The base S^2_j of the bundle π_j is obtained from S^3 by factorizing it by *right* shifts (by elements of the form e^{jt}). Therefore *left* shifts by elements of S^3 act on the factor-space S^2_j.

These left shifts preserve the e^{jt}-right-invariant metric on S^3 and, consequently, preserve the natural metric on the sphere S^2_j. Since the sphere S^3 is connected, the left shifts act on the sphere S^2_j as motions.

All motions of the sphere S^2_j are obtainable in this way from left shifts on S^3, since any element of S^3 can be carried by left shifts into any other, and that means that any direction tangent to the sphere S^2_j may be carried into any other.

Consequently, it is enough to verify Lemmas 1–3 for a specially chosen element of S^3 and it is natural to choose the quaternion 1.

Lemma 4. *Let $\varkappa$ be an imaginary quaternion of length 1, orthogonal to j. Then the projection π_j maps the great circle $\{e^{\varkappa t} : t \in \mathbb{R} \mod 2\pi\}$ of the sphere S^3 onto a great circle of the sphere S_j^2 as a double covering.*

Proof of Lemma 4. Consider the circle of quaternions $e^{\varkappa t} = \cos t + \varkappa \sin t$, $t \in \mathbb{R}$, as a curve in the j-complex plane. Choose in this plane the j-complex Hermitian-orthogonal basis $(1, \varkappa)$. A point with the complex coordinates $(z_1 = x_1 + iy_1, z_2 = x_2 + iy_2)$ in this basis is the quaternion $q = x_1 + y_1 j + x_2 \varkappa + y_2 \varkappa j$.

In the affine coordinate $w = z_2/z_1$ on the projective line $\mathbb{C}\mathrm{P}^1 = S_j^2$ the projection π_j of the circle in $\mathbb{C}^2$ defined above is given by the formula $w = \tan t$, and this proves Lemma 4.

We now prove Lemmas 1 and 3. In the case of Lemma 1 the direction of motion of the point of application on S_j^2 is at all times the projection of the vector field $\mathbf{i}$. Consequently, the direction of motion positively co-orients the moving contact element, remaining orthogonal to the direction of the motion. Meanwhile the point of application describes a great circle with speed 2 by Lemma 4.

In the case of Lemma 3 the direction of motion of the point of application on S_j^2 is at all times the projection of the vector field $\mathbf{k}$. Consequently, the direction of motion is at all times tangent to the moving contact element. Meanwhile the point of application describes a great circle with speed 2 by Lemma 4.

Lemma 5. *The complex j-structure on the sphere $S_j^2 = \mathbb{C}\mathrm{P}^1$ defines on it the orientation in which the frame $(\pi_{j_*}\mathbf{i}, \pi_{j_*}\mathbf{k})$ is positive.*

Proof of Lemma 5. Consider the vectors i and k of the fields $\mathbf{i}$ and $\mathbf{k}$ at the point 1 of the sphere S^3. We compute their projections on the sphere S_j^2 in the affine coordinate $w = z_2/z_1 = x + iy$ on the sphere S_j^2, constructed as in the proof of Lemma 4, in which we take $\varkappa$ as i.

The projections of the vectors i and k are $\partial/\partial x$ and $\partial/\partial y$, respectively (at the point $w = 0$). In fact, by the formulae of the proof of Lemma 4 with $z_1 = 1, z_2 = w$ we find that $q = 1 + xi + yk$. On varying the real parameters x and y near 0 we have

$$\partial q/\partial x\big|_{0,0} = i = \mathbf{i}(1), \qquad \partial q/\partial y\big|_{0,0} = k = \mathbf{k}(1),$$

as was asserted. Lemma 5 has been proved.

Lemma 3 is proved at the same time, since the direction $\pi_{j_*}\mathbf{k}$ of the motion of the point of application is obtained from the direction $\pi_{j_*}\mathbf{i}$ that co-orients the element by turning through an angle $\pi/2$ towards the side defined by the given j-complex structure.

Lemma 2 of § 7 follows from Lemma 2 of § 6 and Lemma 5, since the direction of turning 'from $\mathbf{i}$ to $-\mathbf{k}$', obtained in § 6, is opposite to the direction of turning of the given j-complex structure.

§ 8. The action of right shifts on left-invariant fields

It turns out that left-invariant contact structures on the sphere S^3 are mapped into each other by multiplication by appropriate quaternions on the right.

We denote by $R_q : S^3 \to S^3$ multiplication on the right by the quaternion q of length 1.

Theorem. *Multiplication of S^3 on the right by e^{jt} maps the field $\mathbf{j}$ to itself while the fields $\mathbf{i}$ and $\mathbf{k}$ are turned through an angle $2t$ in the plane generated by them (from $\mathbf{i}$ towards $\mathbf{k}$):*

$$\left(R_{e^{jt}}\right)_* \mathbf{i} = \mathbf{i}\cos 2t + \mathbf{k}\sin 2t.$$

Proof. The vectors of the fields $\mathbf{i}$ and $\mathbf{k}$ are orthogonal to the direction of the field $\mathbf{j}$. Under the isometries $R_q : S^3 \to S^3, q = e^{jt}$, preserving the direction $\mathbf{j}$ of the fibres of the bundle $\pi_j : S^3 \to S^2_j$, the translated vectors remain orthogonal to the directions of the fibres.

The projection to S^2_j of a vector of the field $\mathbf{i}$ transported by the isometry $R_{e^{jt}}$ does not change under change of the parameter t, while the point of application moves along the fibre (in the direction $\mathbf{j}$) with speed 1.

The projection to S^2_j of a vector of the field $\mathbf{i}$, applied at this moving point, turns with angular velocity -2 from $\pi_{j*}\mathbf{i}$ towards $\pi_{j*}\mathbf{k}$, by Lemma 2 of § 7. Consequently, a vector transported by the shifts $R_{e^{jt}}$ turns with angular velocity $+2$ relative to $\mathbf{i}$ in the direction towards $\mathbf{k}$.

Corollary 1. *Multiplication of S^3 on the right by $e^{\pi j/4}$ turns the contact i-structure of S^3 into the k-structure (and the k-structure into the i-structure).*

The vector field $\mathbf{i}$ then becomes the vector field $\mathbf{k}$, while $\mathbf{k}$ becomes $-\mathbf{i}$.

Corollary 2. *Multiplication of S^3 on the right by $e^{\pi j/4}$ turns the bundle $\pi_i : S^3 \to S^2_i$ into the bundle $\pi_k : S^3 \to S^2_k$ and π_k into π_i. The induced maps of the 2-spheres $S^2_i \to S^2_k$ and $S^2_k \to S^2_i$ are isometries. The first of these respects the orientations of these spheres given by the complex i- and k-structures, respectively, while the second changes them. Their compositions are the antipodal involutions of the spheres S^2_i and S^2_k.*

§ 9. The duality of j-fronts and k-fronts of i-Legendrian curves

Let L be an i-Legendrian curve (that is, an integral curve of the i-structure on S^3).

Proposition 1. *Multiplication of S^3 by e^{it} on the right sends an i-Legendrian curve into an i-Legendrian curve.*

Proof. This multiplication is an isometry of S^3. It preserves the field $\mathbf{i}$ and therefore the contact structure orthogonal to it. Therefore integral curves of the $\mathbf{i}$-structure are mapped to integral curves.

Proposition 2. *The fronts of the i-Legendrian curves L and $L \cdot e^{it}$ on the sphere S_j^2 are equidistant (at a distance $2t$ in the direction from $\pi_j L$ towards $\pi_j(L \cdot e^{it})$).*

Proof. This is Lemma 2 of § 7.

Proposition 3. *Cusps of the j-front $\pi_j L$ of the i-Legendrian curve L on S_j^2 correspond to points of inflection to the k-front $\pi_k L$ on S_k^2 and, conversely, the points of inflection on S_j^2 correspond to cusps on S_k^2.*

Proof. At a cusp L has the j-direction, that is, it is tangent to $\{z \cdot e^{jt}\}$. Under the k-projection the circle $\{z \cdot e^{jt}\}$ projects to the great circle on S_k^2 with its normal contact elements. Tangency of the *first* order (between L and $\{z \cdot e^{jt}\}$) in the space $ST^*S_k^2$ is transformed after projection to S_k^2 into tangency of the *second* order of the projections, which implies a point of inflection of the k-front.

Proposition 4. *The front $\pi_j(L \cdot e^{it})$ of the shifted i-Legendrian curve $L \cdot e^{it}$ is congruent to the front $\pi_\varkappa L$ of the original i-Legendrian curve, where $\varkappa = j \cos 2t - k \sin 2t$.*

Proof. Multiplication on the right by e^{it} rotates the fields **j** and **k** in their plane with angular velocity 2 from **k** towards **j** (by the Theorem of § 8) with cyclic permutation of the units. The field **j** is sent by such a rotation into the field $\boldsymbol{\varkappa}$.

Theorem. *The k-front of an i-Legendrian curve on S_k^2 is congruent to the $\pi/2$-equidistant of its j-front on S_j^2.*

Proof. Setting $t = \pi/4$ in Proposition 4 we obtain the congruence of $\pi_j(L \cdot e^{i\pi/4})$ and $\pi_k L$. By Proposition 2, $\pi_j(L \cdot e^{i\pi/4})$ is the $\pi/2$-equidistant of the front $\pi_j L$.

Proof of the triality theorem from § 6. The assertion concerning the dual front has just been proved. The assertion concerning the derivative is obtained from that proof by cyclic permutation of the imaginary units i, j and k.

In fact, multiplication of S^3 on the right by e^{kt} for $t = \pi/4$ takes the field **j** to the field **i**, and the field **i** to the field $-$**j**. Consequently it takes the projection π_i to the projection π_j. The induced isometry $S_i^2 \to S_j^2$ takes the curve $\pi_i L$ to the curve $\pi_j(L \cdot e^{k\pi/4})$. The latter curve on the sphere S_j^2 is the derivative of the front $\Gamma = \pi_j L$ (by Lemma 3 of § 7). The theorem is proved.

Remark. It is not unprofitable to remark that the isometry $S_k^2 \to S_j^2$ constructed in the proof (induced by multiplication on the right by $e^{i\pi/4}$) respects the complex orientations of the spheres S_k^2 and S_j^2, while the isometry $S_i^2 \to S_j^2$ (induced by multiplication on the right by $e^{k\pi/4}$) does not.

CHAPTER III

QUATERNIONS AND CURVATURES

Here the algebra of quaternions is used to obtain additional information about spherical curves. First of all we consider the calculation of the geodesic curvatures of curves on the standard two-dimensional sphere. These results are used below in the proof of a generalized Gauss–Bonnet theorem and formulae for the Maslov index.

§ 10. The spherical radii of curvature of fronts

Consider the front $\Gamma = \pi_j L$ on the sphere S_j^2 of an i-Legendrian curve $L \subset S^3$. The *centre of spherical curvature* of the front at one of its points is the point of intersection of infinitesimally close great circles orthogonal to the front at the point. There are two such centres and they are antipodal.

Definition. The *spherical radius of curvature* of the front at a point is the distance from this point to the centre of spherical curvature along the normal that co-orients the front.

This radius is defined modulo π.

Theorem. *If an i-Legendrian curve L in S^3 has at a given point the direction*

$$\varkappa = \mathbf{j}\cos\theta + \mathbf{k}\sin\theta,$$

then its j-front $\pi_j L$ has at the corresponding point of the sphere S_j^2 the spherical radius of curvature $\theta \mod \pi$.

Proof The centre of curvature of the front at the given point lies on the caustic of the family of its equidistants: it is a singular point of the corresponding equidistant, lying on the normal to the front at that point.

On multiplication of the i-Legendrian curve L on the right by e^{it} its j-front is turned into its $2t$-equidistant (Proposition 2 of § 9). This equidistant is singular when the i-Legendrian curve $L \cdot e^{it}$ is tangent to the directions $\pm\mathbf{j}$ of the fibres of the bundle π_j.

A vector $\varkappa$, orthogonal to $\mathbf{i}$, is turned by multiplication on the right by e^{it} in the moving plane of the fields $\mathbf{j}$ and $\mathbf{k}$ through an angle $2t$ in the direction from $\mathbf{k}$ towards $\mathbf{j}$ (by the Theorem of § 8). Therefore the direction of the curve $L \cdot e^{it}$ (at the given point of L) becomes tangent to the fibre ($\theta = 0 \mod \pi$) after turning the vector $\varkappa$ through an angle $\theta \mod \pi$.

For such a choice of $t (= \theta/2)$ the front of the curve $L \cdot e^{it}$ is $\theta \mod \pi$-equidistant from the original front $\pi_j L$. Consequently, the distance to the centre of curvature of the front along its co-orienting normal is equal to $\theta \mod \pi$.

The theorem just proved allows us to consider the radii of curvature of the j-fronts of i-Legendrian curves, without worrying about their smoothness.

Definition. The *reduced radius of spherical curvature* of a front at one of its points is the minimum of the absolute values of its radii of spherical curvature, $\min |n\pi + \theta|$. This radius does not exceed $\pi/2$.

Example. The reduced radius of spherical curvature of a parallel of latitude is equal to the distance to the nearest pole along a meridian.

From the theorems proved there follow:

Corollary 1. *The sum of the reduced radii of spherical curvature at corresponding points of a front and its dual front is equal to* $\pi/2$.

Corollary 2. *The points of spherical inflection of fronts equidistant from the given one lie on the derivative curve of the caustic.*

In fact, at a point of the caustic the radius of curvature of the corresponding front is equal to 0, while at a point of inflection of the front it is equal to $\pi/2$. Moreover the caustic touches the ray (orthogonal to the front) at a cusp of the front, lying on the caustic.

The theorems proved above enable one to relax the regularity of fronts and caustics in these elementary assertions, requiring only that the i-Legendrian curve is immersed in S^3.

§ 11. Quaternions and caustics

Consider an i-Legendrian curve L on S^3. To it there corresponds the j-front $\pi_j L$ on the sphere S^2_j. The equidistants of this front are the projections of the i-Legendrian curves $L \cdot e^{it}$.

Consider the caustic of this family of equidistants. The family of i-Legendrian curves $L_t = L \cdot e^{it}$ defines a torus immersed in S^3:

$$T^2 = \{l \cdot e^{it} : l \in L, \quad t \in \mathbb{R} \mod 2\pi\}.$$

Definition. The *caustic of the j-projection of an i-Legendrian curve* is the set of critical values of the restriction $p : T^2 \to S^2_j$ of the projection $\pi_j : S^3 \to S^2_j$ to the torus T^2.

Theorem. *The caustic of the j-projection of an i-Legendrian curve is the j-projection of a k-Legendrian curve.*

Proof. Fibre the torus T^2 by i-Legendrian curves L_t with fixed t. Such an i-Legendrian curve has at the point $l \cdot e^{it}$ the j-direction for some $t(l)$, to be precise for $t(l) = \theta(l)/2 \mod \pi$, by the Theorem of § 10.

Definition. The curve $\{l \cdot e^{it(l)} : l \in L\}$ is the *j-critical curve of the i-Legendrian curve L*.

Lemma 1. *The j-critical curve of an i-Legendrian curve L is the set of critical points of the projection $p : T^2 \to S^2_j$.*

Proof of Lemma 1. The tangent plane of the torus at each point is generated by a vector of the field $\mathbf{i}$ and a vector which is orthogonal to it and tangent to the curve L_t. This plane contains the $\mathbf{j}$-direction if and only if the curve L_t is tangent to it.

The surprising resemblance of one-dimensional caustics to one-dimensional wave fronts is clarified by the following lemma.

Lemma 2 (basic). *The j-critical curve of an i-Legendrian curve is k-Legendrian.*

Proof of Lemma 2. Consider an i-Legendrian curve, having the **j**-direction at a given point. We introduce a local parameter s on this curve, reducing to 0 at this point. A point of the curve corresponding to a small value of s has the form

$$l(s) = z + asz \cdot j + o(s), \qquad a \neq 0.$$

The angle of the tangent vector of this i-Legendrian curve with the **j**-direction is given by the formula

$$\theta(s) = bs + o(s),$$

since for $s = 0$ this angle is zero.

Consequently, the corresponding point of the j-critical curve has the form

$$\begin{aligned} l(s) \cdot e^{i\theta(s)/2} &= (z + asz \cdot j + o(s)) \cdot (1 + ibs/2 + o(s)) \\ &= z + s(az \cdot j + (b/2)z \cdot i) + o(s). \end{aligned}$$

The vector in the brackets that is tangent to the j-critical curve at the point z is a real linear combination of the vectors $z \cdot j$ and $z \cdot i$ of the fields **j** and **i** at the point z. It is orthogonal to the vector of the field **k**, as the lemma asserts.

The theorem is proved at the same time.

Thus a caustic and front on the sphere S_j^2 are the j-projections of k-Legendrian and i-Legendrian curves. Therefore the symmetry between the imaginary units allows us to extend to caustics what we already know about fronts.

§ 12. The geodesic curvature of the derivative curve

Here we relate the geodesic curvature of the derivative curve to the derivative of the geodesic curvature of the original curve.

Theorem 1. *The geodesic curvature $\varkappa$ of the original curve and the angle α of its derivative curve with the great circle tangent to the original curve are connected by the relation*

$$\varkappa = \tan \alpha.$$

Proof. Consider an infinitely thin spherical triangle, formed by two infinitely close arcs of great circles that are tangent to the original curve (at points ds apart) and of length $\pi/2$, and the derivative curve. The angles of this triangle adjacent to one of the circular arcs are equal to $\varkappa ds$ and α (Fig. 8). The altitude dropped to that side is equal to $(\tan \alpha)ds$, since the distance of its base from the vertex of the angle α is equal to ds. Since this same altitude is equal to $\varkappa ds$ (being the side of a spherical triangle with sides $\pi/2$ lying opposite the angle $\varkappa ds$) one obtains the relation stated in the theorem.

Example. If the original curve is a parallel of latitude with Euclidean radius $\sin\theta$, then $\varkappa = \cot\theta$ and $\alpha = (\pi/2) - \theta$.

Theorem 1 follows from this example, since the *existence* of a universal relation between $\varkappa$ and α is obvious.

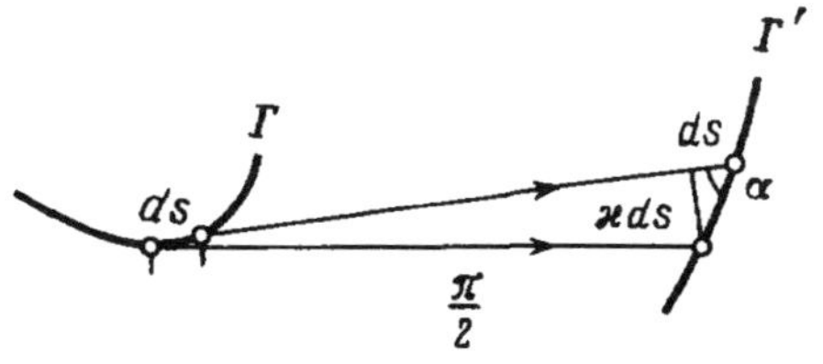

Figure 8

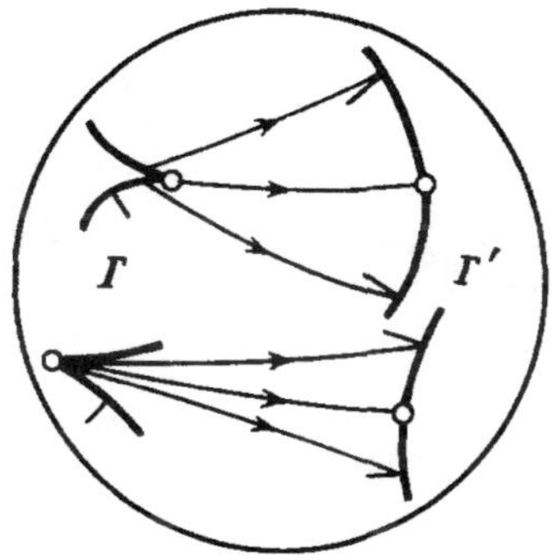

Figure 9

Corollary 1. *The derivative curve of a front is orthogonal to the great circle tangent to the front if and only if this circle is tangent to the front at a cusp* (Fig. 9).

Corollary 2. *The derivative curve of a front is tangent to the great circle tangent to the front if and only if this circle is tangent at a point of inflection* (Fig. 10).

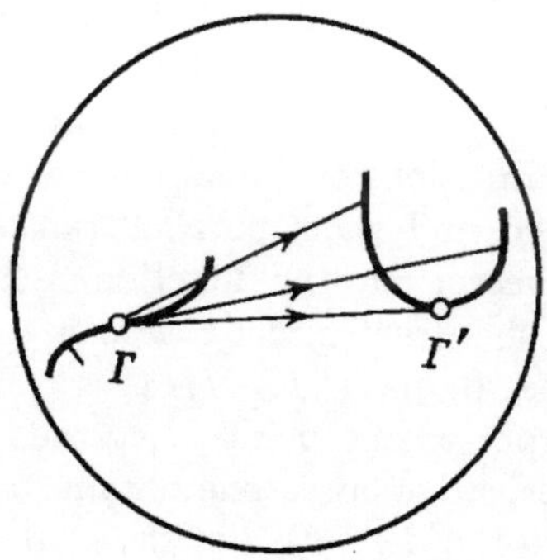

Figure 10

Both these corollaries allow one to estimate from below the number of cusps and points of inflection of the front (in terms of the increase of the angle α along the derivative curve).

Theorem 2. *The great circles touching the original front form a framing on its derivative curve that is parallel with respect to the standard Riemannian metric on the sphere.*

Proof. Consider two points of the original front at a small distance s from each other. We transport in parallel the tangent to the great circle touching the original front at the first point, from a point I of this circle on the derivative curve of the front along the derivative curve to a point II of the circle touching the original curve at the second point (Fig. 11).

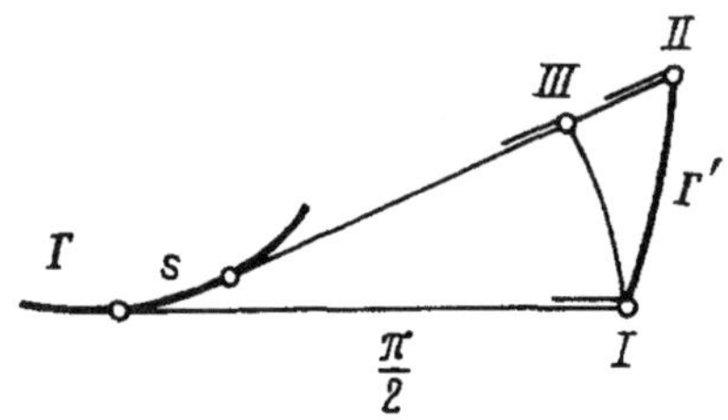

Figure 11

We replace the transfer along the segment (I, II) of the derivative curve by transfer along a path (I, III, II) along the trajectory orthogonal to the family of great circles touching the original front up to a point III on the second tangent circle, and then along this circle touching the front to the point II on the derivative curve.

The area of the triangle (I, II, III) has order s^2 and so the result of transfer along the broken path (I, III, II) differs from the result of transfer along the original curve by $O(s^2)$.

The path (I, III) has geodesic curvature 0 at the point I, since the distance from the point I to the centre of curvature of the path is equal to $\pi/2$. At a point of this path at a distance x from I the distance to the centre of curvature is equal to $\pi/2+O(x)$. Therefore the geodesic curvature at this point is $O(x)$. The distance (I, III) is $O(s)$. Therefore the total angle of turn of a vector carried along the path (I, III) with respect to the directions of the normals to the path (I, III) is $O(s^2)$. The path (II, III) is geodesic, and under parallel transport of a vector along this path the angle between the vector and the direction of the path does not change.

Finally, the direction of the great circle touching the front carried from the point I to the point II along the path (I, III, II) is transformed to the direction of the circle touching the front at the point II, turned through an angle $O(s^2)$. Under transfer along the derivative curve one obtains a result differing from this by another $O(s^2)$. Therefore the derivative of the angle between the transported vector and the directions of the great circles is equal to zero, and that means that the framing is parallel.

Corollary. *The framing of the curve on the sphere S_j^2, obtained by $\tilde{\pi}_j$-projection from the k-vectors along a j-Legendrian curve, is parallel with respect to the Riemannian metric of the sphere S_j^2.*

In fact, we apply Theorem 2 to the front Γ on S_j^2 of an i-Legendrian curve L on $S^3/\pm 1$. The derivative curve of the front is the $\tilde{\pi}_j$-projection of the image L' of the curve L under the action of multiplication on the right by $e^{k\pi/4}$. The framing by the directions of the great circles is the framing by the projections $\tilde{\pi}_{j*}$ of vectors of the field $\mathbf{k}$ along the curve L'. The curve L' is j-Legendrian if L is i-Legendrian, since multiplication on the right by $e^{k\pi/4}$ maps the field $\mathbf{i}$ to the field $\mathbf{j}$.

Thus by Theorem 2 for such a j-Legendrian curve L' its framing by the k-vectors projects onto S_j^2 as a parallel framing.

Any j-Legendrian curve may be obtained from an appropriate i-Legendrian curve by multiplication on the right by $e^{k\pi/4}$. On applying Theorem 2 to the front Γ of this i-Legendrian curve we have the corollary.

Remark. The quaternion units may be permuted. For example, along the projection of an i-Legendrian curve in S_i^2 the projections of the j-vectors (and also of the k-vectors) are parallel.

Theorem 2 enables one to compute explicitly the geodesic curvature of the derivative curve. In fact, this curvature, by Theorem 2, is equal to the derivative of the angle α between the derivative curve and the direction of the framing great circle with respect to the natural parameter l (arc length) of the derivative curve.

Theorem 3. *The geodesic curvature $K = d\alpha/dl$ of the derivative curve is expressed in terms of the geodesic curvature $\varkappa(s)$ of the original curve by the formula*

$$K = \frac{d\varkappa/ds}{(\sqrt{1+\varkappa^2})^3},$$

which can also be written in the form

$$K = \cos^3\alpha\, d\varkappa/ds = d\sin\alpha/ds.$$

Corollary. *The points of inflection ($K = 0$) of the derivative curve correspond to the vertices ($d\varkappa/ds = 0$) of the original curve.*

Proof of Theorem 3. Consider once again the infinitely thin spherical triangle (with sides $\pi/2$, $(\pi/2) - ds$ and dl and the angle α). From this triangle we find that the length dl of the segment of the derivative curve is $ds/\cos\alpha$. But from Theorem 1 we get

$$d\alpha = \frac{d\varkappa}{1+\varkappa^2} = \cos^2\alpha\, d\varkappa.$$

These relations give for $d\alpha/dl$ the expression stated in the theorem.

Example 1. For a parallel of latitude $\varkappa = \text{const}$ and its derivative curve is the equator ($K \equiv 0$).

Example 2. Near an ordinary (semicubical) cusp the square of the radius of curvature of the front is proportional to the distance to the cusp.

In this case $\varkappa(s) \sim 1/\sqrt{s}$, $d\varkappa/ds \sim 1/(\sqrt{s})^3$. The value of K remains bounded as $s \to 0$, as should be the case for an immersed curve. Therefore the K is positive for positive cusps and negative for negative cusps.

Example 3. Near a degenerate cusp of a front ($x \sim p^a, y \sim p^{a+1}$) the curvature $\varkappa$ has order $s^{(1-a)/a}$. In this case $\varkappa(s)$ has order $s^{(a-2)/a} \sim p^{a-2}$ (p is a regular parameter along the derivative curve).

Corollary. *The geodesic curvature of the derivative curve of a front has a zero of order $\mu - 1$ at a point corresponding to a cusp of the front of multiplicity μ.*

Here $\mu = \alpha - 1$: at an ordinary cusp (of multiplicity 1) $a = 2$, $\mu = 1$ and the geodesic curvature of the derivative curve at the origin is not zero.

We conclude that, at a cusp of the original front of multiplicity $\mu, \mu - 1$ vertices of the front vanish. For $\mu = 2$ after a small perturbation the front has a pair of cusps, and the vanishing vertex lies halfway between them.

The formulae obtained lead to new generalizations of the four-vertex theorem (for spherical curves), but I do not elaborate on this here.

§ 13. The derivative of a small curve and the derivative of curvature of the curve

The formulae of the preceding section can be used for the study of plane curves also.

If the original curve is concentrated around the pole on the sphere, then its derivative is concentrated near the equator. The geodesic curvature of a small curve on the sphere placed near the pole is asymptotically equal to the curvature of the plane curve obtained by projecting it to the tangent plane at the pole. The curvature of the derivative curve which is close to the equator also admits a simple asymptotic expression. By combining these two formulae we obtain a rather remarkable expression for the derivative of the radius of curvature of a *plane* curve, given by the following known theorem (see for example the text-book by Santalo).

Consider a curve Γ on the Euclidean plane with Cartesian coordinates (x, y), the parameter on which is the azimuth φ of the normal equator (so that $(\cos\varphi)dx + (\sin\varphi)dy \equiv 0$ along the curve). We denote the radius of curvature of the curve by R and we let $A = x\sin\varphi - y\cos\varphi$ (this measures the distance from the origin to the normal to the curve). We shall denote the derivatives with respect to the parameter φ by dashes.

Theorem.

$$R' = A + A''.$$

Remark. From this formula there follows at once the four-vertex theorem for a convex plane curve, namely: *R' has no fewer that four zeros.*

In fact $-A$ is the derivative by φ of the supporting function $B = x\cos\varphi + y\sin\varphi$ of the curve, so that

$$R' = -(B' + B''').$$

A function of the form $B' + B'''$ has no fewer than four zeros on the circle by the theorems of Sturm (Kellogg and Tabachnikov...), asserting that the Fourier series

$$F(\varphi) = \sum_{n \geqslant m} a_n \cos(n\varphi) + b_n \sin(n\varphi)$$

has no fewer zeros than the number $2m$ of zeros of the first of its harmonics (see [2], [3], [4], [5] for details).

Proof of the theorem. The formula to be proved is a limiting case of Theorem 3 of § 12. Let ε be a small parameter. Consider the small curve $\varepsilon\Gamma$ on the Euclidean plane, tangent to the unit sphere at the North pole. Its curvature is $\varkappa/\varepsilon$, where $\varkappa = 1/R$ is the curvature of Γ. The natural parameter on $\varepsilon\Gamma$ is εs, where s is the natural parameter on Γ.

Project $\varepsilon\Gamma$ to the sphere (for example, along the axis through the pole, or from the centre, or from the South pole). On the sphere we obtain a small curve Γ_ε concentrated near the North pole. The geodesic curvature $\varkappa_\varepsilon$ of this curve is asymptotically equivalent to $\varkappa/\varepsilon$ as $\varepsilon \to 0$, while the natural parameter s_ε is asymptotically equivalent to εs. Consequently the expression of Theorem 2 of § 12 for the geodesic curvature of the derivative of Γ_ε takes the form

$$\frac{d\varkappa_\varepsilon/ds_\varepsilon}{(\sqrt{1+\varkappa_\varepsilon^2})^3} \sim \frac{\varepsilon^{-2}\,d\varkappa/ds}{\varepsilon^{-3}\varkappa^3} = -\varepsilon\frac{dR/ds}{\varkappa} = -\varepsilon\frac{dR/ds}{d\varphi/ds} = -\varepsilon\frac{dR}{d\varphi}. \tag{1}$$

Consider now the derivative of the small curve Γ_ε (Fig. 12). As $\varepsilon \to 0$ this derivative curve tends to the equator. It smoothly depends on ε. Let us introduce on the sphere the coordinates of longitude ψ ($x = \cos\psi, y = \sin\psi$) and latitude λ ($z = \sin\lambda$).

Figure 12

Lemma 1. *The derivative of Γ_ε for small ε is given locally by the equation*

$$\lambda = \Lambda(\psi, \varepsilon), \qquad \Lambda(\psi, 0) = 0.$$

In fact, consider the great circle tangent to Γ_ε at the point corrresponding on Γ to the point whose normal has the azimuth φ. This circle intersects the equator at the point with longitude $\psi = \varphi + (\pi/2) + O(\varepsilon)$. The distance along the circle from the point of tangency to the equator is equal to $(\pi/2) + \varepsilon A(\varphi) + O(\varepsilon^2)$. The expression for λ now follows from the implicit function theorem, since the circle constructed differs from the meridian of longtitude $\varphi + (\pi/2)$ by a quantity of order ε.

Lemma 2. *The geodesic curvature of a smooth curve close to the equator*

$$\lambda = \Lambda(\psi, \varepsilon) = \varepsilon M(\psi) + o(\varepsilon)$$

is given by the formula

$$K = \varepsilon(M + M_{\psi\psi}) + o(\varepsilon).$$

Proof. It is sufficient, for example, to apply the Gauss–Bonnet formula to a small curvilinear quadrilateral bounded by the equator, two nearby meridians and our curve close to the equator.

The area of the quadrilateral is proportional to M, while the difference of the angles between our curve and the meridians is proportional to $M_{\psi\psi}$ (in a first approximation with respect to ε). From this one obtains the required expression for the geodesic curvature of the only side of the quadrilateral that is not a geodesic.

Comparison of Lemmas 1 and 2 leads at once to an asymptotic expression for the geodesic curvature of the derivative of the curve Γ_ε at the point corresponding to the point on Γ where the normal has the azimuth φ:

$$K(\varepsilon) = \varepsilon\big[A(\varphi) + A_{\varphi\varphi}(\varphi)\big] + o(\varepsilon).$$

Comparing this formula with formula (1) we obtain (for the appropriate choice of sign of the geodesic curvature)

$$dR/d\varphi = A + A_{\varphi\varphi},$$

as had to be proved.

Remark. The resemblance of the passage from the small curve to its derivative used here with the contactomorphism

$$ST^*\mathbb{R}^n \to J^1(S^{n-1}, \mathbb{R})$$

of the space of co-oriented contact elements of Euclidean space to the space of the 1-jets of functions on the sphere used in [4] deserves special study.

CHAPTER IV

THE CHARACTERISTIC CHAIN AND SPHERICAL INDICES OF A HYPERSURFACE

The index of a point with respect to a closed plane curve is not conformally invariant. It changes if one takes for the point at infinity some other point of the Riemann sphere.

It turns out that one can adjust the indices of all points (deducting from them a suitable number) so that the adjusted indices are conformally invariant.

In this way, with each component of the complement to a curve on the sphere one can associate its spherical index. The components of the complement, furnished with this index as a multiplicity, form a 2-chain, whose boundary is the original curve.

This remarkable 2-chain is called the characteristic chain of the curve. In an analogous manner one can define the characteristic n-chain of an oriented $(n-1)$-dimensional hypersurface on an even-dimensional sphere and, more generally, of a wave front on an even-dimensional sphere.

Integrals over the characteristic chain appear in the Gauss–Bonnet formula for immersed curves, in the formula for the Maslov index, and so on. Unfortunately the simplicity and naturalness of the constructions of this and the following chapters is somewhat dimmed by the need to pay due heed and attention to endless signs and orientations.

§ 14. The characteristic 2-chain

Consider a smooth immersion of an oriented circle in general position on the oriented 2-sphere. The characteristic 2-chain of such an immersion is the chain consisting of the closures of the regions of the complement to the image with special (integral or semi-integral) coefficients. The curve itself is the boundary of its characteristic chain. The coefficients are chosen in the following way.

Select an auxiliary point B on the sphere outside the immersed curve. We regard it as the 'point at infinity' and identify the complement of this point with the oriented plane. The immersed curve on the sphere transforms to an immersed curve in the plane. This curve has a well-defined index (the number of complete turns of the normal as one travels right round the curve). We denote this by i_B (it depends on the choice of B).

For each point A of the plane outside the curve so obtained there is defined the winding number of the curve around A. Since this number depends not only on A but also on the choice of B, we denote it by i_A^B. The number i_A^B depends on the components of the complement to the curve in which A and B lie (while i_B depends on the component of B).

Theorem 1. *The difference*

$$\tilde{\imath}(A) = i_A^B - (i_B/2)$$

is a conformally invariant function on the complement to the curve on the sphere: it does not depend on the choice of the point B at infinity.

Example 1. Consider the equator on the oriented sphere. We orient it as the boundary of the Northern hemisphere.

This means that the frame (the outward normal of the Northern hemisphere, the vector orienting the equator) positively orients the sphere.

The choice of orientation and co-orientation of a hypersurface in an oriented manifold is said to be *correct* if on placing after the co-orienting normal an orienting frame one orients the manifold positively.

In the Northern hemisphere $\tilde{\imath}(A) = 1/2$, while in the Southern hemisphere $\tilde{\imath}(A) = -1/2$ (Fig. 13, see p. 32). A correctly co-orienting vector is directed to the side of smaller values of i.

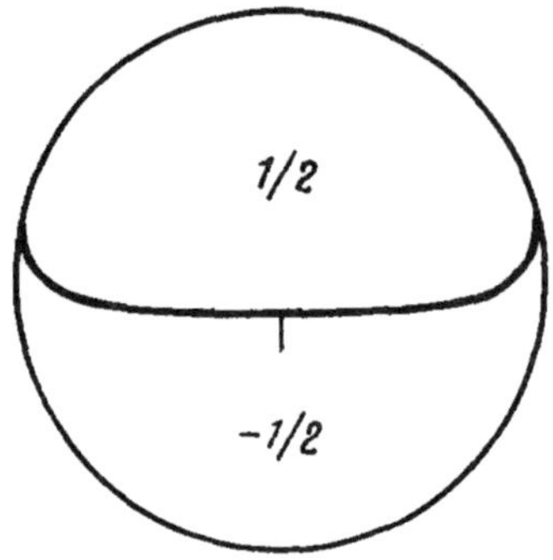

Figure 13

Example 2. Consider a figure of eight on the sphere (with correct orientation and co-orientation). The figure of eight divides the sphere into three regions (Fig. 14). One of these (R) is bounded by both loops, the second region (P) is bounded by the loop with co-orientation directed into the region, while the third region (Q) is bounded by the loop with co-orientation directed outside the loop. Calculations show that $\tilde{i}(P) = -1$, $\tilde{i}(Q) = 1$, $\tilde{i}(R) = 0$.

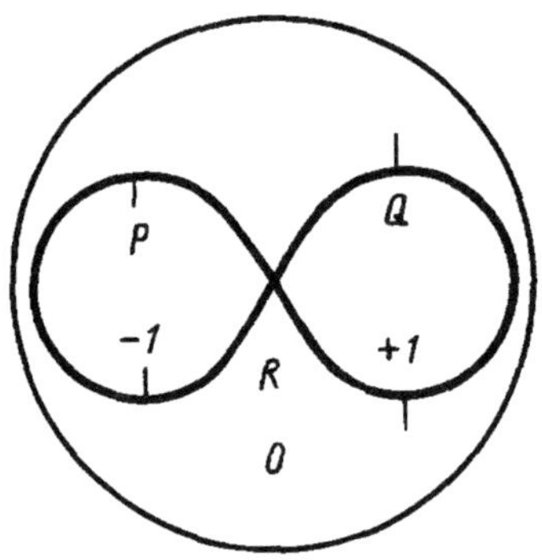

Figure 14

Definition. The *characteristic chain* of an immersed curve in general position on the sphere is the 2-chain consisting of the closures of the regions of the complement with the coefficients $\tilde{i}(A)$.

Suppose that we choose the point B to lie in the same component of the complement of the immersed curve in which A lies. Then $i_A^B = 0$. This proves the following corollary.

Corollary. *The following formula holds for the characteristic chain*:

$$\tilde{i}(A) = -i_A/2.$$

Although this formula appears to be simpler, in practice it is easier to do calculations using the formula of Theorem 1, since in it the point at infinity is fixed.

Let us consider now the lift of the immersion $S^1 \to S^2$ to the Legendrian curve of the normals formed from the correctly co-oriented contact elements of the sphere S^2 (tangent to the immersed curve):

$$\gamma : S^1 \to ST^*S^2 \approx \mathbb{RP}^3 \approx S^3/\pm 1.$$

I remark that the bundle $p : ST^*S^2 \to S^2$ doubly covers the Hopf bundle $S^3 \to S^2$.

'The curve of normals' $\gamma(S^1)$ lies in $\mathbb{RP}^3$ and is doubly covered in S^3 by a curve $\tilde{\gamma}$. The covering curve consists of one or two components, depending on the parity of the index of the corresponding curve on the plane. If the index i_B is even, as for the figure of eight, then there are two components, while if it is odd, then there is one (the parity of the index does not depend on the choice of B, it is opposite to the parity of the number of double points on the immersed curve).

Theorem 2. *The linking coefficient of the Legendrian curve $\tilde{\gamma}$ in S^3 covering the curve of normals, with the fibre of the Hopf bundle lying over the point $A \in S^2 \backslash f(S^1)$, is equal to*

$$2\tilde{i}(A) = 2i_A^B - i_B,$$

independently of the parity of the index ($i_B \mod 2$).

We denote by F the fibre of the bundle $ST^*S^2 \to S^2$ (with its natural orientation).

Theorem 3. *Consider the 2-chain σ in ST^*S^2 whose boundary is the curve of normals γ if the index of the original curve is even, and the curve $\gamma - F$ if it is odd.*

The projection $p_*\sigma$ of the chain σ on S^2 is the characteristic 2-chain $\tilde{i}$ of Theorem 1 (independently of the chioce of σ) for even index ($i_B \mod 2$), while $p_*\sigma = \tilde{i} + (1/2)$ for odd index.

The proofs of Theorems 1–3 are given in § 18 in the more general situation of wave fronts on even-dimensional spheres. But since some facts that we require remain true also for spheres of any dimension, I begin with these more general facts.

§ 15. The indices of hypersurfaces on a sphere

Consider an oriented smoothly immersed hypersurface M^{n-1} on an oriented sphere S^n.

Definition. The *correct co-orientation* of the hypersurface M is the field of unit normals ν such that the frame (normal, tangent frame positively orienting M) positively orients the sphere.

Remark. If M is the boundary of a region (with its usual orientation), then the normal ν is the outward normal of this region.

Consider two points A and B on the sphere outside M. On taking B to be the point at infinity and A to be the origin we may identify $S^N \backslash B$ with $\mathbb{R}^n$ and A with $0 \in \mathbb{R}^n$. The image of M in $\mathbb{R}^n$ will be denoted by M_A^B. The space $\mathbb{R}^n$ and the hypersurface M_A^B immersed in it are oriented.

Definition. The *index* $i_A^B(M)$ of the hypersurface M on the sphere relative to the pair of points (A, B) is the degree of the map of the hypersurface M_A^B to the sphere S^{n-1} by the rays from $A \approx 0$:

$$i_A^B(M) = \deg(s_A^B : M_A^B \to S^{n-1}), \qquad s_A^B(x) = x/|x|$$

(the sphere S^{n-1} is oriented here as the boundary of the ball $|x| \leqslant 1$ in $\mathbb{R}^n$).

Theorem 4. *The index of a smoothly immersed hypersurface on the sphere with respect to a pair of points of the complement depends only on the components of the complement to which these points belong. The function i_A^B on the sphere of a pair of regions of the complement to the hypersurface is skew-symmetric, and may be represented in the form*

$$i_A^B = \varphi(B) - \varphi(A),$$

where φ is a function of the region of the complement.

Consider an n-chain c on S^n, consisting of the closures of regions of the complement of a hypersurface M with coefficients

$$c(A) = i_A^B,$$

where B is an arbitrary fixed point of the complement. The sphere and the components here are oriented.

Theorem 5. *The boundary of the n-dimensional chain c is the original oriented hypersurface:*

$$\partial c = M^{n-1}.$$

Consider a point B outside the image of the hypersurface M on the sphere. Taking B to infinity we identify $S^n \backslash B$ with $\mathbb{R}^n$, as before.

Definition. The *index $i_B(M)$ of the immersed hypersurface M with respect to the point B* is the degree of the Gauss map

$$i_B(M) = \deg(s_B : M^{n-1} \to S^{n-1}),$$

where $s_B(x)$ is the normal vector of length 1 at x which co-orients the hypersurface M^{n-1} correctly.

Theorem 6. *On an even-dimensional sphere the indices of a hypersurface relative to points and pairs are connected by the relation*

$$2i_A^B = i_B - i_A.$$

In particular, the parity of all the indices i_A is the same.

Theorem 7. [6] *On an odd-dimensional sphere the indices of a hypersurface relative to all points of its complement are the same and equal to half the Euler characteristic of the immersed hypersurface:*

$$i_B(M) = \chi(M)/2.$$

The proofs of Theorems 1–7 will be given below, in § 18.

§ 16. Indices as linking coefficients

Consider the manifold $E^{2n-1} = ST^*S^n$ of co-oriented contact elements of the sphere. As with any contact manifold defined by a global contact form α, it has a canonical co-orientation $\alpha \wedge (d\alpha)^{n-1}$.

[An orienting frame can be obtained, for example, by adjoining to a frame orienting the base S^n a frame orienting the fibre S^{n-1}. The orientation of the fibre may be obtained from the orientation of the base as the orientation of the boundary of a small disk of radius r. The outward normals of the disk form the fibre, as $r \to 0$. Here, as usual, the cotangent vectors are identified with the tangent vectors by means of a metric.]

Thus in E^{2n-1} one can consider the indices of intersection of transversal chains of complementary dimension (and also of cycles of complementary dimension).

In particular, we can intersect Legendrian $(n-1)$-dimensional chains with n-dimensional ones.

Definition. The *Legendrian manifold of normals* L^{n-1} of an immersed oriented hypersurface M^{n-1} in the oriented sphere S^n is the manifold of correctly co-oriented contact elements of the sphere tangent to the immersed hyperpsurface M^{n-1}.

If the original immersed manifold M^{n-1} is in general position in S^n, then the Legendrian submanifold L^{n-1} of normals is embedded in E^{2n-1}.

The submanifold L^{n-1} of normals is homologous to the fibre F_b^{n-1} over an arbitrary point $b \in S^n \backslash M^{n-1}$ with some coefficient (since the cyclic group $H_{n-1}(E)$ is generated by $[F]$): there is an n-chain σ such that

$$\partial\sigma = L - kF_b, \qquad k \in \mathbb{Z}.$$

[Here we use the homology groups of E^{2n-1}:
if n is even, then the group is $\mathbb{Z}_2$ in dimension $n-1$ and 0 in dimension n;
if n is odd, then it is $\mathbb{Z}$ in dimensions $n-1$ and n (and, of course, always $\mathbb{Z}$ in dimensions 0 and $2n-1$);
the $\mathbb{Z}_2$-homology groups are $\mathbb{Z}_2$ in dimensions 0, $n-1$, n and $2n-1$.]

Theorem 8. *The index of an immersed hypersurface M relative to a pair of points is equal to the difference between the indices of intersection of the chain σ with the fibres over these points, to be precise*

$$i_A^B = \sigma \cap F_A - \sigma \cap F_B.$$

The non-uniqueness of the choice of σ has no effect on the difference.

Now suppose that the dimension of the ambient sphere S^n is even.

[6] On this matter see also the work of Tabachnikov [15] and the results of White discussed there on Legendrian links in a contact manifold of dimension $4n+1$.

Theorem 9. *On an even-dimensional sphere the index of intersection of the chain σ with the fibre F_A is equal to*

$$\sigma \cap F_A = -i_A/2 + \varepsilon/2,$$

where $\varepsilon = 0$ if the index i_A (and therefore also any index i_B) is even, while $\varepsilon = 1$ if the index i_A is odd.

In particular, the projection of the n-chain σ on the sphere S^n does not depend for even n on the choice of σ, and the boundary of this projection is the original oriented hypersurface M^{n-1}.

Remark. Theorem 9 makes clear why the indices

$$\widetilde{i}(A) = i_A^B - (i_B/2) = -i_A/2$$

can be interpreted as linking coefficients. In fact, for a hypersurface of even index on an even-dimensional sphere under arbitrary choice of the point B

$$\widetilde{i}_A = \sigma \cap F_A.$$

Since in this case $\partial\sigma = L$, Theorem 2 means that $\widetilde{i}_A$ *is the linking coefficient of the Legendrian manifold L of normals of M with the fibre F_A over the point A* (this linking coefficient is defined as the index of intersection of the chain σ filling out L with this fibre).

The proofs of Theorems 8 and 9 are carried out in § 18 with the help of the following constructions.

§ 17. The indices of hypersurfaces on a sphere as intersection indices

Consider two distinct points A and B of the sphere S^n. For this pair of points we construct the n-chain in the manifold E^{2n-1} of co-oriented contact elements of the sphere (to intersect the Legendrian submanifolds with it).

Definition 1. The *cylinder V_A^B over the segment* (A, B) is the n-dimensional submanifold (with boundary) in E^{2n-1} of all contact elements of the sphere applied at the points of the segment (A, B) smoothly embedded in the sphere.

By the segment (A, B) we understand here any smooth embedding of a segment in the sphere joining A to B (for example, a geodesic segment).

The cylinder V_A^B carries a natural orientation.

Definition 2. The orientation of the cylinder V_A^B is counted as *positive* if its intersection index with the Legendrian sphere formed from the outward normals of a small disk on S^n with centre at A is equal to $+1$.

Theorem 10. *The index of an immersed oriented hypersurface M^{n-1} with respect to a pair of points A and B of the oriented sphere S^n is equal to the intersection index of the Legendrian manifold L of normals of the immersed hypersurface M with the cylinder over a segment (A, B):*

$$i_A^B(M) = L \cap V_A^B.$$

Remark. Here it does not matter whether the hypersurface M is correctly oriented, provided only that it is the projection on S^n of the Legendrian manifold L. In fact, changing the orientation of M changes also the orientation of L and therefore also the signs on both the left- and the right-hand sides of the equality.

Theorem 11. *The boundary of the cylinder V_A^B is equal to the difference of the coherently oriented fibres over the points A and B, to be precise*

$$\partial V_A^B = (-1)^n(F_A - F_B).$$

Consider a point B on S^n. For this point we construct the n-chain in the manifold E^{2n-1} of co-oriented elements of the sphere (to intersect Legendrian manifolds with it).

Definition 3. A *B-section* V_B is the n-chain in E obtained from any constant vector field on $\mathbb{R}^n$ under stereographic projection with centre B, $\mathbb{R}^n \to S^n \setminus B$.

Remark. The stereographic projection sends a constant field on $\mathbb{R}^n$ to a manifold diffeomorphic to an open ball of dimension n. Choosing an appropriate diffeomorphism we get a smooth map which extends to the *closed* ball. On the boundary we then obtain a map of the sphere $S^{n-1} \subset \mathbb{R}^n$ to the sphere F_B (the fibre over B), given by the 'folding' map

$$\omega \to e - 2(\omega, e)\omega, \qquad (*)$$

where e is a constant vector of length 1.

The folding map $(*)$ sends any vector ω of the unit sphere S^{n-1} to the reflection of the constant basis vector e in the mirror orthogonal to ω.

One can easily prove the following lemma.

Lemma. *The folding map $(*)$ is realizable geometrically as the stereographic projection of a sphere from its centre onto another sphere of the same radius passing through this centre.*

Example. For $n = 2$ the folding map $(*)$ of a circle on a circle is the ordinary double covering '$\varphi \to 2\varphi$' (Fig. 15).

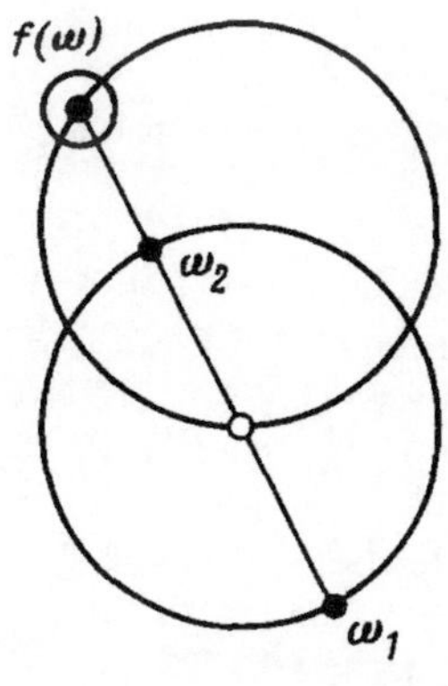

Figure 15

The map of a *closed* ball to E^{2n-1} described above defines an n-chain V_B. Its orientation is induced from the orientation of the base $\mathbb{R}^n \approx S^n \setminus B$.

Theorem 12. *The index of a hypersurface M^{n-1} with respect to B on S^n is equal to the index of intersection of the Legendrian manifold L^{n-1} of normals of the immersed hypersurface M^{n-1} with the B-section V_B:*

$$i_B(M) = L \cap V_B.$$

Remark. Here it does not matter whether the mainfold M is correctly oriented, provided only that it is the projection of a Legendrian manifold L to the sphere S^n. For a change in the orientation of M implies a change in the orientation of L and change in the signs on both the left- and the right-hand sides of the equality.

Theorem 13. *The boundary of the B-section V_B is the image of the sphere S^{n-1} by the folding map $(*)$, to be precise (taking account of the orientations)*

$$\partial V_B = -(1 + (-1)^n) F_B.$$

In other words, on even-dimensional spheres $\partial V_B = -2F_B$, while on odd-dimensional spheres $\partial V_B = 0$.

The geometrical Theorems 10–13 are verifiable directly, from the definitions. For example in Theorems 10 and 12 the contributions of each point of intersection in the left- and in the right-hand side of the equality is the same. It is only necessary to follow accurately through all the orientations. Theorems 1–9 are derived below from these four geometrical theorems.

§ 18. Proofs of the index theorems

Proof of Theorem 6. Let n be even. From Theorems 11 and 13 we have

$$\partial(2V_A^B - V_B + V_A) = 0.$$

Therefore from Theorems 10 and 12 it follows that

$$2i_A^B - i_B + i_A = 0,$$

that is, Theorem 6.

Proof of Theorem 4. From Theorem 11 it follows (for arbitrary n) that

$$\partial(V_A^B + V_B^A) = 0, \qquad \partial(V_A^B + V_B^C + V_C^A) = 0.$$

Applying Theorem 10 we conclude that

$$i_A^B = i_C^B - i_C^A.$$

By this Theorem 4 is proved.

Proof of Theorem 5. For a fixed point B the index $i_A^B(M)$ can be computed as the flow of the radial field in $\mathbb{R}^n \approx S^n \setminus B$ with centre at the point A. The change of this flow corresponding to transversal positive intersection of the hypersurface M by the point A is equal to -1. This proves Theorem 5.

Suppose now that the dimension of the sphere S^n is even.

Proof of Theorem 7. Consider M as an immersed hypersurface in general position in $\mathbb{R}^n \approx S^n \setminus B$. Construct two B-sections, corresponding to the constant fields e_1 and $-e_1$ in $\mathbb{R}^n$. On calculating the degree of the Gaussian map over the points $\pm e_1 \in S^{n-1}$ we obtain

$$i_B = m_0^+ - m_1^+ + \cdots + m_{n-1}^+,$$
$$i_B = m_0^- - m_1^- + \cdots + m_{n-1}^-,$$

where $m_k^\pm$ is the number of points where the co-oriented normal has the direction $\pm e_1$ and for which the restriction of the coordinate function $\pm x_1$ has a Morse critical point of index k.

The restriction of the function x_1 to M has therefore in all $m_k = m_k^+ + m_k^-$ critical points of index k. Thus, $\chi = m_0 - m_1 + \ldots + m_{n-1} = 2i_B$, and Theorem 7 is proved.

Proof of Theorem 8. For n-chains σ and τ in E^{2n-1}

$$(\partial\sigma) \cap \tau = (-1)^n \sigma \cap (\partial\tau).$$

Let $\partial\sigma = L - kF_C$, $\tau = V_A^B$. Then $(\partial\sigma)\cap\tau = i_A^B$ by Theorem 10, since $F_C \cap V_A^B = 0$ (the segment (A, B) may be chosen so as not to pass through C). On the other hand,

$$\sigma \cap (\partial\tau) = (-1)^n(\sigma \cap F_A - \sigma \cap F_B),$$

according to Theorem 11. Finally,

$$i_A^B = (\partial\sigma) \cap \tau = (-1)^n \sigma \cap (\partial\tau) = \sigma \cap F_A - \sigma \cap F_B,$$

which proves Theorem 8.

Proof of Theorem 9. Let n be even. Put $\tau = -V_A$. Then $2\tau = F_A$ by Theorem 13. The cycle $L - \varepsilon F_B$ is homologous to zero, so there is an n-chain σ with boundary $\partial\sigma = L - \varepsilon F_B$. Now $(\partial\sigma) \cap \tau = \sigma \cap (\partial\tau)$. But we have

$$(\partial\sigma) \cap \tau = (L - \varepsilon F_B) \cap V_A = \varepsilon F_B \cap V_A - L \cap V_A,$$

while also $\sigma \cap (2\tau) = 2\sigma \cap F_A$. According to Theorem 12, $L \cap V_A = i_A(M)$. Besides this, $F_B \cap V_A = 1$ due to the choice of the orientation of the fibre F_B and the A-section V_A. Thus, $\varepsilon - i_A = 2\sigma \cap F_A$, which proves Theorem 9.

Proof of Theorem 3. This is a particular case of Theorem 9 (the independence of the chain $p_*\sigma$ from the choice of σ follows also from the fact that $H_n(ST^*S^n) = 0$ for even n).

Proof of Theorem 1. This is a particular case of Theorem 6.

Proof of Theorem 2. We use Theorem 9. Denote by $\hat{\sigma}$ the inverse image of the chain σ under the double covering $S^3 \to ST^*S^2$. The fibre of the Hopf bundle over A is the double covering $\hat{F}_A$ of the fibre F_A of the bundle $ST^*S^2 \to S^2$ over A. We have $\partial\hat{\sigma} = \hat{\gamma} - \varepsilon\hat{F}_C$. Denote by D_C a 2-chain in S^3 with boundary $\partial D = \hat{F}_C$. Then the 2-chain $\hat{\sigma} - \varepsilon D_C$ has boundary $\widetilde{\gamma}$. Therefore the linking coefficient sought for is

$$l(\hat{\gamma}, \widetilde{F}_A) = (\hat{\sigma} - \varepsilon D_C) \cap \hat{F}_A = \hat{\sigma} \cap \hat{F}_A - \varepsilon D_C \cap \hat{F}_A.$$

According to Theorem 9, $\hat{\sigma} \cap \hat{F}_A = 2\sigma \cap F_A = \varepsilon - i_A$. The linking coefficient of the fibres $\hat{F}_C$ and $\hat{F}_A$ of the Hopf bundle is equal to 1, and therefore $D_C \cap \hat{F}_A = 1$. Thus $l(\hat{\gamma}, \hat{F}_A) = -i_A$. But, by Theorem 6, $-i_A = 2i_A^B - i_B$. And so Theorem 2 is proved.

§ 19. The indices of fronts of Legendrian submanifolds on an even-dimensional sphere

Let L^{n-1} be an oriented Legendrian immersed submanifold of the contact manifold ST^*S^n of co-oriented elements of the oriented sphere S^n.

Consider the front of L (its projection to the sphere) as an oriented and co-oriented (singular) hypersurface of the sphere.

Example. Let L be the manifold of normals of an oriented hypersurface immersed in the sphere. Then for the front we get this immersed hypersurface with its regular co-orientation.

Remark. If one changes the *orientation* of L, then for the oriented and co-oriented front one obtains the same co-oriented hypersurface but with changed orientation. In such a case the co-orientation of the oriented immersed hypersurface forming the front is not correct.

In the general case the front has singularities. In such a case there is no relation between the orientation and co-orientation of the front and the orientation of the ambient sphere. Therefore in transferring the preceding results to fronts one needs to omit the correctness of the co-orientation everywhere (replacing it by the agreement of the orientations of the Legendrian manifold L and its front). With this change all the results of the previous sections hold.

Let A and B be points of the complement of a front in the sphere. Identifying B with the point at infinity, we get in $\mathbb{R}^n$ an oriented and co-oriented front M_B^{n-1}.

Definition. The *index* i_B is defined as the degree of the Gauss map $L^{n-1} \to S^{n-1}$ (associating with a point of the Legendrian manifold the direction of the normal which co-orients the front in $\mathbb{R}^n$).

The *index* i_A^B is defined as the degree of the radial map $L^{n-1} \to S^{n-1}$ (associating with a point of the Legendrian manifold the direction of the vector from A to the point of the front).

The quantity

$$\widetilde{i}(A) = i_A^B - (i_B/2)$$

in this new situation is as before a locally constant function on the complement of the front that does not depend on B (it is equal to $-i_A/2$, as in the case of an immersion).

Example. The values of the function $\widetilde{i}(A)$ for some fronts with two cusps are shown in Fig. 16.

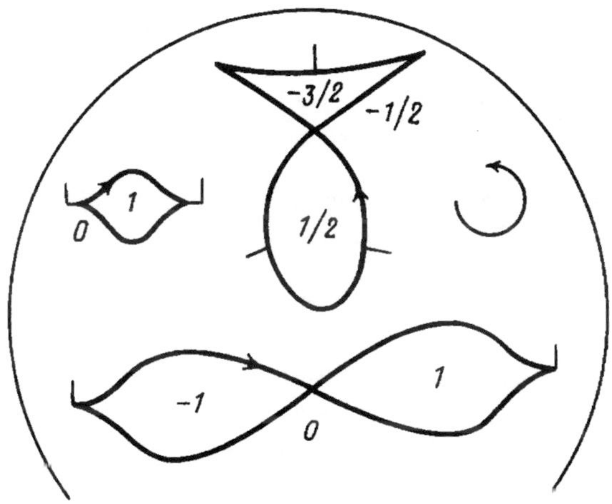

Figure 16

Definition. The *characteristic n-chain* of an oriented and co-oriented front in an even-dimensional oriented sphere is the n-chain consisting of the oriented closures of the components of the complement of the front with coefficients $\widetilde{i}(A)$.

Remark. On changing the orientation of the ambient sphere the characteristic n-chain *does not change.* One can say that an oriented and co-oriented front determines the orientations as well as the multiplicities of the regions of the complement.

On change of the co-orientation of a front the characteristic n-chain also does not change.

On change of the orientation of a front (that is, of the Legendrian manifold corresponding to it) the characteristic chain changes sign: its boundary is always the oriented front.

From these properties of the characteristic chain it is evident that it is possible to define it also for (oriented and co-oriented) fronts on a projective space of even dimension.

To be precise, for the two-sheeted cover of a projective space by the sphere the full inverse image of the characteristic chain of a front on the projective space is the characteristic chain of the full inverse image of the front on the sphere.

Example. The characteristic 2-chain of an oriented circle on the projective plane is the disk bounded by this circle (with its orientation chosen so that its boundary is the given oriented circle). On the Möbius strip complementary to the disk $\widetilde{i} = 0$. All this does not depend on the co-orientation of the front.

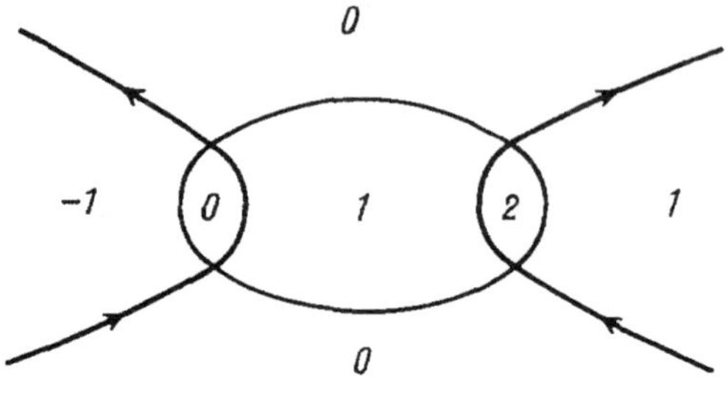

Figure 17

The characteristic 2-chain of the front consisting of a confocal ellipse and hyperbola is shown in Fig. 17.

Theorem 14. *Theorems* 1–6 *and* 8–13 *of this chapter remain true for fronts.*

Even in the proofs practically nothing has to be changed (if we define the indices as mentioned above and define the orientation of the front M independently from its co-orientation as the orientation of its parametrizing Legendrian manifold L).

The proofs of Theorems 10, 12 and 5 are the most crucial. Consider, for example, the assertion of Theorem 12:

$$i_B(M) = L \cap V_B.$$

The proof consists in calculating the contributions of points of transversal intersection of L with V_B both to the left-hand side and to the right. It is asserted that both these contributions are equal for each point separately.

In terms of the Euclidean space $\mathbb{R}^n \approx S^n \setminus B$ the manifold $V_B \subset ST^*\mathbb{R}^n$ is describable as a constant generic (co)vector field v. The contribution to the right-hand side is given only by those points of L where the co-orienting normal to M has directon v.

By choosing the field v to be generic we can arrange that the corresponding points of the front M are non-singular. At these and only at these points the vector co-orienting the front has the direction v.

In computing the number $i_B(M)$ as the degree of the Gauss map $L^{n-1} \to S^{n-1}$ it is enough to count the signs of all these inverse images of the direction v.

At these points the front is a smooth oriented and co-oriented hypersurface. If its co-orientation is correct for the given orientation, then such a point gives the same contribution to both sides of the equality in Theorem 12 (its proof therefore consists in this verification of the coincidence of the contributions).

If the co-orientation of M is not correct for the given orientation, then we change (locally) the *orientations* of L and M. Then the co-orientation becomes correct. The equality of the contributions to the left- and right-hand sides will then hold at the point under consideration. But the contributions both on the left and on the right have changed sign under the change of the orientations. This means that they were equal to start with. By this the assertion of Theorem 12 is proved for the case of fronts.

The assertions of Theorems 10 and 5 are proved in exactly the same way.

The assertion of Theorem 7 for fronts in S^3 or in $\mathbb{R}^3$ is clearly not true: the index i_B depends on the choice of point B.

In the case of S^2, where the front is one-dimensional, all the indices and characteristic chains of generic fronts may be computed also by means of the following construction of averaging the indices of smoothings of the front.

Apart from self-intersections, the only singular points of a generic front are semicubical cusps. There are two ways of smoothing a cusp (Fig. 18): in the one case near the cusp there appears a point of self-intersection, while in the other case there does not. In either case a co-orientable front becomes non-co-orientable, while a non-co-orientable front becomes co-orientable.

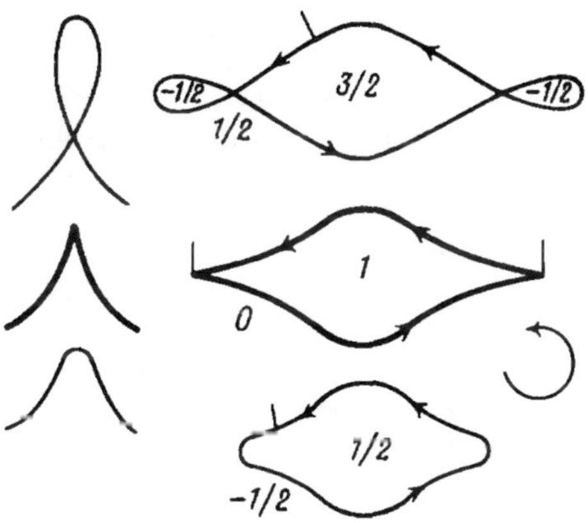

Figure 18

On smoothing in one way or the other all the cusps of a co-orientable front we obtain two new co-orientable fronts (since the number of cusps of a co-orientable front is even). Note that both the fronts so obtained are oriented (like the original front away from its cusps).

Theorem 15. *For the computation of the indices i_B, i_A^B and $\tilde{i}_A^B$ is enough to smooth the original front near all its cusps both ways and to take the arithmetic mean of the values of the corresponding indices for the two immersed curves so obtained.*

Example. Fig. 18 shows the characteristic chains of both smoothings of the simplest curve with two cusps.

Remark. One can form the mean of all the 2^N possible smoothings resulting from smoothing N cusps—the result will be just the same.

Proof. It is sufficient to prove the assertion for plane curves. We denote by $\varphi(t)$ the azimuth of the co-orienting contact element of the original front at the point corresponding to t on the Legendrian curve. For the smoothed curves we denote the corresponding functions by φ_+ and φ_-. Choose the values of the azimuth so that at one non-singular point of the front all three functions have the same value.

On passing through the nearest cusp the differences $\varphi_+ - \varphi$ and $\varphi_- - \varphi$ each change by π but in opposite directions. The same happens on passing through the next cusp. Finally on returning to the original point the difference $\varphi_+ - \varphi$ has changed by an (even) multiple of π in the one direction, and the difference $\varphi_- - \varphi$ in the other. This means that

$$2 \cdot (\text{the increase in } \varphi) = (\text{the increase in } \varphi_+) + (\text{the increase in } \varphi_-),$$

that is, for the index i_B of the original front we have the expression

$$2i_B = i_B(+) + i_B(-)$$

in terms of the indices of both the smoothings.

The number of circuits of the front around different points of its complement in the plane does not change at all for a sufficiently small smoothing:

$$i_A^B = i_A^B(+) = i_A^B(-).$$

Therefore the characteristic chain also coincides with the arithmetic mean of the smoothed chains outside a small neighbourhood of the front:

$$2\widetilde{i}_A = \widetilde{i}_A(+) + \widetilde{i}_A(-).$$

In particular, from this we conclude that even for a front with cusps

$$\widetilde{i}_A = -i_A/2,$$

so that for the indices of fronts with cusps one has conformal invariance, the formula of Theorem 6, and so forth.

CHAPTER V

EXACT LAGRANGIAN CURVES ON A SPHERE AND THEIR MASLOV INDICES

The simplest exact Lagrangian curve on the symplectic sphere is an embedded curve, dividing the area of the sphere in half. Here we calculate the Maslov index of an exact immersed Lagrangian curve: it is expressed in terms of the area of the characteristic chain of the immersed curve, introduced in Chapter IV.

Our construction leads to an unusual version of the Gauss–Bonnett formula and to proofs of the theorems on the dual and derivative curves, formulated in § 1 (Theorems 5 and 9 of § 1 and Theorem 2 of § 3), and to the theorems of duality between area and length.

§ 20. Exact Lagrangian curves and their Legendrian lifts

Consider a smoothly immersed curve on the standard sphere S^2 of radius 1.

Definition. A curve is said to be *exact* if the area that it bounds is a multiple of 2π (half the area of the sphere).

Remark. The area bounded by a curve is the integral of the area form over the 2-chain whose boundary is the curve.

Such a 2-chain is not unique: it is defined up to the addition of the entire sphere. Therefore the area mentioned is defined modulo 4π, which means that the fact of it being a multiple of 2π is well defined. Moreover the distinction between exact curves, the one bounding an area 2π (mod 4π) and the other an area 0 (mod 4π), is also well defined.

Example. An embedded curve is exact if and only if it divides the area of the sphere in half.

A figure of eight is exact if and only if the absolute values of the areas of both the regions bounded by a loop are the same or differ by 2π.

A parallel of latitude, described p times, is exact if and only if it bounds a disk of area $2\pi k/p$, with k an integer.

Consider now the S^1-bundle $\widetilde{\pi}_i : S^3/+1 \to S^2_i$ from § 6.

Theorem. *Exact Lagrangian curves on the sphere S^2_i are precisely the projections on that sphere of i-Legendrian curves lying in the manifold $S^3/\pm 1$ with its standard contact i-structure.*

Proof. The contact i-structure provides the principal bundle $\widetilde{\pi}_i$ with an S^1-connection. A curve in the space of the bundle is i-Legendrian if and only if it is the lift of its projection on the base as an integral curve of the connection. This S^1-connection is invariant with respect to rotations of the sphere. Let us compute its curvature.

For future calculations it is necessary to introduce on $S^3/\pm 1$ a basis for the differential 1-forms.

Definition. The *form* α_i is defined by its values on the vector fields $\mathbf{i}, \mathbf{j}$ and $\mathbf{k}$:

$$\alpha_i(\mathbf{i}) = 2, \qquad \alpha_i(\mathbf{j}) = 0, \qquad \alpha_i(\mathbf{k}) = 0$$

(the coefficient 2 is introduced for convenience, bearing in mind its appearance in Lemmas 1, 2 and 3 of § 7).

The forms α_j and α_k are defined by cyclic permutation of the quaternionic units, so that

$$\alpha_j(\mathbf{j}) = 2, \qquad \alpha_k(\mathbf{k}) = 2.$$

Since the fields $\mathbf{i}, \mathbf{j}$ and $\mathbf{k}$ are left-invariant, all three constructed forms are also left-invariant.

The basic fact in what follows is the following lemma.

Lemma 1. *The differential of the form α_i is the lift of the standard area form ω_i from the complex sphere S_i^2:*

$$d\alpha_i = \widetilde{\pi}_i^* \omega_i.$$

Proof. The differential of α_i in the direction of the field $\mathbf{i}$ is equal to zero. By the homotopy formula

$$0 = L_{\mathbf{i}}\alpha_i = i_{\mathbf{i}}(d\alpha_i) + d(i_{\mathbf{i}}\alpha_i) = i_{\mathbf{i}}(d\alpha_i),$$

so $i_{\mathbf{i}}\alpha_i = 2$ is constant. Conseqently, the form $d\alpha_i$ vanishes when evaluated on the vector $\mathbf{i}$ tangent to a fibre.

This means that the 2-form is the lift of some 2-form from the factor-space S_i^2.

The left invariance of α_i implies that this 2-form on S_i^2 is invariant under rotations of the sphere S_i^2. This means that it is proportional to the standard area form ω_i, that is,

$$d\alpha_i = c\widetilde{\pi}_i^* \omega_i.$$

It remains only to compute the constant c.

For the computation of c we use the following fact.

Lemma 2. *The integral of the form α_i along a fibre of the bundle $\widetilde{\pi}_i : S^3/\pm 1 \to S_i^2$, oriented by the field* $\mathbf{i}$, *is equal to 2π.*

Proof. In the notations of Lemma 2 of § 7 (where the matter concerns the sphere S_j^2, so the quaternion units must be permuted cyclically) a full revolution is accomplished in time 2π (and in the direction opposite to the direction of the complex structure, if the bundle is identified with $ST^*S_i^2$ in the usual way). The time stated is the integral of the form along the fibre oriented by the field $\mathbf{i}$.

Lemma 3. *Consider a vector field with one singular point B on the complex sphere S^2 as a 2-chain σ, oriented by the complex orientation of the sphere. Then $\partial\sigma = -2F_B$, where F_B is the fibre over B (with the orientation given by the direction of the complex-positive rotation of the sphere).*

Proof. This is a particular case of Theorem 13 of § 17 (actually that theorem is obvious). The index of the only singular point of the field is equal to $+2$. Consider the field on the boundary circle of the disk covering this sphere except for a small neighbourhood of the point B. This boundary of the big disk goes around B along a small circle oriented negatively (with respect to the direction of the complex rotation around B). This means that as one goes round this circle a vector of the field turns so that it completes two full revolutions in the negative direction, from which Lemma 3 follows.

Conclusion of the proof of Lemma 1. We compute the integral of the form $d\alpha_i$ over the chain σ. By Stokes' formula

$$\iint_\sigma d\alpha_i = \int_{\partial\sigma} \alpha_i = \int_{-2F} \alpha_i = 4\pi.$$

In fact, the integral of the form α_i over the fibre oriented by the direction of the field $\mathbf{i}$ is equal to 2π. Moreover the direction of the field $\mathbf{i}$ is opposite to that of the complex rotation of the sphere S_i^2 (by Lemma 2 of § 7). Thus $-F$ is the fibre oriented by the direction of the field $\mathbf{i}$. We see that

$$\iint_{S_i^2} c\omega_i = \iint_\sigma d\alpha_i = 4\pi$$

and consequently $c = 1$.

Remark. It is useful to remark that the left-invariant forms α_i, and so on, satisfy the equations

$$d\alpha_i = -\alpha_j \wedge \alpha_k$$

and so on.

In fact, the forms on either side are left-invariant and reduce to zero on the vectors of the $\mathbf{i}$-direction. Therefore it is enough to verify the coincidence only on the pair $(\mathbf{j}, \mathbf{k})$.

On this pair the left-hand side has the value -4, since the complex orientation of the sphere S_i^2 is the orientation from $\mathbf{k}$ to $\mathbf{j}$ (by Lemma 5 of § 7). The right-hand side also has the value -4.

Of course, one can do all this also by direct coordinate calculations, simply multiplying quaternions and explicitly calculating the components of the forms at the point 1, neglecting quantities of the second order of smallness.

Actually all these computations are simply the computations of the structural constants of the group $SO(3)$. But I have given them in full, so as to be sure about the correctness of those cursed signs.

Proof of the theorem. Let L be an i-Legendrian curve on $S^3/\pm 1$ (so that α_i reduces to zero on L). Consider a 2-chain σ with boundary $L - \varepsilon F$ (where $\varepsilon = 0$ if the curve L is homologous to 0 and equal to 1 if not). Here F is a fibre of the bundle $\widetilde{\pi}_i : S^3/\pm 1 \to S_i^2$ with the orientation given by the complex structure of the sphere (so that $\int_F \alpha_i = -2\pi$).

By Stokes' formula

$$\iint_\sigma d\alpha_i = \int_L \alpha_i - \int_{\varepsilon F} \alpha_i = \varepsilon 2\pi.$$

By Lemma 1, $d\alpha_i = \widetilde{\pi}_i^* \omega_i$ and consequently

$$\iint_\sigma d\alpha_i = \iint_\sigma \widetilde{\pi}_i^* \omega_i = \iint_{\widetilde{\pi}_{i*}\sigma} \omega_i$$

is the area bounded by the projection of L on the sphere S_i^2.

Thus this area is an integral multiple of 2π.

Conversely, start with an arbitrary closed Lagrangian curve on S_i^2. It is covered by a segment of a Legendrian curve L', ending above its initial point. Join the end point to the initial point by a segment Δ of the fibre F_B. We get a closed curve $L \cup \Delta$ in $S^3/\pm 1$. We construct a 2-chain σ with boundary $L - \varepsilon F$ just as before.

Now for the area bounded by the original Lagrangian curve (mod 4π) we have the expression

$$\iint_{\widetilde{\pi}_{i*}\sigma} \omega_i = \iint_\sigma d\alpha_i = \int_L \alpha_i - \int_{\varepsilon F} \alpha_i = \int_\Delta \alpha_i + 2\pi\varepsilon.$$

If the original Lagrangian curve is exact, then the area on the left, and that means also the integral of the form α_i over the segment Δ, is a multiple of 2π. This means that the beginning and end points of the segment coincide, and the theorem is proved.

§ 21. The integral of a horizontal form as the area of the characteristic chain

Consider a closed Legendrian curve L in the space of co-oriented contact elements of a standard sphere S^2 of radius 1. We set about the computation of the integral of a horizontal contact form along this curve. First I recall some notations from § 6.

Consider two bundles over the standard sphere S^2_j,

$$\widetilde{\pi}_j : S^3/\pm 1 \to S^2_j; \qquad p\colon ST^*S^2_j \to S^2_j.$$

The fibres of the first bundle are the orbits of multiplication by $\{e^{jt}\}$ on the right. The fibres of the second bundle are the circles consisting of the co-oriented contact elements applied at a point of the sphere.

Definition. The *natural i-identification* of both these bundles is the map associating with a point z of $S^3/\pm 1$ the contact element on S^2_j orthogonal to the projection of the vector of the field $\mathbf{i}$ at z (co-oriented by this projection).

By means of the identification we carry over to the manifold of the co-oriented contact elements the differential 1-form $\alpha_j = \alpha$ (defined in § 20 in terms of $S^3/\pm 1$).

This 1-form is horizontal: planes $\alpha = 0$ are orthogonal to fibres of the bundle. It is left-invariant and invariant with respect to multiplication by e^{jt} on the right. Its integral along the fibre F of our bundle (oriented by the complex structure of the sphere) is equal to -2π (see § 20).

Main theorem. *The integral of the horizontal form α along a Legendrian curve L is equal to the area of the characteristic 2-chain of the front of this curve*:

$$\int_L \alpha = \iint_{c(p_* L)} \omega,$$

where ω is the standard area form on the complex sphere S^2_j.

Proof. We construct a 2-chain σ with boundary $\partial\sigma = L - \varepsilon F$ (where $\varepsilon = 1$ or 0). By Stokes' formula

$$\iint_\sigma d\alpha = \int_L \alpha - \varepsilon \int_F \alpha = \int_L \alpha + 2\pi\varepsilon.$$

According to Theorem 3 of § 14 the projection of the chain σ to the sphere is $p_*\sigma = c + \varepsilon S_j^2/2$. By the theorem of § 20 the differential of the form α is the area of the projection: $d\alpha = p^*\omega$. Thus we find that

$$\iint_\sigma d\alpha = \iint_\sigma p^*\omega = \iint_{p_*\sigma} \omega = \iint_c \omega + \varepsilon 2\pi.$$

Consequently, and finally,

$$\int_L \alpha + 2\pi\varepsilon = \iint_c \omega + 2\pi\varepsilon,$$

as was asserted.

Remark. The standard i-identification may be changed here to an arbitrary τ-identification, where the quaternion τ is orthogonal to j (for example, a k-identification).

§ 22. A horizontal contact form as a Levi-Civita connection and a generalized Gauss–Bonnet formula

The contact structure $\alpha_j = 0$ in $S^3/\pm 1$ determines a connection in the S^1-bundle $\tilde{\pi}_j : S^3/\pm 1 \to S_j^2$.

Theorem 1. *The contact connection $\alpha_j = 0$ is transformed into a Levi-Civita connection of the bundle of (co)tangent vectors of unit length on the sphere S_j^2 under the natural i-identification of the bundles*

$$\tilde{\pi}_j : S^3/\pm 1 \to S_j^2, \qquad p: ST^*S_j^2 \to S_j^2$$

(cotangent vectors are identified with tangent vectors by means of a metric on the sphere).

Here, as in § 21, the i-identification may be changed into an arbitrary τ-identificaton, where the quaternion τ is orthogonal to j.

Proof. The vectors of the fields **i** and **k** at each point of $S^3/\pm 1$ generate the plane $\alpha_j = 0$. An integral curve of the field **k** is i-identified in $ST^*S_j^2$ with the motion of the co-orienting vector of the normal along the geodesic everywhere orthogonal to it, while the field **i** is everywhere tangent to it (by Lemmas 3 and 1 of § 7). Both these motions are parallel transports in the sense of the Levi-Civita connection (according to its definiton). The theorem is proved.

Definition. The geodesic curvature of a co-oriented curve is considered to be *positive* if the curve bends away from the geodesic tangent to it towards its co-orienting vector.

Example. The parallel of latitude, at a distance θ from the pole of the standard sphere of radius 1 and co-oriented by the direction to the pole, has geodesic curvature $\varkappa = \cot\theta$.

From this example we have the following result.

Lemma 1. *The geodesic curvature $\varkappa$ of a co-oriented curve on the sphere of radius* 1 *is connected with the spherical radius of curvature of the curve,* θ, *by the relation* $\varkappa = \cot\theta$.

Consider now a Legendrian curve L in the space $ST^*S_j^2$ of co-oriented contact elements of the sphere S_j^2. Recall that we defined above the horizontal 1-form α_j in the space of the fibration $p : ST^*S_j^2 \to S_j^2$. The sphere S_j^2 is oriented by its complex structure.

Lemma 2. *The value of the horizontal form α_j on any tangent vector of a Legendrian curve L is equal to $\varkappa ds$, where $\varkappa$ is the geodesic curvature of the co-oriented front pL of this curve, while ds is the length of the projection of the vector on the sphere* (*with plus sign if the projection orients the front correctly, and minus if it does not*).

I remind you that the orientation of a co-oriented curve on an oriented surface is *correct* if the frame (co-orienting vector, orienting vector) imparts a positive orientation to the surface.

Proof of Lemma 2. The tangent vector of an i-Legendrian curve is orthogonal to the $\mathbf{i}$-direction. Therefore one can represent it as a combination of the vectors of the fields $\mathbf{j}$ and $\mathbf{k}$:

$$\boldsymbol{\xi} = \mathbf{j}A\cos\theta + \mathbf{k}A\sin\theta, \qquad A > 0.$$

According to the Theorem of § 10, the spherical radius of curvature of the front is θ. According to Lemma 1 the geodesic curvature of the front is $\varkappa = \cot\theta$.

The value of the form α_j on a vector of the field $\boldsymbol{\xi}$ is equal to $2A\cos\theta$ (from the definition of this form). We prove that *the oriented length of a vector of the field* $\boldsymbol{\xi}$ *is equal to* $2A\sin\theta$.

In fact, under the projection p along the j-directon we have

$$p_*\boldsymbol{\xi} = (A\sin\theta)p_*\mathbf{k}.$$

But the oriented length of the projection $p_*\mathbf{k}$ of the vector of the field $\mathbf{k}$ on the sphere S_j^2 is equal to 2 by Lemma 3 of § 7. Thus, $ds(\xi) = 2A\sin\theta$.

Finally we find that

$$\alpha_j(\boldsymbol{\xi}) = 2A\cos\theta = (\cot\theta)2A\sin\theta = \varkappa\, ds(\boldsymbol{\xi}),$$

which is what had to be proved.

Combining Lemma 2 with Theorem 1 and the main theorem of § 21, we obtain the following generalized Gauss–Bonnet formula.

Consider a co-oriented curve K smoothly immersed in the sphere with geodesic curvature $\varkappa$.

Theorem 2. *There is a Gauss–Bonnet formula for immersions*

$$\int_K \varkappa\, ds = \iint_c \omega,$$

where c is the characteristic chain of a correctly oriented curve K and ω is the Gaussian curvature form (that is, for the standard sphere of area 1 simply the area form).

I remind you that a correct orientation of a curve is defined by the condition that the frame (co-orienting vector, orienting vector) positively orients the sphere.

Proof. At first let the metric be standard. By Lemma 2 the left-hand side is then equal to the integral of the horizontal form α along the curve of normals. By the main theorem of § 21 this integral is equal to the double integral on the right-hand side of the formula. So this formula is proved for the standard metric.

Any metric can be obtained from a standard one by a sequence of local deformations. Under a local deformation the equality of the left- and right-hand sides cannot be broken. This follows from the fact that the Gauss–Bonnet formula is true for embedded curves.

Remark. A direct proof of Theorem 2 for an arbitrary metric repeats the proof of the main theorem. The topological equality $p_*\sigma = c + \varepsilon S^2/2$ does not depend on the metric. On the other hand, the relations

$$\alpha(\xi) = \varkappa\, ds(\xi), \qquad d\alpha = p_*\omega, \qquad \int_F \alpha = -2\pi, \qquad \iint_{S^2} \omega = 4\pi$$

are satisfied for the form α of the Levi-Civita connection and its curvature form ω for any metric of the sphere.

Consider now an oriented and co-oriented front on an oriented sphere.

Definition. A point of a front is said to be *positive* if the pair (co-orienting vector, orienting vector) positively orients the sphere.

Example. A semicubical cusp of a front separates its positive branch from its negative one.

Theorem 3. *The following Gauss–Bonnet formula holds for fronts:*

$$\int_K \varkappa\, ds = \iint_c \omega,$$

where c is the characteristic chain of the front K and ds is the element of the oriented length of the front (positive at the positive points of the front and negative at the negative points).

The proof is the same as that of Theorem 2.

Already from Theorem 2 we have an important corollary.

Corollary. *Every exact Lagrangian curve is the derivative of a closed front.*

Proof. Select at a point of the Lagrangian curve the great circle passing through this point. Transport the tangent vector of this circle at this point parallel to itself (in the sense of the standard Riemannian metric on the sphere) along our closed

Lagrangian curve. On returning to the original point we get the initial vector. In fact, the rate of change of the angle of the vector carried parallel to itself along the curve with respect to the tangent vector of the curve is equal to the geodesic curvature of the curve. Therefore for a complete circuit the transported vector turns with respect to the tangent vector through the integral of the geodesic curvature. By Theorem 2 this angle is equal to the area of the characteristic chain. If the original Lagrangian curve is exact, then the angle is a multiple of 2π and the vector returns to its place (and if not, not).

Consider now the great circles intersecting the Lagrangian curve in the directions of the transported vectors. This family of circles has an envelope.

Lemma. *The distance along each of these circles from the original Lagrangian curve to the envelope is equal to* $\pi/2$.

Proof of the lemma. Replace a small segment of our curve by a small great circle arc. The parallelism of the circles constructed above along our curve means that the difference between the angles of this arc with the circles of our family at its ends is of an order of smallness no higher than the first with respect to the length of the arc. But infinitely close great circles intersecting the equator at the same angle intersect at a distance $\pi/2$ from the equator along each of them. This means that infinitely near great circles of our family also intersect at a distance $\pi/2$ along them from the Lagrangian curve.

The derivative of the envelope constructed in this way is our original exact Lagrangian curve. The corollary has been proved.

§ 23. Proof of the formula for the Maslov index

I recall that the Maslov index μ of an oriented and co-oriented front in general position is equal to the difference between the numbers of positive and negative cusps (Fig. 19).

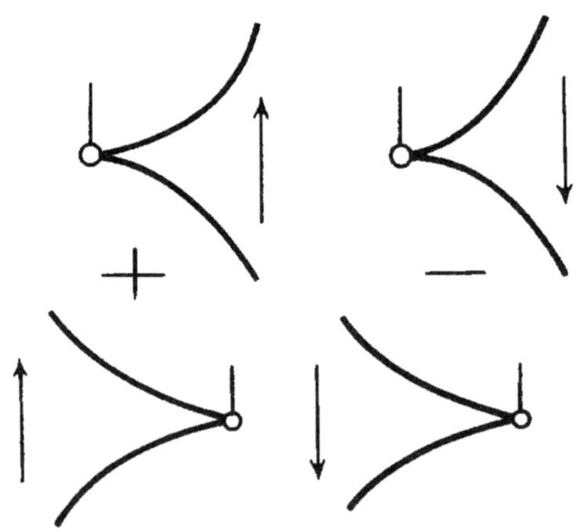

Figure 19

Theorem. *The Maslov index μ of an oriented and co-oriented front on an oriented sphere is connected to the characteristic chain c (that means with the geodesic curvature $\varkappa$ also) of its correctly co-oriented derivative K by the relations*

$$\iint_c \omega = -\pi\mu = \int_K \varkappa\, ds,$$

where ω is the area element of the sphere and ds is the element of length.

Proof. A cusp of the front corresponds to a point where the derivative curve is orthogonal to the family of the framing great circles, with the help of which the derivative has been constructed (see § 12).

So at a generic cusp the angle α between the derivative curve and the framing direction passes through the value $\pm\pi/2$ with non-zero velocity. The sign of the velocity is determined by the sign of the cusp (according to Theorem 3 of § 12 and Example 2 following it).

We conclude that the number μ of cusps, taking account of their signs, is equal to twice the number of turns of the framing vector with respect to the normal vector of the derivative curve.

But the framing vectors form a parallel framing (Theorem 2 of § 12). Consequently the angle of turn of the normal vector with respect to the framing vector after one circuit of the derivative curve is equal to the integral of its geodesic curvature.

The angle of turn of the framing vector with respect to the normal is the opposite, therefore it is equal to $-\int_K \varkappa ds$. Consequently the number of turns is equal to $-\int_K \varkappa ds/(2\pi)$. Therefore twice this number of turns, namely $-\int_K \varkappa ds/\pi$, is equal to the Maslov index μ. Hence the second equality of the theorem is proved.

The first follows from the second by Theorem 2 of § 22.

Remark. The metric of the sphere in the proved theorem is the standard one. Generalizations of this theorem to the case of other metrics are not known to me (the definition of the derivative uses the standard metric).

To verify the sign in the preceding theorem it is necessary to apply it to a front with a non-zero Maslov index.

Example. Consider a system of fronts whose caustic is a parallel of latitude of the standard sphere of radius 1 traversed p times (at a distance $\theta < \pi/2$ from the North pole along meridians). Suppose that the length of the parallel ($2\pi \sin\theta$) is equal to $2\pi/p$. One of the fronts, Γ (for the case $p = 3$), is shown in Fig. 20. The orientation of the sphere is also shown in Fig. 20.

Choose the co-orientation and orientation of the front as shown in Fig. 20. Both the cusps are positive, so that $\mu = 2$. The derivative curve of the front is the ($p = 3$ times traversed) parallel of latitude, at a distance $(\pi/2) - \theta$ from the North pole along meridians. The orientation of the front gives the orientation of the derivative curve Γ', shown in Fig. 20 (see p. 54). The correct co-orientation of this derivative curve is to the side of increasing distance from the North pole.

The geodesic curvature $\varkappa$ of the correctly co-oriented derivative curve is therefore equal to $-\tan\theta$. The integral of this curvature along the derivative curve

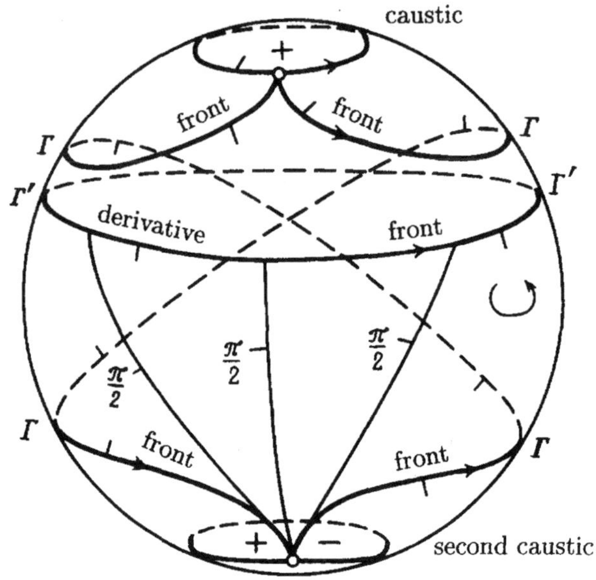

Figure 20

(of length $2\pi p \cos\theta$) is equal to

$$\int_K \varkappa\, ds = -2\pi p \cos\theta \tan\theta = -2\pi p \sin\theta = -2\pi$$

(since the original parallel has length $2\pi/p = 2\pi \sin\theta$).

Our theorem consequently gives $\mu = +2$, as it should.

Remark. For a parallel of latitude K with $z = \cos\theta = k/\pi$, described p times, one would obtain $\mu = 2k$.

§ 24. The area–length duality

Consider an oriented and co-oriented front on the standard oriented two-dimensional sphere of radius 1. The dual front is also oriented and co-oriented.

Theorem 1. *The area of the characteristic 2-chain bounded by the front dual to the given one is equal to the oriented length of the original front.*

Corollary. *The area of the characteristic 2-chain bounded by the given front is opposite to the oriented length of the dual front.*

In fact, the oriented length of the second dual of the front is opposite to the oriented length of the original front, since the antipodal map of the sphere changes its orientation.

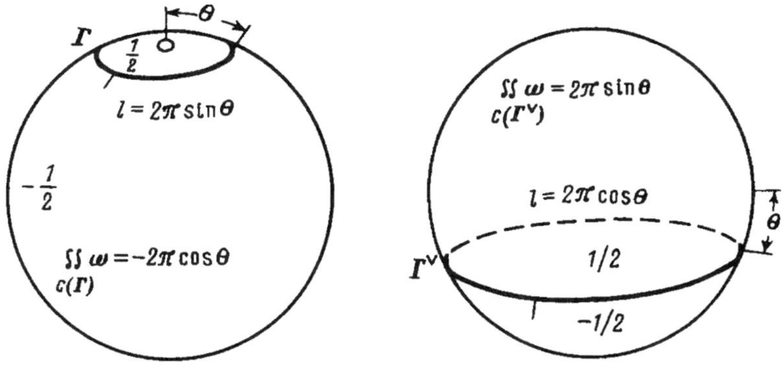

Figure 21

Example. Let the original front be the parallel of latitude at a distance $\theta < \pi/2$ along a meridian from the North pole, co-oriented by the vector directed from the pole (Fig. 21). Take the orientation of this front to be correct (so that the frame (co-orienting vector, orienting vector) positively orients the sphere).

The dual front is the parallel of latitude at a distance $\theta + (\pi/2)$ with the same co-orientation (from the Pole) and with the correct orientation.

The area of the characteristic chain of the dual front is $2\pi \sin\theta$.

The length of the original front is $2\pi \sin\theta$.

The area of the characteristic chain of the original front is $-2\pi \cos\theta$.

The length of the dual front is $2\pi \cos\theta$.

Proof of Theorem 1. Let the original front Γ be the front of an i-Legendrian curve L in $ST^*S^2_j$. Then the dual front $\Gamma^\vee$ is the front of the (also i-Legendrian) curve $L^\vee = Le^{i\pi/4}$ (by Lemma 1 of § 7).

Multiplication by $e^{i\pi/4}$ on the right sends the field $\mathbf{k}$ to the field $\mathbf{j}$ and the field $\mathbf{j}$ to the field $-\mathbf{k}$ (by the theorem of § 8). Therefore

$$\int_L \alpha_k = \int_{L^\vee} \alpha_j, \qquad \int_L \alpha_j = -\int_{L^\vee} \alpha_k.$$

On the other hand, by the main theorem of § 21

$$\int_L \alpha_j = \iint_c \omega, \qquad \int_{L^\vee} \alpha_j = \iint_{c^\vee} \omega.$$

But the oriented length of the projection on S^2_j of the i-Legendrian vector ξ is exactly $\alpha_k(\xi)$ (see the proof of Lemma 2 of § 22). Therefore

$$\int_L \alpha_k = \text{the oriented length of the front } \Gamma,$$

$$\int_{L^\vee} \alpha_k = \text{the oriented length of the front } \Gamma^\vee.$$

The theorem is proved.

I now recall that the derivative curve of the front is dual to the caustic.

Applying Theorem 1 to this pair, we are able to find the length of the caustic. We orient and co-orient the caustic in such a way that the derivative curve of the front (with its orientation defined by the orientation of the front and with the correct co-orientation) is dual to the caustic. Then from Theorem 1 one obtains the following result.

Theorem 2. *The oriented length of the caustic on the standard sphere is equal to an integral multiple of 2π, in fact equal to $-\pi\mu$, where μ is the Maslov index of the front.*

In fact, by Theorem 1 this length is equal to the area of the characteristic chain of the derivative of the front, which is equal to $-\pi\mu$ by the theorem of § 23.

Example. In the situation of Fig. 20 the caustic is the parallel of latitude at a distance θ from the North pole (and described p times). It is co-oriented by the direction towards the pole and oriented incorrectly. Therefore its oriented length is $-2\pi p \sin\theta = -2\pi$, while the Maslov index of the front is $\mu = +2$.

I note also another useful formulation of the area–length duality.

Let $\partial\sigma = L_1 - L_0$ be the homology between two i-Legendrian curves on $\mathbb{R}P^3 \approx ST^*S^2_j$ with fronts Γ_1 and Γ_0 on S^2_j.

Theorem 3. *The area of the projection of the chain σ on the sphere (that is, the difference of the areas bounded by the fronts Γ_1 and Γ_0) is equal to the difference of the oriented lengths of the fronts whose duals are the fronts Γ_1 and Γ_0:*

$$\iint_{\tilde{\pi}_{j*}\sigma} \omega_j = S(\Gamma_1) - S(\Gamma_0) = l(-\Gamma_1^\vee) - l(-\Gamma_0^\vee).$$

The proof is the same as for Theorem 1.

§ 25. The parities of fronts and caustics

Definition. A closed curve in $\mathbb{R}\mathrm{P}^3$ is said to be *even* if it is homologous to zero and *odd* otherwise.

Wave fronts, their derivatives and caustics, equipped with various framings, determine closed curves in $ST^*S^2 \approx \mathbb{R}\mathrm{P}^3$. These curves may be even or odd.

Example 1. All the equidistants of a front, framed with co-orienting normals, have the same parity, since they are mutually homotopic.

Example 2. The derivative curve of a closed front, framed with the directions of the great circles which are parallel along it, has the same parity as the original front.

For moving a distance $\pi/2$ along a tangent can be interpolated by a family of moves by t from the identity map.

A caustic has a natural framing by normals (for which the dual curve of the caustic is the correctly framed derivative curve of the front).

Theorem 1. *The parity of the caustic coincides with the parity of the front if* $\mu \equiv 0 \mod 4$, *and is opposite to it if* $\mu \equiv 2 \mod 4$.

Example 3. In Fig. 20 the caustic is a parallel of latitude described p times. Its parity is the parity of the number p.

The parity of a front (as it is not hard to convince oneself) is the parity of the number $p-1$. In this example $\mu = 2$.

Theorem 1 follows at once from the following fact.

Theorem 2. *The parity of a front coincides with the parity of its derivative curve, framed with the parallel circles, while the parity of the caustic coincides with the parity of the same derivative of the front framed by its normals.*

Both these facts have already been proved above, the first in Example 2 and the second in Example 1.

Proof of Theorem 1. The normal framing of the derivative curve differs from the parallel framing by $\mu/2$ turns (by the theorem of § 23). Each circuit of the framing changes the parity of the curve in ST^*S^2 by 1, since it adds to the homology class of the curve in $\mathbb{RP}^3$ the generator of the group $H_1(\mathbb{RP}^3, \mathbb{Z}) \approx \mathbb{Z}_2$. Therefore the difference of the parities of the caustic and the front is $\mu/2$ (mod 2), as was asserted.

The Maslov index and the parity are (the only) invariants of the regular homotopy of the Legendrian immersion of a circle in the contact manifold ST^*S^2. The Maslov index of the curve is also preserved under the contactomorphisms of this manifold (the author thanks S. L. Tabachnikov and E. Zhir for showing him that this follows from the uniqueness of trivialization of the bundle of contact planes on S^3).

CHAPTER VI

THE BENNEQUIN INVARIANT AND THE SPHERICAL INVARIANT J^+

For immersions of a circle in the plane there is an invariant J^+, defined in [6], that counts positive self-tangencies that occur during perestroikas (see also [5]). It is uniquely defined by the following conditions:

1) on passing through a positive self-tangency (when both the tangent branches have the same orientation) the value of J^+ changes in the same way that the number of double points changes (that is, it increases the value of J^+ or decreases it by 2), while on passing through the remaining perestroikas of codimension 1 (negative self-tangencies and triple intersections) it does not change its value;

2) the invariant is additive for the connected sum:

$$J^+(X+Y) = J^+(X) + J^+(Y);$$

3) the value of J^+ does not depend either on the orientation of the curve or on the orientation of the plane.

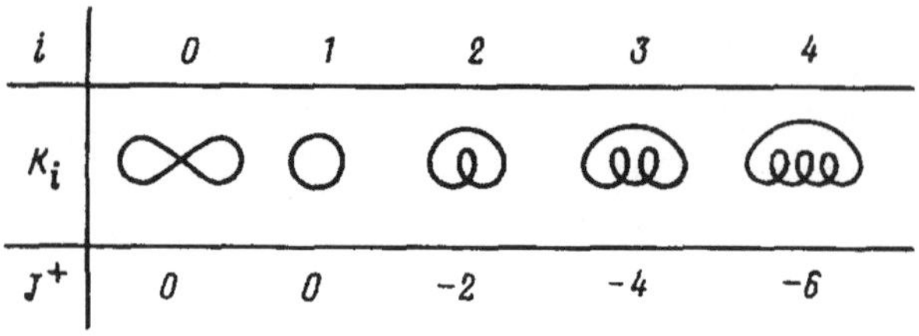

Figure 22

Example. The values of J^+ for the simplest curves are shown in Fig. 22.

The invariant J^+ is not conformally invariant.

Example. The frist and third curves in Fig. 22 are conformally equivalent, but have different values of the invariant J^+.

It turns out that after a small correction the invariant J^+ can be made spherical, that is, conformally invariant.

§ 26. The spherical invariant J^+

Consider on the plane a curve of index i.

Theorem. *The invariant $SJ^+ = J^+ + (i^2/2)$ of the plane curve is conformally invariant (that is, does not depend on which point of the Riemann sphere is chosen as the point at infinity).*

Definition. The quantity SJ^+ is called the *spherical invariant* J^+ of the spherical curve (obtained from the given plane curve by adding to the plane a point at infinity).

Proof of the theorem. Since (by a theorem of Whitney [7]) the space of immersed curves of fixed index i is connected, it is enough to do the following.

1) For one curve of each index to construct a curve conformally equivalent to it but of different index and to verify that the increments in J^+ and $i^2/2$ on passage to the new curve are of opposite sign.

2) To produce just so many such examples that the indices connected by them fall into two classes: even and odd numbers (the parity of the index under conformal transformations does not change, since it is opposite to the parity of the number of points of self-intersection of the curve).

The required examples are shown in Fig. 23.

The verification of the conformal invariance of SJ^+ follows from the identity

$$-2n + (n+1)^2/2 = (n-1)^2/2.$$

The proof is complete.

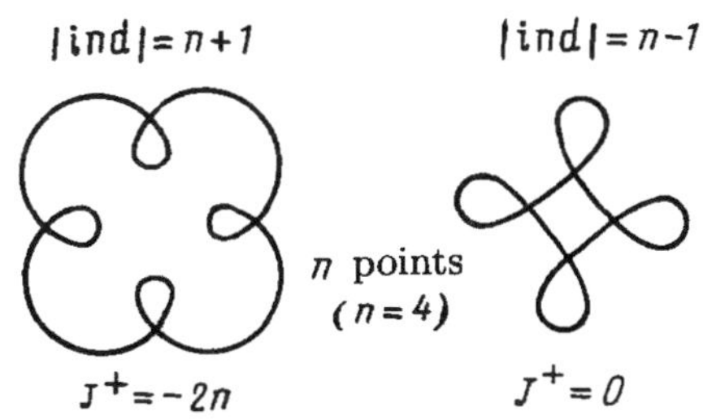

Figure 23

§ 27. The topological meaning of the invariant SJ^+

For each co-oriented front Γ on the sphere S^2 one can define a Legendrian curve $L \subset ST^*S^2 (\approx \mathbb{R}P^3)$; its double cover $\hat{L} \subset S^3$ is a Legendrian curve in the standard contact sphere S^3.

The curve $\hat{L}$ consists either of two components or of one.

Definition. The *Bennequin invariant* $\beta(\hat{L})$ is the linking coefficient $\beta(\hat{L}) = l(\hat{L}, \hat{L}_\varepsilon)$ of the Legendrian curve $\hat{L}$ and its small shift $\hat{L}_\varepsilon$ in the direction transverse to the contact structure.

The orientation of S^3 is chosen here in the ordinary way (that is, in such a way that the linking coefficient of the covering fibres of the bundle $ST^*S^2 \to S^2$ is equal to $+1$).

The Bennequin invariant is an invariant of Legendrian knots but not of immersions: it is defined only for *embedded* Legendrian curves (see [8]).

Theorem 1. *The Bennequin invariant is connected with the spherical invariant* J^+ *by the relation*

$$\beta(\hat{L}) = 2(1 - SJ^+(\Gamma)).$$

In [6] the *linking polynomial* $L(t)$ is defined for an embedded Legendrian curve of index i in $ST^*\mathbb{R}^2$. It belongs to the group ring of the group $\mathbb{Z}_i$. The sum of its coefficients is connected with the invariant J^+ by the relation

$$L(1) + J^+ = 1.$$

In these notations Theorem 1 can be written in the form

$$\beta(\hat{L}) = 2L(1) - i^2.$$

Suppose now that the index $i = 2k$ is even. In this case it is possible to take separately the sums of the coefficients of the polynomial L of even and odd degrees,

$$L_{\text{ev}} = (L(1) + L(-1))/2, \qquad L_{\text{odd}} = (L(1) - L(-1))/2.$$

The covering Legendrian curve $\hat{L}$ in S^3 in this case consists of two components, $\hat{L}_1$ and $\hat{L}_2$.

Theorem 2. *There are two sharpenings of the formula of Theorem* 1:

$$\beta(\hat{L}_1) = \beta(\hat{L}_2) = L_{\text{ev}} - k^2,$$
$$l(\hat{L}_1, \hat{L}_2) = L_{\text{odd}} - k^2.$$

The proofs of Theorems 1 and 2 are obtained by rather direct but heavy calculations of indices of intersection, and I do not dwell on them. The role of the number $i^2/2$ has been discussed before by Tabachnikov in [15].

CHAPTER VII

PSEUDO-FUNCTIONS

The motivation behind all the above research was the wish to promote some generalizations of Morse theory relevant to 'multivalued functions' (see [9], [2], [3], [4], [5], [6], [16], [17]).

We begin with the simplest example: *a function on the circle has no fewer than two critical points.*

What we shall be concerned with is the generalization of this 'Morse inequality' to objects that are more general than functions—the so-called *quasi-functions of Chekanov* (which are, in fact, Legendrian submanifolds, whose Lagrangian curves lie on the surface of a cylinder, and according to the inequality, have no fewer than two points of intersection with its equator) and the pseudo-functions of Givental' (the Lagrangian curves of which lie on the surface of a sphere, and, according to the inequality, intersect each great circle at least twice).

Other generalizations of elementary Morse theory for functions on the circle are provided by the theorem of Sturm (Kellogg, Tabachnikov and others), according to which *the number of zeros of the Fourier series*

$$\sum_{n \geqslant N} a_n \cos(nx) + b_n \sin(nx)$$

on the circle is not less than the number $(2N)$ *of zeros of its first harmonic* (see [2], [3], [4], [5]).

For $N = 1$ this is precisely the theorem of Morse, since a function with mean value 0 is the derivative of a function on the circle. In this way the theorem of Sturm and others is a generalization of the Morse inequality to the case $N > 1$.

In the case $N = 1$ the Morse inequality has been transferred to the 'multivalued' cases of the quasi-functions and pseudo-functions of Chekanov and Givental'.

In the case $N > 1$ only the first steps have been taken in the direction of such transfer.

These first steps, and the conjectures to which they lead, are given below. The generalization of Sturm's theory to functions of more than one variable is a yet more interesting consequence of our constructions, but I do not concern myself with that here.[7]

[7] There is, by the way, an interesting preprint by M. E. Kazaryan on umbilical characteristic classes (Bochum, 1994).

§ 28. The quasi-functions of Chekanov

For any function $f : M \to \mathbb{R}$ there is an exact Lagrangian submanifold $L \subset T^*M$ (formed from the differentials of f at all points). The critical points of the function are intersections of L with the zero section. Suppose that M is compact. The Morse inequalities estimate from below the number of points of intersection of L with the zero section.

The first attempts to generalize the Morse inequalities are related to replacing the Lagrangian section L by Lagrangian submanifolds that need not be sections.

Example. Let $M = S^1$ be the circle. Then $T^*M = \mathbb{R} \times S^1 = \{p, q \mod 2\pi\}$—the cylinder, and L is a curve homotopic to the zero section. The exactness condition consists in the fact that the area between L and the zero section is zero:

$$\int_L p\,dq = 0.$$

If such a curve L has no self-intersections, then it must intersect the zero section twice (for example, if it were above the zero section everywhere then the integral would be positive).

However, if the curve can intersect itself, then the intersection with the zero section can vanish. I described the simplest example of such a thing in 1965 in [9] (Fig. 24)

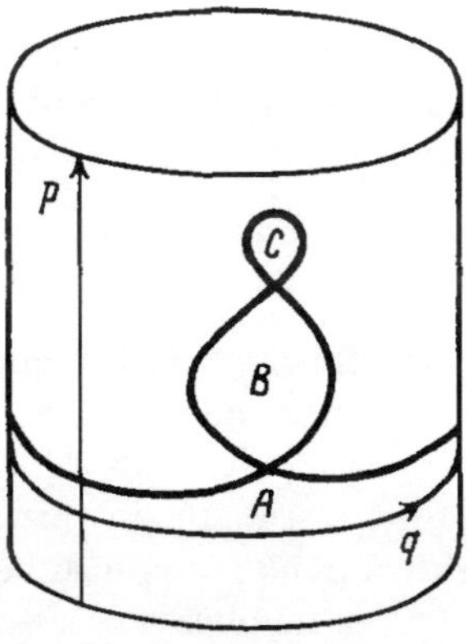

Figure 24

The exactness condition in this example takes the form of a relation between the areas, $A + C = B$. It is easy to construct a curve with $p > 0$ satisfying this condition. For such a curve the 'generalized Morse inequality' (the number of points of intersecton with the zero section $\geqslant 2$) is not true.

I conjectured in 1965 that, for embedded exact Lagrangian submanifolds in the space of the cotangent bundle of a manifold that are Hamiltonially isotopic to the zero section, the number of points of intersection with the zero section is bounded below by the sum of the Betti numbers of the manifold. This conjecture (after the proof by Conley and Zehnder [10] of an analogous conjecture on the fixed points of exact symplectomorphisms of the torus) has been proved by many people: Chaperon [11], Laudenbach and Sikorav [12], and others.

Chekanov [13] noted that the result remains true even for some self-intersecting immersed exact Lagrangian submanifolds, which he called quasi-functions.

Definition. A *quasi-function* on a compact manifold M^m is a Legendrian embedding of M^m in the contact space $J^1(M, \mathbb{R}) = (T^*M) \times \mathbb{R}$ of 1-jets of functions on M^m, which can be joined to the zero section by a regular homotopy consisting of Legendrian embeddings.

The projection of a Legendrian quasi-function embedded in $J^1(M, \mathbb{R})$ is an exact Lagrangian immersion of M in T^*M. This immersion is homotopic to the zero section, but in general has self-intersections.

Chekanov's theorem [13], [16]. *The number of points of intersection of the Lagrangian projection of a quasi-function to the space of the cotangent bundle T^*M with the zero-section (the number of quasi-critical points of the quasi-function) is not less than the sum of the Betti numbers of the manifold M (if one takes multiplicities of points into account).*[8]

Example. Let $M = S^1$. Then the number of points of a quasi-function where $p = 0$ is not less than two.

In particular, the curve of Fig. 24 cannot be the Lagrangian projection of a quasi-function.

This means that for any regular homotopy of the zero section to the curve of Fig. 24 there is a moment when the curve has a double point, breaking it into two loops, along each of which $\int p dq = 0$.

For the simplest homotopy the proof of Chekanov's theorem even for curves on a cylinder is not known.

Remark. Of course, in Chekanov's theorem for the number of quasi-critical points of a quasi-function there were obtained more precise lower estimates than by the sum of the Betti numbers. For many manifolds M this number is not less than the Morse number; the usual lower estimates in this situation for the number of geometrically distinct quasi-critical points have also been obtained.

However, the conjecture that these numbers are always not less than the minimal number of critical points of a function on M (taking account of multiplicities or not) has been neither proved nor disproved.

§ 29. From quasi-functions on the cylinder to pseudo-functions on the sphere, and conversely

The following useful simple lemma is due to Archimedes.

Consider the cylinder $x^2 + y^2 = 1$ wrapped round the sphere $x^2 + y^2 + z^2 = 1$. The *projection of Archimedes* projects the complement to the poles of the sphere onto the surface of the cylinder by horizontal radii $x = e^t x_0$, $y = e^t y_0$, $z = z_0$).

[8]The proof of this important theorem was for a long time missing from the literature, until in 1994 there appeared a preprint by M. Chaperon on this theme (Paris-7).

Lemma. *The Archimedean projection $S^2 \setminus S^0 \to S^1 \times (-1, 1)$ is a symplectomorphism.*

This lemma enables us to transfer quasi-functions to the sphere (less its poles). Adjoining the poles, we arrive at the following concept.

Definition. A *pseudo-function* (on the circle) is an exact i-Lagrangian immersion, which is the projection of an i-Legendrian embedding of a circle in $ST^*S_i^2$, belonging to the component of a Legendrian embedding that projects to the embedding of the equator, in the space of all Legendrian embeddings.

In other words, we deform the equator of the sphere to a smoothly immersed curve so that

1) the curve at all times bounds an area of half the sphere ($2\pi \bmod 4\pi$),
2) during the deformations neither of the loops of the curve into which a double point divides it bounds the area of half the sphere.

Givental's theorem [14]. *A pseudo-function intersects a great circle of the sphere in no fewer than two points.*

Remark. If a pseudo-function has no self-intersections, then what has to be proved is obvious. The example of Fig. 24 allows us to construct an immersed circle in the Northern hemisphere, bounding an area of half the sphere. From Givental's theorem it follows that the immersed curve of Fig. 24 not only does not determine a quasi-function, but is not even a pseudo-function.

Example. If a pseudo-function is obtained from the equator by a homotopy not passing through the poles, then this homotopy projects to the cylinder and there one obtains a quasi-function. For this quasi-function one has, therefore, an assertion close to Chekanov's theorem. If, on the contrary, the pseudo-function passes through the poles during the deformation, then the projection to the cylinder fails to determine a quasi-function. For example, in Fig. 25 (see p. 64) we show one such perestroika. As a result of this perestroika on the surface of the cylinder $S^1 \times (-1, 1)$ one obtains a closed contractible curve, bounding the area 2π.

Such a curve is far from being a quasi-function, but it must nevertheless intersect the zero section at least twice. If the curve is embedded, then this is obvious, since the area of the cylinder whether above or below the zero-section is equal to 2π. Givental's theorem asserts that the intersection points cannot vanish even for curves that have self-intersections, if in the process of deformation none of the double points forms on the curve a loop bounding an area of half the sphere.

Remark. The theorem of Givental' presented here was proved by him in the much more general setting of exact Lagrangian embeddings of $\mathbb{RP}^n$ in $\mathbb{CP}^n$ (our case of spherical curves corresponds to taking $n = 1$).

§ 30. Conjectures concerning pseudo-functions

Symplectic and contact generalizations of the classical four-vertex theorem lead to the following result (see [4]).

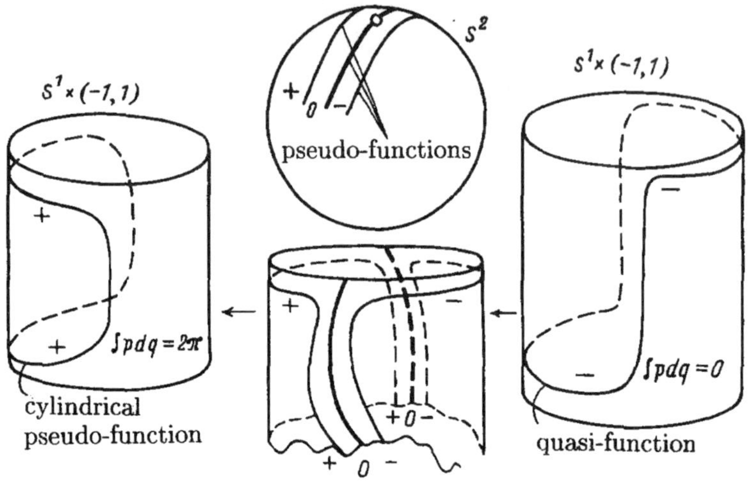

Figure 25

The tennis-ball theorem. *Consider a curve embedded in the standard sphere and bounding an area of half the sphere. Such a curve has no fewer than four points of spherical inflection.*

Using Fig. 24 it is easy to construct an example of a curve, immersed in the standard sphere and obtained from the equator by regular homotopy in the class of immersions bounding the area of half the sphere, that has only two points of spherical inflection.

We have already seen that the curve of Fig. 24 cannot be a pseudo-function.

Conjecture 1. *The number of points of spherical inflection of a pseudo-function is not less than four.*

One can formulate an analogous conjecture concerning quasi-functions on the cylinder.

Definition. A *point of spherical inflection of a quasi-function* is a point of spherical inflection of the projection of its Lagrangian curve to the sphere by rays from the centre.

If a quasi-function is given locally by the equation $p = f(q)$, then the point q of spherical inflection is determined from the equation $f'' + f = 0$.

Conjecture 2. *A quasi-function has no fewer than four points of spherical inflection.*

Theorem. *If the Lagrangian projection of a quasi-function has no points of self-intersection, then it has no fewer than four points of spherical inflection.*

Proof (the author is grateful to B. A. Khesin, whose idea is used here).

Suppose that the spherical projection has fewer than four points of inflection. Then this curve on the sphere can be confined to a hemisphere (see [4]). This hemisphere cannot be bounded by a vertical meridian, since a quasi-function has points with arbitrary values of q.

Thus the Lagrangian curve lies in the half-cylinder $z > ax + by$ (or $z < ax + by$). But $\{(x, y, z - ax - by)\}$ is also a quasi-function. Since everywhere $z \neq ax + by$ we obtain a contradiction to Chekanov's theorem (in its elementary form, since our Lagrangian curve has no points of self-intersection). The theorem is proved.

These conjectures and results can be applied, for example, to the study of the vertices of curves on the Euclidean plane.

Consider a plane co-oriented and smoothly immersed curve of index 1 with no double oriented normals. We associate with it a quasi-function in the following (usual) way.

With the point (x, y) with the co-orienting normal direction φ we associate the 1-jet at the point φ of the function with the value $z = x \cos \varphi + y \sin \varphi$ and the value of its derivative $p = -x \sin \varphi + y \cos \varphi$. If locally $p = A(\varphi)$ and the radius of curvature of the original curve is $R(\varphi)$, then

$$R' = A + A'',$$

according to § 13. This means that the vertices of the original curve (where $R' = 0$) are the points of spherical inflection of the qausi-function so obtained. If the original curve has no double normals with the same orientation, then the Lagrangian curve does not intersect itself. According to the preceding theorem the original plane curve has no fewer than four vertices.

Consider an immersed curve of index 1 on the Euclidean plane.

Conjecture 3. *If the curve of normals of our curve belongs to the component of the curve of normals of a circle in the space of Legendrian embeddings $S^1 \to ST^*\mathbb{R}^2$, then the original curve has no fewer than four vertices.*

Example. The curve of Fig. 26 has two vertices in all. But this curve has $J^+ \neq 0$, and therefore its curve of normals does not lie in the component of the curve of normals of a circle.

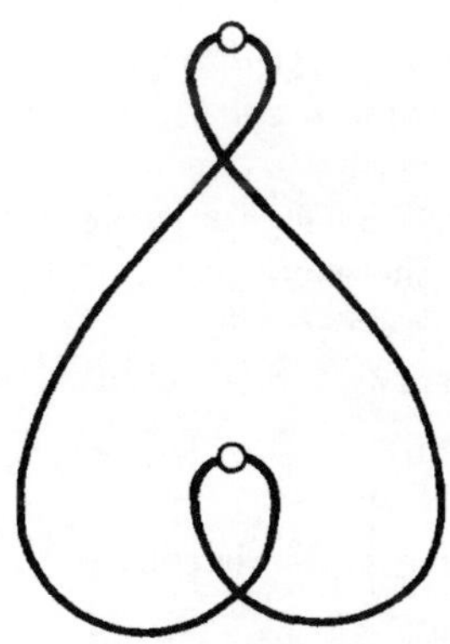

Figure 26

Returning to perestroikas of quasi-functions by passing pseudo-functions through the poles, consider on the surface of the cylinder $|p| \leqslant 1$ a closed contractible smoothly immersed curve, bounding the area 2π and lying in the component of a non-self-intersecting curve in the sense that one can join it to a curve with no self-intersections by a regular homotopy in the class of Lagrangian immersions bounding the area 2π and such that no double point of an intermediate curve forms on it a loop of area 2π (for safety's sake perhaps one should prohibit loops of areas that are multiples of 2π as well).

Is it true that such a Lagrangian curve has no fewer than four points of spherical inflection?

Theorem. *If a contractible Lagrangian curve of area 2π on the cylinder $|p| < 1$ has no points of self-intersection, then it has no fewer than four points of spherical inflection.*

Proof. In the opposite case the projection to the sphere from the centre is confined to an open hemisphere, and the original curve is confined to an open half-cylinder. But since this Lagrangian curve bounds the area 2π it cannot be confined to a half-cylinder of area 2π.

The theorem is proved.

§ 31. Space curves and Sturm's theorem

We now state a geoemtrical problem, leading to the case of an arbitrarily large number of zeros in a theorem of Sturm type.

Consider an immersion of the circle into a Euclidean subspace of dimension $2n$ in a Euclidean space of dimension $2n + 1$. The highest torsion of such a curve is identically equal to zero.

Consider an arbitrary small (along with its derivatives) spatial perturbation of this plane immersion. The number of flat points of the generic perturbed immersion (that is, the number of zeros of the highest torsion) is finite. How small can it be?

Theorem. *The least number of flat points is equal to $2n+2$ for all curves in $\mathbb{R}^{2n+1}$ whose projections into $\mathbb{R}^{2n}$ satisfy the following convexity condition: the number of points of intersection with any $(2n-1)$-dimensional hypersurface in $\mathbb{R}^{2n}$ does not exceed $2n$ (taking multiplicities into account).*

Example. We call a *standard curve* (or *generalized ellipse*) in the space of dimension $2n$ a curve given by the equations

$$x_1 = a_1 \cos t, \quad \dots, \quad x_n = a_n \cos(nt);$$
$$y_1 = b_1 \sin t, \quad \dots, \quad y_n = b_n \sin(nt),$$

where $a_1, \dots, b_n \neq 0$. This curve in $\mathbb{R}^{2n}$ is convex.

Theorem. *The number of flat points of a small perturbation of a standard curve in $\mathbb{R}^{2n+1}$ is no fewer than $2n + 2$.*

Example. For $n = 1$ the standard curve is an ellipse. The assertion consists in saying that a ***perturbation of the ellipse which is small along with its derivatives has no fewer than four flat points.***

This result is true also for curves in odd-dimensional projective spaces.

Analogous results are true also for curves in even-dimensional projective spaces having convex projections or obtainable by small perturbations from convex curves lying in odd-dimensional hyperplanes.

As an example of a convex curve in $\mathbb{R}P^{2n-1}$, take the image of the immersion

$$\begin{aligned} x_1 &= \cos t, \quad \ldots, \quad x_n = \cos(2n-1)t; \\ y_1 &= \sin t, \quad \ldots, \quad y_n = \sin(2n-1)t \end{aligned}$$

(the coordinates being homogeneous, with $t \in \mathbb{R} \bmod \pi$). The corresponding theorems on Chebyshev systems of sections of a non-trivial one-dimensional bundle over the circle will be published in another place.

From these theorems it follows, for example, that a curve projectively dual to a convex curve is convex, and that convex curves in $\mathbb{R}P^m$ form a connected set (M. Shapiro, M. Kazaryan).

Remark. The appearance of numbers of flat points linearly increasing with the dimension of the space leads one to conjecture that the question is related to the Morse theory of a projective space (or more correctly to its product with the circle). How vigorously one can deform a convex curve is not known even for the case of curves in $\mathbb{R}P^2$, where the question reduces to an extension of the theorem of Möbius (on three points of inflection of a curve non-homologous to zero) to curves with self-intersections. For a discussion of the theorem of Möbius see [18].

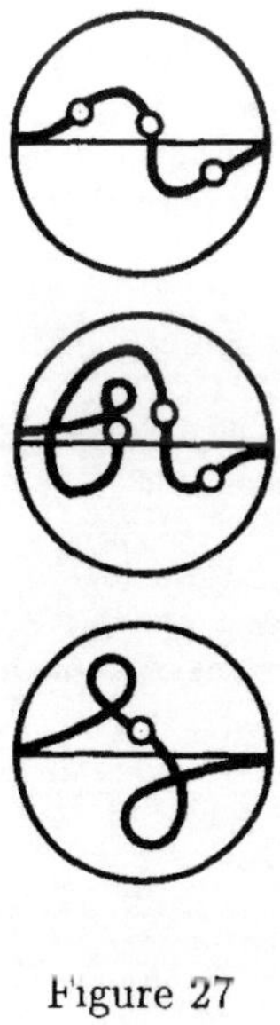

Figure 27

Conjecture. *A non-contractible immersion of a circle in the real projective plane has no fewer than three points of inflection if it can be joined with the standard embedding of the projective line in the class of immersions without dangerous self-tangencies (in which both branches have the same direction).*

Remark. The imposed condition of avoiding dangerous self-tangencies means the preservation under deformation of a type of Legendrian knot. It is easy to construct an immersion with one point of inflection, which can be obtained from the standard embedding of the projective line by once passing through a dangerous self-tangency (Fig. 27).

Bibliography

[1] V. I. Arnold, "Lagrangian and Legendrian cobordisms II", *Funktsional. Anal. i Prilozhen.* **14**:4 (1980), 8–17; English transl. in *Functional Anal. Appl* **14** (1980).

[2] S. L. Tabachnikov, "Around four vertices", *Uspekhi Mat. Nauk* **45**:1 (1990), 229–230; English transl. in *Russian Math. Surveys* **45**:1 (1990).

[3] V. I. Arnol'd, "Sur les propriétés topologiques des projections Lagrangiennes en géométrie symplectique des caustiques", in: Cahiers mathématiques de la decision, vol. 9320, CEREMADE, 1993, pp. 1–9.

[4] V. I. Arnol'd, "The topological properties of Legendrian projections in the contact geometry of wave fronts", *Algebra i Analiz* **6**:3 (1994), 1–16; English transl. in *St. Petersburg Math. J.* **6** (1994).

[5] V. I. Arnol'd, "Topological invariants of plane curves and caustics", University Lecture Series 5, Amer. Math. Soc., Providence, RI 1994, pp. 1–60.

[6] V. I. Arnold, "Plane curves, their invariants, perestroikas and classifications", *Preprint ETH, Zürich, May 1993*; "Singularities and curves", *Adv. Soviet Math.* **21** (1994), 33–91.

[7] H. Whitney, "On regular closed curves in the plane", *Compositio Math.* **4** (1937), 276–284.

[8] D. Bennequin, "Entrelacements et équation de Pfaff", *Astérisque* **107–108** (1983), 87–162.

[9] V. I. Arnol'd, "Sur une propriété topologique des applications globalement chaotiques de la mécanique classique", *C.R. Acad. Sci.* Paris **261** (1965), 3719–3722.

[10] C. Conley and E. Zehnder, "The Birkhoff–Lewis fixed point theorem and a conjecture of V. I. Arnol'd", *Invent. Math.* **73** (1983), 33–49.

[11] M. Chaperon, "Quelques questions de géométrie symplectique d'aprés entre autres, Poincaré, Arnold, Conley et Zehnder", *Astérisque* **105–106** (1983), 231–249.

[12] F. Laudenbach and J.-C. Sikorav, "Persistence d'intersection avec la section null au course d'une isotopie hamiltonienne dans un fibré cotangent", *Invent. Math.* **82** (1985), 349–358.

[13] Yu. V. Chekhanov, "Legendrian Morse theory". (The reference given in the Russian text is incorrect.)

[14] A. B. Givental', "Nonlinear generalization of the Maslov index", in: Theory of singularities and its applications, *Adv. Soviet Math.* **1** (1990), 71–103.

[15] S. L. Tabachnikov, "The computation of the generalised Bennequin invariant by means of fronts", *Funktsional. Anal. i Pilozhen* **22**:3 (1988), 89–90; English transl. in *Functional Anal. Appl.* **22** (1988).

[16] V. I. Arnol'd, "First steps in symplectic topology", *Uspekhi Mat. Nauk* **41**:6 (1986), 3–18; English transl. in *Russian Math. Surveys* **41**:6 (1986).

[17] V. I. Arnol'd, "Invariants and perestroikas of fronts in the plane", *Trudy Mat. Inst. Steklov.* **209** (1995), 1–60.

[18] V. I. Arnol'd, "The branched covering $\mathbb{C}P^2 \to S^4$, hyperbolicity and projective topology", *Sibirsk. Mat. Zh.* **29**:4 (1988), 36–47; English transl. in *Siberian Math. J.* **29** (1988).

Steklov Institute of Mathematics

Received 25/OCT/94
Translated by I. R. PORTEOUS
Typeset by $\mathcal{A}\mathcal{M}\mathcal{S}$-TEX

Remarks on eigenvalues and eigenvectors of Hermitian matrices, Berry phase, adiabatic connections and quantum Hall effect

V.I. Arnold

Abstract. The singularities of the stratification of the space of the Hermitian matrices according to the multiplicities of the eigenvalues are described as an informal complexification of the previous study of the space of the real symmetric matrices. The degeneration of the spectral sequence associated to this stratification provides some strange combinatorial identities. The eigenvector bundles over the manifold of the Hermitian matrices with simple spectra are equiped with the natural connections, describing also the adiabatic approximation to the oscillations of the linear systems defined by the slowly varying skew Hermitian matrices. The curvature of this connection is singular at the codimension 3 variety of the Hermitian matrices having multiple eigenvalues. The resulting jumps of the integrals of the curvature form at the crossings of this variety by the moving surface of integration are responsible for the quantum Hall effect.

§1. Introduction

The space of Hermitian matrices is stratified according to the multiplicities of the eigenvalues. The subvariety formed by those matrices which have multiple eigenvalues has codimension *three* in the space of Hermitian matrices. The singular points of this subvariety correspond to the matrices having more multiple eigenvalues and higher multiplicities. They form the strata of codimension at least 6 in the space of Hermitian matrices. The simplest properties of this stratification have been described in [1]. In the present paper some new properties of this stratification and some problems and conjectures related to it are described.

The geometry of the space of the Hermitian matrices having no multiple eigenvalues is related to the adiabatic connections on the eigenvector bundles (known in physics under the name of the Berry phase), while the Hermitian matrices having multiple eigenvalues are responsible for the quantum Hall effect. The author is grateful to S.P. Novikov, who has explained to him the relation of the paper [1] to the quantum Hall effect, and is also grateful to A. Vainstein, M. Shapiro, B. Shapiro, A. Givental, V. Vassiliev, and A.I. Neistadt for their useful discussions.

* This article is an expanded version of a lecture read 25 March, 1994, at the conference on Real Algebraic Geometry at the Institut Henry Poincaré, Paris

Originally published in: *Selecta Math.* (*N. S.*) **1**(1), 1–19 (1995), © 1995 Birkhäuser Verlag, Basel.

Reprint by Springer Nature Switzerland AG 2022 with kind permission

§2. The natural stratification of the space of Hermitian matrices

The strata are labeled by the vectors of the multiplicities of the ordered eigenvalues $\lambda_1 < \lambda_2 < \dots$. Our goal is the study of the singularities of the closures of these strata, of their adjacencies to each other, of the behavior of the natural fibrations of the eigenvectors over these strata and of the natural connections on these fibrations at the boundaries of the strata.

All the results might be formulated for the Lie algebra of the skew-Hermitian matrices, but I formulate them in the Hermitian case to preserve the parallelism with the stratification of the space of the real symmetric matrices.

It is technically convenient to consider the matrices whose traces vanish. To avoid the noncompactness, it suffices to consider the stratification of the sphere of those Hermitian matrices having trace zero, whose norms are equal to one.

Denote by $\mathcal{F}_0 \subset \mathcal{F}_1 \subset \mathcal{F}_2 \subset \cdots$ the filtration of the sphere by the closed subvarieties $\mathcal{F}_p$ of the Hermitian matrices having at most $p+1$ different eigenvalues (and having trace zero and norm one).

This filtration generates a spectral sequence, convergent to the (co)homologies of the sphere.

Proposition. *The spectral sequence of the filtration $\{\mathcal{F}_p\}$ degenerates at the second term.*

The proof [2] of this proposition, found by M. Shapiro and A. Vainstein (to whom I had communicated this proposition as a conjecture) will be published elsewhere.

Example. Consider the Hermitian matrices of order $n = 4$. The corresponding sphere has the dimension $n^2 - 2 = 14$. I shall denote the strata like $(\lambda_1 < \lambda_2 = \lambda_3 < \lambda_4)$ by the symbols like $(12|34)$. Our stratification of the sphere consists of $2^{n-1} - 1 = 7$ strata (corresponding to the vertices of a cube of dimension $n-1 = 3$). The adjacencies of these strata (and their real codimensions in S^{14}) are shown in Fig. 1.

Every stratum of $\mathcal{F}_p \setminus \mathcal{F}_{p-1}$ is fibered over an open simplex of dimension p (formed by the eigenvalues). Each fiber consists of the Hermitian matrices having the given eigenvalues with given multiplicities. These fibers are also shown in Fig. 1. They are the complex flag manifolds. Indeed, an Hermitian matrix is unambiguously defined by its eigenvalues and by their eigenspaces, which are pairwise Hermitian orthogonal. The set of the eigenspaces V_i (ordered by the order of the eigenvalues λ_i on the real axis) is bijectively defined by the flag of their sums $W_i = V_i + V_{i+1} + \cdots$.

For instance, the manifold of the complete flags in $\mathbf{C}^4$ is fibered over $\mathbf{CP}^3$ with a fiber which is fibered over $\mathbf{CP}^2$ with fiber $\mathbf{CP}^1$. Hence I denote the variety

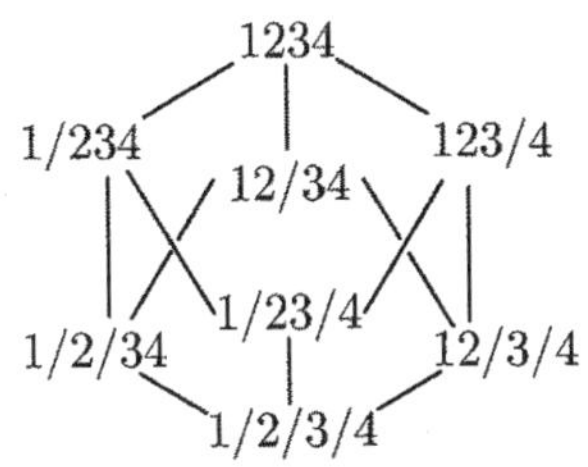

$\operatorname{codim}_{\mathbf{R}} = 0$	$p = 3$	$\dim_{\mathbf{C}} \mathbf{CP}^{3,2,1} = 6$	1356531
$\operatorname{codim}_{\mathbf{R}} = 3$	$p = 2$	$\dim_{\mathbf{C}} \mathbf{CP}^{3,2} = 5$	123321
$\operatorname{codim}_{\mathbf{R}} = 6$	$p = 1$	$\dim_{\mathbf{C}} \mathbf{G}_4^2 = 4$	11211
$\operatorname{codim}_{\mathbf{R}} = 8$	$p = 1$	$\dim_{\mathbf{C}} \mathbf{CP}^3 = 3$	1111
$(\operatorname{codim}_{\mathbf{R}} = 15)$	$(p = 0)$		

Fig. 1 Stratification of the space of Hermitian matrices of order 4

of the complete flags in $\mathbf{C}^4$ by $\mathbf{CP}^{3,2,1}$. Similarly, $\mathbf{CP}^{3,2}$ is the manifold of the flags $\mathbf{C}^3 \supset \mathbf{C}^2$ in $\mathbf{C}^4$, while G_4^2 denotes the Grassmannian (of the $\mathbf{C}^2$ in $\mathbf{C}^4$).

The rightmost column in Fig. 1 shows the (even-dimensional) Betti numbers of the corresponding flag manifolds. For instance, the Betti numbers of $\mathbf{CP}^2$ would be shown as (111).

The homologies of the flag manifolds are isomorphic to the homologies of the corresponding direct products of Grassmannians (for instance, $H_\bullet \mathbf{CP}^{3,2} \approx H_\bullet(\mathbf{CP}^3 \times \mathbf{CP}^2)$).

This follows, for instance, from the fact that the complex Schubert cells forming the spaces, the bases and the fibers of the corresponding fibrations are even dimensional, and hence all the boundary operators in the corresponding complexes are vanishing (and all the spectral sequences degenerate).

Each stratum is not only a trivial fibration, but is also canonically trivialized: an Hermitian matrix is unambiguously defined by the sets of its different eigenvalues and of the corresponding eigenspaces.

The initial term of our spectral sequence consists of the homologies of the spaces (a stratum with a fixed p modulo the strata with smaller values of p), equivalent to the Thom's spaces of the trivial fibrations over the flag manifolds with fibers of real dimension p.

These homologies are isomorphic to the homologies of the corresponding flag manifolds (with shift of the real dimension by p). The Betti numbers of the unions of the Thom's spaces corresponding to the strata of Fig. 1 are shown in Fig. 2.

The differential of the spectral sequence is defined by the following construction. The direct product of a cycle of a flag manifold with the boundary of the corresponding simplex of the eigenvalues is naturally embedded into the space of Hermitian matrices. This direct product lies in the union of the strata having smaller values of p than the initial stratum. The differential of the spectral sequence

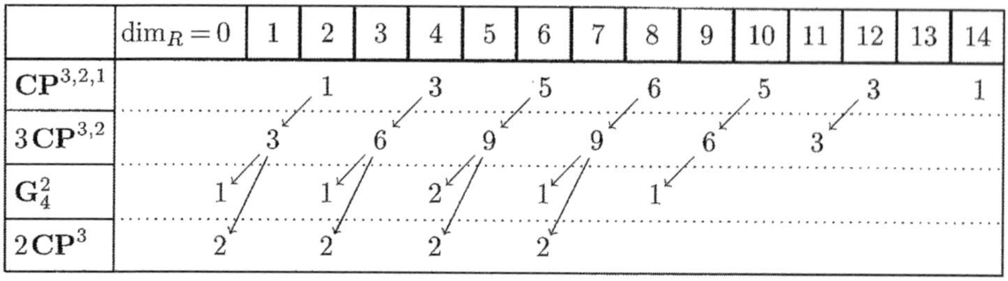

Fig. 2 Differentials of the spectral sequence

sends the initial cycle to the part of the boundary over the faces of codimension 1 in the simplex.

The main observation is that the remaining part of the boundary is a chain whose dimension is smaller than the dimension of the boundary of the product of the cycle of the flag manifold with the simplex of the eigenvalues.

The degeneration of the spectral sequence means that the homologies of the differential defined above vanish (except in dimensions 0 and $2^n - 2$, where the homologies of the sphere appear).

For instance, in Fig. 2 this degeneration manifests itself in the vanishing of the Euler characteristics of the complexes, denoted by the arrows. For instance, $6 - 9 + (1 + 2) = 0$, and so on.

REMARK. Similar calculations for arbitrary values of n generate a large set of strange combinatorial identities. These identities, corresponding to the strata of small codimensions, are written explicitly below.

§3. The small codimension strata

The complexes shown in Fig. 2 are separated by the following characteristics of the complexity of the degeneration.

DEFINITION. The *defect* of a stratum is the difference between the (complex) dimension of the manifold of the complete flags and the complex dimension of the manifold of the flags corresponding to the given stratum (which is the manifold of the Hermitian matrices with given eigenvalues of given multiplicities).

LEMMA 1. *The defect of the stratum corresponding to a_i eigenvalues of each multiplicity $i = 2, 3, \ldots,$ is equal to*

$$q = a_2 + 3a_3 + 6a_4 + \cdots + (i(i-1)/2)a_i + \cdots.$$

Proof. The complex dimension of the manifold of the complete flags in $\mathbf{C}^n$ is equal to $(n-1)+\cdots+1$. The complex dimension of the manifold of the flags, corresponding to one multiple eigenvalue of multiplicity i, is equal to $(n-1)+\cdots+i$. In this case the defect is equal to $1+\cdots+(i-1)=i(i-1)/2$.

The contributions of different eigenvalues to the defect are independent. The summation of these contributions provides the above formula for the defect.

REMARK. This formula may also be written in the form

$$q=\sum_j \frac{n_j(n_j-1)}{2}$$

(summation over the different eigenvalues λ_j of multiplicities n_j).

DEFINITION. *The Young diagram* of a stratum corresponding to a_i eigenvalues of each multiplicity i is the diagram with a_i lines of lengths $i-1$ (for each $i=2,3,\ldots$).

The lengths of the lines are the numbers of the *lost eigenvalues*. The defect of a stratum depends only on its Young diagram (which, however, does *not* define the stratum).

LEMMA 2. *The strata, whose defects are smaller or equal to 4 correspond to the seven simplest Young diagrams, shown in Fig. 3. The codimension of the union of all the strata, having the higher values of the defect, is equal to 14.*

Each Young diagram corresponds to several strata, since the diagram does not take into account the ordering of the multiple eigenvalues on the real axis. The number of strata, corresponding to each diagram, is shown in the rightmost column of Fig. 3.

The manifold of complex flags, corresponding to a given stratum, depends only on its Young diagram.

DEFINITION. The *Poincaré polynomial* of a stratum X is the polynomial

$$b(t)=1+b_2(X)t+b_4(X)t^2+\cdots,$$

where b_{2i} is the Betti number of the corresponding flag manifold. These Poincaré polynomials are easily computable, since the homologies of the corresponding fibrations into the Grassmannians are isomorphic to those of the direct products and the homologies of the Grassmannians are isomorphic to their Schubert cells chains groups.

Young Diagram	Stratum	codim_R	defect$q$	number of strata
$\varnothing$	$1,\ldots,n$	0	0	1
$\square$	$1,\ldots,i\|i+1,\ldots,n$	3	1	$n-1$
	$1,\ldots,i\|i+1,\ldots,j\|j+1,\ldots,n$	6	2	$\binom{n-2}{2}$
	$1,\ldots,i\|i+1,\ldots,j\|j+1,\ldots$ $\ldots,k\|k+1,\ldots,n$	9	3	$\binom{n-3}{3}$
	$1,\ldots,i\|i+1,\ldots,j\|j+1,\ldots$ $\ldots,k\|k+1,\ldots,m\|m+1,\ldots,n$	12	4	$\binom{n-4}{4}$
	$1,\ldots,i\|i+1\|i+2,\ldots,n$	8	3	$n-2$
	$1,\ldots,i\|i+1\|i+2,\ldots,j\|j+1,\ldots,n$ or $1,\ldots,i1i+1,\ldots,j\|j+1\|j+2,\ldots,n$	11	4	$2\binom{n-3}{2}$

Fig. 3 Simplest strata of the space of Hermitian matrices

EXAMPLE. For the manifold of the complete flags in $\mathbf{C}^n$

$$b(t)=\prod_{i=1}^{n}\frac{1-t^i}{1-t}=1+(n-1)t+\cdots$$

REMARK. The Poincaré polynomials are reciprocal (the Poincaré duality). For instance, in the example above

$$b(t)=t^N\left(1+\frac{n-1}{t}+\cdots\right).$$

This is the form in which I shall use the Poincaré polynomials of the strata of our stratification: the small codimension homology groups will be compared in small codimension strata.

The first terms of the Poincaré polynomials of the seven simplest strata of small defects are shown in Fig. 4 (provided that n is not too small). The points mean the coefficients that will not be used in the sequel.

Young diagram	b_0	b_1	b_2	b_3	b_4	Number of strata
$\varnothing$	1	$n-1$	$\binom{n}{2}-1$	$\binom{n+1}{3}-n$	$\binom{n+2}{4}-\binom{n+1}{2}$	1
	1	$n-2$	$\binom{n-1}{2}$	$\binom{n}{3}-1$	$\dots$	$n-1$
	1	$n-3$	$\binom{n-2}{2}+1$	$\dots$	$\dots$	$\binom{n-2}{2}$
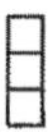	1	$n-4$	$\dots$	$\dots$	$\dots$	$\binom{n-3}{3}$
	1	$\dots$	$\dots$	$\dots$	$\dots$	$\binom{n-4}{4}$
	1	$n-3$	$\binom{n-2}{2}$	$\binom{n-1}{3}$	$\dots$	$n-2$
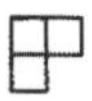	1	$\dots$	$\dots$	$\dots$	$\dots$	$2\binom{n-3}{2}$

Fig. 4 The first Betti numbers of the simplest strata

§4. The combinatorial identities

These identities generalize the identity $5-9+2+2=0$ and other similar identities of Fig. 2.

To write these identities, I identify the complex p-dimensional homology of every complex flag manifold with the homology of the dual dimension (using the Poincaré duality). Now the differential of the spectral sequence acts from the space of the complex p-(co)dimensional homologies of the disjoint union of the complex flag manifolds with a given Young diagram to the space of the complex p'-(co)dimensional homologies of the disjoint union of the complex flag manifolds whose Young diagrams are directly adjacent to the given diagram and for which $p+q=p'+q'$ (where q is the defect of the initial Young diagram and q' the defect of the directly adjacent one).

Here a stratum is *directly adjacent* to a given one if it is adjacent and the number of its distinct eigenvalues is one less than the number of distinct eigenvalues of the given stratum. The area of a directly adjacent diagram is thus larger than the area of the initial Young diagram by one.

Example. The Young diagram "..." is directly adjacent to the diagram ":", since $(\lambda_1=\lambda_2, \lambda_3=\lambda_4)$ may degenerate into $(\lambda_1=\lambda_2=\lambda_3=\lambda_4)$.

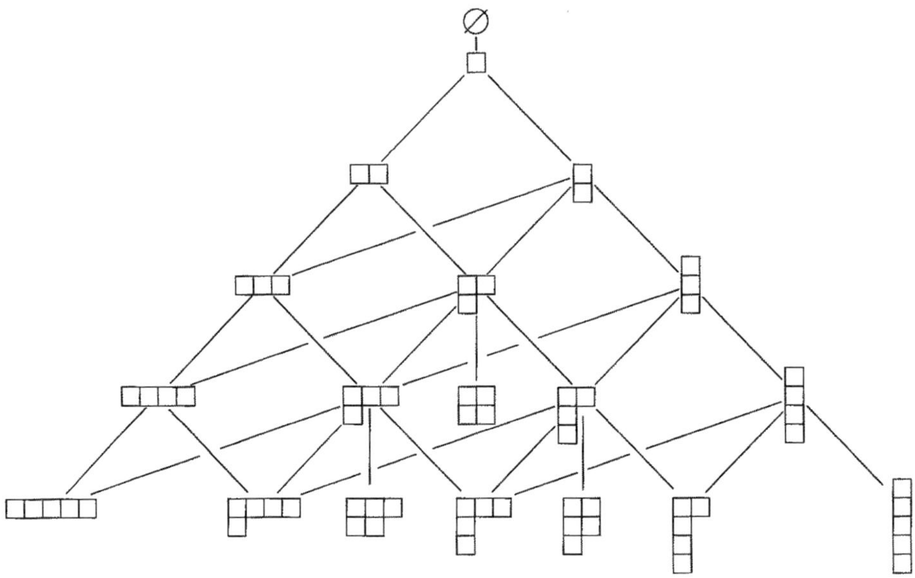

Fig. 5 The direct adjacencies of the simplest Young diagrams

The general rule is: to obtain a directly adjacent diagram from a given diagram one should replace two (possibly void) lines of the lengths a and b by a line of the length $a + b + 1$.

The list of the direct adjacencies of the simplest Young diagrams is shown in Fig. 5.

REMARK. The conservation of the quantity $p + q$ follows from the geometrical construction of the differential. Indeed, let A be the complex dimension of the manifold of the complete flags. The complex dimension of the initial manifold of degenerate flags is then $A - q$ (according to the definition of the defect q). The real dimension of a cycle of complex codimension p in this flag manifold equals $2(A - q - p)$. Let s be the number of distinct eigenvalues for the initial stratum. The product of the cycle with the simplex corresponding to this stratum defines a relative cycle of real dimension $2(A - q - p) + s$. The part of its boundary over a simplex of dimension $s' = s - 1$ (which is the face of the initial simplex) has the real dimension $2(A - q - p) + s - 1$. This boundary corresponds (in the directly adjacent stratum of defect q') to a complex cycle of such codimension p', that its product with the simplex of dimension s' has the same dimension, as the boundary that we have considered:

$$2(A - q' - p') + s' = 2(A - q - p) + s - 1,$$

hence $p' + q' = p + q$.

Stratum	$p+q=4$	$p+q=3$	$p+q=2$	$p+q=1$	$p+q=0$
$\varnothing$	$\binom{n+2}{4}-\binom{n+1}{2}$	$\binom{n+1}{3}-n$	$\binom{n}{2}-1$	$n-1$	1
$\square$	$(n-1)\left(\binom{n}{3}-1\right)$	$(n-1)\binom{n-1}{2}$	$(n-1)(n-2)$	$n-1$	
	$\binom{n-2}{2}\left(\binom{n-2}{2}+1\right)$	$\binom{n-2}{2}(n-3)$	$\binom{n-2}{2}$		
	$\binom{n-3}{3}(n-4)$	$\binom{n-3}{3}$			
	$\binom{n-4}{4}$				
	$(n-2)(n-3)$	$n-2$			
	$(n-3)(n-4)$				

Fig. 6 The spectral sequence of the small codimension homologies

Comparing the Betti numbers of the flag manifolds with the above information, we obtain the dimensions of the spaces of our spectral sequence, shown in Fig. 6.

Each box in Fig. 6 represents the dimension of the space of the cycles of complex codimension p in each flag manifold of defect q with a given Young diagram (for $p+q\leq 4$ and for the values of n which are not too small).

The differential acts vertically downstairs, into the homologies with the same value of $p+q$ of the manifolds of flags whose Young diagrams are directly adjacent to the given diagram (and hence whose areas are larger by one than that of the initial Young diagram).

The degeneration of the spectral sequence implies the vanishing of the Euler characteristic in each column (except the rightmost one). This vanishing provides the combinatorial identities: *the alternating sum of the products of the Betti numbers of the flag manifolds with the numbers of the strata having the corresponding Young diagrams is equal to zero (the alternating signs being defined by the parities of the areas of the Young diagrams).*

Thus, we obtain an infinite sequence of strange combinatorial identities, starting from

$$\binom{n}{2}-1+\binom{n-2}{2}=(n-1)(n-2),$$

$$\binom{n+1}{3}-n+\binom{n-2}{2}(n-3)+(n-2)=(n-1)\binom{n-1}{2}+\binom{n-3}{3},$$

$$\binom{n+2}{4} - \binom{n+1}{2} + \binom{n-2}{2}\left(\binom{n-2}{2}+1\right) + \binom{n-4}{4} + (n-2)(n-3) =$$
$$(n-1)\left(\binom{n}{3}-1\right) + \binom{n-3}{3}(n-4) + (n-3)(n-4).$$

I do not know any bijective proof of any of these identities.

REMARK 1. The consideration of the cycles of small dimension leads to a *different* chain of combinatorial identities.

REMARK 2. (A.B. Givental). Representing the (co)homologies of the Grassmannians and of the flag manifolds in terms of the symmetric functions, one might interpret the degeneration of our spectral sequence as a set of relations between the symmetric functions of different symmetry types.

§5. The informal complexification

Any notion in real geometry has an informal "complex version" (see [3]). The table below lists some examples.

The real version.	The complex version.
$\mathbf{R}$	$\mathbf{C}$
π_0	π_1
S^0	S^1
$\mathbf{Z}_2 = \mathbf{Z}/2\mathbf{Z}$	$\mathbf{Z}$
$\mathbf{R}^n \setminus 0$	$\mathbf{C}^n \setminus 0$
π_{n-1}	π_{2n-1}
$O(n), SO(n)$	$U(n), SU(n)$
Morse theory	Picard-Lefschetz theory
Stiefel-Whitney classes	Chern classes
$\mathbf{RP}^n$	$\mathbf{CP}^n$
$S^1 = \mathbf{RP}^1$	$S^2 = \mathbf{CP}^1$
the Möbius S^0-bundle	the Hopf S^1-bundle
a manifold with boundary	a 2-ramified covering
orientation	homotopy class of Hermitian framings

Consider from this point of view the stratifications of the spaces of the real symmetric and of the complex Hermitian matrices of order n.

In the real case the variety of the matrices having double eigenvalues has codimension 2, and its complement is not simply-connected (see [1]). To describe the fundamental group of the space of the symmetric matrices with simple spectra, choose a matrix with no multiple eigenvalues and choose a normed eigenvector for each of the eigenvalues. The normed eigenvectors are defined up to their signs, forming thus the sphere S^0.

Choose a closed loop in the space of the symmetric matrices having no multiple eigenvalues, starting from the chosen matrix and returning to it. Continuing the normed eigenvectors over this loop, we shall return to the initial matrix, in general, with a new frame of the eigenvectors. We thus obtain the *monodromy representation* of the fundamental group of the manifold of the symmetric matrices having simple spectra onto the group $\mathbf{Z}_2^{n-1}$ of the reversals of an even number of the eigenvectors. This representation is a faithful representation of the fundamental group ([1]).

On each two-dimensional local transversal to each of the $n-1$ strata $\lambda_i = \lambda_{i+1}$ of codimension 2 in the space of the real symmetric matrices the variety of the matrices having multiple eigenvalues is represented by a point. The circle centered at this point in linked with it on the transversal. It represents a generator of the fundamental group of the manifold of the real symmetric matrices having simple spectra.

The bundle of the normed eigenvectors over this circle is nontrivial. The continuation along this circle changes the signs of the i-th and of the $(i+1)$-th eigenvectors. In other terms, the restrictions of the i-th and of the $(i+1)$-th normed eigenvector bundles to this circle are the Möbius S^0-bundles. (The Möbius S^0-bundle is the fibration of the boundary of the Möbius band over its central circle).

Now we informally complexify all the above description (borrowed from [1]), replacing the real symmetric matrices by the Hermitian ones.

The real codimension of the variety of the Hermitian matrices having a multiple eigenvalue equals 3. Its transversal at a regular point is a 3-dimensional real space. On the local transversal to each of the $n-1$ strata $\lambda_i = \lambda_{i+1}$ the variety of the Hermitian matrices having a multiple eigenvalue is represented by a point. The two-sphere S^2, linked with this point in $\mathbf{R}^3$, represents one of the generators of the second homology (or homotopy) group of the manifold of the Hermitian matrices having simple spectra.

The normed eigenvector, corresponding to a simple eigenvalue of an Hermitian matrix, is defined up to its phase (up to its multiplication by a complex number whose modulus is equal to one). The S^1-bundle of the normed eigenvectors over the sphere S^2 (linked with the variety of the Hermitian matrices having a multiple eigenvalue) is nontrivial. This S^1 bundle (informally complexifying the Möbius S^0-bundle over S^1) is the Hopf S^1-bundle over S^2, as an easy calculation shows.

EXAMPLE. Consider the space $\mathbf{R}^3$ of the traceless Hermitian matrices of order 2,

$$A = \begin{pmatrix} x & z \\ \bar{z} & -x \end{pmatrix}, \qquad z \in \mathbf{C}.$$

The only matrix having a multiple eigenvalue in this family is the zero matrix. The linked sphere has the equation $x^2 + z\bar{z} = 1$. The eigenvector bundle over the complement to the origin of $\mathbf{R}^3$ has a *natural connection*: One transports a normed eigenvector of a given matrix into the closest normed eigenvector of an infinitesimally deformed Hermitian matrix.

This natural connection is the informal complexification of the monodromy of the eigenvector bundle of the real symmetric matrices. The curvature of this connection is a closed two-form (with real values) on the space $\mathbf{R}^3$ of the traceless Hermitian matrices of order $n = 2$ (singular at the variety of the matrices having a multiple eigenvalue, which consists in the case $n = 2$ of one point 0.)

A direct computation shows that the integral of this curvature form over the linked sphere equals 4π. The singularity of the 2-form at the origin is Newtonian. A closed 2-form in $\mathbf{R}^3$ corresponds to a divergence-free vector field. In our case the vector of this field is directed along the radius vector and its length is inversely proportional to the square of the distance to the origin.

These calculations show, among other things, that each of the n curvature 2-forms of the eigenvector bundles' natural connections over the manifold of the traceless Hermitian matrices of order n with simple spectra has the Newtonian polar singularities on the corresponding strata of the variety of the matrices with multiple eigenvalues.

These remarkable characteristic 2-forms (whose sum vanishes identically) generate the cohomology ring of the manifold of the traceless Hermitian matrices with simple spectra. It would be interesting to understand the relation of the ring, generated by these 2-forms to the above cohomology ring: do there exist nontrivial linear combinations of the products of these forms, which are homologous to zero, being not identically zero?

It would also be interesting to understand the "residues" of these singular characteristic forms on the different strata of our stratification and their relations to the characteristic forms of the eigenspace bundles over these strata. One might try to represent the characteristic forms as the integrals of their "residues" along different strata.

It would be interesting to find the real symmetric matrix versions of the geometry of the Hermitian matrices stratification described in sections 1–3.

§6. The eigenvector bundles in physics: the adiabatic connection

I shall now briefly describe two (missed) applications of the informal complexification of the constructions of [1] to physics: they lead directly to the adiabatic connections on the bundles of the eigenvectors (known in physics as the Berry phase theory [4], [5]) and to the (integer) quantum Hall effect theory.

The manifold of the Hermitian matrices of order n having no multiple eigenvalues is the base space of n eigenvector bundles. The natural connections on these bundles, described above (in section 4) coincide as we shall see with the *adiabatic connections*, which we shall now define.

Consider a "quantum" system with a slowly varying Hamiltonian

$$\frac{dz}{dt} = iA(\epsilon t)z, \qquad \epsilon \ll 1$$

where A is an Hermitian matrix, depending on a slowly varying parameter ϵt. Suppose that only one mode is excited initially. In the *adiabatic approximation* the resulting oscillations are described as the forever-one-mode oscillations, with slowly varying eigenvector and eigenfrequency. In this approximation one neglects the fast oscillations, perturbing the averaged adiabatic approximating solution. This approximation is reasonable for the time intervals of order $t \sim 1/\epsilon$.

The phase of the adiabatic approximating oscillation (with respect to the naturally transported initial eigenvector) is defined as the integral of the momentary (slowly varying) frequency, $S(t) = \int_0^t \lambda(\tau)\, d\tau$, where $\lambda(\tau)$ is the momentary eigenvalue of A at moment τ.

The adiabatic approximation describes the slow transport of an eigenvector along a path in the manifold of the Hermitian matrices having simple spectra. This averaged adiabatic transportation defines a connection on the eigenvector bundle. I shall call it the *adiabatic connection*.

PROPOSITION. *The adiabatic connection coincides with the natural connection on the eigenvector bundle.*

I was unable to find a proof of this (essentially known, compare [5]) fact in the literature. The following rigorous proof, based on the modern technology of the averaging theory of classical mechanics has been proposed by A.I. Neistadt. Diagonalize the matrices by a parameter-dependent linear (unitary) change of variables. In the new coordinates the matrices of the transformed time-dependent system will be only slightly non-diagonal, provided that the parameter varies slowly. Diagonalize them in the first approximation by a small unitary time-dependent change of variables. In these new coordinates the adiabatic approximation is just the solution of the diagonalized system (the perturbation being of second order only). This calculation of the adiabatic approximation implies the coincidence of the adiabatic connection with the natural one.

Here, however, I shall give in detail an uglier line of reasoning, calculating the following: the components of both connections in a time-independent Hermitian orthonormal frame $(v_1, \ldots, v_n)$ of the eigenvectors of the unperturbed momentary Hermitian operator; the corresponding eigenvalues $(\lambda_1, \ldots, \lambda_n)$; and the matrix elements $b_{k,m}$ of the perturbing Hermitian operator B (in the basis of the eigenvectors of A).

Let $v_m + \epsilon w_m + \cdots$ be the eigenvectors of the perturbed operator $A + \epsilon B$ (we transport v_m using the natural connection; the dots mean the higher-order terms in the perturbation parameter $\epsilon \to 0$). The corrections w_m to the eigenvectors are described by their components $u_{k,m}$, namely $w_m = \sum u_{k,m} v_k$, where $k \neq m$ in the sum.

The absence of the term with $k = m$ in this linear combination means that the connection is natural (the correction being orthogonal to the original vector).

The first order correction is given by the following classical formula of perturbation theory.

LEMMA 1. *The components of the correction ϵw_m to the eigenvector v_m are*

$$\epsilon u_{k,m} = \frac{\epsilon b_{k,m}}{\lambda_m - \lambda_k} \qquad \textit{for } k \neq m, \tag{1}$$

the correction to the eigenvalue $\lambda_m(\epsilon) = \lambda_m(0) + \epsilon\delta_m + \cdots$ being $\delta_m = b_{m,m}$.

Proof: Compare the first order terms in ϵ in the identity

$$(A + \epsilon B)(v_m + \epsilon w_m + \cdots) = (\lambda_m + \epsilon\delta_m + \cdots)(v_m + \epsilon w_m + \cdots).$$

The resulting matrix u is skew-Hermitian (B being Hermitian) and depends on B linearly. Formula (1) defines a *connection* on the eigenvector bundle. It is the natural connection.

Now we start to calculate the adiabatic connection. We have to find an approximate solution (in first approximation with respect to the small parameter ϵt) of the linear equation with slowly varying coefficients

$$dz/dt = i(Az + \epsilon t Bz + \cdots), \qquad z(0) = v_m.$$

The perturbation parameter is now ϵt and the dots mean the higher order terms (like $(\epsilon t)^2$ and so on).

To calculate the adiabatic approximation for small ϵt we have to average the deviation of the solution from the unperturbed solution $z_* = e^{i\lambda_m t} v_m$ along many periods of the unperturbed motion, taking ϵ so small that the product ϵt will remain small.

Let the perturbed solution be $z = z_* + \epsilon x + \cdots$, and let x_k be the components of the vector $x = \sum x_k v_k$.

LEMMA 2. *The components of the correction to the solution of the unperturbed system are*

$$\epsilon x_k = \frac{\epsilon b_{k,m}}{\lambda_m - \lambda_k}\left[t e^{i\lambda_m t} + i\frac{e^{i\lambda_m t} - e^{i\lambda_k t}}{\lambda_m - \lambda_k}\right] \qquad \textit{for } k \neq m,$$

$$\epsilon x_m = \epsilon i b_{m,m} e^{i\lambda_m t} t^2/2.$$

Proof: Vector x verifies the equation in variations

$$dx/dt = i(Ax + tBz_*), \qquad x(0) = 0, \qquad z_* = e^{i\lambda_m t}v_m.$$

Seeking the solution in the form $x = e^{iAt}C(t)$, where $C(t) = \sum c_k(t)v_k$, we find the coefficients c_k by the quadratures:

$$dc_k/dt = ite^{i\alpha t}b_{k,m}, \qquad \alpha = \lambda_m - \lambda_k, \qquad c_k(0) = 0.$$

We use now the evident

LEMMA 3. *For $\alpha \neq 0$ one has*

$$\int_0^t \tau e^{i\alpha\tau}\, d\tau = \frac{t}{i\alpha}e^{i\alpha t} + \frac{e^{i\alpha t} - 1}{\alpha^2}.$$

This formula implies, for $k \neq m$, the expression

$$c_k = \left(\frac{t}{\alpha}e^{i\alpha t} + i\frac{e^{i\alpha t} - 1}{\alpha^2}\right) b_{k,m},$$

while for $k = m$, one finds

$$c_m = it^2 b_{m,m}/2.$$

These formulae prove Lemma 2.

To separate the adiabatical drift from the fast oscillations around it in the expression of Lemma 2, consider the large values of t, which are the integer multiples of the period of the fast oscillations of frequency λ_m (this stroboscopical trick is equivalent to the averaging). Of course, we choose a very small value of ϵ, thus the value of ϵt remains small. In this averaged approximation we neglect the second (oscillating) term in the expression of Lemma 2 comparatively to the first (growing) term. Thus, we obtain for $k \neq m$ the adiabatic approximations to the components of the correction to the eigenvector v_m in the form

(2) $$\epsilon x_k = \frac{\epsilon t b_{k,m}}{\lambda_m - \lambda_k}, \qquad \text{while } \epsilon x_m = i\epsilon b_{m,m} z_* t^2/2$$

The last correction (complex proportional to v_m) in Lemma 2 corresponds exactly to the slow change in the frequency $\lambda(t) = \lambda_m + \epsilon t b_{m,m} + \cdots$. Indeed, the adiabatic approximation phase is $S(t) = \lambda_m t + \epsilon b_{m,m} t^2/2 + \cdots$. This phase provides the approximate solution, whose projection on the direction v_m is (in the first approximation) equal to

(3) $$e^{iS(t)}v_m = z_*(1 + i\epsilon b_{m,m}t^2/2 + \cdots).$$

The correction term has the value x_m given by Lemma 2. The formulae (1)–(3) show that the adiabatic connection coincides with the natural connection. It might be interesting to derive the connections on other bundles from the suitable adiabatic approximations.

REMARK. The relations of the adiabatic approximation to the bundle connections were known in the 1920's or even earlier. For instance in the classical textbook by F. Klein on the Higher Geometry [11], the Levi-Civita connection in Riemannian geometry is interpreted (following Radon) as the slow transportation of an oscillator (like a pendulum) along the Riemannian configuration manifold. However, these ideas, as far as I know, have never been presented in a mathematically rigorous form.

In the modern physical literature, the quasiclassical symplectic geometry version of the adiabatic connection is known as the *Berry phase* (and *Hannay angles*) theory.

Particular (but quite instructive) cases of these connections had been discovered in the 1930's in optics (S.M. Rytov, see [6], [7]) and in the 1940's in the then classified works in gyroscopes theory (A.Yu. Ishlinkii, see [8]).

Consider a thin curved optical fiber, fixed at the two end points, where its directions are also fixed. Consider a beam of the polarized light, entering the fiber at one end point and leaving it at the other. The leaving beam defines the polarization direction at the end point.

Deform the fiber, preserving the endpoints and the end directions. The direction of the polarization of the leaving beam will turn at some angle. This angle has been calculated by Rytov.

Consider the Gauss map, sending each point of the optical fiber to the direction of the fiber at this point. One obtains a curve on the unit sphere. The two curves, corresponding to the initial position of the fiber and to the deformed position, connect the same two points on the sphere. *The spherical area of the domain, bounded by these two curves, equals the rotation angle of the polarization direction.*

It is clear that this Rytov formula calculates the transport (corresponding to a connection) along a closed loop as the integral of the curvature of a connection (along a two-chain, bounded by the loop). The physicists call this natural connection "the inertia of the polarization direction".

In the gyroscopes theory one slowly moves the rotating body along a closed loop (say, transporting the gyroscope along the surface of the earth). The angle of the resulting additional rotation equals to the spherical area, bounded by the projection of the loop to the unit sphere. This Ishlinksii's formula is also the calculation of the transport corresponding to a connection in terms of its curvature (see [8]).

The quantum mechanics manifestation of the natural (or of the adiabatic?) connection is the well-known Aharonov-Bohm effect, showing that a quantum particle feels the magnetic *potential* rather than the magnetic force: it's phase depends on the magnetic potential even when the particle moves in the region where the magnetic field vanishes.

§7. The characteristic number in physics: the quantum Hall effect

The topological nontriviality of the eigenvectors bundle over the manifold of the Hermitian matrices with no multiple eigenvalues manifests itself in physics as the quantum Hall effect (see [9]).

Consider a 2-surface in the space of Hermitian matrices. If the surface is generic, all the corresponding matrices have only simple eigenvalues (since the codimension of the variety of the matrices with multiple eigenvalues is equal to 3).

The i-th normed eigenvectors of the matrices of a generic two-parameter family form an S^1 bundle over the base surface. If the surface is closed (or under other convenient boundary conditions) the Chern numbers of these bundles are defined. These characteristic numbers of the eigenvector bundles can be calculated as the integrals of the curvature forms of the natural connections along the base surface.

Suppose now, that the surface depends on one more parameter (I shall call it "the time", while physically it might be quite different — say, an exterior electric potential or the magnetic field). Then the moving surface will momentary cross (at some moments) the variety of the Hermitian matrices having multiple eigenvalue. This crossing will generically be transversal.

At this transversal crossing the Chern numbers of two eigenvector bundles will change by 1 (whether they will increase or decrease can be calculated using the natural co-orientation of the variety of the Hermitian matrices having multiple eigenvalues at its regular points).

Indeed, to obtain the surface immediately after the critical moment from the surface just before this moment it suffices to consider a small sphere S^2 linked with the crossing point (corresponding to a matrix having a multiple eigenvalue) in the three-dimensional parameter space. The bundle of the eigenvectors over this sphere is the standard Hopf bundle and its Chern number equals 1. The Chern number of the eigenvector bundle over the connected sum with this sphere of the surface just before the crossing differs by 1 from the Chern number of the bundle over the initial surface (since the integral is additive). The Chern numbers of the eigenvector bundle over the connected sum surface and over the surface immediately after the transversal crossing are equal. Indeed, these two surface are isotopic in the three-dimensional parameter space (and thus their eigenvector bundles are isomorphic). Thus the jump of the Chern number at a transversal crossing is equal to 1.

The sequence of the changes of the Chern numbers of the eigenvector bundles by 1, corresponding to the moving surface (to a one-parameter family of surfaces in the space of Hermitian matrices) is observable in the physical experiments as the (integer) quantum Hall effect (see [9]).

Hence, the theory of this effect might be considered as the (missed) informal complexification of the study [1] of the monodromy of the eigenvectors of the symmetric matrices depending on the parameters.

These examples show that the informal complexification (for which there is still no formal definition) is a useful tool, deserving more attention. It would be interesting, for example, to understand what kind of mathematical structure is the complexification of the homology theory.

It is closer to the topic of the present article to study the unitary group manifold as the informal complexification of the real Lagrangian Grassmannian. The unitary group manifold compactifies the space of Hermitian matrices (defining the "Hermitian generating functions") in the same sense in which the real Lagrangian Grassmannian compactifies the space of the real symmetric matrices (defining the quadratic generating functions in the real symplectic geometry).

This leads immediately to the Hermitian version of the Lagrange manifolds and to the corresponding problems in symplectic topology, waiting for the courageous researchers. (The resulting theory should be compared with the Givental's quaternionic version [10] of the Lagrange manifolds theory.)

References

[1] V.I. Arnold. *Modes and quasimodes.* Funkt. Anal. and its Appl., **6:2** (1972), 94–101.

[2] M. Shapiro, A. Vainstein. *Stratification of Hermitian matrices, the Alexander mapping and the bundle of eigenvspaces.* Preprint, Weizmann Inst., 1994, 6pp.

[3] V.I. Arnold. *On some problems in singularity theory.* Trudy Seminare S.L. Sobolev, Novosibirsk, vol. 1 (1976), pp. 5–15, (in Russian; English translation in: Singularities, Proc. Symp. Pure Math., vol. 40^1, AMS, 1983, pp. 57–69).

[4] M.V. Berry. *Quantal phase factors accompanying adiabatic changes.* Proc. Roy. Soc. London, vol. A-392 (1984), pp. 45–57.

[5] Y. Aharonov, J. Anadan. Phys. Rev. Lett. Vol. 58 (1987), pp. 1593.

[6] S.M. Rytov. *On the transition from wave to geometric optics.* Doklady Ac. of Sc. USSR, **18:2** (1938), 263–266.

[7] V.V. Vladimirskii. *On the rotation of the polarization plane along a curved ray.* Doklady Ac. of Sc. USSR, **31:3** (1941), 222–225.

[8] A.Yu. Ishlinskii. *Mechanics of Special Gyroscopical Systems.* Ukrainian Acad. of Sc., Kiev, 1952 (see also: A.Yu. Ishlinskii. Orientation, gyroscopes, and inertial naviagation. Nauka, Moscow, 1976, page 195).

[9] Y. Collin de Verdière. *Effect quantique de Hall.* Seminaire de Theorie Spectrale et Geometrie, no. 11 (1993), Institut Fourier, Grenoble.

[10] A.B. Givental. *Quaternionic Sturn theory and symplectic geometry.* Lectures at Strasbourg (June 1991) and at Bologna (September 1991).

[11] F. Klein. *Einleitung in die höhere Geometrie.* 3^{d} Aufl., Berlin, 1926.

V.I. Arnold
Steklov Mathematical Institute
42 Vavilova St.
Moscow 117966, GSP-1/Russia

and

University Paris-Dauphine
Place du Maréchal
de Lattre de Tassigny
F-75775 Paris Cedex 16

Problems on singularities and dynamical systems

V.I. Arnold

This paper more or less reproduces the 1991 list of problems of the Singularity Seminar at Moscow State University. Some of the previous lists (compiled every year) have been published in [1–4]. I.G. Petrovsky once said that he had succeeded in doing something in mathematics more because he had not been aware of many things than because he knew something. However, he added, it was always very important to know, that something is not known.

FAKE $\mathbb{R}^4$ AND DYNAMICAL SYSTEMS

Is it possible to write down explicitly a vector field in $\mathbb{R}^5$ *(say, with polynomial or trigonometric polynomial components), whose orbit space is a fake* $\mathbb{R}^4$*?*

Let us consider the product $X \times \mathbb{R}$, where X is a manifold, homeomorphic to $\mathbb{R}^4$ but not diffeomorphic to it. The product is diffeomorphic to $\mathbb{R}^5$. Let us consider the vector field $v = (0, 1)$ on the product space. The orbit space of v is X. What are the restrictions on the components of the vector field v in the usual coordinates of $\mathbb{R}^5$, imposed by the fact that X is fake? Can these components be polynomials or trigonometric polynomials?

LIMIT CYCLES AND ELLIPTIC CURVES

The elliptic curve, associated with a real limit cycle of a holomorphic plane vector field, is the local orbit space in the complex neighbourhood of the cycle.

Originally published in: *Developments in Mathematics: The Moscow School.*
Editors: V.I. Arnold and M. Monastyrsky [M.I. Monastyrskii].

How large can the set of elliptic curves be in the orbit space of polynomials vector fields, or of polynomial mappings; or of algebraic correspondences with fixed discrete invariants (genera, degrees)?

Let us consider a mapping of the complex line, which is biholomorphic on one boundary circle of a (non-circular) annulus and which sends it to the other boundary circle. Identifying each boundary point with its image, we obtain an elliptic curve. This elliptic curve may be thought of as representing the (local) orbit space of the mapping.

Deforming the initial circle (and its image) we shall not change the orbit space and hence the modulus of the elliptic curve, provided that the mapping is fixed.

If the mapping depends holomorphically on parameters, the modulus of the elliptic curve will depend on these parameters holomorphically up to the moment when the topology of the local orbit space changes. This change will not occur until the image of the initial circle touches this circle. However, deforming the initial circle, we can continue the elliptic curves to a larger domain of the parameter space (Fig 7.1).

How large is the natural boundary (and the Riemann surface) of the analytic continuation of the elliptic curve, corresponding to the standard mapping

$$z \to z + a + b \sin z \qquad z \in \mathbb{C} \bmod 2\pi?$$

We start from the elliptic curve, corresponding to $b = 0$, $\operatorname{Im} a > 0$

$$\mathbb{C}/(2\pi\mathbb{Z} + a\mathbb{Z})$$

and we continue it analytically in a and b. Or we can start from the annulus, bounded by the real axis and its image ($|b| < \operatorname{Im} a$) $\mathbb{C} \bmod 2\pi$.

Figure 7.1 *The analytic continuation of an elliptic curve in the space of orbits.*

The elliptic curve is

$$\mathbb{C}/(2\pi\mathbb{Z} + \omega\mathbb{Z})$$

The 'modulus' ω is holomorphic in a and b at least for $|b| < \operatorname{Im} a$; $\omega(a, 0) = a$ for $\operatorname{Im} a > 0$. The formulae for the calculation of the 'complex rotation angle' ω have been discussed in [5].

It would be interesting also to study the boundary of the domain of definition of the elliptic curve for the 'semistandard' mapping $z \to z + i\omega + ibe^{iz}$.

We associate with a real limit cycle of a holomorphic vector field the complex Poincaré mapping on the transverse line and then apply the preceding constructions to obtain an elliptic curve. However, for the general definitions of a circle and an elliptic curve in the orbit space we do not need any global transversal, hence these notions are well defined for the global foliation of the complex phase space into complex phase curves.

PSEUDOPERIODIC TOPOLOGY AND FERMI SURFACES

A function of two variables defines a plane vector field, whose vector is tangent to the level lines—the Hamiltonian vector field. Thus, the Hamiltonian vector fields on a 2-plane are integrable. This is also true for Hamiltonian vector fields on other 2-surfaces, provided that the Hamiltonian function is single valued.

A Hamiltonian vector field on a two-dimensional torus may be defined also by a multivalued function on the torus, whose differential is a well-defined closed 1-form. In this case the dynamics, defined by this 'integrable' Hamiltonian system, may have peculiar ergodic and topological properties (mixing and so on), in spite of the existence of the 'explicit' formulae for the solutions of the Hamiltonian differential equations.

Similarly, $n - 1$ multivalued functions on an n-dimensional torus define a generalized Hamiltonian (or Nambu) system, which is explicitly integrable, but whose phase curves may have globally a very peculiar behaviour.

This way we obtain interesting divergence-free periodic vector fields in ordinary Euclidean 3-space. In spite of the explicit 'integrability' of the equations of the field lines, very little is known about their global topological properties, even in the simple but

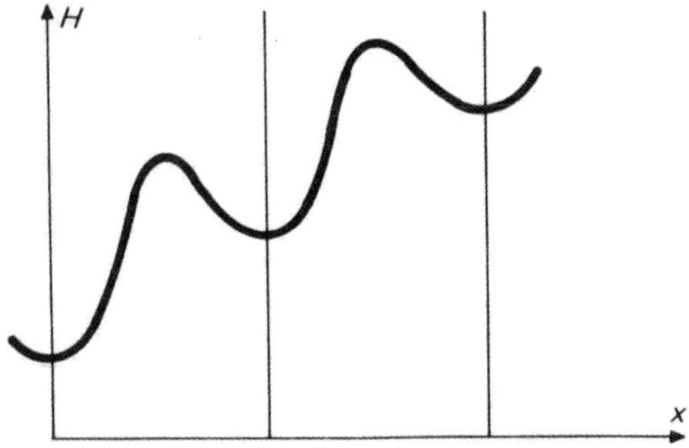

Figure 7.2 *A pseudoperiodic function.*

physically meaningful case of the irrational plane sections of the Fermi surfaces in solid-state physics.

To formulate the problems of pseudoperiodic topology I need the following definition. A mapping $H:\mathbb{R}^n \to \mathbb{R}^k$ is called **pseudoperiodic** (with respect to the lattice $\mathbb{Z}^n$) if

$$H(x+t) \equiv H(x) + \omega(t) \qquad \forall x \in \mathbb{R}^n, t \in \mathbb{Z}^n$$

where $\omega:\mathbb{Z}^n \to \mathbb{R}^k$ is a monomorphism. In other words, H is a sum of a linear mapping and of a periodic mapping, and the linear mapping ω sends the basis of $\mathbb{R}^n$ into n completely incommensurable (rationally independent) vectors of $\mathbb{R}^k$ (Fig. 7.2).

We shall call the level sets of quasiperiodic mappings **pseudoperiodic varieties.**

Let us consider an analytic pseudoperiodic mapping $H:\mathbb{R}^3 \to \mathbb{R}^2$, and let $c \in \mathbb{R}^2$ be a non-critical value. Then the pseudoperiodic curve $H^{-1}(c)$ is smooth.

Can the pseudoperiodic curves $H^{-1}(c)$ have more than one unbounded component?

It is easy to see that at least one unbounded component exists. The number of unbounded components of pseudoperiodic curves is exactly one for the generic pseudoperiodic function of two variables. For generic pseudoperiodic functions in n variables each of the sets $H > c$ and $H < c$ has exactly one unbounded component (see [6]).

However, very little is known about quasiperiodic varieties of higher codimension.

A submanifold of the Euclidean space $\mathbb{R}^n$ is called **$\mathbb{Z}^r$-periodic** if it is invariant under the translations by the vectors of an integral sublattice $\mathbb{Z}^r$ ot $\mathbb{R}^n$.

Let us consider an irrational (affine) generic plane section of a $\mathbb{Z}^3$-periodic surface (**Fermi surface**) in $\mathbb{R}^3$. S.P. Novikov has asked the following question:

Is it true that every unbounded component of this curve belongs to a finite neighbourhood of some straight line?

According to A. Zorich [7], this is true for an open set in the space of planes. It is perhaps wrong for some 2-plane sections of $\mathbb{Z}^4$-periodic surfaces in $\mathbb{R}^4$ (since the sections might be 5-symmetric and having unbounded components; see [8]).

Suppose that a connected generic r-dimensional $\mathbb{Z}^r$-periodic submanifold is given in $\mathbb{R}^n$. Let us cut it by a generic (completely irrational) affine hyperplane. The number of unbounded components of the complement of the hyperplane is exactly two (see [6]). The 'quasiperiodic intersection theory' should generalize this answer ('two') to the case of the intersection with an affine space of higher codimension.

Remark 7.1

Part of the proofs in [6] depend on an interesting generalization of the Sard lemma, which has been recently proven by Yu. Chekanov (unpublished).

Let us consider a C^∞-smooth divergence-free vector field in Euclidean n-space.

Let us consider the closed unit ball, and one of the connected components of the subset of the ball where the vector field does not vanish (Fig. 7.3).

Lemma 7.1

The flux of the field through the intersection of the above component with the boundary of the ball vanishes.

In the case $n=2$ the flux is equal to the sum of the increments of the flow function. This sum vanishes, since each component of

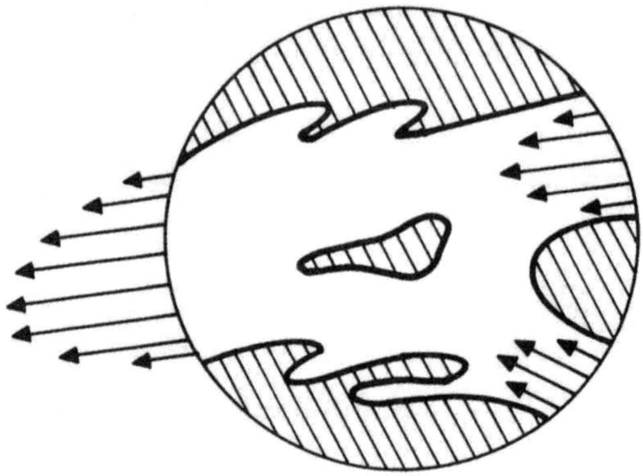

Figure 7.3 *The generalized Sard lemma.*

the boundary inside the ball consists of critical points of the function, and hence the function is constant along this component, according to Sard's lemma.

The above generalization of Sard's lemma to $n-1$ forms in n-space suggests the possibility of generalizations to k-forms.

Remark 7.2

The results of [6] implies that the Hamiltonian system with a multi-valued Hamiltonian function on a 2-torus is ergodic outside

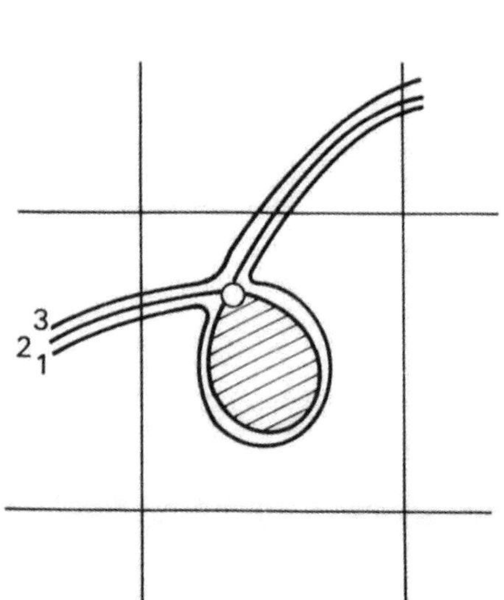

Figure 7.4 *A flow on the torus and the equivalent special flow.*

the union of the 'stagnation domains' (whose number on the torus is generically finite). Moreover, the dynamics outside the stagnation domain is that of the 'special flow' over a circle rotation with a logarithmically singular time function (Fig. 7.4).

The special flow over a diffeomorphism $A:X\to X$ with time function $f:X\to\mathbb{R}^+$ is the flow defined in the subgraph

$$S=\{x,y:x\in X,0\leqslant y\leqslant f(x)\}$$

by the vector field $\partial/\partial y$, the points $(x,f(x))$ being identified with $(A(x),0)$.

Khanin and Sinai [9] have recently proved that such a flow $\{g^t\}$ is mixing, provided that $A:S^1\to S^1$ is a generic circle rotation and f: $S^1\to\mathbb{R}^+$ is a generic function f with asymmetric logarithmic singularities. Mixing means that

$$\frac{\operatorname{mes}(g^tU)\cap V}{\operatorname{mes} S}\to\frac{\operatorname{mes} U}{\operatorname{mes} S}\frac{\operatorname{mes} V}{\operatorname{mes} S}$$

(the difference tends to zero as $(\log t)^{-1}$).

The asymmetry of the time function f in our case is due to the fact that the orbits passing near a stagnation point from one side of the separatrix visit the neighbourhood of the stagnation point once, and from the other side twice.

It would be interesting to study the ergodic properties of the motion in the generalized integrable pseudoperiodic Hamiltonian system with critical lines on the 3-torus. In the covering Euclidean 3-space this field is defined by two pseudoperiodic Hamiltonian functions:

$$\dot{x}=[\operatorname{grad} f,\operatorname{grad} g]$$

In the general case, a mapping $H:M^n\to N^{n-1}$ and the volume elements μ on M and ν on N uniquely define the generalized Hamiltonian vector field v by the formula

$$i_v\mu=H^*\nu$$

This field is divergence-free ($L_v\mu=0$) and has $n-1$ first integrals ($i_v\mathrm{d}H=0$).

Let us consider in more detail the case of $\mathbb{R}^3$ with $f=(a,x)+F(x)$, $g=(b,x)+G(x)$, for F and G $\mathbb{Z}^3$-periodic. Each orbit belongs to a finite neighbourhood of a straight line $(a,x)=0$, $(b,x)=0$. Suppose that this line is transversal to the planes $z = \text{const}$. In the analytical case (and in the generic C^∞ case) each orbit intersects each plane $z = c$ at a finite set of points and the number of these points is uniformly bounded (the constant does not depend on c or on the orbit).

Hence an orbit corresponding to a non-critical value of (f, g) is either closed or connects $z = \infty$ to $z = \infty$. There are four possibilities $(-+, --, +-, ++)$ for an oriented orbit. If the z-component of the vector product $[a, b]$ is positive, the $(-+)$ curve exists in any non-singular level set $f = \text{const}$, $g = \text{const}$. It is not clear whether the other cases are possible in completely irrational analytic systems.

The $(--)$ and $(++)$ curve, if they do exist, would correspond to a set of zero measure. This is because the corresponding subset of the torus can be decomposed in an infinite sequence of disjoint measurable sets of equal measure.

The $(+, -)$ curve exists in some systems with a rational line $(a, x) = 0 = (b, x)$. They are probably impossible when this line is completely irrational. If they do exist for a set of positive measure of values of (f, g), then they would exist among the level lines for almost every choice of the value (in the Lebesgue measure sense).

A curve of critical values, which is neither closed nor connecting ∞ to ∞, ends at a critical point (provided that the system is analytic). Such critical points, which are the zeros of our vector field, generically form a smooth curve on T^3. In $\mathbb{R}^3$ this curve is covered by a smooth curve, whose components are either closed or unbounded (of rational directions).

Suppose that there are no critical points. In this case the mapping (f, g) can be reduced to the normal form $f = (a, x)$, $g = (b, x)$ by a diffeomorphism of the torus. The vector field v is thus reduced to the form $[a, b]h$, where h is a positive periodic function.

Indeed, a closed non-degenerate 1-form df on T^3 can be reduced to the standard normal form (Laudenbach, Blank, Cerf, ...). Choose a Riemannian metric on $\mathbb{R}^3$. The gradient of the restriction of g to the planes $df = 0$ is nowhere zero. The gradient lines can be continued indefinitely. The restriction of g to such a line is a diffeomorphism onto the real line. Hence each level set $f = \text{const}$, $g = \text{const}$ on $\mathbb{R}^3$ has one component, which is an unbounded curve. Now it is easy to construct a T^2 transversal to the field v on T^3. The linearizing coordinates can then be written explicitly.

If there exist critical points some of the orbits are perhaps closed in $\mathbb{R}^3$. Each closed orbit of the field v in $\mathbb{R}^3$ is linked (mod 2) with the critical curves, provided that this orbit bounds a finite domain on the surface $f = \text{const}$. This follows from the fact that the Euler characteristic of the domain is then odd. There exist systems in 3-space having critical points but which have no arbitrary small closed orbits.

The linking number mod 2 is always equal to the class of the framing of the orbit by the frame $(\nabla f, \nabla g, v)$ in the group $SL(3)$. This follows from the fact that both numbers mod 2 are equal to the linking number of the curve $(\nabla f, \nabla g)$ in $\mathbb{R}^6$ with the variety V^4 of collinear pairs of vectors in $\mathbb{R}^3$.

I know of no example of a closed orbit totally unlinked to the critical curve, but there exist examples when the linking number is even.

BIFURCATION COMPLEXITY AND SKOLEM'S THEOREM

Let us consider a local holomorphic mapping of the complex 2-plane to itself, preserving the origin, and two holomorphic curves containing the origin.

How fast a rate of growth is possible for the intersection multiplicity of the image of the first curve under the nth iteration of the mapping with the second curve when n is growing (provided that all these multiplicities are finite)?

Example 7.1

$A(x, y) = (x, y^2), X = \{y = x), y = \{y = 0\}$.

Then $A^n X = \{y = x^{2^n}\}$. Hence the multiplicity (2^n) can grow exponentially.

If all the images $A^n X$ are smooth an exponential upper bound is proven in [10]. The multiplicities also cannot grow faster than some exponent, provided that everything is algebraic (see [11], [12]). However, for general holomorphic curves the problem is open, even when the mapping is polynomial and finite-to-one.

If the mapping is locally biholomorphic, the situation is quite different. In this case the multiplicities are uniformly bounded [10].

Moreover, one can replace the 2-plane by $\mathbb{C}^m$ and the two curves by two holomorphic submanifolds X^k and Y^l, containing the fixed point 0 of our biholomorphic mapping A.

In the case of submanifolds of complementary dimensions $(k+l=m)$ one can replace the intersection multiplicity by the linking number of the intersections of A^nX and of Y with a small sphere, centred at the origin.

In the case of hypersurfaces $(k=l=m-1)$ one can consider the order of tangency (the Milnor number). In the general case one may consider any abstract Milnor number defined by the following construction.

Let us consider the space J of (infinite) jets of pairs of submanifolds of dimensions k, l at the origin of $\mathbb{C}^m$. Let us fix a filtration

$$J \supset \Sigma^1 \supset \Sigma^2 \supset \cdots$$

where each Σ^i is an algebraic subvariety (defined by a finite system of polynomial equations on a finite set of the Taylor coefficients). The generalized Milnor number of a pair of submanifolds is the maximal value of i such that the pair of jets of the submanifolds belongs to Σ^i (thus Σ^i consists of pairs whose generalized 'multiplicity' of intersection is at least i).

The boundedness of any generalized Milnor numbers of the pairs (A^nX, Y), for all $n \in \mathbb{Z}$ is proved in [10]. This proof depends on the Skolem theorem on the zeros in recurrent sequences.

Let

$$f(n) = \sum_{i=1}^{m} c_i \lambda_i^n$$

be a finite exponential sum with complex coefficients. Then the set of its integer roots

$$\{n \in \mathbb{Z} : f(n) = 0\}$$

is a finite union of arithmetic progressions.

One can replace here the exponential sum f by a limit of such exponential sums, that is, by a similar sum, where the c_i are polynomials in n of degree smaller then m_i and where $(\sum_{m_i=m}$ (in other words, by any recurrent sequence of order m).

Remark 7.3

If the λ_i are non-resonant ($\prod \lambda_i^{n_i} \neq 1$ for any non-zero integers n_i), then the set of roots is finite.

Does there exist a uniform bound for the numbers of roots of non-resonant recurrent sequences of order m?

This seems to be unknown even for third-order recurrent sequences, where the bound might be equal to 6.

The algebraic curves in the projective λ-plane defined by the vanishing of the corresponding generalized Van der Mond determinants, $\det|\lambda_i^{n_j}| = 0$, intersect each other non-trivially only for very special choices of the triples n_j.

Remark 7.4

The intersection multiplicities, discussed above, contain as a special case the bifurcation multiplicity of periodic points in dynamical systems.

Let $g: M \to M$ be an analytical diffeomorphism having a fixed point 0. Let us consider the cartesian square $M \times M$, and the mapping

$$A = (\mathrm{id}, g): M \times M \to M \times M, \quad A(x, y) = (x, g(y)).$$

This mapping transforms the diagonal $\{y = x\}$ into the graph of g (Fig. 7.5). Its nth iterate A^n transforms the diagonal into the graph of g^n. Hence the multiplicity of the intersection of the nth image of the diagonal with the diagonal at the fixed point $(0, 0)$ is equal to the multiplicity of the periodic point 0 of period n for the mapping g.

The boundedness of these bifurcation multiplicities follows also

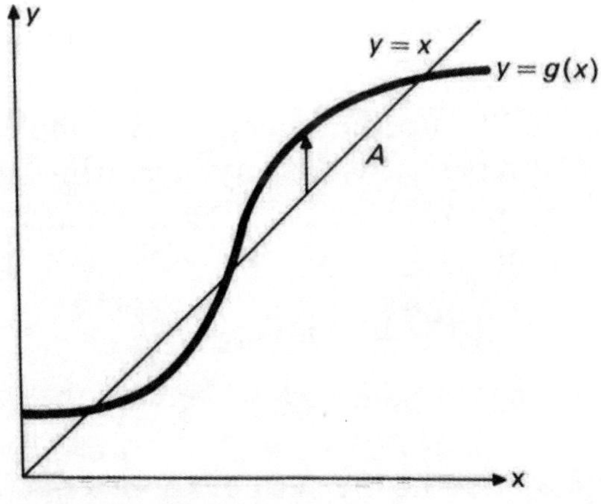

Figure 7.5 *Fixed points as intersection points.*

from the theorem of M. Shub and D. Sullivan [13] on bounding the Poincaré indices of the iterates of a diffeomorphism.

If g is not invertible, we can guess that the multiplicity of the fixed point 0 of g^n cannot grow with n faster than exponentially, but this is not proven (unless everything is algebraic).

The above local upper bounds of the asymptotics of the complexity of the intersections and bifurcations in dynamical systems are the counterparts of the following global results.

Let M^m be a compact smooth manifold, X^k and Y^l compact submanifolds, $A:M\to M$ a diffeomorphism. We move X, applying A n times, and we intersect it with Y. We obtain the intersection

$$Z_n=(A^nX)\cap Y$$

and we try to bound its 'topological complexity' $|Z_n|$ as a function of time n (Fig. 7.5).

If everything is algebraic, the topological complexity cannot grow faster than exponentially:

$$|Z_n|<ce^{\lambda n}.$$

The arguments of [11] prove this for such invariants as the sum of the Betti numbers, the characteristic numberss, the Morse number, and hence the numbers of the generators and relations of the fundamental group (see [10], [12], [14]). Perhaps one can even bound the lengths of the relations. The number of possible manifolds Z_n verifying these bounds is probably finite for each n (up to homotopy equivalence, homeomorphism, diffeomorphism and even isotopy of embeddings). It would be interesting to know how fast the above numbers of types can grow with n (provided that the set of types is indeed finite).

In the smooth (C^∞) case the intersection complexity may grow arbitrarily fast. More precisely, for any increasing sequence $a_n\to\infty$ there exists a quadruple (M^2, A, X^1, Y^1) and an increasing sequence of integers n_i such that

$$|(A^{n_i}X)\cap Y|>a_{n_i}\qquad \forall i$$

where $|\cdot|$ is the number of points and where all the intersections are transversal [12].

Similar examples exist for the periodic points of diffeomorphisms of manifolds of dimension $m\geqslant 2$ and for the periodic orbits of vector fields on manifolds of dimension $m\geqslant 3$ ([15], 16]).

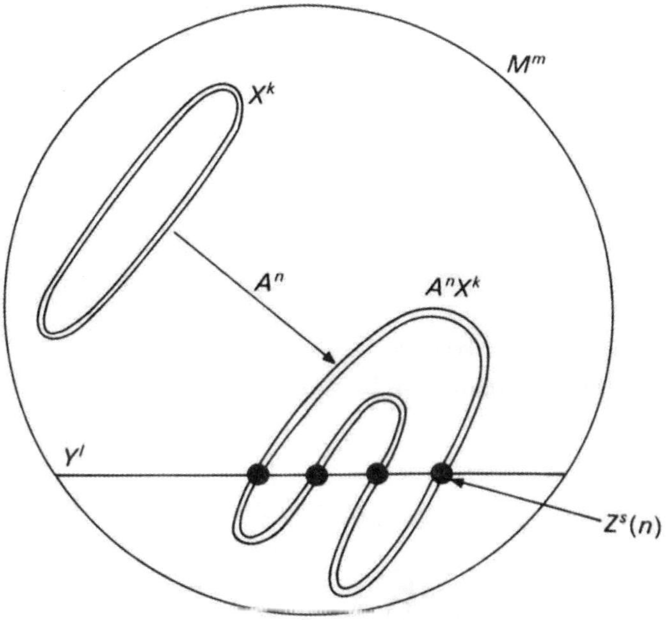

Figure 7.6 *Dynamics of the intersections.*

The exponential upper bound holds, however, with probability 1 for many measures of the topological complexity (see [14]). Let us consider a quadruple (M^m, A, X^k, Y_t^l) where $A{:}M \to M$ is a diffeomorphism and Y_t is a family of submanifolds of dimension l, depending generically on a parameter $t \in T^p$ (Figure 7.6). Then, for almost all t (in the sense of Lebesgue measure on T) the topological complexity of $Z_n(t) = (A^n X) \cap Y_t$ grows at most exponentially:

$$|Z_n(t)| \leqslant Ce^{\lambda n},$$

provided that the number of parameters p is sufficiently large ($p \geqslant m^2$).

Remark 7.5

One needs here the following surjectivity: the tangent spaces of the submanifolds Y_t at all their points and for all values of the parameter t should cover the neighbourhood of the manifold for tangent spaces of Y_0 in the bundle N of l-subspaces of the tangent spaces of M (the derivative of the corresponding mapping $Y_0 \times T \to N$ should be surjective at any point of $Y_0 \times 0$). This surjectivity property holds for generic families, provided that the dimension p of the parameter space is sufficiently high.

One can choose as a measure of the complexity, $|Z|$, in the preceding theorem

1. the volume V_s of dimension $s = \kappa + l - m$,
2. the number of components b_0,
3. the sum of Betti numbers b,
4. the Morse number μ,
5. any characteristic number p, or
6. the total absolute curvature K.

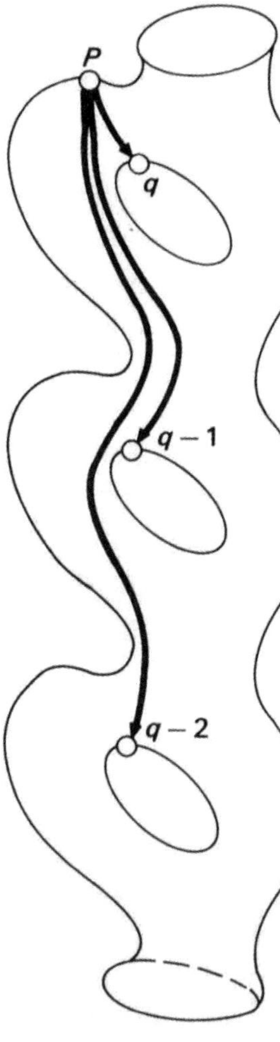

Figure 7.7 *Instantons from p to q − n.*

($K(Z)$ is the s-volume of the manifold of tangent planes of Z in the Grassmann bundle of s-planes of tangent spaces of M).

It is not known whether one can bound exponentially the lengths of the relations in the fundamental group (the numbers of generators and of relations are bounded by the Morse number, hence by the total absolute curvature, hence by an exponent of n).

It is not known whether the number of periodic orbits is bounded exponentially with probability 1 in the preceding sense (it is probably true). The results hold, probably, also for smooth mappings which ae not diffeomorphisms, but this has not beeen proven.

Among other probable generalizations of the above results I mension the following two:

1. (S.P. Novikov). Let us consider a cyclic covering of a compact manifold and a generic pseudoperiodic smooth Morse function f on the covering space (whose differential is lifted from the initial compact manifold).

   We denote by $+1$ the action of $\mathbb{Z}$ on the covering space, and we suppose that $f(x+1) \equiv f(x)+1$. Let p be a critical point of index i and q of index $i-1$. Let us consider the 'instantons' connecting p to $q-n$ (the orbits of grad f, Fig. 7.7).

Is the number of instantons connecting p to $q-n$ bounded by an exponent of n?

2. Let us consider a hyperbolic fixed point of a plane mapping. We fix a neighbourhood of the hyperbolic point. We define the **order** of its homoclinic point as the number of points of the orbit of this homoclinic point which do not belong to the fixed neighbourhood of the hyperbolic point (Fig. 7.8).

Is the number of homoclinic orbits of order n bounded by an exponent of the order?

Probably both exponential bounds hold with probability 1 in generic C^∞-systems.

In the analytic case there can be no bound to the complexity of intersection. This was proven in 1991 by O. Kozlovsky. He considers an analytical torus mapping

$$A(x,y) = (x+\alpha, y+f(x)) \bmod 1$$

and the circles $X = Y = \{y=0\}$.

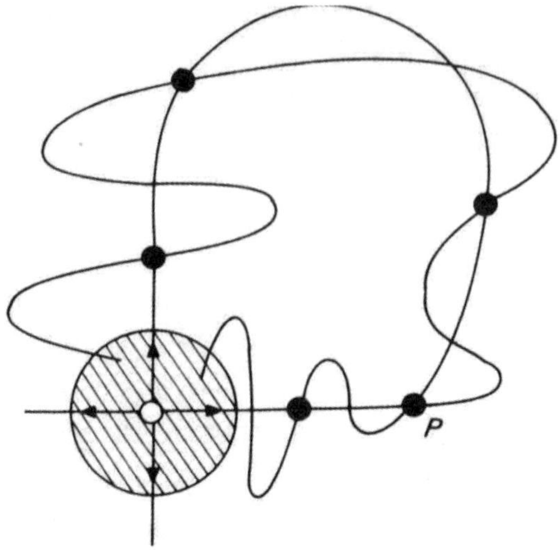

Figure 7.8 *Two homoclinic orbits of order 5.*

If one chooses the irrational number α and the hyperlacunary trigonometric series f cleverly, the number of intersection points $|(A^n X) \cap Y|$ will grow with n faster than any given sequence (for some increasing sequence n_i of values of n).

One conjecture is that the situation is similar for the periodic points of analytic diffeomorphisms and for the periodic orbits of analytical vector fields. It might even be the case for a polynomial vector field:

Does there exist for any increasing sequence $a_n \to \infty$ a polynomial vector field on $\mathbb{R}^m$ (entering the spheres $|x| = r > 1$), which has more than a_{n_i} periodic orbits of period at most n_i for an increasing sequence $n_i \to \infty$ (provided that all the periodic orbits are non-degenerate)?

QUASICONVEX HYPERSURFACES AND PROJECTIVE TOPOLOGY

Let us consider a smooth hypersurface in the real projective space P^n and two linked projective subspaces P^k and P^l, where $k + l = n - 1$.

Definition 7.1

The hypersurface is **quasiconvex** (with respect to P^k and P^l) if any straight line connecting a point of P^k with a point of P^l intersects it transversally and at exactly two points.

Example 7.2

Let $k=0$, $l=n-1$. The boundary of a convex body in the affine space $\mathbb{R}^n = P^n - P^{n-1}$, containing P^0, is quasiconvex with respect to P^0 and P^{n-1}.

Definition 7.2

A connected smooth hypersurface in the real projective space is **locally hyperbolic** if its second quadratic form is nowhere degenerate.

Example 7.3

If the form is positive definite, the surface is locally quadratically convex. A one-sheeted hyperboloid in P^3 is locally hyperbolic.

It is easy to see that closed connected (locally quadratically) convex hypersurfaces in projective spaces are the ordinary convex hypersurfaces of their affine parts (see, for example, [17]).

Is it true that closed connected locally hyperbolic non-convex surfaces in P^3 are quasiconvex?

This question may be considered as a projective version of the Hilbert problem on the complete surfaces of negative curvature in the Euclidean 3-space.

Is the set of closed connected locally hyperbolic non-convex surfaces in P^3 connected? Does any such surface admit a convex plane section?

The geometry of locally hyperbolic surfaces in affine and projective spaces provides many interesting generalizations of the maximum principle of the theory of partial differential equations, the function's graph being replaced by hypersurfaces whose topology might be non-trivial.

The peculiar topological properties of smooth submanifolds of locally projective spaces deserve special study. I mention here the recent classification of non-degenerate curves in S^n due to M.Z. Shapiro (1991)

An immersion $S^1 \to S^n$ is non-degenerate (positive), if the osculating flag is nowhere degenerate (if the Frenet frame defines a fixed orientation of S^n).

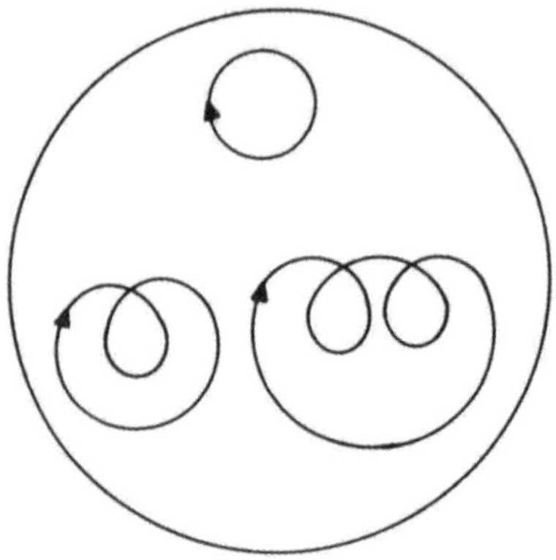

Figure 7.9 *Three classes of the convex curves on the 2-sphere.*

The number of regular homotopy classes of positive immersions of oriented manifolds $S^1 \to S^n$ is equal to 3 if n is even and to 2 if n is odd. The three classes of locally convex curves on S^2 (discovered by J. Little [18]; see also B.Z. Shapiro [19]) are represented in Fig. 7.9. This is one of the first theorems of projective topology; see the analysis of the inflections of projective curves and the 4-vertex theorem in [17] for more results in this almost unexplored domain.

4-VERTEX THEOREM AND 4-UMBILICAL POINT THEOREM REVISITED

The 4-vertex theorem says that a closed convex plane curve has at least four vertices (extremal points of the curvature). The 4-umbilical point theorem says that a generic convex surface has at least four points where both principal curvatures coincide. Both theorems seem to be connected to symplectic and contact topology rather than to Euclidean metrical geometry, involved in their standard formulation. That is, they provide lower bounds for the number of singular points of a special kind of the normal Lagrange mappings defined by convex curves and surfaces in Euclidean spaces. These singularities are well defined in the general case, and it is very interesting to see which properties of Lagrange projections imply such bounds.

Let us start with the following elementary generalization, going back to Jacobi's dynamics [20]. Consider any Riemann metric on a sphere S^n (say, that of a generic ellipsoid in $\mathbb{R}^{n+1}$), and fix a generic point 0 on the sphere. The conjugate points of the geodesics issuing from 0 form a hypersurface of dimension $n-1$ in S^n—the **caustic**.

The problem is to find a lower bound for the number of singularities of these caustics.

Jacobi claimed that such a caustic on the surface of a triaxial ellipsoid has exactly four cusps. But he did not prove this statement. The corresponding last geometrical paper in the last volume of his collected works [21] remained unfinished.

Is it true that the caustic, formed by the rth *conjugate points along all the extremals issuing from a given point of a 2-sphere, has at least four cusps (provided that the caustic is generic)?*

Is it true that the caustic, formed by the rth *conjugate points along all the extrema issuing from a given point of a 3-sphere, has at least four D_4-singularities (provided that the caustic is generic)?*

In the case of a positive curvature metric on a 2-sphere, it is clear that the components of the critical set of the normal mapping (of the mapping exp from the (co)tangent space of the sphere at our point of the sphere) corresponding to the first, the second conjugate point and so on are disjoint, and hence the caustic can be decomposed into the first branch, the second one and so on. It is easy to prove that the first branch has at least four cusps. It seems that this is also true for the other branches, but I do not think it has ever been proven.

Whether the caustics on the higher-dimensional spheres can be decomposed into compact branches is less clear (even for the case of the ellipsoid in $\mathbb{R}^4$, where everything can be computed explicitly).

CRITICAL POINTS AND LAGRANGIAN SINGULARITIES

The symplectic generalization of the 4-vertex theorem interprets the vertices as the cusps of the Lagrangian projection of a Lagrangian submanifold of the symplectic phase space—the cotangent bundle of the plane. That is, this Lagrangian manifold is formed by all the unit vectors of the lines normal to the curve, defining the same orientation of the line as the interior normal vector (provided that we identify the tangent and the cotangent vectors using the Euclidean metric of the plane).

The cusps of the projection of this Lagrangian submanifold to the

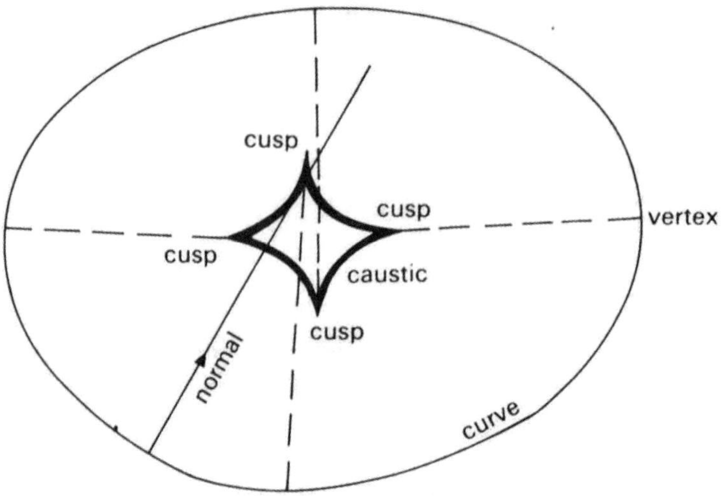

Figure 7.10 *The caustic of a plane convex curve.*

plane are the centres of curvature of the initial generic curve at the vertices. These cusp points are the singular points of the caustic, formed by the centres of curvature at all points (Fig. 7.10).

The Lagrangian manifold formed by the unit vectors is diffeomorphic to a cylinder. The unit vectors, based outside a large disc containing the caustic, form two collars (semi-cylinders) on the Lagrangian cylinder. These collars are projected to the plane diffeomorphically.

The middle part of the cylinder between the collars is projected with singularities. The set of its critical values is the caustic; the vertices correspond to the cusps of this projection.

Is it possible to replace the middle part of our cylinder by another Lagrangian embedding in such a way that the resulting mapping of an embedded Lagrangian cylinder to the plane has no cusps (and coincides with the initial one on the collars)?

One can associate with each boundary collar some index of a loop in the space of functions on the circle, having no singularities more complicated than x^3. The difference of the two indices corresponding to the two collars would imply the existence of the cusp singularities in the middle [22]. However, for our cylinder the indices of both collars are equal. It seems, therefore, that there exists at least an immersion of a Lagrangian cylinder with no cusps and with given boundary conditions on both collars.

Do there exist such embedded Lagrangian cylinders?

Our Lagrangian cylinder is optical (belongs to a hypersurface $p^2 = 1$, which is quadratically convex in each fibre of the cotangent bundle).

Is it possible to replace our cylinder by an optical immersed (embedded) Lagrangian cylinder, whose projection to the plane has no cusps, provided that the boundary collars remain unchanged?

Apart from applications to problems of Lagrangian singularities, the topological study of spaces of functions with mild singularities has other applications.

The topological invariants of the space of Morse functions on a given compact manifold (or the spaces of functions whose critical points belong to some less restrictive class of mildly degenerate singularities) are, in principle, interesting invariants of the smooth manifold.

Are the homotopy types of these function spaces defined by the topological type of the manifold, or do they really depend on the smooth structure?

It is interesting also to study such spaces as the space of non-negative functions (equipped with the filtration by the number of roots), the boundary of the space of non-degenerate symplectic or contact structures on a manifold (equipped with the filtration by the number of degeneration points).

Let us consider a Morse function on a connected compact manifold. We can push all the critical points to a small ball B^n centred on a point of the manifold (by a diffeomorphism). The restriction of the function to the neighbourhood of the boundary of this ball defines there a Lagrangian (or Legendre) collar—the set of the first differentials of the function or of its 1-jets at the points of the spherical layer.

Example 7.4

Suppose that the initial manifold was a 2-torus. Then the Lagrangian collar cannot bound inside the ball a Lagrangian section which intersects the zero section less than three times.

Indeed, otherwise we obtain a function on T^2 with less than three critical points.

Can one reconstruct a differential manifold knowing its Lagrangian collar? Which manifolds admit functions, coinciding on balls, containing all the critical points?

One can replace here the functions by the exact Lagrangian submanifolds which do not intersect the zero section of the cotangent bundle.

Let us consider a family of smooth functions as a function defined on the total space of a smooth fibration (with compact base and fibres).

Can one give a lower bound to the numbers of degenerate critical points (of different types) of the restrictions of the function to the fibres in terms of the topology of the fibration?

Such bounds might be considered as generalized Morse inequalities, since they would measure the discrepancy of the Morse complexes of the fibres, induced by the non-triviality of the fibration.

I shall end with a nice problem, due to B. Teissier. Let us consider a smooth real function in a neighbourhood of a critical point of finite multiplicity, and suppose that the index of the gradient vector field at the critical point vanishes.

Does there exist a function with no real critical points in a versal deformation of the given function?

A similar question can also be formulated for the Lagrangian collar defined by the restriction of the function to the neighbourhood of a sphere centred at the critical point.

Does this collar bound a Lagrangian disc (or another Lagrangian manifold, embedded in the cotangent bundle space of the ball) which does not intersect the zero section and which is bounded by the given collar?

ACKNOWLEDGEMENT

The author thanks Mrs Le Bronnec for rapidly deciphering a difficult manuscript.

REFERENCES

1. V.I. Arnold, Some unsolved problems in singularity theory, *Tr. Sem. S.L. Sobolev po kubaturn. formulam*, Novosibirsk, 1 (1976), 5–15 (in Russian).
2. V.I. Arnold, On some problems in singularity theory. In *Geometry and Analysis*, Papers dedicated to the memory of V.K. Patodi, Proc. Indian Ac. Science, **99**:1 (1981), 1–9.
3. V.I. Arnold, On some problems in singularity theory. In *Singularities*, Proc. Symp. Pure Math., vol. 40, Amer. Math. Soc. 1983, 57–69.
4. V.I. Arnold, Ten problems. In V. I Arnold, ed., *Theory of Singularities and its Applications* Advances in Soviet Mathematics, Vol. 1; Amer. Math. Soc., 1990, 1–8.
5. V.I. Arnold, Small denominators I. Mapping the circle onto itself, *Izv. Akad. Nauk SSSR*, Ser. Math., **25** (1961), 21–86; Math. USSR Izv., 1965.
6. V.I. Arnold, Topological and ergodic properties of closed differential 1-forms, *Functional Anal. Appl.*, **25**:2 (1991), 1–12.
7. A. Zorich, The problem of S.P. Novikov on the semiclassical motion of an electron in a homogeneous magnetic field, closed to a rational field, *Uspekhi Math. Nauk* **39**:5 (1984), 235–236; Russian Math Surveys, **39**:5 (1984) 287–288.
8. V.I. Arnold, *Huygens and Barrow, Newton and Hooke: pioneers in Mathematical Analysis and catastrophe theory, from evolvents to quasicrystalls*, Birkhäuser, Basel, 1990.
9. K. Khanin and Ya. G. Sinai, Mixing in integrable Hamilton systems, *Functional Anal. Appl.*, **26** (1992).
10. V.I. Arnold, *Majoration of Milnor Numbers of Intersections in Holomorphic Dynamical Systems*, preprint 652, Utrecht University, April 1991, 1–9.
11. M. Artin and B. Mazur, On periodic points, *Ann. Math.*, **81** (1965), 82–99.
12. V.I. Arnold, Dynamics of Intersections. In P. Rabinowlitz and E. Zehnder, eds, *Analysis et cetera, Research Paper Published in Honor of Jürgen Moser's 60th Birthday*, Academic Press, NY. 1990, 77–84.
13. M. Shub and D. Sullivan, A remark on the Lefschetz fixed point formula for differentiable maps. *Topology*, **13** (1974), 189–191.
14. V.I. Arnold, Dynamics of Complexity of Intersections, *Bol. Soc. Bras. Math.*, **21**, 1 (1990), 1–10.

15. E. Rosales, On the growth of the number of periodic points of dynamical systems, *Functional Anal. Appl.*, **25**:4, (1991), 14–22.
16. E. Rosales, On the growth of the number of long periodic solutions of differential equations, *Functional Anal. Appl.*, **26** (1992).
17. V.I. Arnold, Ramified covering $\mathbb{CP}^2 \to S^4$, hyperbolicity and projective topology, *Sib. Math. J.*, **29**, 5, (1988), 36–47.
18. J. Little, Nondegenerate homotopies of curves on the unit 2-sphere, *J. Diff. Geom.*, **4** (1970), 339–348.
19. B.Z. Shapiro, *Disconjugate Linear Ordinary Differential Equations, Flag Varieties and Classification of Symplectic Leaves in Several Kac–Moody and Gelfand–Dikii Algebras*, Matematiska Institutionen, Stockholms Universitet, 1990 (esp. 43–47).
20. C.G.J. Jacobi, *Vorlesungen über Dynamik*, G. Reimer, Berlin, 1884 (reprinted Chelsea, USA, 1969).
21. C.G.J. Jacobi, Über die Curve welche alle von einem Punkte ausgehenden geodätischen Linien eines Rotations-ellipsoides berühnt. In *Gesammelte Werke*, G. Reimer, Berlin, 1891, 72–87.
22. V.I. Arnold Spaces of functions with mild singularities, *Functional Anal. Appl.*, **23**:3 (1989), 1–10.

On some problems in the theory of dynamical systems

V.I. Arnold

To Jean Leray who was my scientific advisor at Sorbonne in 1965,
with admiration and best wishes

Translation of: Sur quelques problèmes de la théorie des systèmes dynamiques. *Topological Methods in Nonlinear Analysis* (*Journal of the Juliusz Schauder Center*), **4**, no. 2 (1994), 209–225.[1]

Translated by Nikolai Kruzhilin.

"If I did something in mathematics, it was not because of knowing much, but rather because I was ignorant of many things", I.G. Petrovskiĭ, one of my teachers in mathematics, said. "However, it was always very important", he added, "to know that this or that problem had not been solved yet". To the best of my knowledge, the eight problems that follow are open.

1. Exotic spaces $\mathbb{R}^4$ and dynamical systems

Can one encounter exotic spaces $\mathbb{R}^4$ (differentiable manifolds that are homeomorphic but not diffeomorphic to the linear space $\mathbb{R}^4$) in the conventional analysis?

I propose a natural construction which, in principle, describes all the exotic spaces $\mathbb{R}^4$ by explicit formulae [1].

Consider a vector field in the linear space $\mathbb{R}^5$.

The orbit space of such a (suitably selected) field is diffeomorphic to an arbitrary prescribed exotic space $\mathbb{R}^4$.

Indeed, the Cartesian product of an exotic space $\mathbb{R}^4$ by $\mathbb{R}$ is diffeomorphic to the linear space $\mathbb{R}^5$. Thus each exotic space $\mathbb{R}^4$ can be defined by five differentiable functions of five variables, the components of the vector field.

Problem. *Can one obtain an exotic space* $\mathbb{R}^4$ *starting from a vector field whose components are polynomials? trigonometric polynomials? analytic functions? elementary functions? Can one write explicitly such a vector field?*

[1] In the French version, the list of references contains some errors (in particular, the references [24–26] are missing) due to a typographic flaw. In the present translation, the correct references are restored by using the Russian version of the paper in: Vladimir Igorevich Arnold. Selected-60. Technical editor: M.B. Sevryuk. PHASIS, Moscow, 1997, pp. 533–551.

Translation of: *Topological Methods in Nonlinear Analysis* **4**(2), 209–225 (1994),

Translated by Nikolai Kruzhilin

2. Pseudoperiodic topology

Pseudoperiodic topology is motivated by problems in the quantum physics of solids and the geometry of quasicrystals, by investigations of closed 1-forms that are not exact, and by studies of slow mixing and "stochastic webs" in integrable and nearly integrable Hamiltonian systems.

DEFINITION. A *pseudoperiodic mapping* is the sum of a linear map and a periodic map.

EXAMPLE. The expression

$$f(x, y) = ax + by + \sin x + \cos y$$

defines a pseudoperiodic function in $\mathbb{R}^2$ (with the period lattice $2\pi\mathbb{Z}^2$).

DEFINITION. A *pseudoperiodic variety* is the inverse image of a point under a pseudoperiodic mapping.

EXAMPLE. The intersection of a periodic surface in $\mathbb{R}^3$ (for instance, a Fermi surface) and an irrational plane (orthogonal to the "magnetic field") is a pseudoperiodic curve studied by S.P. Novikov [23].

Consider a pseudoperiodic (but not periodic) curve in $\mathbb{R}^n$ (with the fixed period lattice $\mathbb{Z}^n$). *Assume that the rank of the linear part of the corresponding mapping is maximal* (equal to $n - 1$). Then it is obvious that the curve contains an infinite branch (lying at a finite distance from a certain straight line).

PROBLEM. *Is the noncompact component of such a pseudoperiodic curve always unique?*

REMARK 1. The conjecture of the uniqueness of an infinite branch has not been proven even for a generic linear part of the pseudoperiodic mapping (for instance, for a linear part satisfying the standard Diophantine inequalities which only fail for linear maps constituting a set of zero Lebesgue measure).

REMARK 2. The example of planar curves ($n = 2$) shows that there can be infinitely many *compact* components.

REMARK 3. The same example shows that in the *periodic* case there can be an arbitrary (odd) number of infinite branches.

REMARK 4. In the case $n = 2$ (or more generally, for pseudoperiodic *hypersurfaces* of dimension $n - 1$ in $\mathbb{R}^n$), there can be only one noncompact component [2].

REMARK 5. If the pseudoperiodic mapping $f : \mathbb{R}^n \to \mathbb{R}^{n-1}$ that defines our curve has no singular points (for instance, this occurs when the periodic perturbation is small compared with the main linear part), then each curve $f^{-1}(c)$ consists of (no more than) a single component. In this case the mapping f is "rectifiable": it can be reduced to its linear part by a diffeomorphism of the torus $\mathbb{R}^n/\mathbb{Z}^n$ (see [3]; for $n = 2$ this result is due to A.N. Kolmogorov [20]).

REMARK 6. Recently I.A. Dynnikov (a third-year student of the Moscow State University) proved the uniqueness conjecture for the infinite component in the case of planar pseudoperiodic curves in $\mathbb{R}^3$ (when one of the two pseudoperiodic functions defining the curve is linear).

EXAMPLE. The planar curve given by the equation

$$y = ax + F(x, y, bx)$$

(where F is a function with period 1 in each of its three arguments) contains a unique infinite branch if a is an irrational number.

REMARK 7. I.A. Dynnikov also proved the conjecture due to S.P. Novikov [23] on the intersections of Fermi surfaces with planes. Novikov's conjecture stated that *each connected component of such a curve lies in a finite neighborhood of a certain straight line*. This conjecture has been proven for generic planes (satisfying the standard Diophantine conditions). On the other hand, Tsarëv and Dynnikov constructed counterexamples to this conjecture for exceptional planes (which form a set of measure zero).

3. Numbers of periodic points and of limit cycles

DEFINITION. A point x is said to be *periodic of period n* for a mapping A if this point is a fixed point of the iteration of A:

$$A^n x = x.$$

Let $A : M \to M$ be an analytic diffeomorphism of a compact analytic manifold (for example, of the torus T^2).

PROBLEM. *Is the number of periodic points of period n majorized by an exponential function of n?*

Here we assume that the periodic points are nondegenerate (that is, 1 is not an eigenvalue of the derivative of A^n at x). Generic diffeomorphisms A do not possess degenerate periodic points.

REMARK 1. The example of the mapping of the torus T^2 given by the matrix $\left(\begin{smallmatrix} 2 & 1 \\ 1 & 1 \end{smallmatrix}\right)$ shows that the number of periodic points of period n can grow no more slowly than an exponential function of n (this also holds for generic mappings).

REMARK 2. M. Artin and B. Mazur proved that for algebraic mappings, the number of points of period n is majorized by $Ce^{\lambda n}$ [13].

PROBLEM. *Is the number of periodic orbits of a period no larger than T (in a compact ball in $\mathbb{R}^n$) of a polynomial vector field majorized by an exponential function of T?*

REMARK 3. By contrast, for C^∞ manifolds and mappings *no* majorization whatsoever is possible [17].

CONJECTURE. *The number of periodic points of a C^∞ mapping grows almost always no faster than an exponential function of the period.*

Here “almost always” means “for almost all values of the parameters (in the sense of Lebesgue measure) in any generic family of mappings depending on sufficiently many parameters”.

Remark 4. This definition of genericity, which allows one to neglect phenomena that only occur in exceptional cases (from the viewpoint of measure theory) in finite-parameter families, is *quite different* from the definitions used by experts in topology and probability theory.

Topologists, who use the concept of category, would neglect a Cantor set of measure equal to 99% of the total measure of the manifold in question.

Probabilists consider measures in function spaces that are concentrated on not very smooth functions (like a Brownian motion).

The above definition of “almost always” introduced by Kolmogorov [21] is well adapted to the theory of dynamical systems.

4. Asymptotic topological complexity of intersections

This problem, which generalizes Problem 3, is motivated by the theory of solitons (see [23]).

Consider two compact submanifolds X^k and Y^ℓ of a compact manifold M^m. Let $A : M \to M$ be a differentiable mapping. Consider the successive images of X under the iterations A^n of A. To measure their complexity (growing with time n) we consider their intersections with the fixed manifold Y:

$$Z(n) = (A^n X) \cap Y.$$

This intersection Z is generically a smooth manifold of dimension $s = k + \ell - m$.

The problem consists in exploring the asymptotic behavior of the topological complexity $|Z(n)|$ *of* $Z(n)$ *as a function of time n.*

Example 1. If X and Y have complementary dimensions, then $Z(n)$ is generically a finite set and $|Z(n)|$ is the number of its points.

Example 2. Let $g : \Omega \to \Omega$ be a generic dynamical system. Consider $M = \Omega \times \Omega$, $A = (\mathrm{Id}, g)$, $X = Y =$ the diagonal of M. In this case $Z(n)$ is the finite set of pairs (x, x), where x is a periodic point of g of period n. Thus Problem 3 is a special case of Problem 4.

Remark 1. If (A, M, X, Y) are real *algebraic* mapping / manifolds, then the topological complexity grows no faster than an exponential function of time:

$$|Z(n)| \le C e^{\lambda n}.$$

This was proven in [4, 5] for such measures of topological complexity as the Betti numbers, characteristic numbers, etc.

Now consider the case of C^∞ manifolds and mappings.

Problem. *Is the topological complexity of the intersection* $Z(n)$ *almost always majorized by an exponential function of time* n*?*

REMARK 2. When A is a diffeomorphism, this conjecture was proven in [5] for the following measures of topological complexity: the number of connected components, Betti numbers, Morse numbers, characteristic numbers, Riemannian volume, absolute total curvature, etc.

The proofs are based on the Borel–Cantelli lemma (the Chebyshev inequality) in probability theory, on the one hand, and on the majorization of the Betti numbers by the absolute total curvature due to Chern and Lashof [14], on the other hand.

REMARK 3. One can also majorize the number of generators and the number of relations of the fundamental group. However, I cannot prove any majorization for the lengths of the relations.

REMARK 4. "Almost always" in [5] means the following: for almost all values of the parameter t in generic families of the fixed manifolds Y_t, one has

$$|Z_t(n)| \leq C(t)e^{\lambda n},$$

provided that the number of parameters $p = \dim\{t\}$ is sufficiently large. The "tomographic" arguments in [5] use values of p of an order of m^2.

REMARK 5. The result in [5] probably remains valid almost always if A or X, rather than Y, depends on parameters.

REMARK 6. There exist C^∞ and *even analytic* examples where $|Z(n)|$ grows faster than any prescribed function (for some increasing sequence n_i of moments of time).

The example given by O.S. Kozlovskiĭ [22] (a first-year student of the Moscow State University) is particularly instructive. In this example, M is the torus $T^2 = \{(x, y) \bmod 2\pi\}$, $X = Y$ is the circle $y = 0$, and A is an analytic diffeomorphism of the form

$$(x, y) \mapsto (x + 2\pi\lambda,\ y + f(x)).$$

Choose an irrational number λ and a very lacunar series

$$f(x) = \sum_i f_{M_i} \sin M_i x.$$

The possibility to obtain an arbitrary growth rate of $|Z(n_i)|$ is ensured by the existence of irrational numbers λ for which the elements of the continued fractions increase arbitrarily fast.

5. Growth of the Milnor numbers in holomorphic dynamics

This is a local version of Problem 4. Consider two germs of holomorphic curves passing through the origin of the plane $\mathbb{C}^2$:

$$(X, 0) \hookrightarrow (\mathbb{C}^2, 0) \hookleftarrow (Y, 0)$$

and a germ of a holomorphic mapping $A : (\mathbb{C}^2, 0) \to (\mathbb{C}^2, 0)$ leaving the origin fixed.

One moves X by the iterations of A and studies the intersections of $A^n X$ with Y.

DEFINITION. The *Milnor number* $\mu(n)$ is the multiplicity of the intersection of the curves $A^n X$ and Y at the origin.

Example. $A(x, y) = (x, y^2)$, $X = \{(x, x)\}$, $Y = \{(x, 0)\}$. The images $A^n X$ are the "parabolas" $y = x^{2^n}$, and therefore $\mu(n) = 2^n$.

Return to the general case of arbitrary (A, X, Y).

Problem. *Are the Milnor numbers $\mu(n)$ majorized by an exponential function of time n?*

Here we assume that A is of finite multiplicity and that $A^n X$ and Y are distinct for each n.

Remark 1. Exponential majorization was proven in [6] for the case of smooth curves $A^n X$ (as in the example).

Remark 2. It seems that such exponential majorization also holds for holomorphic subvarieties $(X^k, 0)$ and $(Y^\ell, 0)$ of $(\mathbb{C}^m, 0)$ of arbitrary dimensions and for the generalized Milnor numbers defined by the following construction.

Consider an algebraic filtration $V_1 \supset V_2 \supset V_3 \supset \dots$ of the space $V_1 = J^\infty$ of the infinite jets of pairs of holomorphic mappings

$$f : (\mathbb{C}^k, 0) \to (\mathbb{C}^m, 0), \qquad g : (\mathbb{C}^\ell, 0) \to (\mathbb{C}^m, 0)$$

at the origin.

The varieties V_i are algebraic subvarieties of J^∞, in other words, each of these varieties is defined by polynomial equations (imposed on a finite number of the Taylor coefficients). However, this finite number depends on i.

Definition. The *generalized Milnor number* $\mu(f, g)$ is the greatest integer i for which the pair (f, g) belongs to V_i.

Example. The local versions of the topological (or, more generally, discrete) invariants of the intersections considered in Problem 4 are generalized Milnor numbers (associated with very special filtrations).

Conjecture. *The generalized Milnor numbers $\mu(n)$ of the pairs $(A^n X, Y)$ are majorized by an exponential function of n (provided that A is a mapping of finite multiplicity and all these Milnor numbers are finite).*

Remark 3. This conjecture does hold for diffeomorphisms A. In addition, in this case *the sequence of generalized Milnor numbers $\mu(n)$ is bounded by a constant.* The proof of this fact [6] is based on the following theorem due to Skolem.

Let $f(n) = a_1 f(n-1) + \dots + a_m f(n-m)$ be a recurrent sequence. Then the set of the values of n for which $f(n) = 0$ is the union of a finite set and a finite collection of arithmetic progressions.

This theorem in number theory, which was proven by a logician, belongs in fact to the theory of dynamical systems.

Remark 4. The results in [6] generalize a theorem due to M. Shub and D. Sullivan [25]: the values of the Poincaré index of iterations of a C^1 diffeomorphism at its fixed point are bounded (under the condition that all these indices are finite). In both the results, the sequences of the values (of the generalized Milnor number and of the index) are periodic starting from some moment.

The theorem of Shub and Sullivan was generalized by D. Fried [15] who considered the complex Atiyah–Bott indices in place of the Poincaré index.

6. Infinitesimal version of Hilbert's 16th problem and Abelian integrals

Hilbert's 16th problem has two parts: the question of the ovals of algebraic curves and the question of the limit cycles of polynomial vector fields.

In between these two problems, there is also an intermediate question: examining the limit cycles born from the level curves of a first integral.

This problem contains, as a special case, investigating the number of real zeros of the periods of Abelian integrals as a function of parameters. Suppose that a polynomial vector field in the plane admits a first integral whose level curves are cycles (which fill at least a certain annulus in the plane).

Consider small polynomial perturbations (of a fixed degree) of this vector field. The places where limit cycles are born are given in the first approximation by the zeros of a certain integral (discovered by Poincaré) along the unperturbed closed orbits (these are the level curves of the first integral).

PROBLEM. *Is the number of zeros of the Poincaré integral bounded (by a constant depending only on the degree of the perturbations)?*

EXAMPLE. The generalized Lotka–Volterra system

$$\dot{x} = x(a + bx + cy), \qquad \dot{y} = y(d + ex + fy)$$

admits sometimes a first integral

$$I = x^{\alpha} y^{\beta} z^{\gamma},$$

where z is an (inhomogeneous) linear function of x and y. The unperturbed orbits $I = \text{const}$ are algebraic or non-algebraic curves, depending on the coefficients $(a, \dots, f)$. The problem is to figure out *whether the number of unperturbed orbits that give rise to limit cycles is bounded by a function of the degrees of the perturbations.*

REMARK 1. For all values of the coefficients of the initial system, this number is probably bounded by the same constant.

However, it has not even been proven that the number of zeros of the Poincaré integral is bounded in the case of hypergeometric integrals (in the sense of Gelfand, Varchenko, Aomoto, and others) corresponding to individual generalized Lotka–Volterra systems (or even to non-generalized Lotka–Volterra systems where $b = f = 0$).

REMARK 2. In the special case where the unperturbed system is Hamiltonian with a polynomial Hamilton function H, the problem can be stated in terms of the periods of Abelian integrals. Consider the complete Abelian integral

$$I(h) = \oint (P\,\mathrm{d}x + Q\,\mathrm{d}y) \tag{1}$$

along an oval of the algebraic curve $H(x, y) = h$. The polynomials $P(x, y)$ and $Q(x, y)$ describe the infinitesimal variation of the Hamiltonian vector field, and I is the Poincaré integral.

The problem consists in bounding from above the number of real zeros of the function I for all the polynomials (P, Q) of a fixed degree.

This problem of a uniform bound for the numbers of real zeros of complete Abelian integrals has been solved by A.N. Varchenko [26] and A.G. Khovanskiĭ [19] by two distinct methods. However, none of these proofs presents a bound itself for the number of real roots of I.[2]

Remark 3. In the even more special case of the system with the cubic Hamilton function

$$H(x, y) = y^2 + x^3 - x, \tag{2}$$

the precise bound for the number of zeros of the (this time, elliptic) Poincaré integrals was found by G.S. Petrov, a first-year student of the Moscow State University [24].[3]

The family of functions (1) obtained for all the polynomials P and Q (of a bounded degree) forms a linear space of some finite dimension N. Consequently, there exist perturbations (P, Q) such that the Poincaré integral $I(h)$ has at least $N-1$ real zeros (which can moreover be placed at arbitrary points h_i).

Definition. An N-dimensional linear family of functions of one variable is called a *Chebyshev family* if each function (not identically equal to zero) in this family has less than N zeros.

If this is the case, the linear differential equation for which the functions of the family are solutions is called a *disconjugate equation*.

Theorem (Petrov). *The family of the Poincaré integrals $I(h)$ corresponding to the cubic Hamilton function* (2) *is a Chebyshev family.*

Here the perturbations (P, Q) that define the integral (1) are arbitrary polynomials of degree at most n. In this case the family of integrals (1) has dimension n. Thus the maximal number of real zeros of the Poincaré integrals (1) in the case (2) is equal to the perturbation degree minus 1.

Remark 4. Petrov's results show that the Picard–Fuchs equation (of the corresponding Gauss–Manin connection) is disconjugate. The symplectic Sturm theory (see [7]) suggests that this property of disconjugacy must be connected with a variational principle for this linear Picard–Fuchs equation, whose (quadratic) Lagrangian must be positive definite.

In this case theorems in symplectic geometry would yield the property of Lagrangian disconjugacy: the number of moments of non-transversality of a moving Lagrangian plane and a fixed plane would be minimal.

---

[2] In the Russian version of the paper (see the beginning of this text), this sentence is revised as follows: However, none of these proofs presents an explicit bound from above; what has been proven is just the *existence* of such a bound for the number of real roots of the function I. (Editors' remark)

[3] His real name is G.S. Petrov-Tan′kin. (Editors' remark)

For a hyperelliptic Hamilton function $H = y^2 + p(x)$, this program was implemented by A.B. Givental [16]. He established the property of Lagrangian disconjugacy of the corresponding Picard–Fuchs equation.

Unfortunately, Lagrangian disconjugacy does not imply directly any bound for the number of zeros of particular solutions of the linear Hamiltonian equation (it only provides a bound for the number of zeros of certain determinants built from several solutions).

7. Materialization of resonances in holomorphic dynamics

It has been well known since Poincaré's works on celestial mechanics that resonances (commensurabilities of frequencies) imply the divergence of the series describing perturbed motions (because of "small denominators", which vanish when an exact resonance occurs).

The divergence of the Taylor series of the function $\arctan x$ for $|x| > 1$ is explained by the presence of a complex singularity at the point i. This singularity, which presents a topological obstruction to convergence, materializes divergence which looks mysterious otherwise.

The idea of materialization of resonances consists in finding topological obstructions to the convergence of series of the perturbation theory in the behavior of orbits of the perturbed system in the complex phase space.

One can find many examples of the implementation of this program in [8–10]. The proofs of the divergence of series of the perturbation theory given by Poincaré and Siegel can be viewed as indications of the presence of periodic orbits (with quite long periods) in complex neighborhoods of the real phase space.

Here I state my old (1958) conjecture of this type, which apparently still remains open (at least formally).

Consider a holomorphic mapping of a neighborhood G of a circle S^1 (embedded in the complex plane $\mathbb{C}$) to another neighborhood of the same circle,

$$A : (G, S^1) \to (G', S^1).$$

Assume that A induces a diffeomorphism of the circle S^1 that is conjugate to a rotation R_λ through angle $2\pi\lambda$ by means of a conjugating diffeomorphism B holomorphic in a neighborhood of the circle:

$$A = BR_\lambda B^{-1}.$$

Assume that the "Poincaré rotation number" λ is irrational.

Remark 1. According to another conjecture of 1958 proven by M. Herman [18], a holomorphic diffeomorphism B exists if the rotation number λ of A satisfies the standard Diophantine conditions (which fail only for a set of λ that is of Lebesgue measure zero).

Example. A holomorphic function f in a neighborhood of the real axis that is 2π-periodic and real on this axis defines the diffeomorphism $x \mapsto x + f(x)$ of the circle $\{x \bmod 2\pi\}$ provided that $f'(x) > -1$ on the real axis.

Assume that the maximal annulus M (diffeomorphic to $S^1 \times \mathbb{R}$) in which the mapping A is conjugate to a rotation lies in the neighborhood G of the circle S^1 together with its boundary ∂M.

In this case, the conjecture of materialization of resonances can be formulated as follows.

PROBLEM. *Is it true that each neighborhood of any point of the boundary ∂M contains a periodic point of A?* [4] *Is this true at least generically?*

REMARK 2. The periodic points of A lie outside the neighborhood of the circle where A is conjugate to the irrational rotation R_λ. So these points prevent the convergence of series of the perturbation theory and materialize resonances.

REMARK 3. In the case where the conjugating diffeomorphism B is not holomorphic, one would expect the existence of periodic points in each neighborhood of any point of the circle S^1.[5]

REMARK 4. The conjectures on an approximation of the conjugation boundary by periodic points can be formulated for the local case of a fixed point of a holomorphic mapping $A : (\mathbb{C}, 0) \to (\mathbb{C}, 0)$ whose derivative at the origin is equal to $e^{2\pi i\lambda}$. One expects that, generically, there are periodic points in a neighborhood of the boundary ∂M of the maximal disc M in which A is holomorphically conjugate to a rotation.[6]

REMARK 5. For the exceptional irrational rotation numbers λ, J.-C. Yoccoz and R. Pérez-Marco have constructed polynomial[7] mappings A for which the periodic points do not approximate the boundary ∂M of the annulus or disc of holomorphy of the conjugating diffeomorphism.

REMARK 6. On the other hand, for generic rotation numbers λ, they have proven the approximation conjecture for polynomial and rational mappings.[8]

REMARK 7. Apparently, this problem remains open (at least formally) even for mappings defined by trigonometric polynomials, for example, for mappings of the form $x \mapsto x{+}a{+}b \sin x$, for which the approximation conjecture was originally stated in 1958.

REMARK 8. There is still much to be done in the general program of materialization of resonances. For instance, I refer to the recent works by M.B. Mishustin on normal forms of neighborhoods of holomorphic curves on holomorphic surfaces.

REMARK 9. Quite recently J. Moser studied perturbations of holomorphic curves lying on the surface of the torus $\mathbb{C}^2/\mathbb{Z}^4$ under a pseudoholomorphic perturbation of the holomorphic

[4] In the Russian version of the paper, this question is revised as follows: *Is it true that each neighborhood of any point of the boundary ∂M contains a point of a periodic orbit of the mapping A, this orbit lying in an arbitrarily small neighborhood of the boundary?* (Editors' remark)

[5] In the Russian version of the paper, this remark is revised as follows: In the case where the conjugating diffeomorphism B is not holomorphic, one would expect the existence of points of periodic orbits, that lie in an arbitrarily small neighborhood of the circle S^1, in each neighborhood of any point of this circle. (Editors' remark)

[6] In the Russian version of the paper, this sentence is revised as follows: One expects that, generically, there are points of periodic orbits, that lie in an arbitrarily small neighborhood of the boundary ∂M of the maximal disc M in which A is holomorphically conjugate to a rotation, near any point of the boundary. (Editors' remark)

[7] In the Russian version of the paper, the word "polynomial" is absent. (Editors' remark)

[8] In the Russian version of the paper, this remark is revised as follows: On the other hand, for generic rotation numbers λ, the approximation conjecture has not been verified even in the case of polynomial and rational mappings. (Editors' remark)

structure of the torus. There is no doubt that for this problem, one can develop a theory of materialization of resonances parallel to the theory of bifurcations of elliptic curves on holomorphic surfaces constructed in [9].

8. Analytic insolvability and geometric insolvability in the chaos theory

It seems that chaotic behavior of dynamical systems makes quite precise questions on their structure undecidable.

Here I present several old conjectures (see [11, 12]) on undecidability of classical problems in the theory of dynamical systems, such as the problems of stability of equilibrium points, integrability in quadratures of ordinary differential equations, and complete integrability of Hamiltonian systems in mechanics. My definitions of insolvability are very different from the ones used in algebra and in the theory of algorithms developed by logicians.

In the spirit of the geometry of manifolds, the solvability of a problem should not depend on the coordinates used to introduce the data of the problem (the right-hand sides of differential equations, the Hamilton functions, etc.).

Definition 1. A problem is said to be *geometrically unsolvable* if the class of problems obtained from this problem by diffeomorphic changes of the coordinates and parameters contains no problem analytically solvable (in the sense of Definition 2 below).

Example. Consider a vector field depending on parameters in a neighborhood of an equilibrium point. The problem of stability of this point for this family is geometrically unsolvable if this problem is analytically unsolvable for each family equivalent to the given one (in the sense of local diffeomorphisms depending on parameters).

Remark 1. The problem of integration of a differential equation is geometrically solvable in a neighborhood of a nonsingular point of the corresponding vector field, although this problem may be analytically unsolvable in a neighborhood of a singular point, even for polynomial fields.[9]

The definition of analytic solvability of local analytic problems which we mentioned above (and which we will describe below) is very loose. For example, all the Diophantine equations are solvable in this sense. Thus analytic insolvability in this sense is a much stronger property than algorithmic insolvability.

As the data of a local analytic problem, I consider a finite number of Taylor series *in a fixed coordinate frame*. In the stability problem for an equilibrium point, these are the series of the components of the vector field at this point. The space of their Taylor polynomials of a given degree is then endowed with distinguished coordinates. The definition of analytic solvability describes the allowed operations yielding a solution.

We start by defining functions, varieties, and mappings *admissible* in the arithmetic spaces $\mathbb{R}^N$ and $\mathbb{C}^N$. These are the minimal sets of varieties and "mappings" possessing the following properties:

---

[9] In the Russian version of the paper, the second part of this remark is revised as follows: although this problem may be analytically unsolvable for some fields. On the other hand, this problem is probably geometrically unsolvable in a neighborhood of a singular point even for polynomial vector fields. (Editors' remark)

1. the arithmetic spaces $\mathbb{R}^N$ and $\mathbb{C}^N$ (for any N) are admissible manifolds;
2. any rational mapping is admissible;
3. the image and the inverse image of an admissible variety under an admissible mapping are admissible;
4. the intersection and the union of a pair of admissible varieties are admissible, and so are the complements and the direct product;
5. the composition of two admissible mappings is admissible;
6. a mapping (which can even be multi-valued) is admissible if and only if its graph is an admissible variety;
7. if $f(x, y)$ is an admissible function, then its derivative with respect to x and its primitive function (whose branch is fixed by the initial values on an admissible subvariety) are admissible.

REMARK 2. The primitive function in question is given by complex integration. Hence the graph of this multi-valued function is an admissible variety.

Thus, analytic continuation is an admissible operation in this theory, and so are all the arithmetic operations, differentiation, integration, substitutions, and taking the inverse mapping.

REMARK 3. One can develop the concept of analytic solvability based on real integration. This theory is different from the one discussed here. For example, the function sin on the real axis is admissible if complex integration is allowed, but it does not belong to the minimal set of admissible functions if one only allows real integration (although the restriction of this function to any finite interval of the real axis is always an admissible function).

If the function sin is admissible, so is the set of integers. Consequently, the set of integer values of the parameter y for which a polynomial $f(x, y)$ has an integer root x is admissible, whatever the given polynomial f with integer coefficients.

For simplicity consider a *binary* local analytic problem, that is, a problem with only two possible solutions, "yes" or "no" (for instance, the stability problem for an equilibrium point). For each value of k, the space of Taylor polynomials of degree k of the data for a binary local problem falls into three parts:

A_k: The Taylor polynomials for which the answer is "yes" (independently of the terms of higher degrees).

B_k: The Taylor polynomials for which the answer is "no" (independently of the terms of higher degrees).

C_k: The Taylor polynomials for which the answer depends on the terms of higher degrees.

DEFINITION 2 (see [11]). A binary local analytic problem is *analytically solvable* if

(i) the sets A_k, B_k, C_k are admissible varieties (for each k),

(ii) the codimension of the variety C_k tends to infinity with the degree k of the Taylor polynomials.

PROBLEM. *Is the stability problem for an equilibrium point of a vector field in $\mathbb{R}^n$ geometrically solvable?*

Remark 4. There is an old conjecture (see for instance [12]) that this problem is probably neither analytically nor even geometrically solvable (at least for $n \geq 3$).

I think that the sets A_k, B_k, C_k are not admissible even for very small values of k, maybe already for $k = 2$ (for $k = 1$ these sets are admissible for any n according to Lyapunov's classical theorem).

Remark 5. I think that also the problems of integrability in quadratures and complete integrability of Hamiltonian systems are neither analytically nor even geometrically solvable (in some sense close to that described above for binary local problems).

Remark 6. It is even possible that there exist completely integrable systems whose integrability is nevertheless *not decidable by means of analytic operations* applied to the data of the problem (namely, to the expression of the Hamilton function in the standard coordinates (p, q) in the phase space). I do not see any reason why this cannot occur, for instance, for polynomial Hamilton functions.

Remark 7. I hope that the proof of insolvability (in the sense described above) of such problems as the stability problem is not possible without a real progress in understanding the behavior of dynamical systems: this is not a problem in logics.

Following the Russian tradition, the above problems are formulated in the least general form possible. It is easy to generalize them. However, I believe that it is concrete problems that remain the main — and maybe the only — source of inspiration for mathematicians and of all general theories.

References

1. V.I. Arnold. Problems on singularities and dynamical systems. In: *Developments in Mathematics: the Moscow School* (Eds. V.I. Arnold and M.I. Monastyrsky). Chapman & Hall, London, 1993, 251–274.
2. V.I. Arnold. Topological and ergodic properties of closed 1-forms with incommensurable periods. *Funktsional. Anal. i Prilozhen.* **25** (1991), no. 2, 1–12 (in Russian); the English translation: *Funct. Anal. Appl.* **25** (1991), no. 2, 81–90.
3. V.I. Arnold. Polyintegrable flows. *Algebra i Analiz* **4** (1992), no. 6, 54–62 (in Russian); the English translation: *St. Petersburg Math. J.* **4** (1993), no. 6, 1103–1110.
4. V.I. Arnold. Dynamics of intersections. In: *Analysis, et cetera. Research papers published in honor of Jürgen Moser's 60th birthday* (Eds. P.H. Rabinowitz and E. Zehnder). Academic Press, Boston, MA, 1990, 77–84.
5. V.I. Arnold. Dynamics of complexity of intersections. *Bol. Soc. Brasil. Mat.* (*N. S.*) **21** (1990), no. 1, 1–10.
6. V.I. Arnold. Bounds for Milnor numbers of intersections in holomorphic dynamical systems. In: *Topological Methods in Modern Mathematics. Proceedings of the symposium in honor of John Milnor's sixtieth birthday* (*Stony Brook, NY, 1991*) (Eds. L.R. Goldberg and A.V. Phillips). Publish or Perish, Houston, TX, 1993, 379–390.
7. V.I. Arnold. Sturm theorems and symplectic geometry. *Funktsional. Anal. i Prilozhen.* **19** (1985), no. 4, 1–10 (in Russian); the English translation: *Funct. Anal. Appl.* **19** (1985), no. 4, 251–259.
8. V.I. Arnold. Remarks on singularities of finite codimension in complex dynamical systems. *Funktsional. Anal. i Prilozhen.* **3** (1969), no. 1, 1–6 (in Russian); the English translation: *Funct. Anal. Appl.* **3** (1969), no. 1, 1–5.

9. V.I. Arnold. Bifurcations of invariant manifolds of differential equations and normal forms of neighborhoods of elliptic curves. *Funktsional. Anal. i Prilozhen.* **10** (1976), no. 4, 1–12 (in Russian); the English translation: *Funct. Anal. Appl.* **10** (1976), no. 4, 249–259.
10. V.I. Arnold. *Supplementary Chapters to the Theory of Ordinary Differential Equations.* Nauka, Moscow, 1978 (in Russian). The English translation: *Geometrical Methods in the Theory of Ordinary Differential Equations.* Springer, New York–Berlin, 1983.
11. V.I. Arnold. On local problems of analysis. *Vestnik Moskov. Univ. Ser. I Matem. Mekh.* 1970, no. 2, 52–56 (in Russian); the English translation: *Moscow Univ. Math. Bull.* **25** (1970), no. 1–2, 77–80. See also: V.I. Arnold. *Collected Works. Volume II. Hydrodynamics, Bifurcation Theory, and Algebraic Geometry (1965–1972).* Springer, Berlin, 2014, 193–196.
12. V.I. Arnold. Dynamical systems and differential equations. In: *Mathematical Developments Arising from Hilbert Problems* (Ed. F.E. Browder) (Proc. Symp. Pure Math., vol. 28). Amer. Math. Soc., Providence, RI, 1976, 59–60.
13. M. Artin, B. Mazur. On periodic points. *Ann. of Math. (2)* **81** (1965), no. 1, 82–99.
14. Sh.-sh. Chern, R.K. Lashof. On the total curvature of immersed manifolds. *Amer. J. Math.* **79** (1957), no. 2, 306–318; On the total curvature of immersed manifolds. II. *Michigan Math. J.* **5** (1958), no. 1, 5–12.
15. D. Fried. Periodic points of holomorphic maps. *Topology* **25** (1986), no. 4, 429–441.
16. A.B. Givental. Sturm's theorem for hyperelliptic integrals. *Algebra i Analiz* **1** (1989), no. 5, 95–102 (in Russian); the English translation: *Leningrad Math. J.* **1** (1990), no. 5, 1157–1163.
17. E. Rosales-González. On the growth of the numbers of periodic orbits of dynamical systems. *Funktsional. Anal. i Prilozhen.* **25** (1991), no. 4, 14–22 (in Russian); the English translation: *Funct. Anal. Appl.* **25** (1991), no. 4, 254–262; On the growth of the number of long-period solutions of differential equations. *Funktsional. Anal. i Prilozhen.* **26** (1992), no. 2, 29–35 (in Russian); the English translation: *Funct. Anal. Appl.* **26** (1992), no. 2, 99–105.
18. M.-R. Herman. Sur la conjugaison différentiable des difféomorphismes du cercle à des rotations. *Inst. Hautes Études Sci. Publ. Math.* No. 49 (1979), 5–233.
19. A.G. Khovanskiĭ. Real analytic manifolds with the property of finiteness and complex Abelian integrals. *Funktsional. Anal. i Prilozhen.* **18** (1984), no. 2, 40–50 (in Russian); the English translation: *Funct. Anal. Appl.* **18** (1984), no. 2, 119–127.
20. A.N. Kolmogorov. On dynamical systems with an integral invariant on a torus. *Doklady Akad. Nauk SSSR (N. S.)* **93** (1953), no. 5, 763–766 (in Russian); the English translation in: *Selected Works of A.N. Kolmogorov. Vol. I. Mathematics and Mechanics* (Ed. V.M. Tikhomirov) (Mathematics and Its Applications (Soviet Series), vol. 25). Kluwer Acad. Publishers, Dordrecht, 1991, 344–348.
21. A.N. Kolmogorov. The general theory of dynamical systems and classical mechanics. In: *Proceedings of International Congress of Mathematicians (Amsterdam, 1954)*, Vol. 1. North-Holland, Amsterdam, 1957, 315–333 (in Russian); the English translation in: *Selected Works of A.N. Kolmogorov. Vol. I. Mathematics and Mechanics* (Ed. V.M. Tikhomirov) (Mathematics and Its Applications (Soviet Series), vol. 25). Kluwer Acad. Publishers, Dordrecht, 1991, 355–374.
22. O.S. Kozlovskiĭ. The dynamics of intersections of analytic manifolds. *Dokl. Akad. Nauk* **323** (1992), no. 5, 823–825 (in Russian); the English translation: *Russ. Acad. Sci. Dokl. Math.* **45** (1992), no. 2, 425–427.
23. S.P. Novikov. Quasiperiodic structures in topology. In: *Topological Methods in Modern Mathematics. Proceedings of the symposium in honor of John Milnor's sixtieth birthday (Stony Brook, NY, 1991)* (Eds. L.R. Goldberg and A.V. Phillips). Publish or Perish, Houston, TX, 1993, 223–233.
24. G.S. Petrov. On the number of zeros of complete elliptic integrals. *Funktsional. Anal. i Prilozhen.* **18** (1984), no. 2, 73–74 (in Russian); the English translation: *Funct. Anal. Appl.* **18** (1984), no. 2, 148–149; Elliptic integrals and their non-oscillation. *Funktsional. Anal. i Prilozhen.* **20** (1986), no. 1, 46–49 (in Russian); the English translation: *Funct. Anal. Appl.* **20** (1986), no. 1, 37–40; The Chebyshev property of elliptic integrals. *Funktsional. Anal. i Prilozhen.* **22** (1988), no. 1, 83–84 (in Russian); the English translation: *Funct. Anal. Appl.* **22** (1988), no. 1, 72–73; The problem of the number of zeros of an elliptic integral is a semi-algebraic problem. *Mat. Zametki* **44** (1988), no. 3, 393–401 (in Russian); the English translation: *Math. Notes* **44** (1988), no. 3–4, 699–703; Complex zeroes of an elliptic integral. *Funktsional. Anal. i Prilozhen.* **23** (1989), no. 2, 88–89 (in Russian); the English translation: *Funct. Anal. Appl.* **23**

(1989), no. 2, 160–161; Non-oscillation of elliptic integrals. *Funktsional. Anal. i Prilozhen.* **24** (1990), no. 3, 45–50 (in Russian); the English translation: *Funct. Anal. Appl.* **24** (1990), no. 3, 205–210.

25. M. Shub, D. Sullivan. A remark on the Lefschetz fixed point formula for differentiable maps. *Topology* **13** (1974), no. 2, 189–191.
26. A.N. Varchenko. Estimation of the number of zeros of an Abelian integral depending on a parameter, and limit cycles. *Funktsional. Anal. i Prilozhen.* **18** (1984), no. 2, 14–25 (in Russian); the English translation: *Funct. Anal. Appl.* **18** (1984), no. 2, 98–108.

Mathematical problems in classical physics

V.I. Arnold

Mathematics is the name for those domains of theoretical physics that are temporarily unfashionable. The advantage of being unfashionable is that it presents the possibility of the rigorous and deep investigation of well-established mathematical models. A sleeping "physical theory" can be formulated as a chain of statements having exact mathematical meaning of mathematical conjectures. Such conjectures can then be proved or disproved. In many cases the mathematical problems arising this way are very difficult, and progress is rather slow. It is much easier to obtain a new result in an unexplored domain. Hence, most researchers carefully avoid any thinking on the old classical problems.

On the other hand, most of the new developments in physics are due to the exploitation by physicists of the theories developed by the mathematicians in previously unfashionable domains. Thus, it is useful to compile from time to time the lists of sleeping problems in unfashionable domains—just to know that the problems are still open.

The inclusion of a problem in the list that follows does not reflect its objective importance; rather, the choice is based on my personal taste.

1.1 Differential Invariants and Functional Moduli

Consider any local classification problem: We classify some objects (for example, functions, fields, varieties, or mappings), and we call two objects equivalent if one can be reduced to the other by a clever change of variables.

Example 1. The classification of the Riemannian (or Einstein) metrics f at a neighborhood of a point of the space up to the local diffeomorphism of the space preserving the point.

Example 2. The classification of the vector fields f on a manifold at a neighborhood of a zero point of the field up to the local diffeomorphism of the manifold preserving the point.

Example 3. The classification of the smooth mappings $f : M^m \longrightarrow N^n$

Originally published in: *Trends and Perspectives in Applied Mathematics*.
Editor: L. Sirovich (Appl. Math. Sci. 100). Springer, New York, 1–20 (1994),
Reprint by Springer Nature Switzerland AG 2022

at a neighborhood of a point x of M up to the local diffeomorphisms of M and of N preserving x and $f(x)$.

Example 4. The classification of the Hamilton vector fields f at a neighborhood of a zero point of the Hamilton field up to the local symplectomorphisms preserving this point.

Example 5. The local classification of differential equations $y'' = f(x, y, y')$.

Example 6. The classification of the germs of hyperkählerian structures on a manifold of dimension $4n$ up to the local diffeomorphisms.

The classical normal forms, to which one can reduce these objects by the action of the corresponding infinite-dimensional group, contain arbitrary functions. In most cases these functions depend on fewer variables than the initial objects.

However, these descriptions seem to depend on the special choice of the algorithms of the reduction. Is there any intrinsic, coordinate-free meaning in the assertion that the answer "depends on m arbitrary functions of n variables"? The arbitrary functions, intrinsically associated to the objects of our classification, are called the *functional moduli.*

The problem might be formulated formally as follows [1]. We start with some functional space of objects that we wish to classify (or of their Taylor series at a point). A group of diffeomorphisms acts on this space.

We fix an integer k and we consider the k-jets (the Taylor polynomials of degree k) of our objects. They form a finite-dimensional manifold. The action of the group of diffeomorphisms defines the action of the corresponding finite-dimensional Lie group on the finite-dimensional space of jets.

Consider the dimension of the orbits space at the k-jet of a given object f. We call this dimension of the moduli space of the k-jet of f *the moduli number of the k-jet.* We denote it by $m(k)$. To describe all these moduli numbers together we form *the Poincaré series of moduli numbers*

$$M(t) = \sum_{k=0}^{\infty} m(k)t^k .$$

If the objects depend regularly on q arbitrary functions of p variables as of moduli, the Poincaré series is equal to a rational function, namely, $M(t) = q/(1-t)^p$.

Problem. *Is it true that the Poincaré series of moduli numbers in most of the local problems of analysis are rational functions?*

For instance, is this the case in Examples 1–6 for almost any f (any f that does not belong to some set of codimension infinity in the space of Taylor series)?

In a sense, the situation here is similar to that of the Hilbert theorem on the finiteness of the bases of the ideals of analytic function germs. The difference with this classical theory is that our problem belongs in differential algebra rather than commutative algebra.

The finiteness theorem in differential algebra has been at least formulated by Tresse, a student of Lie [2]. He claims that all the differential invariants are generated by a finite number of invariant functions and of invariant vector fields (as differential polynomials). However, it is not clear what the exact assumptions are for the validity of this theorem. Its proof, as far as I know, has not yet been absorbed by modern mathematics and has not been read with the modern criteria of mathematical rigor. The questions above on the Poincaré series provide some test for the maturity of the theory of differential invariants.

1.2 Logarithmical Asymptotics and Wave Fronts

The asymptotical expansions of many functions in mathematical physics contain logarithms of the coordinates as well as the ordinary monomials: The terms of such expansions have the form

$$(x_1^{n_1}\log^{m_1} x_1)(x_2^{n_2}\log^{m_2} x_2)\cdots(x_\kappa^{n_\kappa}\log^{m_\kappa} x_\kappa),$$

where the coordinates x_i are small.

For smooth or analytical functions and mappings there is a theory of the sufficient jets in the homogeneous and quasi-homogeneous gradings (see, e.g., [3]). This theory permits neglecting the "higher-order terms" of the Taylor expansion at the origin when the topological (or even analytical) structure of the object (function, variety, mapping) is defined by the terms of the smallest order. For instance, the behavior of a function at a neighborhood of a critical point is defined by the second-order terms, provided that they form a nondegenerate quadratic form ("the Morse lemma").

Is it possible to construct a similar theory for the expansions that contain the logarithmical terms?

Of course, the logarithms do not change the respective strengths of two ordinary monomials if their strenghts are different (say, $x^2\log^3 x$ is weaker than x). However, $x_1\log x_1$ is stronger than x_2 (in the homogeneous grading).

The sufficient jets theory should give the conditions on the systems of monomials, including the logarithms under which the addition of the higher-order terms will not affect the topological or combinatorial structure of the object.

Example. x^2+ (higher-order terms) has a minimum at the origin, even if the higher-order terms contain the logarithms.

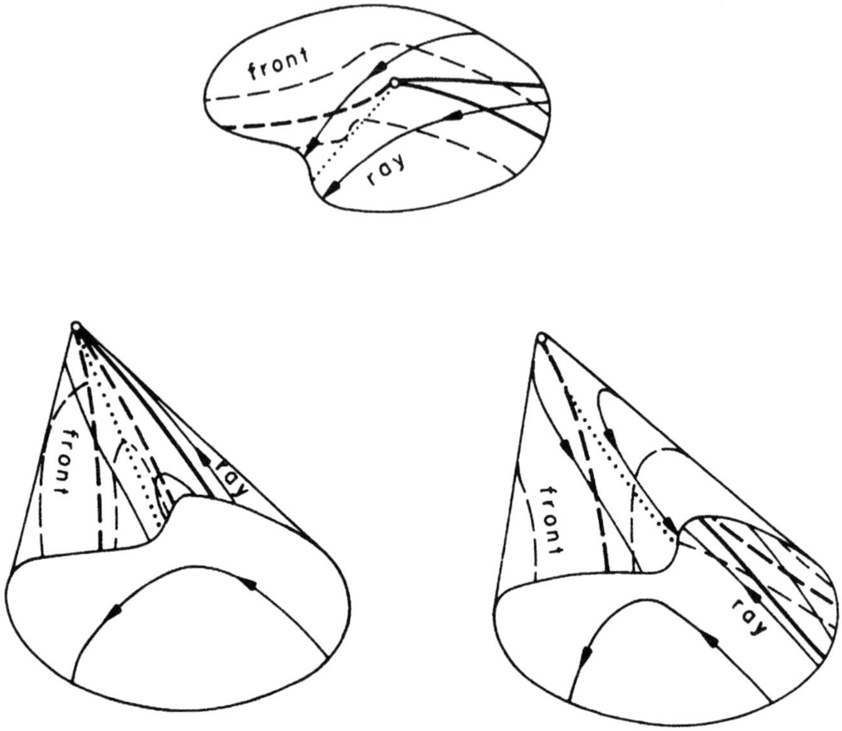

FIGURE 1. Wave front bifurcations at a double eigenvalue.

As an example where such a theory might provide meaningful information on the physical phenomena, I shall mention the ***transformation of the waves, governed by variational principles*** in inhomogeneous nonstationary media.

At the level of the geometrical optics, the evolution of the wave fronts and the generic singularities of the ray systems are described by the series containing the logarithms. These series are described in [4], where one can also find descriptions of the shapes of the singularities of wave fronts and their generic bifurcations (Fig. 1). In fact, Fig. 1 represents the generic (non-Legendrian) projection of the corresponding Legendrian surface from the five-dimensional contact manifold to the three-dimensional space, while the wave front bifurcations are described by the Legendrian projection at this surface to the three-dimensional space-time. These figures are based on the study of the "main" parts of the expansions, containing the logarithmical terms.

It is not proved that the truncation of the expansions will not change the patterns of the singularities and bifurcations. Although this seems to be rather inprobable, the absence of mathematical tools in dealing with this test problem shows how weak our understanding is of the topological structure of functions with more than one variable whose asymptotical expansions contain logarithms.

1.3 Hydrodynamical Attractors

Observations suggest that when the Reynolds number becomes large the motion of the fluid becomes chaotic and turbulent. In spite of many mathematical works, there are no theorems to explain these phenomena. I shall discuss some old conjectures, discussed in the 1950s (see [5]) and maybe earlier but still open.

Consider the Navier–Stokes system of equations of motion for a viscous uncompressible fluid. The equation for the vorticity field $\mathbf{r}$ is

$$\partial \mathbf{r}/\partial t + \{\mathbf{v}, \mathbf{r}\} = \nu(\Delta \mathbf{r} + \mathbf{R}),$$

$$\text{curl } \mathbf{v} = \mathbf{r}, \quad \text{div } \mathbf{v} = 0,$$

where $\{\cdot,\cdot\} = \text{curl}(\cdot \wedge \cdot)$ is the Poisson bracket, Δ the Laplacian, ν the viscosity (a small parameter), and $\nu\mathbf{R}$ the curl of the exterior force. Suppose for simplicity that the periodic boundary conditions (the domain of the motion being the 3-torus $T^3 = \mathbb{R}^3/\mathbb{Z}^3$) are imposed and that the field $\mathbf{R}$ have the same period.

Conjecture 1. *There exists a minimal attractor whose dimension grows indefinitely when the viscosity vanishes* ($\dim \longrightarrow \infty$ *for* $\nu \longrightarrow 0$).

Conjecture 2. *The minimum of the dimensions of the attractors of the Navier–Stokes system grows indefinitely when the viscosity vanishes.*

Remark 1. The exterior force field is chosen proportional to the viscosity ν to make the scale of the attractors independent on the Reynolds numbers.

Remark 2. A.N. Kolmogorov (1958) suggested that the assertion of Conjecture 2 might be wrong, but he had no doubts about Conjecture 1, even for the flows on the 2D torus with the exterior field pumping the first harmonic (say, for $\mathbf{R} = \sin y$).

Remark 3. The Lorenz-type attractors correspond to similar but different models. Both the Kolmogorov and Lorenz systems can be written in the form of (infinite) systems of ordinary differential equations with quadratic functions on the right-hand side:

$$X_i^{\cdot} = \Sigma a_{jk}^i X_j X_k - b_i X_i + c_i \ .$$

The first (quadratic) term corresponds to the inertial term $\{\mathbf{v}, \mathbf{r}\}$ in the Navier–Stokes system. It has the energy integral, which we can write as $E = \Sigma X_i^2/2$ (choosing the scales of the variables X_i appropriately),

$$E^{\cdot} = 0 \quad \text{if} \quad b_i = c_i = 0 \ .$$

The energy balance is due to the dissipation provided by the viscosity (the linear term), and to the pumping of energy provided by the exterior force (the constant term):

$$E^{\cdot} = -\Sigma b_i X_i^2 + \Sigma X_i c_i \ .$$

Thus, the energy is dissipated outside some ellipsoid of the phase space and is growing inside it.

For the Lorenz model the pumping region of the phase space is the interior of a quadratic cone. It is an infinite domain. This higher instability of the motion in the Lorenz-type models makes it possible to prove numerically (and even rigorously) the existence of higher-dimensional attractors in such models.

For the initial Navier–Stokes system (even in dimension 2) there still are no such rigorous results.

1.4 Fast Dynamo and Stochastization Problems

Consider the linear equation of ideal magnetohydrodynamics, where the velocity field $\mathbf{v}$ of the incompressible fluid is known and is stationary (time independent). The unknown magnetic field $\mathbf{B}$ is stretched by the movement of fluid particles, whereas a slow magnetic diffusion dissipates the magnetic energy:

$$\frac{\partial \mathbf{B}}{\partial t} + \{\mathbf{v}, \mathbf{B}\} = \mu \Delta \mathbf{B}, \quad \operatorname{div} \mathbf{B} = 0,$$

where the magnetic viscosity μ is a small parameter and $\{\ ,\ \}$ is the Poisson bracket of vector fields.

For simplicity, we suppose that the boundary conditions are periodic (the "star" or "planet" whose magnetic field growth is modeled is replaced by the 3D torus $\mathbb{R}^3/2\pi\mathbb{Z}^3$). The following question has been formulated by Ja.B. Zeldovich.

Does there exist a divergence-free velocity field $\mathbf{v}$ *such that the magnetic field* $\mathbf{B}$ *is exponentially growing in time for some initial field* $\mathbf{B}_0$?

Such a field $\mathbf{v}$ is called a fast kinematical dynamo, provided that the increment of the magnetic field is minorated by a positive constant, when

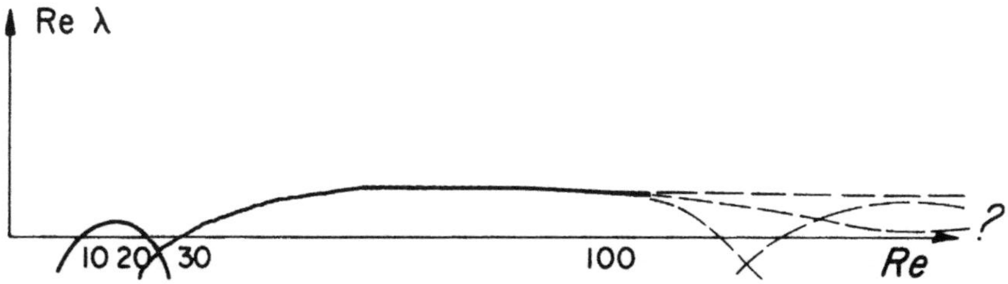

FIGURE 2. Increment of the fast dynamo problem as a function of the magnetic Reynolds number.

the magnetic Reynolds number $1/\mu$ grows to infinity:

$$\mathbf{B} = e^{i\lambda t}\mathbf{B}_0, \quad \operatorname{Re}\lambda(\mu) > C > 0, \quad \mu \longrightarrow 0\ .$$

Does there exist any fast kinematical dynamo divergence-free vector field $\mathbf{v}$?

The increment of the magnetic field is due to the stretching of the fluid particles by the flow. Hence, it is natural to suppose that the fast dynamo vector field $\mathbf{v}$ should have a positive Liapunov exponent on a set of initial points with positive measure.

This old conjecture is at present neither proved nor disproved. M.M. Vishik has proven that a vector field $\mathbf{v}$ cannot be a fast dynamo unless it has a point where the Liapunov exponent is positive. Since most velocity fields have a positive Liapunov exponent point, this theorem excludes practically no candidate divergence-free field; they might be fast kinematical dynamo fields.

The numerical experiments with the so-called *ABC* flows (which are the eigenvectors of the curl operator on T^3) show a dynamo effect (with an increment of the same order of magnitude as the empirical Liapunov number) for the magnetic Reynolds numbers $1/\mu$ in the intervals between 10 and 20 [6] and between 30 and at least 100 [7] (Fig. 2). There is a gap between 20 and 30. It is not clear whether at 100 the asymptote has been attained. One may speculate that there will be an infinite sequence of gaps for $1/\mu \to \infty$ or an asymptotic decline of the increment $\operatorname{Re}\lambda(\mu)$ to zero, or $\liminf \operatorname{Re}\lambda(\mu) > C > 0$.

Numerically this is a first eigenvalue problem for matrices of the order of many millions, even for reasonable Reynolds number (hundreds). The physically meaningful magnetic Reynolds numbers are of the order of magnitude 10^8. The corresponding matrices are (and will remain) too large for any computer.

The eigenfunctions, corresponding to the growing mode, show some peculiar concentration in the neighborhoods of the invariant manifolds of the stagnation points of the flow. This observation suggests the possibility of a short wave "quasi-classical" asymptotic analysis of the concentration phenomena.

There is a simplified model of the dynamo problem where the "diffusion" term is neglected and the "physical" dynamo mechanism has been suggested by Ya. B. Zeldovich [8] and A.D. Sacharov.

Consider a three-dimensional ball ("a star") and a divergence-free vector field in it ("a magnetic field").

A diffeomorphism of the ball to itself, preserving the volume elements, models the stationary motion of the conductive fluid inside the star.

Does there exist such a diffeomorphism whose iterations, applied to the initial vector field, will make field energy grow exponentially with the number of iterations?

If $\mathbf{B}_0$ is the initial magnetic field and g the diffeomorphism, it is required that for all n

$$\iiint (g_*^n \mathbf{B}_0)^2 d^3x \geq Ce^{\lambda n}, \quad \lambda > 0 .$$

The solution, proposed by Zeldovich and Sacharov, is the following construction. We start with a solid torus $S^1 \times D^2$ embedded in the ball. We take it out and stretch S^1 twice, contracting D^2 in such a way that the volume element is preserved. Then we fold the new solid torus twice (using the two-fold covering $S^1 \longrightarrow S^1$; Fig. 3) and we put the resulting solid torus in its place.

The energy of the longitudial field in the solid torus (directed along the S^1 component) grows exponentially under the iterations of the above construction since the field is stretched twice by the longitudial stretching of the solid torus.

Unfortunately, this construction is not a diffeomorphism of the ball. Making it smooth, we lose control over the stretching at a finite portion of the solid torus. These small portions make the proof of the exponential growth of the energy impossible. In any case we have here the same difficulty that is well known in the theory of the stochastization of the Hamiltonian dynamical systems at the homoclinic points; although the existence of the Bernoulli-type chaos has been known from the time of the classical works of Alexeev [9], *it is still unknown whether the corresponding invariant set of the phase space has positive or zero measure.*

Remark. The following idea (from M. Herman) might solve the problem. Consider the elliptic curve as a double covering of a sphere ramified at four points. The Anosov automorphism of the torus (say, $(x, y) \to (2x+y, x+y)$) iterated six times induces on the sphere a diffeomorphism with positive Liapunov exponents. Blowing up the four fixed ramification points, one

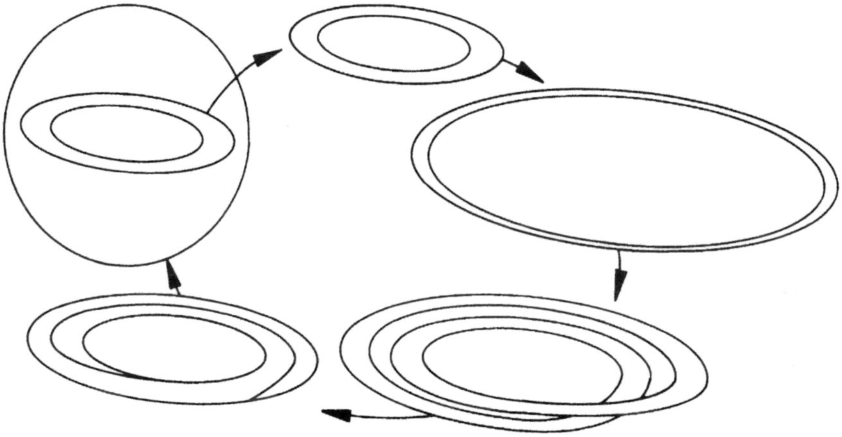

FIGURE 3. The Zeldovich–Sacharov topological dynamo mechanism.

obtains a diffeomorphism of a disc with three holes with almost everywhere positive Liapunov exponents. It may be performed inside the ball.

1.5 Minimal Magnetic Field

Consider a divergence-free vector field on a Riemannian 3-manifold (a magnetic field inside a star). Volume-preserving diffeomorphisms (motions of the perfectly conductive incompressible matter inside the star) transport the field (the field is "frozen").

Given an initial divergence-free vector field, we seek the diffeomorphism that minimizes the energy of the field. From the physical point of view one supposes that the Lorentz force produced by the field will move the matter, unless the magnetic field energy attains the minimal possible value. Part of the magnetic energy will thus be transformed to the kinectic energy of the motion. Then it will be dissipated by the hydrodynamical viscosity. Finally, the movement will stop when the energy of the magnetic field attains the minimum.

From the point of view of a mathematician, we search for the infimum rather than for a minimum, since the minimum over the group SDiff of volume-preserving diffeomorphisms might not be attained.

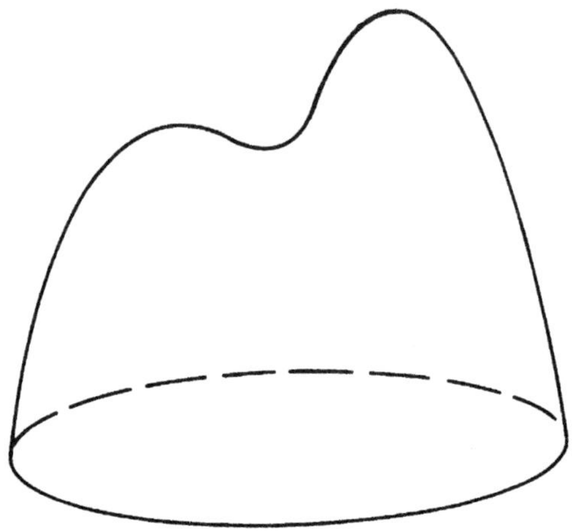

FIGURE 4. The graph of the function whose Dirichlet integral should be minimized.

If the minimum is not attained, the sequence of minimizing diffeomorphisms may converge to a nonsmooth and even discontinuous measure-preserving mapping, and the corresponding field may also become very bad.

It is not clear, however, how patalogical they really are even in the following simple two-dimensional model [10].

Consider a smooth function u_0 on the disk $x^2 + y^2 \leq 1$.

Problem. *Find the infimum of the Dirichlet integral*

$$I[u] = \int\int\left(\left(\frac{\partial u}{\partial x}\right)^2 + \left(\frac{\partial u}{\partial y}\right)^2\right)dx\,dy$$

over the set of all smooth functions u, which can be obtained from u_0 by the area-preserving diffeomorphism of the disk to itself.

Suppose, for simplicity, that u_0 vanishes at the boundary and has exactly one maximum inside the disk. Then the minimum is obtained on the function, which depends only on the distance to the center of the disk and whose smaller values sets $\{x, y : u(x, y) \leq c\}$ have the same area as those of the initial function u_0 (this follows essentially from the isoperimetric inequality and Schwartz inequality).

If the initial function has more critical points (say, two maxima and a saddle point; Fig. 4), the situation is not so simple. Some numerical experiments suggest a minimizer with a singular line, but I do not know of any theorem dealing with this case.

The variational problems of this type are also relevant for the study of the steady flows of ideal (incompressible, inviscous) fluid. The Euler equations in both problems formally coincide if we replace the magnetic field in the magnetic field problem by the vorticity in the hydrodynamics. In the two-dimensional case, this steady Euler equation says that the unknown function and its Laplacian should be dependent:

$$J(u, \Delta u) = 0,$$

where $J(f, g)$ is the Jacobian (in the hydrodynamical case, u is the stream function).

The situation in the three-dimensional problem is even worse than in the two-dimensional case above. There exist topological reasons to have *no* minimizing diffeomorphisms. The topological properties of the extremal vector fields (verifying the variational Euler equations) are very special. These extremal fields are either integrable (having nested systems of invariant tori) or force-free (being the eigenvectors of the curl operator). Most of the divergence-free initial vector fields are not diffeomorphic to any extremal field. It seems that the infimum is not attained even when we replace the diffeomorphisms with homeomorphisms. The problem of the mathematical description of the minimizing sequences in this case is wide open; no conjectures exist, as far as I know. It is only known that the infimum might be zero if the initial field lines are not linked (A.D. Sacharov 1973, M. Freedman 1991). The infimum is positive if the initial field's helicity is positive [10], or if there exist at least two linked field lines [11].

The difficulty of three-dimensional hydrodynamics (for instance, the difficulty of the problem of the existence and uniqueness of solutions of the 3D Navier–Stokes equation) seems to be related to the same geometrical difference between the topological properties of the orbits of the coadjoint representation of the group SDiff of measure-preserving diffeomorphisms in two and three dimensions, that is, responsible for the difference between the 2D and 3D extremal magnetic fields that we discussed above. Namely, in two dimensions, there exists an infinite sequence of "enstrophy-type integrals," which are functionals on the space of fields, invariant under the coadjoint representation. Thus, the orbits are closed subvarieties of infinite codimension in the space of 2D vector fields. In three dimensions no such integrals besides the helicity are known. They probably do not exist, and the generic orbits are probably dense in the level varieties of the helicity functional in some space of 3D fields [12].

1.6 Gravitational Shock Waves

Consider the dust-like gravitating matter in the usual Euclidean 3-space (we neglect the expansion of the universe and relativistic effects). Suppose that the initial distribution of particles is described by a smooth positive density function and that the initial velocity field is smooth and potential (there are physical reasons for the last assumption: the growth rate of the nonpotential perturbations is smaller than that of the potential ones; see [13]). The model of noninteracting particles shows that the singularities of the density function will be formed at the places of faster particles bypassing the slower ones. Computer experiments show that these singularities will also exist in the case where each particle is moving in the gravitational field formed by the whole distribution of particles.

These singularities form a very peculiar cellular structure. The density singularities form "caustic" surfaces. These caustic surfaces decompose the space into *voids*—regions of small density of particles. The caustics, forming the walls of the voids, themselves have singularities at some lines, where the density is even higher than at the generic points of the caustics. These lines, in turn, have singular points where the density is higher than along the lines. The whole system of caustics evolves with time and at some moments the bifurcations occur. At the moment of these bifurcations the order of the singularity of the density function is unusually high.

For the model of noninteracting particles, this picture is well founded in the mathematical theory of Lagrange singularities [3]. A Lagrange submanifold of a phase space is the mathematical equivalent of the potential flows of physicists. The time evolution of phase points is described by the Hamiltonian differential equations. The phase flow of a Hamiltonian system transforms Lagrange submanifolds into Lagrange submanifolds. However, even in the case where the initial Lagrange submanifold was a section of the cotangent bundle of the physical space (to each physical point one velocity vector corresponds) the image of this Lagrange manifold transported by the Hamiltonian phase flow may no longer be a section (Fig. 5). This means that the velocity field may become multivalued. Several flows of particles with different velocity fields are able to travel through "the same physical point" at the same time, provided that their total densities are not too large.

The caustics are the sets of critical values of the projections of the Lagrangian submanifolds to the physical space. The hierarchy of the singularities of caustics is well understood in the theory of Lagrangian singularities and explains the cellular structure described above.

The computer experiments with gravitationally interacting particles show essentially the same general patterns of singularity formation, but the mathematical description of these phenomena is much less complete.

From the mathematical point of view, the problem consists of the study of the so-called Vlasov–Poisson system of equations. This system defines the evolution of the density of phase points in the phase space. The motion of the phase points is governed by the Hamiltonian system of ordinary differential equations. The Hamiltonian function is the sum of the standard kinetic energy and of the potential energy defined by the gravitational potential of the distribution of the masses in the physical space. The last distribution is obtained from the density in the phase space by the integration along the velocity or moment spaces (along the fibers of the cotangent bun-

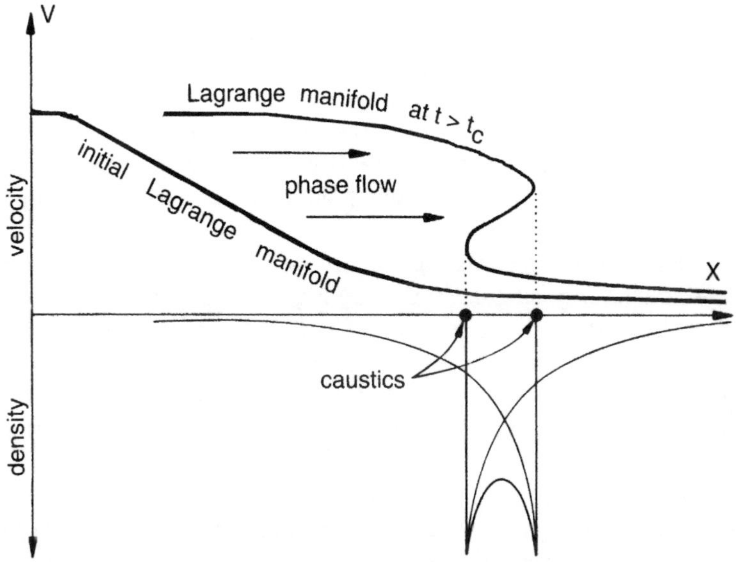

FIGURE 5. Caustics and multivalued velocity field, generated by the dust-like matter evolution.

dle). This is a closed system of equations for the unknown time-dependent phase space density. The system contains the differential equations of the transport and the Poisson equation for the potential. The right-hand side of the Poisson equation is the integral of the phase space density along the fibers.

The initial condition is a distribution along a Lagrangian submanifold in the phase space, representing a potential vector field. Such a distribution also defines the density of particles in the physical space. We suppose that this initial distribution is generic among those concentrated at Lagrange sections of the cotangent bundle space.

Problem. *Describe the singularities of the caustics hypersurfaces and of the density of particles in the physical space after the formation of the first caustics.*

A Lagrange submanifold will remain Lagrangian under the evolution defined by the Vlasov–Poisson equation. However, after the formation of the caustics it will no longer be a manifold but rather a variety with singularities over the caustics. Indeed, the singularities of the projection of a Lagrange manifold or variety to the physical space generate the singularities of the density of particles at the points of the caustics in the physical space. The formation of these density singularities makes the gravitational potential and hence the Hamilton function singular. Thus, the phase flow mappings are no longer diffeomorphisms over the points of the caustics. Hence, the initially smooth Lagrange submanifold in the phase space is transformed by the phase flow to a variety having singularities over the points of the caustics.

One guesses that these singularities are rather weak when compared with the generic Lagrange singularities (only higher-order terms of Taylor series are discontinuous at the caustics) and hence that the topological pattern should be the same as in the case of noninteracting particles. This has not yet been proved, however, for even the simplest case of one-dimensional dust-like matter, where the gravitational potential of a particle is the modulus of the distance from this particle [14]. In this case, one expects the semicubical cusp singularity of the newborn caustic in the two-dimensional space-time (the distance between the two newborn caustic) points in the physical 1D space to be of the order $(t - t_c)^{3/2}$, where t_c is the caustics formation moment. The density singularity at the caustics should be of the order $(-1/2)$.

One expects that the corresponding Lagrange variety in the phase 2-plane has four singular points (Fig. 6). Two of them correspond to the points where the variety has a vertical tangent. This verticality provides the physical space density singularity. The two other singular points of the Lagrange variety are weaker singularities at the intersections with the fibers over the caustical points.

In spite of these singularities of the Lagrange variety, the topological patterns of the generic caustics and their generic bifurcations as well as the singularities of the density seem to be essentially the same as in the case where the particle interaction is not taken into account and so where the Lagrange manifolds remain smooth.

Is it true that the topological structure of the singularities of the solution of the Vlasov–Poisson equations with generic initial distributions concentrated on generic smooth Lagrange sections of the cotangent bundle is the same as for the case of the Vlasov equation (where the gravitational interaction is not taken into account)? Does the density singularity scale the same way at the neighborhood of the points of the caustics and of their singularities as for the case with noninteraction particles?

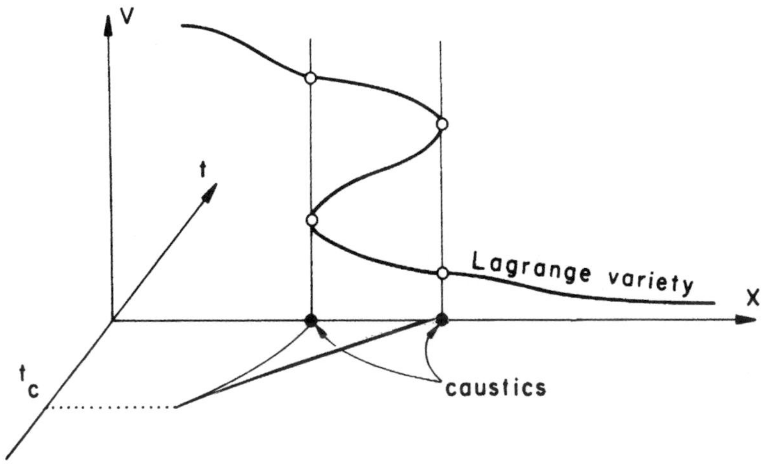

FIGURE 6. The four singularities of the moving Lagrange variety at the caustics.

1.7 Oscillating Integrals

An oscillating integral is the function of the form

$$I(h) = \int e^{iF(x)/h} a(x) dx$$

where $h \longrightarrow 0$ is a small parameter, the phase F being a real smooth function in $\mathbb{R}^n$. We shall suppose for simplicity that *amplitude* a is finitely supported.

According to the stationary phase principle the main contribution to the integral is due to the critical points of the phase function. If the phase function F is generic, the critical points are nondegenerate (Morse) points; in this case the integral scales like $h^{n/2}$ (Fresnel: see [3] and [4]).

If the phase function depends on the parameters, some non-Morse singularities may appear at certain values of the parameters even for generic families of phase functions $F(x, y)$ (where y is a p-dimensional parameter). Such special values of the parameter are called the *caustical values.* They indeed form the caustic of the Lagrangian submanifold defined by the stationary phase condition

$$\{(y, z) : \partial F/\partial x = 0, \quad z = \partial F/\partial y\}$$

in the cotangent bundle space of the parameters' space. At the caustical values of the parameters the integral is larger than at the ordinary ones:

$$I(h) \sim h^{n/2-\beta(y)},$$

where β is called the *oscillation index* and $\sim$ means that we neglect the logarithmical multipliers $\ell n^{\gamma} h$.

The oscillation indices depend on the singularities of the caustics. For the generic points of a generic caustic, $\beta = 1/6$ (the Airy function case). For the ordinary cusps of the generic caustics, $\beta = 1/4$. To every type of singularity of projections of Lagrange submanifolds to the base of the cotangent fibration there corresponds a singularity index. The hierarchy of these singularities is studied in the general theory of Lagrange singularities, and the indices occurring in the generic families depending on p parameters have been computed for p up to 10 (see [3] and [4]).

Physically $F(x, y)$ is the optical distance between the source point x and the observation point y, h is proportional to the wavelength, and the integral I describes the light brightness at the observation point.

For any number p of parameters there is a worse singularity occurring unavoidably in generic p-parameter families—that having the maximal oscillating index β. Thus, there is a function $\beta(p)$–the maximal oscillating index for generic p-parameter families. If the number n of variables x in the integral is also fixed, the maximal oscillating index in p-parameter families may be different: $\beta_n(p) \le \beta(p) = \beta_\infty(p)$.

Problem. *Calculate the asymptotical behavior of the maximal oscillation indices $\beta_n(p)$ and $\beta(p)$ occurring in the generic p-parameter families of oscillating integrals of functions of n variables.*

The natural conjecture is that the worst phase functions are $x_1^3 + \cdots + x_n^3 + x_{n+1}^2 + \cdots$ with a suitable $n(p)$ in the case when n is not bounded a priori and $x_1^k + \cdots + x_n^k$ with a suitable $k(p)$ in the case when n is bounded. These conjectures lead to the following conjectures:

$$\beta(p) \sim \sqrt{2p}/3 \quad (p \longrightarrow \infty),$$

$$\beta_n(p) \sim n/2 - n/\sqrt[n]{n!p} + \cdots \quad (p \longrightarrow \infty)\,.$$

1.8 Hamiltonian Chaos

Consider a generic analytic Hamiltonian system close to an integrable one:

$$H = H_0(p) + \varepsilon H_1(p, q, \varepsilon),$$

where the perturbation H_1 is 2π-periodic in the angle variables $(q_1, \ldots, q_n)$ and where the nonperturbed Hamiltonian function H_0 depends on the action variables $(p_1, \ldots, p_n)$ generically. Let n be greater than 2.

Conjecture 1 ([15, 16]). *For any two points p', p'' on the connected level hypersurface of H_0 in the action space there exist orbits connecting an arbitrary small neighborhood of the torus $p = p'$ with an arbitrary small neighborhood of the torus $p = p''$, provided that $\varepsilon \neq 0$ is sufficiently small and that H_1 is generic.*

Conjecture 2 ([15, 16]). *A stationary point of a generic analytic Hamiltonian system is Liapunov unstable, provided that the quadratic part of the Hamiltonian function at the stationary point is not positive or negative definite.*

The first thing one has to do in this direction is study the averaged systems at double resonances. These are Lagrange systems on a two-dimensional torus $\{q_1, q_2 \bmod 2\pi\}$ whose Lagrange functions have the form $L = T - U$, where $U(q)$ is a 2π-periodic function and $T = a\dot{q}_1^2 + 2b\dot{q}_1\dot{q}_2 + c\dot{q}_2^2$, a translation-invariant quadratic form (which might be definite or indefinite).

The problem is to find the double asymptotic orbits whose projections to the torus are positive and negative asymptotic to two periodic solutions of arbitrary two-homology classes [16].

In the elliptical case (when T is, say, positive definite) one can use the Jacobi metric. If the total energy is higher than the maximum of the potential energy U, this is an honest Riemannian metric on the torus. For each point and each integer homology class there exists a geodesic starting at that point and asymptotic to a cycle of the given homology class [17].

When the total energy is equal to the maximum of the potential energy, one can hope to find geodesics that are asymptotic to the maximal point in one direction and to a periodic orbit of any given homology class in the other. Two such orbits form a chain connecting two periodic orbits belonging to different homology classes. Perturbing this chain, one can hope to obtain a doubly asymptotic orbit.

In the case of Lorenzian kinetic energy, the Jacobi metric is not Riemannian but rather Finslerian. In this case, one should first prove the existence of periodic orbits in a given homology class. It seems that the contemporary technique of the variational calculus in the large has no ready methods for this problem.

It seems also that the problem of the *minoration of the number of the closed orbits of the motion of a charged particle in a magnetic field orthogonal to the surface along which the particle moves* is still not solved (conjecturally the number of the periodic orbits is generically at least $2g + 2$ on a surface of genus g for any given initial energy). Mathematically, this is the problem on the closed curves with prescribed positive geodesic curvature at any point of a Riemannian surface. In the higher-dimensional variant the magnetic field force is replaced by a closed two-form. These problems can be approached by the methods of the modern symplectic topology; however, up to now, only partial results have been obtained.

There are also many interesting problems in the domain of the pseudoperiodical integrable Hamiltonian systems, where the chaotic behavior has been observed only numerically. Although these problems are also related to the quasi-crystal geometry and to that of the Fermi surfaces in solid-state physics, they can be considered as chaos problems in Hamiltonian systems with 1 or 3/2 degrees of freedom. The corresponding numerical works are known under the name of *stochastic webs* theory describing the phase plane of a 1D particle in an electromagnetic wave field (Sagdeev, Zaslavski, and others; see, e.g., [18]). Here I quote only one particular problem in these systems.

Consider q vectors $(\mathbf{k}_1, \ldots, \mathbf{k}_q)$, based at the origin of the Euclidean plane, connecting the origin to the vertices of a regular q-gon. Consider the sum of q equal harmonic waves with these wave vectors. If $q \neq 1, 2, 3, 4, 6$ (say, for $q = 5$), the sum is not a periodic function but a quasi-periodic one.

Example.

$$H(\mathbf{r}) = \sum_{j=1}^{5} \cos(\mathbf{k}_j, \mathbf{r}), \quad q = 5 .$$

Problem. *Do all the closed components of the level lines $H = h$ that bound domains containing the origin lie in a finite neighborhood of the origin?*

Do there exist unbounded phase curves of the Hamiltonian system defined by the Hamiltonian function H?

Similar problems are also open for iterations of the plane mapping $T = AB$ where A is a $2\pi/q$ rotation and $B(x, y) = (x, y + a \sin x)$. Numerical experiment show, it seems, that unbounded orbits exist for the arbitrary small values of a.

I shall end with an old question on general (non-Hamiltonian) dynamical systems [1, 19]:

Is the problem of stability of a stationary point of a vector field whose components are polynomials with integer coefficients algorithmically decidable?

References

[1] V.I. Arnold, *Geometrical Method in the Theory of Ordinary Differential Equations.* Springer-Verlag, New York, 1983.

[2] A. Tresse, Sur les invariants différentiels des groupes continus des transformations. *Acta Mathematica* **18**, 1–88 (1894).

[3] V.I. Arnold, S.M. Gusein-Zade, and A.N. Varchenko, *Singularities of Differentiable Maps.* Vol. 1, Birkhäuser, Boston, 1985; Vol. 2, 1988.

[4] V.I. Arnold, *Singularities of Caustics and Wave Fronts.* Kluwer, Boston, 1991.

[5] V.I. Arnold, Kolmogorov's attractors. *Proc. Roy. Soc.*, A **434**, No. 1890, 19–22 (1991).

[6] V.I. Arnold and E.L. Korkina, Growth of the magnetic field in a three-dimensional flow of uncompressible fluid. *Vestn. Mosc. Univ. Ser. Math. Mech.* (1983), no. 3, 43–46 (translated as *Moscow Univ. Math. Bulletin*).

[7] U. Frish and D. Galloway, A numerical investigation of magnetic field generation in a flow with chaotic streamlines. *Geophys. Astrophys. Fluid Dyn.* **29**, no. 1, 13–18 (1984).

[8] Ya. B. Zeldovich, *Collected Works.* Nauka, Moscow, 1986.

[9] V.M. Alexeev, Quasirandom dynamical systems. *Math. USSR-Sb.* **5**, 73–128 (1968); **6**, 505–560 (1968); **7**, 1–43 (1969).

[10] V.I. Arnold, The asymptotic Hopf invariant and its applications. In *Materialy Vsesoysnoi Shkoly po Differentsialnym Uraneniyam Besconechnym Chislom Stepenei Svobody Dilizhane*, Armenian Acad. of Sc.,1973; Erevan, 1974, pp. 229–256 [English translation: *Selecta Math. Sov.* **5**, 327–345 (1986)].

[11] M.H. Freedman and Z.-X. He, Links of tori and energy of incompressible flows. *Topology* **30**, 283–287 (1991).

[12] V.I. Arnold and B.A. Khesin, Topological methods in hydrodynamics. *Ann. Rev. Fluid Dyn.* **24**, 145–166 (1992).

[13] V.I. Arnold, S.F. Shandarin, and Yu.B. Zeldovich, The large-scale structure of the universe. *J. Geophys. Astrophys. Fluid Dyn.* **20**, 111–130 (1982).

[14] A.A. Roitvarf, On the motion of continuous media in a force field with a root singularity. *Vestn. Mosc. Univ.*, no. 1, 65–68, 1987 (translated as *Mosk. Univ. Math. Bull.*).

[15] V.I. Arnold, Small denominators and problems of stability of motion in classical and celestial mechanics. *Russian Math. Surveys* **18**, no. 6, 85–193 (1963).

[16] V.I. Arnold, Stability problem and ergodic properties of classical dynamic systems. In *Proceedings of International Congress of Mathematicians*, Nauka, Moscow, 1966, pp. 387–392.

[17] E.V. Gaidukov, Asymptotic geodesics on a Riemannian manifold nonhomeomorphic to the sphere. *Sov. Math. Dokl.*, **7**, no. 4, 1033–1035 (1966).

[18] V.I. Arnold, Huygens, Barrow, Newton and Hooke, pioneers in mathematical analysis and catastrophe theory. In *Evolvents to Quasicrystals*, Birkhäuser Verlag, Basel, 1990.

[19] V.I. Arnold, Problèmes résolubles et problèmes irresolubles analytiques et géométriques. To appear in *Formes et Dynamique, Renaissance d'un Paradigme. Hommage à René Thom*, Eshel, Paris, 1994.

Solvable and unsolvable analytic and geometric problems

Vladimir Arnold

Steklov Mathematical Institute, Moscow, Russia, and Université Paris-Dauphine, France

Translation of: Problèmes résolubles et problèmes irrésolubles analytiques et géométriques. In: *Passion des Formes. Dynamique Qualitative, Sémiophysique et Intelligibilité. À René Thom. Vol. 1* [Passion for Forms. Qualitative Dynamics, Semiophysics, and Intelligibility. To René Thom. Vol. 1], M. Porte (Ed.), ENS Éditions, Fontenay-St Cloud, 1994, 411–416.

Translated by Emmanuel Ferrand.

> *"To tell about him, one has to be a Frenchman, because only people of this nation manage to explain to others what they themselves do not understand."*
>
> N.S. Leskov, Nesmertel′nyĭ Golovan [Deathless Golovan]

Abstract: Here we discuss conjectures about the insolvability of local problems of analysis and compare them with the insolvability posed by the theory of algorithms.

Keywords: algorithm – integrability – stability – dynamical systems.

What is the most charming in the heretical texts of Thom is his way of keeping his thought free of the chains of logic, and how he is not fearful of sentences without any precise meaning. I have the honor of considering myself, partially, as a student of Thom, because I learned a lot at his IHÉS seminar which I attended throughout 1965. The most important idea I mastered there is, maybe, the following principle (attributed to Thom, and not to his students[1]): "there will always be dumb people to prove theorems". This revolutionary idea deeply changed my mathematical universe — I understood that one can "jump over theories" (not only "over computations", as Galois said). Although I could never get rid of theorems and enjoy the resulting intellectual freedom, I was deeply contaminated by the dream of reckless mathematical fantasies.

I present here, in honor of the jubilee of the great Master, a series of such fantasies in the form of rather vague conjectures.

The exact formulations of these conjectures will not be presented here for the following reasons:

[1] In the Russian version of the paper (in: Vladimir Igorevich Arnold. Selected-60. Technical editor: M.B. Sevryuk. PHASIS, Moscow, 1997, pp. 577–582), these words are revised as follows: attributed to Thom by his students. (Editors' remark)

Translation of: *Passion des Formes. Dynamique Qualitative, Sémiophysique et Intelligibilité. À René Thom. Vol. 1.* Editor: M. Porte. ENS Éditions, Fontenay-St Cloud, 411–416 (1994),

Translated by Emmanuel Ferrand

1) they are too long,
2) the reader can reconstruct them himself/herself,
3) there exist several non-equivalent versions of these conjectures.

I – Geometrically unsolvable problems

The conjectures are dealing with the notions of "solvable" and "unsolvable" analytic problems. Typical examples are a) the problem of stability of motion in a neighborhood of a stationary point of a dynamical system; b) the problem of integrability of a system of differential equations; c) the problem of complete integrability of a system of canonical Hamiltonian equations.

The most vague formulation of the conjectures in question is the statement that all these problems are unsolvable.

The insolvability here is stronger than in the theory of algorithms because, for example, any Diophantine equation is solvable in our sense.[2] Consequently, the conjecture implies more than the absence of an algorithm in the sense of logicians, we speak here of the absence of a more general procedure (an "analytic algorithm" as described below).

In the spirit of the geometry of manifolds, the solvability of a problem should not depend on the coordinate frame used to specify the data of the problem (the right-hand sides of differential equations, the components of vector fields, the Hamilton functions). Hence I propose the following

Definition. *A problem is* geometrically unsolvable *if, among the problems one obtains from this problem by diffeomorphic changes of parameters, there are no analytically solvable problems.*

The definition of the analytic solvability is given below.

Example. The problem of stability of a stationary solution of a system of differential equations depending on parameters is geometrically unsolvable if this problem is analytically unsolvable for any family equivalent (via local diffeomorphisms depending on the parameters) to the given one.

Remark. The problem of integrating a differential equation is geometrically solvable in a neighborhood of a nonsingular point of the corresponding vector field, although this problem might be analytically unsolvable. This problem is probably geometrically unsolvable in a neighborhood of a singular point, even for polynomial vector fields.

II – Analytic solvability

We are now going to describe the analytic solvability. In this theory, the coordinate frame is fixed. The *data* of the problem are the functions defining the problem (the components of the vector field, the Hamilton function, etc.). These functions are fixed (as functions of the

[2] The word "Diophantine" was missing in the original. Cf. the last paragraph of § II in this paper and Arnold V.I., Sur quelques problèmes de la théorie des systèmes dynamiques. *Topological Methods in Nonlinear Analysis* **4** (1994), no. 2, 209–225, translated into English in the present volume. (Editors' remark)

coordinates and the parameters). The problem (of stability, integrability, etc.) is *analytically solvable* if the solution (the set of values of the parameters for which the system is stable, integrable, etc.) can be expressed in terms of the data through analytic operations, whose list contains:

1) all arithmetic operations,
2) calculating the derivative,
3) integration,
4) substitutions,
5) inversion.

To be more precise, one has to consider a set of "admissible varieties" and "admissible mappings" possessing the following properties:

1. The arithmetic spaces $\mathbb{R}^n$ and $\mathbb{C}^n$ (for all n) are admissible manifolds.
2. Any rational mapping is admissible.
3. The image and the inverse image of an admissible variety by an admissible mapping are admissible.
4. The intersection, union, and complements of two admissible varieties are admissible.
5. The superposition of two admissible mappings is admissible.
6. If $f(x, y)$ is an admissible function, then its derivative with respect to x as well as its primitive function (whose branch is fixed by the values at an admissible point) are admissible.

Remark. The primitive function in question is given by complex integration; the graph of this multi-valued function is therefore an admissible variety.

Thus, the analytic continuation is, in our theory, an admissible operation.

One can also develop a notion of solvability based on real integration. This theory is different from the one discussed here. For example, the function *sin* on the real line is admissible if complex integration is allowed, but is not contained in the minimal set of admissible functions if only real integration is allowed (even though the restriction of this function to any finite interval of the real line is always an admissible function).

Since the function *sin* is admissible, so is the set of all integers. This implies that the set of integer values of the parameter y such that a polynomial $f(x, y)$ with integer coefficients has an integer root x is admissible, whatever polynomial f is given.

III – Almost solvable problems; conjectures

Return to an analytic problem specified by a choice of the data functions (the components of the vector fields, the Hamilton functions, etc.) depending also on parameters.

Definition. The admissible set of the problem *is the minimal admissible set containing the data functions.*

The problem is analytically solvable *if the solution is an admissible function of the parameters.*

Example. The problem of stability of a stationary solution O of a dynamical system depending on parameters is analytically solvable if the set of the values of the parameters such that the stationary solution O is stable is an admissible variety.

Conjecture 1. There exist a number M and two functions N and D such that the problem of stability of an equilibrium point O of a vector field in $\mathbb{R}^n$ whose components are polynomials of degree d is analytically unsolvable

a) if n and d are larger than M,
b) if $d > 1$ and n is larger than $N(d)$,
c) if $n > 2$ and d is larger than $D(n)$.

The conjecture that this problem does not admit an algorithm, in the sense of logicians, for components with integer coefficients has already been published in (Arnold, 1976).

Conjecture 2. The problem of integrability of a differential equation defined by a vector field in a space of dimension n larger than 1 whose components are polynomials of degree d larger than 1 is analytically unsolvable.

Conjecture 3. The problem of complete integrability of a system of canonical Hamiltonian equations defined by a polynomial Hamilton function of degree d larger than 2 in a space of dimension $2n$ larger than 2 is analytically unsolvable.

Conjecture 4. All the problems of conjectures 1–3 are also geometrically unsolvable.

Remark 1. The *integration* problems are probably also unsolvable. However, the situation cannot be excluded that a family of systems is integrable (in the sense that integrals exist) but not in a solvable way (the integrals are not admissible functions).

For example, the commuting integrals of a Hamiltonian system integrable in the sense of Liouville can be non-admissible functions, i. e. not expressible in terms of the Hamilton function and the admissible operations.

Remark 2. Conjectures 1–4 deal with polynomial systems. This restriction is not natural: we need a definition where the roles of parameters can be played by functions (for example, by the components of a vector field). Formally, such a definition can be formulated without any difficulty. The result is a formalization of the possibility of expressing the answer of the problem (of stability, integrability, etc.) in the form of the action of admissible operations on the data functions.

However, in this case, even the problem of figuring out whether a critical point of an analytic function of one variable at the origin is a local minimum is not solvable (since the order of the necessary derivatives is not bounded *a priori*).

Therefore, I propose the following definition (Arnold, 1970):

Definition. *A problem containing a function as a parameter is* almost solvable *if in the space of functions, there exists a decreasing sequence of exceptional subvarieties of increasing codimensions such that the problem is solvable outside any of those exceptional subvarieties.*

(The answer allows an admissible representation in terms of the data functions, provided that those functions do not belong to one of the exceptional subvarieties; this representation depends on the exceptional subvariety in an arbitrary way.)

The class of solvable problems can be further restrained. One can add that the exceptional subvarieties should be admissible themselves; one can restrict the dependence of this subvariety and of the procedure solving the problem on the number of the subvariety in the sequence.

But I think that such restrictions are not necessary:

Conjecture 5. There are no almost solvable problems among the problems of conjectures 1–4.

I hope that the proofs of these conjectures will rely on the understanding of the dynamical systems rather than on tricks of formal logic, and that they will provide interesting information on the structure of the discriminant varieties and the bifurcation diagrams of dynamical systems, the objects whose study has been initiated by R. Thom (Thom, 1971).

References

1. Thom R., The bifurcation subset of a space of maps. In: *Manifolds—Amsterdam 1970* (Proc. Nuffic Summer School on Manifolds) (Ed. Kuiper N.H.) (*Lecture Notes in Math.*, vol. 197). Springer, Berlin, 1971, 202–208.
2. Arnold V.I., On local problems of analysis. *Vestnik Moskov. Univ. Ser. I Matem. Mekh.* 1970, no. 2, 52–56 (in Russian); the English translation: *Moscow Univ. Math. Bull.* **25** (1970), no. 1–2, 77–80. See also: Arnold V.I., *Collected Works.* Volume II. Hydrodynamics, Bifurcation Theory, and Algebraic Geometry (1965–1972). Springer, Berlin, 2014, 193–196.
3. Arnold V.I., Dynamical systems and differential equations. In: *Mathematical Developments Arising from Hilbert Problems* (Ed. Browder F.E.) (*Proc. Symp. Pure Math.*, vol. 28). Amer. Math. Soc., Providence, RI, 1976, 59–60.

Projective topology

V.I. Arnold

Translation of: *Kvant*, 1995, no. 6, 5–11.
Translated by Nikolai Kruzhilin.

Let us plot a hyperbola in the plane. It has two branches. Consider one of the straight lines, the asymptotes of the hyperbola, to which the branches approach at infinity. We connect the parts of the two branches of the hyperbola that, in going to "infinity" to the left and to the right, approach the asymptote we have selected, by a smooth curve without self-intersections (Fig. 1). One of the first theorems in projective topology, the Möbius theorem, says that there are *at least three inflection points* on the curve obtained. (An *inflection point* of a smooth curve is a point in whose neighborhood the curve goes over from one side of the tangent line to the other, similarly to what we see for the curve $y = x^3$ at the point $x = y = 0$.) Having plotted a couple of dozens of curves you will easily see it yourself that this result is true. However, its proof is by no means easy. In this article I expound the simplest basic concepts of projective topology, which is a rather young area, although its foundations go back to A.F. Möbius' works (the 19th century) which had actually motivated the invention of the very well-known "Möbius band".

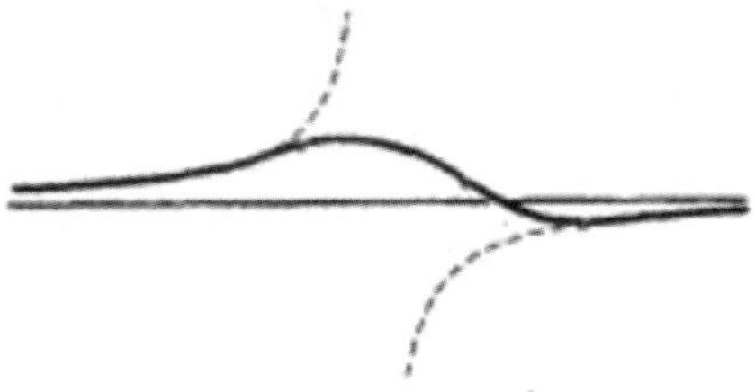

Fig. 1 The Möbius theorem on three inflection points

It turned out recently that this theory is closely related to the so-called Sturm theory of oscillations of solutions of differential equations, to symplectic and contact topology (that is, the topology of the phase spaces in mechanics and the topology of systems of the wave fronts of light rays in optics), to knot theory, to algebraic geometry, and even to quantum field theory. Of course, one cannot cover all these links in this article. However, I hope to explain you what types of problems arise here. Although the statements of these geometric problems are entirely elementary, all the heavy machinery of modern mathematics is (so far?) helpless before them — what we need here is perhaps youthful imagination.

Translation of: *Kvant*, no. 6, 5–11 (1995),
Translated by Nikolai Kruzhilin

Projective plane

D. Hilbert said that, for mathematicians, points can be objects of any kind, for instance, beer mugs. Now we will follow this principle. The plane — which you know very well, of course — is one of the most fundamental concepts in mathematics. However, apart from the conventional plane, mathematicians also consider the so-called projective plane, which is obtained from the conventional one by adding "points at infinity". To distinguish the conventional plane from the projective one, the former is called the affine plane (I do not know why); I explain the origins of the term "projective plane" in what follows.

The easiest way to get used to the projective plane is to rely on the following definition: *points of the projective plane are straight lines passing through the origin O of the three-dimensional space*.

The meaning of this definition is as follows.

Consider the conventional plane $z = 1$ (with coordinates (x, y)) in the three-dimensional space with coordinates (x, y, z) (see Fig. 2). With each point P in this conventional plane, we associate the straight line OP that connects it with the origin O. This line $p = OP$ is unequivocally determined by the point P in the conventional plane; in its turn, it determines P uniquely.

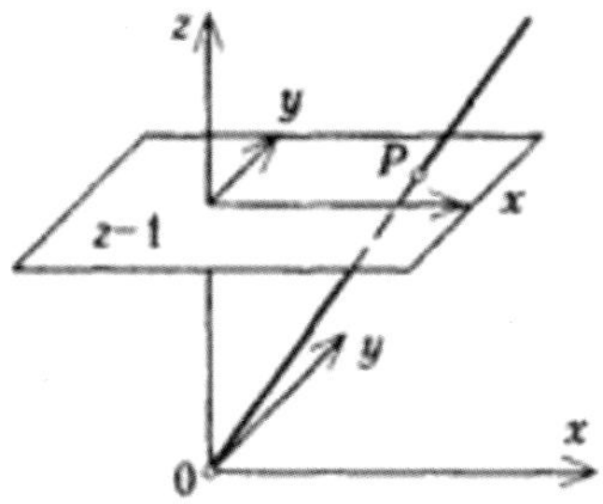

Fig. 2 An affine chart of the projective plane

In this sense our conventional plane is called an *affine chart* of the projective plane. The point P is called the representation of the corresponding point p (that is, of the straight line OP) in the projective plane on the affine chart.

Maps which are most often used in geography represent only a small portion of the Earth surface, and even a hemisphere map covers only half of this surface. Likewise, the affine chart of the projective plane does not represent the whole projective plane on our plane $z = 1$. Indeed, the straight lines parallel to the plane $z = 1$ do not intersect it and have no representations on our chart. Still an affine chart of the projective plane presents a much fuller image of the projective plane than, say, a geographical map of one hemisphere does: almost all the points in the projective plane have their representations on our affine chart. Namely, the points in the projective plane that are not represented on the affine chart form just a circle (why?).

Problems

1. Show that the whole projective plane is covered by three affine charts (which correspond to the planes $z = 1$, $x = 1$, and $y = 1$).

2. Fig. 3 shows how a certain triangle in the projective plane is represented on the affine chart $z = 1$. Draw the same triangle on the affine chart $x = 1$ (the answer, which is depicted in Fig. 7 below, is pretty surprising).

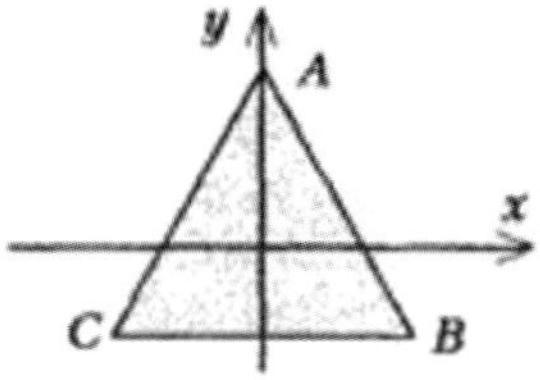

Fig. 3 The representation on an affine chart of a triangle in the projective plane

3. The equation $x^2 + y^2 = 1$ determines a circle on the affine chart $z = 1$. This circle is the representation of a closed curve without self-intersections in the projective plane. Draw the representation of this curve on the affine plane $x = 1$.

Instead of treating points in the projective plane as straight lines passing through the point O in the three-dimensional space, one could also define a point in the projective plane to be a pair of antipodal points on a sphere centered at O (Fig. 4). Indeed, any such line intersects the sphere in two antipodal points, and each pair of antipodes determines a straight line.

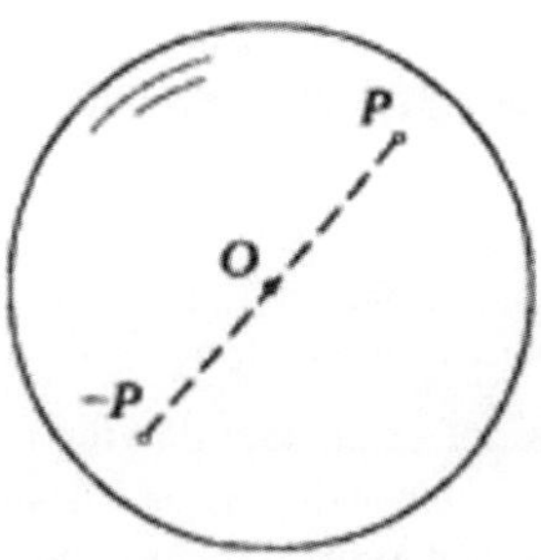

Fig. 4 The projective plane as a sphere with antipodal points identified

Mathematicians say that *the projective plane is obtained from the sphere by gluing the diametrically opposite points*. One could also very well confine oneself to a hemisphere, for instance, the Southern one. In this case we still have to glue the diametrically opposite points lying on the equator (Fig. 5). This could make you think that points on the equator are exceptional, but in fact no singularity arises there: all the points in the projective plane are quite similar (as all the points in the conventional plane or all the points in the sphere are).

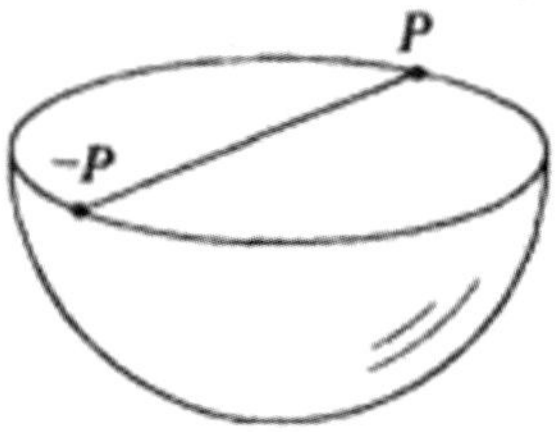

Fig. 5 Gluing the projective plane from a hemisphere

The projective plane is usually denoted by $\boldsymbol{RP}^2$ (the real projective space of dimension 2).

Problem 4. Define and investigate the projective line $\boldsymbol{RP}^1$.

Projective properties

The precise meaning of the "similarity" of all the points in the plane or the sphere is as follows: there exist transformations of the plane (sphere) that preserve all the properties of interest to us. For example, in the case of Euclidean geometry of the plane, the (distance preserving) motions of the plane are such transformations.

Transformations of the plane that take straight lines to straight lines (and therefore preserve the property of being parallel for straight lines) are said to be *affine*. Translations, rotations, reflections, and extensions along some direction are clearly affine transformations. Each affine transformation of the plane (for example, a homogeneous dilation) is a combination of several transformations from this list.

Problems

5. Prove that an affine transformation takes the midpoint of a line segment to the midpoint.
6. Prove that an affine transformation takes the medians of a triangle to the medians.

The last problem provides a proof of the fact that the medians of a triangle intersect at a single point. Indeed, this is obvious for an equilateral triangle, and an arbitrary triangle can be turned into an equilateral one by an affine transformation.

Definition. A transformation of the projective plane onto itself is said to be *projective* if it takes projective lines to projective lines.

Remark. Of course, you have guessed already what a projective line in the projective plane is: it is a curve whose representation on some (and therefore any) affine chart is a straight line. For example, if we use the affine chart $z = 1$, then it contains the representations (given by equations of the form $ax + by = c$) of all the projective lines but one in the projective plane. If we view a point P in the projective plane as the straight line OP in the three-dimensional space, then each projective line in this plane turns out to be a plane in the three-dimensional space that passes through O. On the other hand, if we view the projective plane as a sphere where the diametrically opposite points are identified, then the projective lines are great circles on the sphere.

Problem 7. Prove that each affine transformation of the plane extends to a projective transformation of the projective plane containing it.

The term "projective geometry" has its origins in the following example (which is also called the theory of perspective).

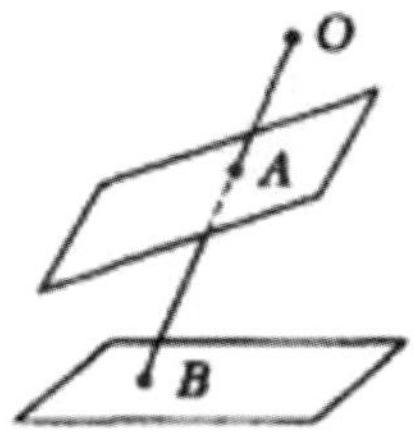

Fig. 6 Perspective transformation

Consider two arbitrary planes in the three-dimensional space and a point O outside them (Fig. 6). A *perspective mapping* of one of these planes to the other is the mapping that, to each point A of intersection of a straight line passing through O with the first plane, assigns the point B of intersection of this line with the second plane. In fact, a perspective mapping is a projection (by means of the rays issued from the center O).

Problems

8. Prove that any perspective mapping is the representation of a projective transformation of the projective plane on an affine chart.

9. Find projective transformations of the plane that take a circle to a) a hyperbola, b) a parabola.

Definition. A property of a geometric figure in the plane is said to be *affine* if it is preserved by affine transformations, and it is said to be *projective* if it is preserved by projective transformations.

Example. Being parallel is not a projective property of straight lines, but it is an affine property.

The branch of science that explores the projective properties of geometric figures is called *projective geometry*.

Remark. Projective transformations of geometric figures in the projective plane can change lengths, angles, areas and so on strongly. However, such properties of the figures as the number of connected components of a figure in the projective plane are not altered (see Fig. 7).

Indeed, the projective transformations of the projective plane are just the affine transformations of the three dimensional space that fix the origin.

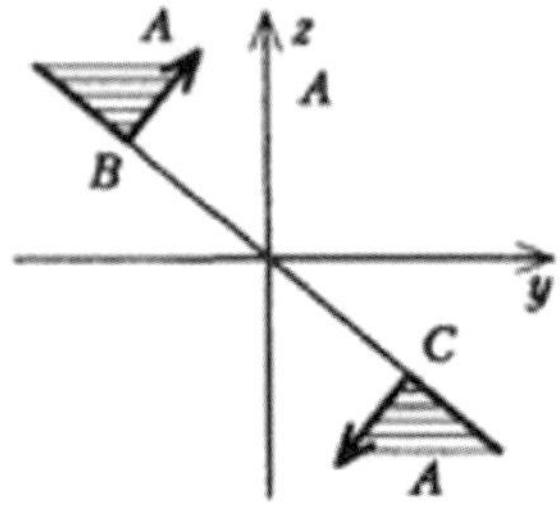

Fig. 7 The representation of the triangle in Fig. 3 on another affine chart

The properties of geometric figures that do not change under continuous (non-tearing) transformations are called *topological* properties, which means ones related to the branch of science exploring locations (from the Greek $\tau\acute{o}\pi o\varsigma$, location). Examples of topological properties are: the number of connected components of a geometric figure or of its complement, the number of branches of a curve that come together at a point, and so on.

Topological properties of the projective plane

In the small the projective plane is indistinguishable from the conventional plane or the sphere. However, its global topological properties are quite different from the properties of the conventional plane and those of the sphere.

For instance, consider a small disc (a neighborhood of a point) in the conventional plane, in the sphere, and in the projective plane.

Problem 10. Prove that in all the three surfaces above, the domains complementary to the disc are topologically distinct (they cannot be transformed one into another by a mapping that is one-to-one and continuous together with its inverse).

Solution. The complement of a disc in the sphere is also a disc. The complement of a disc in the plane is topologically equivalent to an annulus. A disc and an annulus are topologically different. For example, this can be seen from the fact that each closed curve in a disc can be contracted continuously to a point, while an annulus contains a closed curve that cannot be contracted (Fig. 8).

What does the complement of a disc in the projective plane look like?

Theorem. *The complement of a disc in the projective plane is topologically equivalent to a Möbius band.*

Proof. We use the model in which the projective plane is a sphere with the diametrically opposite points identified (Fig. 9). In this model, the complement of a disc in the projective plane is the belt between, say, the two tropics (with the same identification of points). The part of this ring domain that lies in the Western hemisphere contains a unique representative of each point of the ring, apart from the points lying on the meridian separating the hemispheres. The segments between the tropics on the meridians of longitudes 0° and 180° are antipodal and represent the same line segment in the projective plane. Thus, to obtain the complement

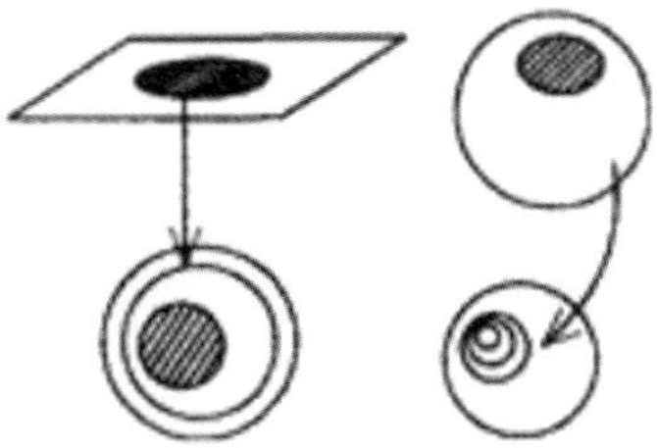

Fig. 8 The complements of a disc in the sphere and in the plane

of a disc in the projective plane, in the "rectangle" lying between the tropics in the Western hemisphere one has to glue the segments of the boundary meridians after turning one of them upside down relative to the other. After such gluing, a Möbius band is produced: this is its standard definition.

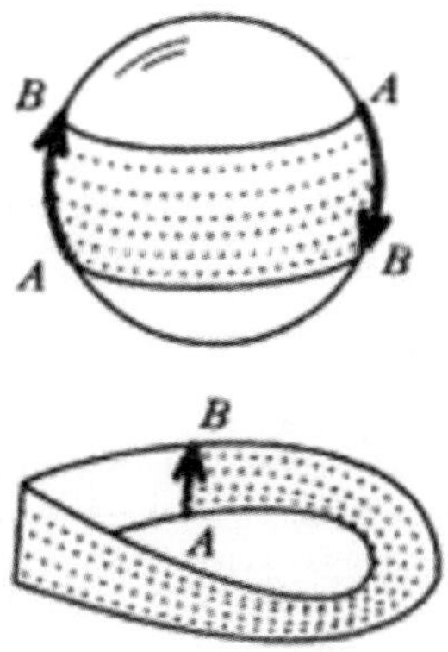

Fig. 9 The complement of a disc in the projective plane

Embeddings of a circle in the projective plane

Consider a closed smooth curve without self-intersections in the projective plane.

Example 1. The boundary circle of a small neighborhood of a point in the projective plane exemplifies such a curve.

Example 2. A projective line (for instance, the line at infinity which we add to the affine plane to obtain the projective plane) is also a closed curve without self-intersections in the projective plane. As far as their intrinsic topological properties are concerned, these two smooth curves are similar (both are circles). However, the positions of these topological circles in the projective plane are entirely different.

Problems

11. Prove that these two closed curves in the projective plane are not smoothly equivalent: there exists no one-to-one transformation of the projective plane into itself that is smooth together with its inverse and takes one curve to the other.

Solution. The first curve can be contracted continuously to a point in the projective plane, whereas the second one cannot (Fig. 10).

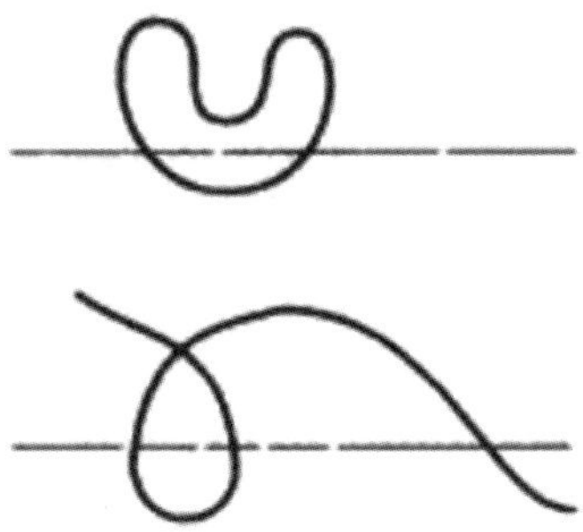

Fig. 10 Two circles lying in the projective plane in a contractible and a non-contractible way

That the second curve is not contractible can be proven, for instance, as follows. Fix a projective line and count the points of intersection of the smooth curve in question with this line; here we assume that all the intersections are generic, that is, without tangencies.

For a curve of the first type the number of intersections is even, and for a curve of the second type it is odd. Indeed, two projective lines intersect at a unique point and at a nonzero angle. When a smooth curve is deformed, the number of points of intersection is changed by two (at an instant of tangency). So the parity of this number is always preserved. Hence, after whatever deformations of the curve, intersections will persist (their number at generic moments of time will remain odd). Thus, a curve of the second type cannot be contracted.

12. Prove that each closed smooth curve without self-intersections in the projective plane can be deformed (in the class of all the curves with these properties) either to a circle bounding a small neighborhood of a point or to a projective line.

Hint. The first case takes place if the number of points of intersection with (some — and then any) generic projective line is even and the second case occurs if it is odd (Fig. 11).

Remark. A curve of the first type is represented by two closed curves on the sphere, and a curve of the second type is represented by a single curve.

13. Prove that an ellipse as well as a hyperbola and a parabola are smooth closed curves without self-intersections in the projective plane. What are the types of these curves (Fig. 12)?

14. Consider the curve in the affine plane that is given by the equation $y = \frac{1}{1+x^2}$. Prove that this affine curve turns to a closed smooth curve without self-intersections in the projective plane after adding a single point "at infinity" (Fig. 13). What is the type of this curve?

15. Consider the curve in the affine plane that is given by the equation $xy(x + y) = 1$. Prove that this affine curve turns to a closed smooth curve without self-intersections in the projective plane after adding three points "at infinity". Determine the type of this curve.

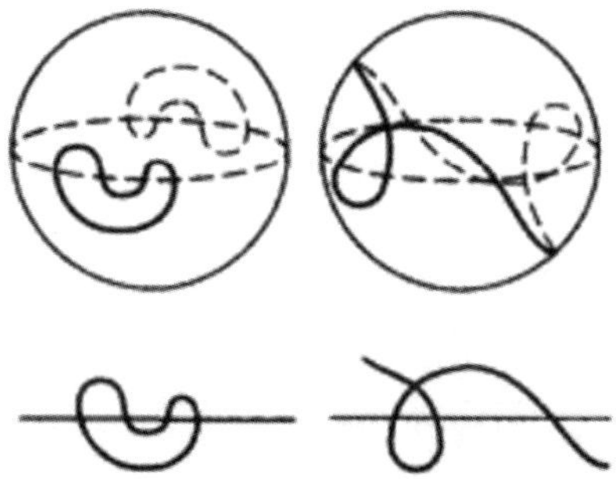

Fig. 11 Two types of curves in the projective plane and their representations on the sphere

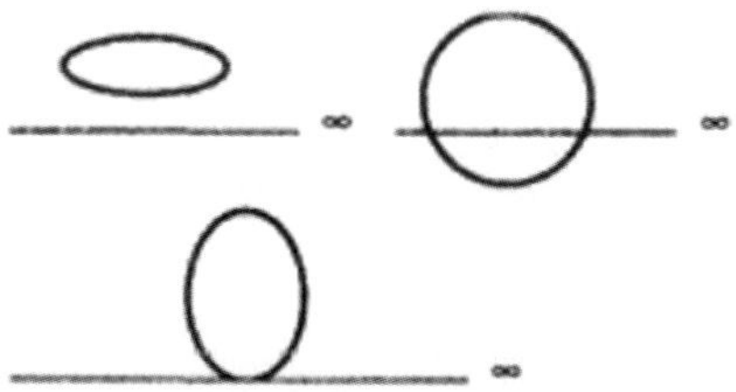

Fig. 12 An ellipse, a hyperbola, and a parabola in the projective plane

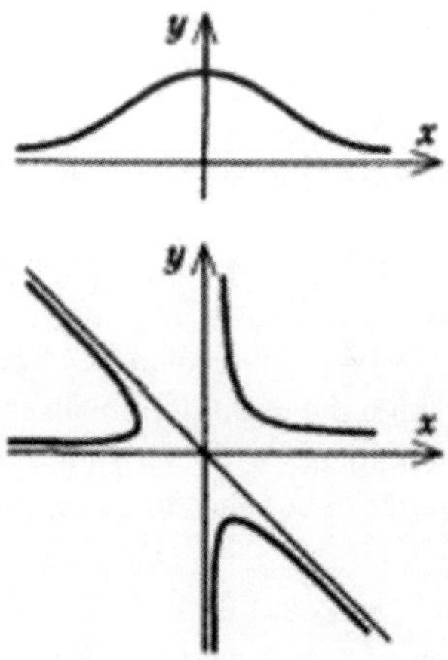

Fig. 13 Cubic curves in the projective plane

16. Consider the curve in the affine plane that is given by the equation $y = x^3$. Prove that this affine curve turns to a closed non-smooth curve without self-intersections in the projective plane after adding a single point "at infinity".

Inflection points of curves

The simplest example of an inflection point of a smooth planar curve is the point O on the graph of the function $y = x^3$ (Fig. 14).

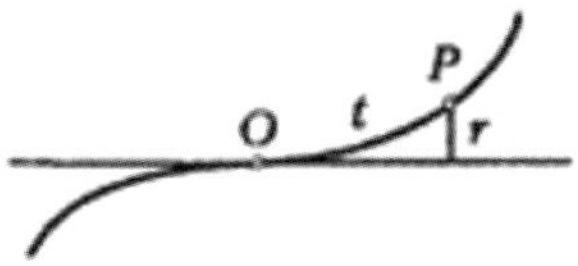

Fig. 14 An inflection point of a planar curve

Definition. An *inflection point* of a smooth planar curve is a point at which the curve has a higher order of tangency than the usual one with its tangent line (that is, an order higher than the order of tangency of a circle with its tangent line).

Let t denote the distance from a point P on the curve to the point O under study (Fig. 14). Then the point O in question is an inflection point if (and only if) the limit of the ratio r/t^2 of the distance r of the point P from the tangent line at the point O under study to the square of the distance t between P and the point O in question is zero:

$$\lim_{t\to 0} \frac{r}{t^2} = 0.$$

In other words, an inflection point is a point where the curvature of the curve vanishes (and the radius of curvature is infinite).

At the first glance, the property of a point on a curve to be an inflection point is related to the Euclidean geometry of the plane: the definition involves distances. However, after some reflection we see that (by contrast with the particular values of the curvature and of the radius of curvature) the property of a point on a curve to be an inflection point is invariant under affine transformations.

Problems

17. Prove that the property of a point on a smooth curve to be an inflection point is invariant even under projective transformations.

Hint. A projective transformation takes projective lines to projective lines. It changes the distances between close representations of close points in the affine plane by a bounded factor. Therefore, under a projective transformation, the tangent lines to a curve go to tangent lines to the transformed curve. The orders of tangency at the corresponding points are also the same.

18. Prove that the curve $y = \frac{1}{1+x^2}$ (Fig. 13) has an inflection point at infinity.

19. Prove that the points at infinity of the cubic curve $xy(x + y) = 1$ (Fig. 13) are not inflection points.

20. Prove that the curve $y^2 = x^3$ has a simplest inflection point at infinity (of the same order as the curve $y = x^3$ has at the origin). Find its tangent line at this inflection point.

Remark. We see that inflection points can be defined for smooth curves (even with self-intersections) both in the affine plane and in the projective plane. Incidentally, we obtain a definition of such points for curves in the sphere as well — recall that the sphere is locally indistinguishable from the projective plane.

In this case great circles tangent to a curve in the sphere play the role of tangent lines.

21. In the (affine) plane, consider a locally smooth curve (with a self-intersection) which has the shape of a figure of eight. This curve has two inflection points. Prove that during a continuous deformation of this curve (in the class of locally smooth curves in the affine plane), we will always obtain curves with at least two distinct inflection points (of course, here a curve is meant to be a smooth mapping of the circle parametrizing the curve into the plane, and it is the inverse images of the inflection points on this circle that are distinct).

22. A figure of eight drawn on the sphere also has two inflection points. Is it true that the curves in the sphere that are obtained from a figure of eight by a continuous deformation (in the class of locally smooth curves in the sphere) have at least two inflection points?

Answer. No, an example is shown in Fig. 15.

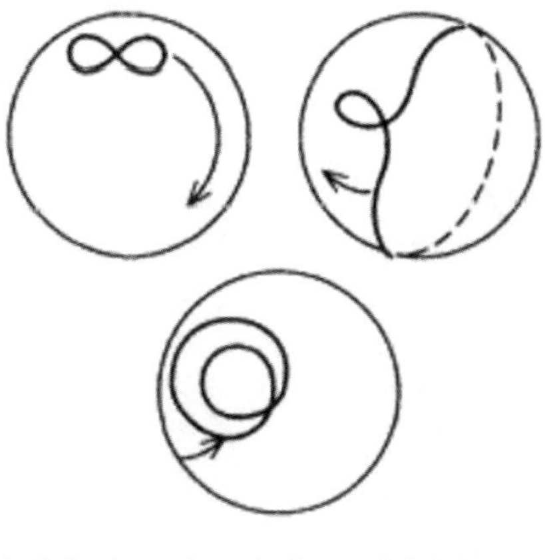

Fig. 15 Disappearing of both the inflection points of a figure of eight in the sphere

23. Find the minimal number of inflection points of a curve of a prescribed topological type (in the plane, in the sphere, and in the projective plane). The solution to this problem is not known yet.

The Möbius theorem on three inflection points

Now everything is ready for the statement of the Möbius theorem.

Theorem. *A smooth closed non-contractible curve without self-intersections in the projective plane has at least three inflection points.*

Example. On the cubic curve $xy(x + y) = 1$ (Fig. 13), all the three inflection points are at infinity, and on the curve $y(x^2 + 1) = 1$, one of these is at infinity, while the other two lie in the affine plane with coordinates (x, y).

Remark. If the curve is cubic (that is, given by an equation of degree three), then there are precisely three inflection points, and all of them lie on one projective line (for the curve $xy(x + y) = 1$ this is the line at infinity, and for the curve $y(x^2 + 1) = 1$ it is parallel to the Ox axis). I know of no simple proof for this result in algebraic geometry. There are many proofs of the Möbius theorem (although he did not prove it himself apparently). Unfortunately, no proof that I know of can be presented in the framework of this article.

Problems

24. Let us deform continuously a projective line in the class of locally smooth curves in the projective plane (self-intersections are allowed). Is it true that there are always at least three inflection points on a curve obtained?

Answer. No, a deformation is shown in Fig. 16. In this figure, the projective plane is symbolically represented as a disc with the boundary circle where the diametrically opposite points are identified. We can view this disc as the image of a hemisphere that produces the projective plane via gluing the diametrically opposite boundary points. In the deformation shown in Fig. 16, at some time instant we obtain a curve with a positive self-tangency (a self-tangency is said to be positive if the two corresponding tangent vectors inducing the orientation of the curve have the same direction).

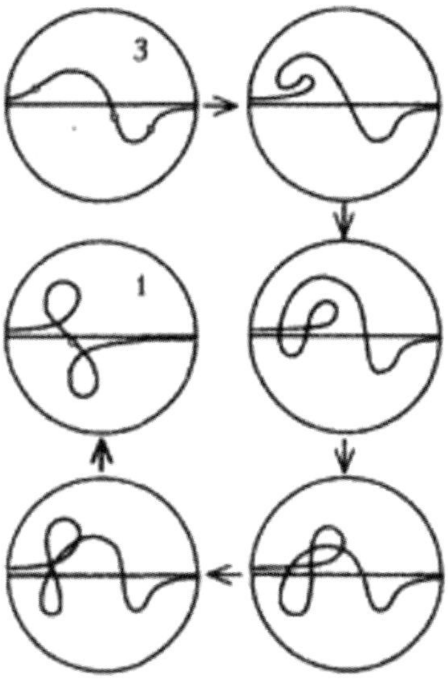

Fig. 16 Removing two inflection points of a curve in the projective plane

25. Do three inflection points survive on a locally smooth curve obtained from a projective line via a continuous deformation during which no positive self-tangency occurred? (Recently D.A. Panov, a student of the Moscow State University, proved that all the deformations during which no positive self-tangencies occur and no more than five inflection points are formed simultaneously at any intermediate instant, preserve three inflection points.)

In all the examples I am aware of where two of the three inflection points annihilate, a positive self-tangency must occur on the way. If we could prove that this is indeed unavoidable, we would obtain a substantial generalization of the Möbius theorem. Perhaps, for such a generalization we have to understand better the deep reasons for the existence of three inflection points stated in the Möbius theorem. The situation with the next result, related to the Möbius theorem, is quite similar.

The tennis ball theorem

Consider a closed smooth curve without self-intersections in the sphere (Fig. 17).

Fig. 17 Four inflection points of the seam of a tennis ball

Theorem. *A closed smooth curve without self-intersections which divides the sphere into two parts with equal areas has at least four inflection points.*

Example. On the surface of a tennis ball, there is a seam which divides the area of the sphere equally in half. There are four inflection points on this seam; they are easily seen in Fig. 17.

Now consider a curve with self-intersections. Can we define when such a curve divides the area of the sphere in half?

In this case the curve divides the sphere into more than two parts, and in finding the "area bounded by the curve" one has to take these parts into account with different coefficients. Namely, the coefficient at some point is taken to be an arbitrary integer; then we increase it by 1 each time we cross the curve in the positive direction. If the sum of the areas of all the parts taken with these coefficients is 2π modulo 4π, then we say that the curve divides the area of the sphere equally in half.[1]

[1] Here the radius of the sphere is assumed to be 1. (The translator's remark)

Problems

26. Prove that the above definition is consistent: whether the sum of the areas is 2π modulo 4π or not, is independent of the choice of the "positive" side of the curve and of the choice of the coefficient at the reference point.

27. Consider a continuous deformation of the equator of the sphere in the class of locally smooth curves (maybe with self-intersections) that divide the area of the sphere in half. Is it true that any curve obtained via such a deformation has at least four inflection points?

Solution. This is not true: the corresponding deformation is shown in Fig. 18 (where the areas of the parts A, B, and C satisfy the relation $A + C = B$).

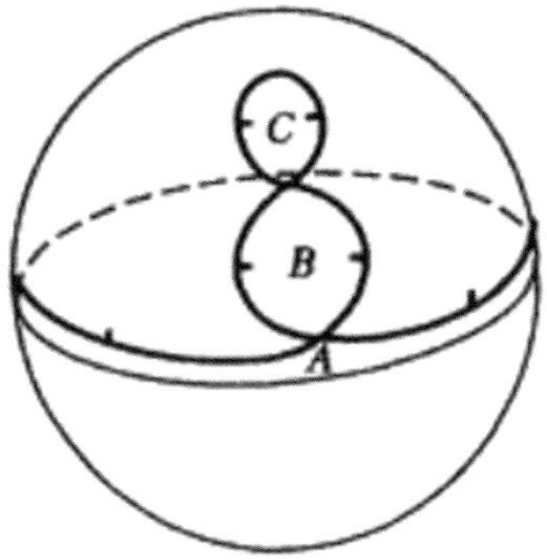

Fig. 18 Removing two inflection points of a curve in the sphere

Conjecture. During such a deformation of the equator, four inflection points persist as long as no closed loop of any curve obtained in the deformation divides the area of the sphere in half.

Remark. Consider another property of closed curves without self-intersections that divide the area of the sphere in half: they intersect each great circle at least twice.

If during the deformation process (in the class of locally smooth closed curves dividing the area of the sphere in half) the curve starts to intersect itself, then this property of intersecting each great circle at least twice can disappear (this is exemplified by the curve in Fig. 18).

However, this property cannot get lost if no loops dividing the area of the sphere in half (and distinct from the whole curve) are formed in the process of deformation.

I know of no simple proof of this elementary geometric fact (non-elementary proofs, using methods of so-called symplectic topology, were given by Yu.V. Chekanov and A.B. Givental).

Suitable "multi-dimensional analogues" of the Möbius theorem have not been proven; moreover, they have not even been stated, nor even as conjectures, nor even for surfaces close to a projective plane in the three-dimensional projective space. Cubic surfaces of this type contain four curves of points with vanishing Gaussian curvature. Can their number be smaller when the surface is not given by a third degree equation?

Projective duality

Consider the set of all the projective lines in the projective plane. In a natural way, one can view this set as (another) projective plane! This (pretty amazing) fact is called the phenomenon of projective duality. It requires considering precisely the projective plane: the straight lines of the affine plane constitute a Möbius band rather than another affine plane (check this).

This phenomenon is called duality because the projective lines in the projective plane obtained (it is said to be dual to the original plane) can be identified with the points of the original plane in a natural way. Therefore, the projective plane dual to the plane dual to the given one is the original projective plane. We cannot obtain a third plane via repeating our construction, we return to the starting plane this way.

The easiest way to understand the definition of duality is by relying on the definition of the projective plane as the set of straight lines passing through zero in the usual three-dimensional vector space $\{(x, y, z)\}$. Consider linear homogeneous functions in this three-dimensional space. They have the form $ax + by + cz$ and constitute the three-dimensional vector space $\{(a, b, c)\}$ themselves. This space is said to be dual to the original vector space.

Problem 28. Prove that the second dual space of a vector space coincides with the original vector space.

Now consider a straight line in the dual space that passes through 0. Points on this line that are distinct from 0 are linear functions $ax+by+cz$ of (x, y, z) proportional to each other (with nonzero coefficients of proportionality). In the original space, such a function determines a plane passing through the origin. This plane is independent of the choice of a point (a, b, c) on the straight line in question in the dual space.

Thus, to each straight line (passing through 0) in the dual space, we assign a plane (passing through zero) in the original space. This natural correspondence is one-to-one. It converts the set of the planes passing through zero in the original space into the projective plane obtained by "projectivization" of the three-dimensional space dual to the original one.

On the other hand, after projectivization of the original space, a plane passing through zero becomes a projective line in the original projective plane. This natural correspondence is also one-to-one.

Thus, we have constructed a natural one-to-one correspondence between the set of the projective lines in the original projective plane and the set of the points of another projective plane, namely, the plane that is projectivization of the vector space dual to the original space.

Problems

29. Consider the set of all the projective lines passing through a fixed point in the projective plane. Prove that they form a projective line in the dual plane.

30. Consider the set of all the tangent lines to a circle in the original plane. Prove that they form a "circle" in the dual plane.

Definition. The set of all the tangent lines to a curve in the projective plane is called the *dual curve* of the original curve.

Problems

31. Consider the curve $y = x^3$ in the projective plane. Prove that in a neighborhood of the point corresponding to the inflection tangent ($y = 0$), the dual curve has a singularity (a "cusp" similar to the singularity of the curve $u^2 = v^3$ at the origin).

32. Prove that the second dual curve of a convex curve (with the curvature vanishing nowhere) coincides with the original curve.

Remark. In fact, this also holds for (not too bad) non-convex curves provided that we define the tangent lines at the points where the curve is not smooth in an appropriate way.

33. Investigate the curves projectively dual to the curves in Fig. 13.
Answer. These are hypocycloids with three cusps (Fig. 19).

Fig. 19 The dual curve of the curve $y = 1/(x^2 + 1)$

The Möbius theorem looks particularly beautiful when stated in the terms of the dual plane.

Consider a projective line in the original projective plane. This is just a point in the dual plane. Now we deform the original projective line. The tangent lines to the resulting smooth curve constitute the dual curve in the dual plane. If the deformed curve is close to a projective line then the dual curve is small (close to a point).

According to the Möbius theorem, the dual curve of the deformed curve has at least three cusps, similarly to the small hypocycloid in Fig. 19.

Remark. One can show that for "most" small deformations of a projective line, the dual curve has precisely three cusps (although for some special deformations, one can obtain dual curves with five, seven, or even an arbitrary odd number of cusps).

Problem 34. Draw the curve dual to the curve $xy(x^2 - y^2)(x - 2y) = C$. For large values of C, the original curve is close to a projective line in the projective plane (namely, to the "line at infinity" of the plane $\{(x, y)\}$).

"Area–length" duality

The geometries of the original and dual planes are fairly amazingly interrelated. We saw above that inflection points of a curve in the original plane correspond to cusps of the dual curve in the dual plane. It turns out that the lengths of curves in one of the two planes dual to each other correspond to the areas of the domains bounded by the dual curves in the dual plane. Of course, when curves are not convex or have self-intersections, one has to take the lengths and areas into account here with appropriate signs.

Remark. The lengths and areas in the projective plane are defined to be equal to the lengths and areas in the corresponding sphere. More precisely, we treat the original three-dimensional space as a Euclidean one and consider the unit sphere there (from which the projective plane is obtained by gluing the diametrically opposite points). Defined in this way, the lengths and areas are not preserved by projective transformations and are not projective characteristics of geometric figures.

Now we can regard the dual projective plane as the same sphere. To do this, to each plane passing through the origin in our Euclidean space, one should assign the orthogonal straight line (also passing through the origin). Mapping the dual plane onto the original one in this way, we get the procedure to measure the lengths, angles, and areas also in the dual plane.

Example. On going over to dual curves, the curves that divide the area of the sphere in half are transformed into curves with zero length (the signs of the lengths of the arcs of the dual curve change each time we pass through a cusp).

Problem 35. Prove that the envelope of the family of the normals to a convex planar curve has length zero.

Consider a smooth oriented curve in a sphere of radius 1.
Extend each tangent great circle by a distance of $\pi/2$ ahead of the tangency point.

Problem 36. Prove that the resulting curve is smooth (this holds even when the original curve has simplest cusps of the type $x^2 = y^3$: we only need the "forward" direction to be continuous at any cusp).

The curve obtained by this construction of extending the tangents by $\pi/2$ is called the *derivative* of the original curve.

Problems
37. Find the derivative of a parallel (circle of latitude) in the sphere.
Answer. This is the equator parallel to this circle.

38. Prove that the derivative of a closed curve without self-intersections in the sphere divides the area of the sphere in half.

This result is dual to the property of the envelope of the family of the normals to have zero length (this property holds for curves both in the plane and in the sphere — think about why).

The "area–length" duality pertains to the branch of mathematics which is called integral geometry. Integral geometry has many applications in mathematics and outside it (for example, in tomography, in probability theory, in the topological theory of characteristic classes, and in quantum field theory).

Some important results in this field were discovered by astrophysicists rather than by mathematicians (in connection with the problem of interpreting the results of measurements through optical and radio telescopes). The topological theory of characteristic classes arose as a by-product of investigations in mathematical statistics. These are further examples of mysterious links which exist between all areas in mathematics, be they very abstract or quite applied.

Problem 39. One throws a bounded planar curve at random onto a lined paper sheet and then computes the mean value, over a large number of independent throws, of the number of intersections with lines on the sheet. Prove that the (limit) mean number of intersections is proportional to the length of the curve (but independent of its shape). Calculate this number for a needle of length 1, which is thrown onto a plane with spacing 1 between neighboring lines ("Buffon's needle problem").
Answer. $2/\pi$ (because for a circle of length π, the number of intersections is always 2).

What is projective topology?

Since F. Klein's time (that is, for more than a century) mathematicians classify various branches of geometry depending on the transformations preserving the corresponding properties of geometric figures. For example, angles and lengths are in the realm of Euclidean geometry, for they are preserved by Euclidean motions (and also by reflections if we do not take orientation into account).

The notions of a median and the barycenter of a triangle pertain to affine geometry, and the notion of an "inflection point", to projective geometry. From this abstract standpoint, topology is a science considering the properties of geometric figures that are preserved by all continuous transformations (which also have continuous inverse transformations), while differential topology is a science considering the properties preserved by smooth transformations (which are differentiable together with their inverse transformations).

For example, a square and a disc in the plane are topologically equivalent but smoothly non-equivalent. It was discovered recently that there exist fake four-dimensional spaces, which are topologically equivalent to the standard four-dimensional space, but not smoothly equivalent to it.

The Möbius theorem shows that the abstract standpoint provides an unreasonably narrow view of the subject. For instance, the statement on three inflection points cannot be included in differential topology because the notion of an inflection point is not invariant under all smooth transformations (but only under projective ones). On the other hand, the Möbius theorem is clearly topological in nature, although this does not fit the Procrustean bed of the abstract definition.

I suggest that the following terminology should be used: apart from the topological invariants, topology (with some prefix or another) explores also all the discrete invariants of the geometric figures considered by this or that geometric theory (responsible for the prefix).

For example, the number of vertices of a convex polygon in the affine plane is a discrete invariant of this convex polygon. The concept of a convex polygon is related to affine geometry, so the number of vertices is an invariant of affine topology.

The number of inflection points of a planar curve is an invariant of projective topology, and the Möbius theorem is a simplest fact in this branch of science, whose development has been greatly delayed because it was not covered by the abstract algebraic nomenclature used for the mathematical sciences.

Questions to V.I. Arnold

Interviewed by Michèle Audin and Patrick Iglésias
in Luminy, France, on December 31, 1991

Translation of: Questions à V.I. Arnold. *Gazette des Mathématiciens*, 1992, no. 52, 5–12.
Translated by Emmanuel Ferrand.

La Gazette: Among all of your results, which one do you consider the most important? How did the idea come to you, how did you find it?

V. Arnold: I do not know what the "importance" of a mathematical result is. It is probably a quantity inversely proportional to its technical difficulty. It could be that the most important thing I made in mathematics is not generally associated with my name. Hence I will just choose a result that I appreciate.

A few months ago, Hironaka told me about a certain method to analyze real algebraic curves, which he liked. This method is based on the topology of manifolds of dimension 4. To an algebraic curve in the complex projective plane, one associates the 4-dimensional manifold that is the double cover of the plane ramified along the curve. Hironaka did not know that it was me who had invented this method [1].

It is not his fault: the habit to attribute a result to the last author who used it, or to the one from whom we learned it, has contributed a lot to the progress of science. I do not give other examples because I am very grateful to the mathematicians and physicists who have used and continued my research. America does not bear the name of Columbus. Nabokov said that one pays the biggest tribute to Tolstoy not by citing the name of Tolstoy but by evoking that of Anna Karenina.

Let's come back to algebraic curves. I remember that it was I.G. Petrovskiĭ, the rector of the Moscow State University and the creator of the theory of real algebraic varieties [2, 3] who asked me to read the thesis of D.A. Gudkov [4]. Gudkov solved the question of Hilbert's 16th problem on the configurations of the ovals of real algebraic curves of degree 6 in the projective plane. In that very difficult thesis, which I never read, I was struck by the congruence modulo 8 conjectured by Gudkov:

$$p - m \equiv k^2 \bmod 8,$$

where p is the number of the ovals of a smooth curve of degree $n = 2k$ that are "located inside" an even number of ovals, while m is the number of the ovals located inside an odd number of ovals, provided that the total number of the ovals is maximal. This maximal value is equal to $g + 1$ where $g = (n - 1)(n - 2)/2$ is the genus of the curve, as was shown by

Translation of: *Gazette des Mathématiciens*, no. 52, 5–12 (1992),
Translated by Emmanuel Ferrand

Harnack [5]. Curves of degree n with $g+1$ ovals exist for all n, and they were called M-curves (M for maximal) by Petrovskiĭ.[1]

Gudkov found all the configurations of the 11 ovals of M-curves of degree 6 (see the figure).

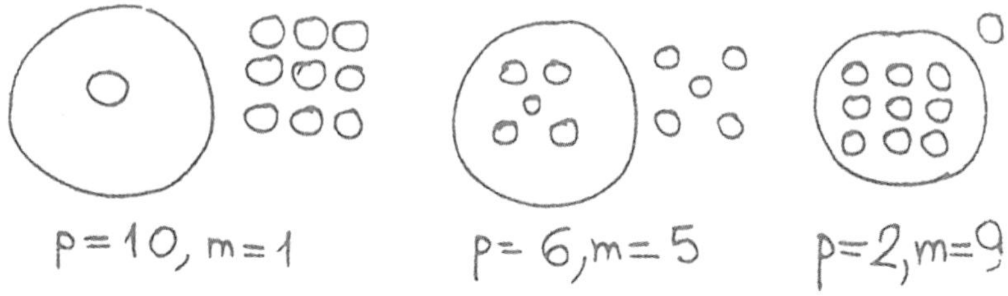

Fig. 1 Eleven ovals of M-curves

Gudkov's congruence had been confirmed for all the M-curves known to that date. But one did not see any relation between oval configurations in the projective plane and arithmetic. I knew from the works by Rokhlin and Milnor that the congruences modulo 16 play a crucial role in the topology of 4-dimensional differentiable manifolds, like the congruences modulo 8 in the theory of quadratic forms (I learned that in Serre's magnificent book *"Cours d'Arithmétique"*[2]). So I immediately started searching for a 4-dimensional manifold. I first thought of the complement of the complex curve in the complex projective plane and of its image in the quotient space $S^4 = \mathbb{CP}^2/\text{conj}$, where conj is the complex conjugation. After a few weeks I began to study coverings (being inspired by then recent papers by Rokhlin, Atiyah, Bott, and Singer).

It was January then and it was very cold in Moscow, and the heating (which is centralized and monopolized by the government) was down (as it happened several times afterwards and is likely to happen again in a near future). My hand could not hold a pen, and I was able to work only while skiing in the forests surrounding Moscow. After a few days I understood that the road to a proof of Gudkov's conjecture was obstructed by a lemma which I could not prove (one can find it in [1]). At that time I had a very young brilliant student, A.N. Varchenko, who used to overcome technical obstacles easily. I gave him this problem among others. Less than in a week, he told me that he had managed to prove the lemma. Once the obstacle was removed, I continued without fear and finished the proof quickly.

[1] It is important here that the ovals of an M-curve (and, furthermore, of any smooth algebraic curve) of an even degree lie two-sidedly in the projective plane, so that each of them bounds an interior domain and an exterior one. (Editors' remark)

[2] J.-P. Serre. *Cours d'Arithmétique.* Presses Universitaires de France, Paris, 1970. Deuxième édition: 1977. English translation: *A Course in Arithmetic.* Springer, New York, 1973. (Editors' remark)

The temperature went up and the heating system was repaired. I was therefore able to verify the proofs by writing them down. Everything was fine except the lemma: Varchenko's proof was incorrect. I could not go back, I had to overcome the obstacle. I did that in several days of intense work, and that's how I proved Gudkov's conjecture modulo 4.

Without Varchenko's incorrect proof, I would never have finished the work since I doubted the lemma, and I would not have had the courage to derive the necessary corollaries from an unproven lemma. Without those corollaries, I would not have overcome the technical difficulties of the proof of the lemma. Thus, the essential source of the final result was an error!

The most important outcome of all this was not Gudkov's conjecture itself but rather the discovery of unexpected links between algebraic geometry of real curves, topology of manifolds of dimension 4, and arithmetic of quadratic forms. From that moment on, the theory developed very quickly. Rokhlin, to whom I described the theory, almost immediately finished the proof of Gudkov's conjecture modulo 8. More recent achievements were presented at International Congresses of Mathematicians by Kharlamov (1978), Viro (1983), Khovanskiĭ (1983), Nikulin (1986), Shustin (1990). Unfortunately, there is no general exposition of the theory in the form of a book (one can nevertheless consult the paper [6] and Kharlamov's commentary in Petrovskiĭ's *Selected Works* in Russian[3]).

In 1979, I wrote [6] that among the 268 282 855 trees describing the configurations of 22 ovals, at least 10 and at most 144 can be realized by M-curves of degree 8. Nowadays these numbers have narrowed to 78 and 91, but not all possible configurations of M curves of degree 8 have been found yet.

The essential idea of [1] was that the complexification of the notion of a manifold with boundary $f(x, y) \geq 0$ is the notion of a ramified double covering $f(x, y) = z^2$. I always wonder what could be "complexified versions" of topological notions derived from the notion of boundary, like those of homology or cobordism. This program has been carried out only for Lagrangian and Legendrian cobordisms in symplectic and contact geometry.

G: You had and have many students and, more generally, many mathematicians have been influenced by your ideas. Could you tell us how you were educated, who were your teachers?

V. Arnold: I am a mathematician in the fourth generation. Even though my father died when I was 11, our house was full of mathematical books. Like most of Russian mathematicians, I participated in mathematical olympiads when I was 14 to 17 years old, with some success (I was usually the second). One evening every week, we had a "kruzhok" (a working group[4]) at the University, where university students suggested problems to school students. Every pupil had to explain solutions to the problems that he had solved during the last week. There I learned more about mathematics than during all my years at the University: the art

[3] V.M. Kharlamov. *Topology of real algebraic varieties.* In: I.G. Petrovskiĭ. *Selected Works. Volume 1. Systems of Partial Differential Equations. Algebraic Geometry.* Nauka, Moscow, 1986, pp. 465–493 (in Russian). Now these *Selected Works* with all the commentaries are available in English translation: I.G. Petrowsky. *Selected Works. Part I. Systems of Partial Differential Equations and Algebraic Geometry.* Gordon and Breach Publishers, Amsterdam, 1996. (Editors' remark)

[4] To be more precise, a hobby group. "Kruzhok" literally means "small circle" in Russian. (Editors' remark)

of reasoning correctly, of oscillating between induction (examples, counter-examples) and deduction (conjectures, lemmas). The Russian mathematical school was very much based on this tradition. We entered the University being researchers already, maybe without erudition (which we despised as "useless" and even "dangerous"), but very ready to climb rocks, perhaps more difficult than those one encounters in the great mountains of serious mathematics.

I entered the Moscow State University in 1954, when the discriminatory restrictions had become less severe for a few years. There was a "kruzhok" for first-year students on the theory of functions of real variables, led by A.G. Vitushkin. One had to discover the whole theory without reading anything, by finding proofs of a series of well-chosen theorems. This was a wonderful schooling, although very difficult. I remember, for instance, Kirillov and Vinberg among the participants. The spirit of honest competition (one should not consult manuals) is a characteristic feature of this somewhat sectarian system. At the end of the year, we arrived at the variational theory of functions of several variables (today called integral geometry, Grassmann manifolds, and characteristic classes).

The algebra course was delivered by E.B. Dynkin. I was stunned that such a great mathematician could give such bad lectures. Later I understood that the algebraists do this on purpose to make their science more hermetic and to raise the authority of the art of moving equation terms from right to left. The course was filled with non-motivated definitions. For example, one had to study determinants without understanding that they are the volumes of parallelepipeds. Groups appeared at the end but as sets endowed with two operations satisfying some axioms one had to learn by heart. It was a great relief to read the linear algebra textbooks by Shilov and Gelfand (where, for example, all the axioms of a Euclidean space were replaced by the single fact that any assertion of elementary geometry that can be formulated in terms of three vectors must be true).

The official education in Moscow was already very much Bourbakized. Dynkin's course was actually the best. The analysis course was at the level of Dedekind's cuts at the very time when in the course of experimental physics, one already used vector analysis (with wrong proofs of the Stokes theorem). Luckily, it was not mandatory to attend the lectures provided that one had successfully passed the exams.

I wrote my first paper while preparing for the algebra exam at the first year. Dynkin proposed a problem on the Sturm method: find a formula for the number of roots of a polynomial in a domain bounded by a rational curve. This problem was solved by two different methods by two students, Kirillov and myself. Apparently our professor thought that the solutions had been suggested by more mature mathematicians. But we passed all the (public) trials of this suspicious inquisitor, and he realized that this was not the case. Morally, after Vitushkin's school, it was completely impossible for us to submit a solution that would not be our own.

Dynkin asked us to write a paper. After 6 or 7 rejected versions, we had to send the paper to a journal. But at that moment I found at home an old book by Cauchy which contained essentially the same results. The paper has never been published.

There was also a competition for the students of the first two years. About ten problems were posted on the wall. I remember a series of problems which ended with a classification of closed ideals of the algebra of continuous functions. We had an impression that we were creating this theory ourselves. Another problem was to trace what is called today (I do not know why) the Mandelbrojt set. It was the only problem we did not solve. We did not have computers in 1955.

At the beginning of my second year at the University, Kolmogorov started a seminar on the theory of approximations of functions for students of my generation. He formulated elementary problems (at the same time aiming, as I understand now, at Hilbert's 13th problem on a representation of functions of three real variables as a superposition of functions of two variables). After that he left for France for several months.

Meanwhile, Dynkin organized a very interesting seminar which attracted the best students. I remember his remarkable style. A theorem is written on the blackboard. One also sees a lemma proven five minutes ago. A student, chosen by Dynkin, is standing in front of the blackboard. He has to prove the theorem. He does not manage. Dynkin therefore starts helping him. He is telling a story. It is a story of a monkey. There are bananas hanging from the ceiling and two sticks. The monkey has to unhook bananas. None of the two sticks is long enough. But one can make a longer stick by using the two sticks together. There are monkeys who are intelligent and achieve this within two minutes. For those who are not it takes much longer. "Are you ready?" he finally asks the student. Nobody was smiling at the seminar.

Later I learned that it was not Dynkin who had invented this style. He copied Gelfand who in turn copied Landau. A large part of Moscow mathematicians has passed through this system. Nevertheless, I remember an excellent talk which I attended there: it was a colloquium style talk on celestial mechanics by V.M. Alekseev.

When Kolmogorov returned from Paris, I told him what I had managed to do with his problems, and he invited me to work with him. I left Dynkin's seminar with a great relief.

Kolmogorov had an enormous authority. He was interested in all sciences and worked in all domains of mathematics except in algebraic geometry and number theory. For almost twenty years he hardly had any contact with foreign mathematicians. In 1963 he told me that his discoveries in topology (especially his four notes in *Comptes Rendus de l'Académie des Sciences* on cohomology[5]) had not been appreciated by the mathematical community. "They use — he told me — the cohomology groups that I introduced (generalizing physical concepts) at the same time as Alexander. However, in my papers, I also introduced a ring structure. I believe that many discoveries could be made by employing this very important structure which is not used by topologists."

This remark of Kolmogorov illustrates the mathematical atmosphere in Moscow, isolated from the outside world. It also shows (as Rokhlin explained to me) the greatness of Kolmogorov who was able to estimate the importance of his discoveries.

In 1956 Kolmogorov was mostly working on problems of information theory. The problem of superpositions is connected with these studies. He found a very simple proof of the Vitushkin inequality:

$$n/p \leq m/q$$

[5] English translations of these four notes of 1936 with references to the French originals are published as papers nos. 32–35 in A.N. Kolmogorov. *Selected Works. Volume I. Mathematics and Mechanics.* Kluwer Academic, Dordrecht, 1991. This volume also contains commentaries to these papers by Kolmogorov himself and G.S. Chogoshvili. (Editors' remark)

provided that each p-differentiable function of n variables is a superposition of q-differentiable functions of m variables. In particular, the number of information units necessary to encode a p-differentiable function of n variables with the ε-accuracy grows like $(1/\varepsilon)^{n/p}$.

For non-differentiable functions, the situation is different. Kolmogorov represented any continuous function of four real variables as a superposition of continuous functions of three real variables. He also proved that to reduce continuous functions of three variables to functions of two variables, it is enough to embed the universal tree in $\mathbb{R}^3$ in such a way that any continuous function on the tree can be representable as the sum of three continuous functions each of which depends on one coordinate only.

I constructed such an embedding. It was my first serious work. Kolmogorov himself immediately wrote the text of my note for *Doklady Akad. Nauk SSSR*, which he praised a lot later. I think that Kolmogorov was right: we always have to encourage the young ones. It was so different from Dynkin's monkeys!

One week later, Kolmogorov decomposed an arbitrary continuous function of n real variables into a superposition of continuous functions of one variable and a single function of two variables, the sum. He was saying later that it was the most difficult of his results, and that he never managed to work ferociously on a single subject for more than two weeks. I am proud that my proof stimulated him: he discovered his theorem by simplifying my constructions.[6]

From this moment on, Kolmogorov treated me as an independent mathematician rather than as a student. He allowed me to work freely at any kind of mathematics by my own choice. I chose to continue my research on the representation of functions on curves as sums of functions of coordinates. For trees the problem had been solved. I therefore considered closed plane curves such as ellipses, for example. This Dirichlet problem for the hyperbolic equation depends on a certain dynamical system on the curve. For an ellipse, this system is a rotation of the circle. For nearby curves one encounters all the difficulties caused by resonances and small divisors, like in celestial mechanics.

I immediately formulated two conjectures on holomorphic diffeomorphisms of the circle. I did not manage to prove them. The first one was proven by Michel Herman 20 years later. As for the second one, J.-C. Yoccoz and R. Pérez-Marco proved recently that it holds for almost all rotation numbers but not for all.

Being unable to prove my conjectures, I considered a simplified model: local diffeomorphisms. I solved the problem in this case by using a Newton-type method invented by Kolmogorov in 1954 (later Malgrange explained to me that Henri Cartan had already employed this method 10 years earlier in his research on what one called later the theory of coherent sheaves).[7] Then I discovered papers by C.L. Siegel who had already dealt with the

---

[6] One can find English translations of Kolmogorov's papers on this subject, accompanied by commentaries by Arnold and Kolmogorov himself, in the volumes of Kolmogorov's selected works recently published in the series "Mathematics and Its Applications" by Kluwer. (Arnold's remark) — See papers nos. 55 and 56 in A.N. Kolmogorov. *Selected Works. Volume I. Mathematics and Mechanics.* Kluwer Academic, Dordrecht, 1991. (Editors' remark)

[7] See papers nos. 52 and 53 in A.N. Kolmogorov. *Selected Works. Volume I. Mathematics and Mechanics.* Kluwer Academic, Dordrecht, 1991 and the commentaries by Kolmogorov himself and Arnold. (Editors' remark)

simplified problem during the war. Kolmogorov who was unaware of them said "we are in a good company".

This was the origin of my first work (1958) on the theory which was later called by physicists the KAM theory.[8]

Let's return to Hilbert's 13th problem. It is unknown whether any algebraic function of three complex variables can be represented as one of the branches of a superposition of algebraic functions of two variables. I always conjectured that this is not possible even for the function $f(x, y, z)$ proposed by Hilbert:

$$f^7 + xf^3 + yf^2 + zf + 1 = 0,$$

and even if one replaces algebraic functions of two variables by multi-valued mappings topologically equivalent to algebraic functions.

Later this conjecture led me to braid groups, the geometry of discriminants, and singularity theory. Recently S. Smale and V.A. Vassiliev used my results on braids in their theory of algorithmic complexity. Vassiliev has derived that the topological complexity of an approximate calculation of a root of a polynomial of degree n lies between n and $n - c \log n$ [7].

G: The French people, in general, believe that there is a Russian style in mathematics. What is your opinion?

V. Arnold: I believe that the styles of N. Hitchin and D. Bennequin are more Russian than those of Suslin and Fedorchuk (the professor who replaced P.S. Alexandrov at the topology chair of the Moscow State University; the chairs in Moscow are practically lifelong). The Russian articles "on one property of one solution of one PDE" are as boring as their counterparts in *Comptes Rendus de l'Académie des Sciences*. In my view, what we call the Russian style is the style of mathematicians who consider mathematics as an indivisible entity, mainly looking for links between most distant branches of mathematics and sciences. I could cite Novikov, Sinaĭ, Manin, Faddeev, Vershik, Kirillov, Fuchs, Khovanskiĭ. Our masters, like Kolmogorov, Petrovskiĭ, Pontryagin, Gelfand, all had very diverse mathematical interests. I believe that the Russian style is a heir of the style of Klein and Poincaré rather than of Hilbert and Bourbaki.

G: You have just spent a few months in Orsay, what is your assessment?

V. Arnold: A. Beauville, L. Illusie, J.-M. Bismut, A. Douady, R. Temam welcomed me very warmly, but the email was not functioning. In Moscow, two days before leaving for Orsay, I waited for two hours to buy some bread. When bread was finally delivered, there was just

[8] KAM is for Kolmogorov–Arnold–Moser. For Arnold's very detailed recollections on the origin and the first steps of KAM theory (with a suitable bibliography), see M.B. Sevryuk. *Translation of the V.I. Arnold paper "From superpositions to KAM theory" (Vladimir Igorevich Arnold. Selected–60, Moscow: PHASIS, 1997, pp. 727–740)*. Regul. Chaotic Dyn. **19** (2014), no. 6, 734–744. All the early works by Arnold including his papers on superpositions of functions and on KAM theory mentioned above are collected in V.I. Arnold. *Collected Works. Volume I. Representations of Functions, Celestial Mechanics and KAM Theory (1957–1965)*. Springer, Berlin, 2009. Here we confine ourselves with the note that Arnold's first work on KAM theory (1958) was published only in 1961 in *Izvestiya Akad. Nauk SSSR, Ser. Matem.* (Editors' remark)

enough of it to serve the queue up to me and two more people. I was very happy. One cannot find this kind of happiness in France.

The mathematics[9] in Orsay remains as hermetic as it was 26 years ago when I stayed one year in Paris. Seminar titles like "End of the proof of the property P" are not helping me much. I would have liked to attend a seminar like the "Bourbaki seminar" every week, but unfortunately there was only one. It was a great pleasure for me to see again Thom, Sullivan, Gromov, Ruelle, and Berger at the IHÉS.

The Orsay library is magnificent. It is the first one where I could see the collection of the *Encyclopaedia of Mathematical Sciences* arranged on a single bookshelf — it seems, as a result of J.-C. Sikorav's efforts — and not scattered according to the names of the authors (like in the library of Congress and therefore in all the libraries in the United States).

I am sorry to say that of the mathematical results I have just learned in France, those that interested me the most are due to Japanese (T. Shintani), English (P. Kronheimer), Russian (P. Zograf, M. Kontsevich), and American (M. Freedman, Z.X. He) mathematicians.

G: In the recent but already very popular short paper [8], you point out that the mathematical culture of Russian students is decreasing, and you suggest some explanations. Isn't there also a drop in the numbers of gifted students choosing scientific fields (in favor of law, management, or economics, for example), like in France?

V. Arnold: This does happen too, but this should not prevent one from providing a good education to those who remain. Some of them are very good. The decline is less visible at the level of Moscow's "normaliens"[10] than at universities in general.

G: A few years ago in another interview you complained about bad relations between Soviet and Western scientists, the latter not citing properly the former. Now, when people can meet more easily, has that changed?

V. Arnold: The personal relations are excellent, but nobody (of whom I talked to) in Orsay has noticed the negative solution of Hilbert's 21st problem (on the realization of monodromy groups by Fuchsian systems) published by A.A. Bolibrukh in Moscow several years ago. Its popularized version published in the US by Anosov has changed nothing.

I have the impression that now, fearing no longer, one will completely neglect the Russian contributions. A book recently published in Paris attributes to Thom and Milnor (1964–65) the classical estimates due to Petrovskiĭ and Oleĭnik (1949, 1951 [9]) of the Betti numbers of algebraic varieties (on which, by the way, the initial proof of Vitushkin's inequality was based). One of the authors told me that he knew the history of the question. He cites, it is true, those Russian mathematicians whom he has met.

G: What can you tell us about the impact on science of the political changes and the economic problems occurring in the Soviet Union? On the Academy of Science, on the independent universities?

---

[9] In the original, "la mathématique" (singular) is used here (as opposed to the more common plural "les mathématiques"), referring to the somewhat bourbakist unifying terminology. (Translator's remark)

[10] Arnold refers here to the students of the rather elitist École normale supérieure in Paris. (Translator's remark)

V. Arnold: I am not aware of the current situation in the Academy of Sciences. The Academy will probably no longer be able to cover travel expenses for trips abroad, and a large part of them will have to be cut. It seems that the libraries will not get any new books or Western journals from 1992 on — there is no enough money even to pay for 1991.

The Communist Party committee was kicked out of the faculty. In its place, a new chair has been created for the local representative of an important institution whose current name I do not know. Since the Brezhnev times, this person has been in charge of choosing those professors who are allowed to travel abroad.

The Independent University of Moscow has been created by a group of enthusiasts (N.N. Konstantinov, M.K. Polivanov, and others). There are two Departments: Mathematics and Theoretical Physics. The students are at the level of Parisian "normaliens". Among the professors, one finds well-known mathematicians like A.A. Kirillov, A.N. Rudakov, Yu.S. Il′yashenko. The difficulties are enormous, the main obstacle being the military service. There is also the funding issue.

G: What is the effect on science of the dismantling of the Soviet Union?

V. Arnold: Nationalist and fascist regimes will discriminate students, researchers, and professors who are "ethnically foreigners", for example, Russians. I hope that the mathematical school of Khar′kov, one of the most important in the former USSR, will survive within Ukraine. The impact of the economic disaster seems to be more dangerous.

G: One has the impression of a real brain drain from the USSR towards the Western countries and Israel. What do you think, what remedy can we consider?

V. Arnold: Either a Pol Pot-type dictatorship or a brain drain towards Latin America.

The communists that were active in mathematical pogroms are now asking me for recommendation letters for universities in the West. They have more information than us and, like rats, they leave the sinking ship.

The salary of a university professor was at the level of 20 dollars per month for several years. Now it is much less. The forecast is that the prices will get about five-fold at the beginning of 1992. The effective salaries will therefore be divided by five. I have been told that the state has confiscated the convertible money paid by the Western editors to Russian authors (as a result of the bankruptcy). I do not know whether this is true; if this is the case, the brain drain will increase, because it was a substantial help for active mathematicians.

G: How do you see the future?

V. Arnold: If we believe the statistics published in Moscow, the longevity in the USSR has decreased by ten years in the last decades. An easy cannibal computation shows that the number of the lost "person-years" is equivalent to the effect of an immediate execution of approximately 80 million inhabitants (a lemma: one loses approximately a half of one's life when one is shot down). This is more than all the other crimes of communists, the losses during the war were about 30 million victims only. Thus, the reduction in the number of eaters, the problem formulated by Lenin in 1918, will soon be solved.

The disaster of the Bolshevik revolution produced in the following years a paradoxical flourishing of mathematics in Russia. A similar miracle can happen again after an atomic civil war or a Kuwait-like scenario, but where nobody will have any interest in defending Paris.

I am therefore quite optimistic about the future of the Russian mathematical school. Personally, I hope to be useful for my country and to avoid the necessity to emigrate, my age being already close to the mean longevity.

References

1. V.I. Arnold. *On the arrangement of ovals of real plane algebraic curves, involutions of four-dimensional smooth manifolds, and the arithmetic of integral quadratic forms.* Funkts. Anal. Prilozhen. **5** (1971), no. 3, 1–9 (in Russian). English translation: Funct. Anal. Appl. **5** (1971), no. 3, 169–176. That translation was not quite satisfactory. Adequate English translation in: V.I. Arnold. *Collected Works. Volume II. Hydrodynamics, Bifurcation Theory, and Algebraic Geometry* (*1965–1972*). Springer, Berlin, 2014, pp. 239–249.
2. I.G. Pétrowsky. *Sur la topologie des courbes planes réelles et algébriques.* Comptes Rendus Acad. Sci., Paris **197** (1933), no. 22, 1270–1272.
3. I.G. Petrovskiĭ, O.A. Oleĭnik. *On the topology of real algebraic surfaces.* Izvestiya Akad. Nauk SSSR, Ser. Matem. **13** (1949), no. 5, 389–402 (in Russian). English translation in: *Translations. Series One. Volume 7. Algebraic Topology.* Amer. Math. Soc., Providence, RI, 1962, pp. 399–417 and in I.G. Petrowsky. *Selected Works. Part I. Systems of Partial Differential Equations and Algebraic Geometry.* Gordon and Breach Publishers, Amsterdam, 1996.
4. D.A. Gudkov, G.A. Utkin. *Topology of Sixth-Order Curves and Fourth-Order Surfaces* (*Concerning D. Hilbert's 16th Problem*). Uchenye Zapiski Gor'kovskogo Universiteta, **87** (1969) (in Russian). This volume contains six papers by Gudkov (pp. 5–153) and three papers by his student Utkin (pp. 154–211). All these papers are translated into English as D.A. Gudkov, G.A. Utkin. *Nine Papers on Hilbert's 16th Problem.* Amer. Math. Soc. Translations. Series 2, **112** (1978).
5. A. Harnack. *Ueber die Vieltheiligkeit der ebenen algebraischen Curven.* Math. Annalen **10** (1876), Nr. 2, 189–198.
6. V.I. Arnold, O.A. Oleĭnik. *Topology of real algebraic varieties.* Vestnik Mosk. Univ. Ser. 1 Mat. Mekh., 1979, no. 6, 7–17 (in Russian). English translation: Moscow Univ. Math. Bull. **34** (1979), no. 6, 5–17; reprinted in: V.I. Arnold. *Collected Works. Volume III. Singularity Theory* (*1972–1979*). Springer, Berlin, 2016, pp. 281–293.
7. V.A. Vassiliev. *Cohomology of braid groups and complexity of algorithms.* Funkts. Anal. Prilozhen. **22** (1988), no. 3, 15–24 (in Russian). English translation: Funct. Anal. Appl. **22** (1988), no. 3, 182–190.
8. V.I. Arnold. *A mathematical trivium.* Uspekhi Mat. Nauk **46** (1991), no. 1, 225–232 (in Russian). English translation: Russ. Math. Surveys **46** (1991), no. 1, 271–278; reprinted in: *Arnold: Swimming against the Tide.* Eds. B.A. Khesin, S.L. Tabachnikov. Amer. Math. Soc., Providence, RI, 2014, pp. 47–55.
9. O.A. Oleĭnik. *Estimates of the Betti numbers of real algebraic hypersurfaces.* Mat. Sbornik **28(70)** (1951), no. 3, 635–640 (in Russian).

In every mathematician, there is an angel and a devil

Vladimir Igorevich Arnold
(professor at the Steklov institute in Moscow)
Interview by Marimar Jiménez

Translation of: En todo matemático hay un ángel y un demonio. *El País* (*Futuro*), June 29, 1994, p. 3.

Translated by Daniel Peralta-Salas.

Regarded as one of the greatest figures of this century in his discipline, this Russian mathematician was conferred the degree of doctor *honoris causa* by the Universidad Complutense de Madrid last week. He urges to overcome the current *suicidal divorce* of mathematics from the rest of the sciences, and points out that some hot topics, such as Fermat's Last Theorem, have little to do with this science.

Vladimir Igorevich Arnold, 57, holds the opinion that French mathematician Henri Poincaré's idea that mathematics and physics are the same thing is relevant again. Arnold, rather than talking about specific aspects, prefers to talk about general topics related to his expertise, on which he has striking opinions for non-experts.

Question: What is the relation between mathematics and catastrophe theory?

Answer: Understanding of mathematics by the general public is minimal, if any at all. Therefore mathematics and other sciences need to do some advertising: cybernetics, fractals, catastrophe theory, ... they all fit within this objective.

Q: Would it be then a less serious part of mathematics?

A: Simply, it is not mathematics, it is a speculation based on it. Catastrophe theory makes use of mathematics.

Q: Why did you get into this world?

A: In Russia, there is a long tradition of mathematical studies since childhood. Five-year-old kids are given mathematical problems that are very easy to formulate, but difficult to solve. An example: There is a barrel of wine and a glass of tea. You take a spoon of wine from the barrel and put it into the tea. Now you have a non-uniform cocktail in the glass. You take a spoon of this cocktail and take it back to the barrel of wine; then there is a foreign liquid in the barrel, as well as in the tea. The problem is: Which of the two places has more foreign liquid: is there more wine in the tea or more tea in the wine?

Q: And what is the solution?

Translation of: *El País* (*Futuro*), June 29, 1994, p. 3,
Translated by Daniel Peralta-Salas

A: The solution is based on the so-called conservation law. The level of liquid in the glass is the same as in the beginning, and the same with the barrel, because the part of tea that was extracted was exactly replaced. Therefore, the amount of foreign liquid in both places is identical. This is real mathematics. There are no numbers here, but the reasoning is mathematical.

Q: What was the first problem you solved?

A: The first one I solved took me a whole day. It was the following: Two old ladies begin walking from two points in the opposite directions. The points can be Alcalá de Henares and Madrid. They start at dawn and meet at some point at noon. Then they continue their trips with the same speed as they had. One arrives in Madrid at 4 pm, the other arrives in Alcalá at 9 pm. The problem is to find out at what time the sun rose that day.

This problem can be solved without algebra. It is like one of the old problems of merchants in the XVIII century. In those days algebra did not exist yet. When I solved it I felt I had discovered something very unusual. Something very special. It is a pleasure difficult to compare with anything else.

Q: How old were you?

A: 12 years old.

Q: Have you solved many important mathematical problems since then?

A: It is an impossible question to answer. It is the same as if you ask a surgeon what is the most important surgery he has done. He does several surgeries every day. But perhaps, the most famous problems associated with me are stability of the planetary motion, impossibility to predict the weather well in advance, and chaos theory. Also the problem of possible configurations of real algebraic curves in optics and in quantum mechanics. The theory I have developed is called quantum catastrophe theory by physicists.

Q: Can you explain your theory?

A: Yes. It is possible to explain everything. In a cup, when the sun is shining, it is very easy to see that the shadow draws a curve, a cusp. If the cup is moved, the cusp does not disappear. It moves as well. The problem is why the cusp appears and why it persists. The explanation from the mathematical viewpoint is the same as why glasses help eyesight. Ultimately, it is the same mathematical theory that explains functioning of the eye lens, as well as colors of the rainbow.

Q: Should all mathematics have practical applications?

A: One never knows when a theory will prove to be applicable. In fact, the most applied parts of mathematics sometimes have been created with a different spirit. It has taken 2000 years to find applications for conic sections of ancient Greeks.

Quantum mechanics is based on the theory of differential equations, which was created without having any applications in mind. The Galois theory is now the main mathematical tool used in cryptography by agencies such as the KGB or the CIA, and it was created without any practical or political idea more than 150 years ago. In my experience, applications always come from 10 to 20 years after the creation; and much later for some.

Q: How do computers influence mathematics?

A: A lot. I have experience with French students in the University of Paris, who asked me to help decide if the fraction 4/7 is bigger or smaller than one, because they had forgotten the calculator at home. Anyway, I think that computers are useful for mathematics in the same way as a ruler is useful for drawing. Just like a straight line made with a ruler by a person who can draw will be better than that made by the best painter, a computer can make a person with average abilities go further than the best genius.

Q: The feeling of non-experts is that mathematics is becoming more and more complicated, stranger and stranger. Even scientists from other areas see this as well.

A: For several decades there has been an attitude within mathematics, which I would call criminal, that led to a divorce from several sciences. By now this suicidal tendency has essentially changed, and the direction of modern mathematics is to return to be the same science as modern physics. But it will take several decades before this gets noticed, especially in teaching children. We are still seeing disasters originated by the scholastic times of mathematics. Mathematics is a name for the part of theoretical physics that currently is out of fashion.

Q: Why did this criminal attitude occur?

A: This tendency has existed since Newton and Leibniz. The latter represented the scholastic trend and Newton represented applications to the real world. This was already 300 years ago. The importance of both trends has been oscillating with some periods. In the words of Hermann Weyl "Inside every mathematician the devil of abstract algebra is fighting with the angel of geometry". For some years, the devil was winning, but now we are in a good condition.

Q: How does a mathematician work?

A: There are many different styles. I solve many problems while I am skiing wearing a swimsuit. I leave my house in Moscow with the problem in my head; after skiing 40 km I return with some solution in mind. The next step is to check the solution with a pencil and paper. Most times the solution is wrong, but then a new problem has been created.

Of course, sometimes you have to do long computations for hours, either with a pencil and paper or with computers.

Q: Do you think the stir that arose with Fermat's theorem was reasonable?

A: I am not interested in this problem. It has an advertising effect, the same as with catastrophe theory. As the great Russian physicist Landau said: Prime numbers are made to be multiplied and are very useful in physics because any number can be decomposed as a product of them. However, some mathematicians are studying the decomposition of numbers as a sum of prime numbers. Why to insist on the desire to add them?

Will Russian mathematics survive?

V.I. Arnold

V. I. Arnold is a professor at the Steklov Mathematical Institute in Moscow.

The disaster of the Bolshevik Revolution produced a paradoxical flourishing of mathematics in Russia. One may hope that the present catastrophic situation will produce a similar miracle after a nice atomic civil war or after the establishment of a dictatorship of Nazi or fundamentalist type. We still have brilliant undergraduate students, and the mathematical high schools still prepare excellent students for the universities. However, the danger of the complete extinction of these sources of good mathematicians is quite real if no help comes.

The two main reasons for the emigration of mathematicians from Russia (and from fSU in general) are the disastrous economic situation and the fear of a new totalitarianism. The economic situation is changing every day from bad to worse. One dollar was equivalent to approximately 125 roubles on the free market in the spring of 1992; in September the price of a dollar was approximately 350 roubles. The inflation seems to grow faster than exponentially. Counted in dollars or in commodities, the salary of a full professor was more or less stable at somewhere between ten and thirty dollars a month. The students' stipends were five or ten times smaller.

Food, rent, and other necessities in the USSR were traditionally so low that it was possible to survive. A student's stipend was sufficient to buy, say, 200 loaves of bread in a month. In 1992, prices (in dollars) have started to approach Western standards. Last summer, the student stipend was still sufficient to buy, I guess, about fifty loaves of bread a month, but the present inflation rate (reaching sometimes 10% in one day) is constantly reducing the effective salaries.

The Russian government is no longer able to support science. The sciences were traditionally supported because of their military usefulness. More than three-quarters of the Soviet economy was (and still is) military. This disproportion ruined the country.

According to the Russian official statistics, men's mean longevity in our country (sixty-one to sixty-two years) is approximately ten years less than in the Western countries (and, some years, less than it was in the USSR twenty years ago). There are cities where the mean longevity is thirty-five years. A small cannibal calculation shows that, given the size of the population in the USSR, the reduction of longevity by ten years produces the same loss as the immediate execution of about 80 million citizens. (Lemma: you lose approximately one-half your life when they shoot you.)

In the present economic situation, the drastic cut in funding for science is a natural (though very incompetent) action. The large-scale brain-drain is an evident result of the economic disaster combined with the permission to visit foreign countries. This brain-drain is equally annoying to the Russian and the American or European science administrations. It is one of the many points where the KGB and the CIA have common interests. Bad relations between our countries are vitally needed for both of them, and both are strong enough to enhance the return to the Cold War situation. The reintroduction of exit visas for Russian scientists is a clear sign of the direction of the changes. We still must return the "foreign" passport upon returning to Russia from abroad. The system is still the same, it is like a faucet; it is open and the water is running, but you can close it at any moment.

The fear of the return to Nazi-type totalitarianism is another reason for emigration, perhaps as important as the economic disaster. For seventy-five years, our country was ruled by secret groups of terrorists. The names they used have been systematically changed, and I do not even know what the present name is (*partija, commissija, upravlenie, commissariat, comitet, ministerstvo, politburo, presidium, soviet*, ...). In any case, the incompetence of the professional killers in most other domains is quite natural.

For many years the terror was implemented through the Communist Party, but twenty million thieves was too much for the ruined economy. They decided to fire the corrupted servants. The Party is no longer ruling officially, but the property is in the same criminal hands and the powerful *nomenclatura* is essentially the same, with minor personnel replacements.

The Party Committee room at the mathematical department at Moscow State University has been given to a newly established chair in mathematics. The chairman has not changed his place: he is the person responsible (over the last twenty years, I think) for the choice of the mathematicians deserving the permission of the Party and of the KGB to attend conferences abroad.

The university had elected a new Rector. The old one

Originally published in: *Notices Amer. Math. Soc.* **40**(2), 104–107 (1993),

Reprint by Springer Nature Switzerland AG 2022 with kind permission

was famous as an anti-Einsteinist. The new one is the former vice-rector, a mathematician, who had implemented the ideas of the former one both in science and in the administration. This Communist had risen to a position of gray eminence at the university through the channels of the Komsomol and of the Communist Party. He had created at the university a large number of "Administrative Councils". These people, approved by him, have finally "democratically" elected him as the new rector. He is responsible for the "gas chambers" system of the entrance examinations to Moscow University (the existence of which, as far as I know, has yet to be admitted by the university administration and has not been condemned by any official institution of Russia). He had replaced the old generation of professors by Communists and is responsible for the degradation of the scientific quality of the faculty.

The proportion of the KGB people in Soviet universities and in the Academy of Sciences is probably similar to that in East Germany (ranging from one in three professors to one in ten professors). Some of the Party and KGB people are real scientists and nice people. By the way, the Communists are the first to emigrate: they are better informed on the scale of the economic disaster and more active than the rank and file slaves.

The corresponding member of the Soviet Academy, Siforov, for many years headed the Moscow Institute for the Transmission of Information. He was one of those who was the most helpful to Moscow mathematicians, supporting the best research in mathematics. People like Dobrushin, Margulis, and Kontsevich (to name only a few) work at his Institute. In his reminiscences, published recently in Moscow, Siforov describes (among other interesting things) his work as a secret agent of the KGB (then under a different name) at a Leningrad scientific institution. His duty was to report the activity of his teacher, Admiral and Academician A.I. Berg. When Berg was finally arrested and sent to the GULAG prison, this secret agent Siforov supported his teacher's family and saved his teacher from starvation by sending food to the prison. When Berg was released (to become the leading figure in the Navy Radio and radar technology), he returned from the GULAG to the house of his student. He was aware of Siforov's role. They have remained close friends.

Be it good or bad, the new administration consists mostly of the Communist Party and KGB people (unless they come from Sverdlovsk). They have the ability, which they are not using, to do a lot of harm.

I received recently a pile (about one meter high) of letters, books, and journals sent to me from abroad ten or twenty years ago. This dangerous correspondence (for instance, Einstein's biography by A. Pais, 1982) has been carefully preserved for me at the *spezkhran* of Moscow University. At least one institution is still working as efficiently as it has for these seventy-five years, and this permits us to be rather optimistic for the future.

I leave to the psychiatrists and to the criminologists the study of the writings of our Mengeles who require the introduction of the official inequality of the "ethnically alien" people in Russia. The idea behind this is the belief that the irrational and mysterious Russian soul (so superior to all others, as any reader of Dostojevsky knows) will then miraculously save Russia (and later the whole Humanity) from the horrors of modern technological civilization. They are right in saying that they are not requiring physical extermination and that it is impossible for someone who cannot read Russian to appreciate the unbelievable affinity of their theories with the Nazi propaganda of the 1920s. However, mathematicians in Russia are able to understand the message, which enhances the panic emigration. The Pamjat broadcasts all over Europe would be impossible under the present conditions in Russia without the use of state-owned equipment.

My generation of compatriots deserves the present disaster: we supported the slavery for too many years and spent our lives working for an evidently criminal state. The younger generation is less guilty, and it is difficult to blame their unwillingness to continue to serve the unrepentant criminals.

What can be done to help to preserve the Russian mathematical school? (I am most familiar with the Russian situation, but the other fSU states have similar problems; all the fascist governments are discriminating against the "ethnic aliens", which in most cases outside Russia means "Russians".) I shall describe here several ideas (keeping in mind that the changing situation may make them meaningless tomorrow).

1. On September 15, 1992, the Mathematical College of the Independent University of Moscow started its third semester. The College was established by a Founding Committee and with the help of a group of scientific advisers (V.I. Arnold, A.A. Beilinson, R.L. Dobrushin, A.A. Kirillov, S.P. Novikov, A.N. Rudakov, Ya.G. Sinai, V.M. Tikhomirov, L.D. Faddeev, A.G. Khovansky, and M.A. Shubin). It is working in close contact with the Mathematical Physics College, founded by M.K. Polivanov and headed at present by O.I. Zavialov. The Mathematical College has recently elected a dean, A.N. Rudakov.

The Founding Committee has hired for the current academic year six professors (A.A. Kirillov, A.N. Rudakov, V.M. Tikhomirov, E.B. Vinberg, A.L. Gorodentsev, and A.B. Sossinsky), twenty-two instructors (including V.A. Vassiliev, S.M. Gusein-Zade, S.K. Lando, Yu.M. Neretin, B.L. Feigin, and D.V. Alekseevsky), and fifteen assistants.

The mathematical syllabus has been drafted by a methodological group (Arnold, Kirillov, and Rudakov). The present number of students in mathematics is sixty-one. They have been admitted after successfully passing difficult examinations. Forty-eight of them are also students at Moscow State University. The Independent University is in the process of registration under the new Russian law on education, passed July 30, 1992, and hence has no military education, which its students are advised to complete elsewhere. At present, the Mathematical College is renting working space at Moscow high school #2, not far from the state university.

The needs have been up to now covered, so to speak, by the selfless enthusiasm of various individuals associated with the college. $2000 has been received via contributions of Professors MacPherson, Khovansky, and Beilinson. The college estimates that its basic needs come to approximately $25,000 a year, which is perhaps underestimated by at least a factor of two.

The interested reader may obtain more details from the Founding Committee (N.N. Konstantinov, nsknkonst@glas.apc.org) or from the dean (A.N. Rudakov, rudakov@systud.msk.su).

2. CyrTUG (Cyrillic TEX Users Group) was born in May 1991. Now there are about fifty individual and twelve institutional members including the Mech.-Mat. faculty of Moscow State University, Keldysh Institute of Applied Mathematics, Ioffe Physics and Techniks Institute (St. Petersburg), Novosibirsk Mathematical Institute of the Siberian Branch of the Academy of Sciences, Nauka Publishers, and Mir Publishers (Moscow). The group is arranging courses, seminars, and meetings and publishes textbooks and other materials for nonprofit goals.

Help might include the permission to publish and distribute some American textbooks in Russian (TEX and mathematics). Second-hand computers might be extremely useful. Financial support and the exchanges of experts (in TEX and in mathematics) would be useful.

For further information, contact the CyrTUG President, Joseph V. Romanovsky (St. Petersburg University, jvr@math.lgu.spb.su) or the Executive Director, I. Makhovaya, CyrTUG, 2 Pervy Rizhsky Per., Moscow 129820 (irina@mir.msk.su).

3. The recent jump (by a factor of 100 and more) of the printing cost has destroyed the long tradition of the Russian translations of best mathematical books by Mir and Nauka. You may help to translate your (or others') books by buying fifty copies of the Russian version for $20 each (which is very cheap). If you are interested, contact A.S. Popov at Mir, Pervy Rizhsky Per.2, Moscow 129820 (asp@mir.msk.su).

There is also a project for producing new editions of the classical series of mathematical books for high schools (*Biblioteka Mathematicheskogo Kruzka, Popular Lectures in Mathematics, Library of Physics and Mathematical Schools, Kvant Library*). This would be an invaluable help to the new generations of mathematicians in Russia, since the books are in many cases real jewels. But one needs $10,000 to print 10,000 copies of a book of 400 pages.

A special editorial board has been formed at Mir to choose which books should be rereleased. If you have an idea for how this can be accomplished, contact Popov at the address above.

4. Something should be done to help directly the mathematical high schools. Second-hand computers could be very important, since children are still studying "informatics" with no computers in Russian schools. Another idea is to give grants ($500 each) to the teachers of the students with best results at the mathematical olympiads (about twenty of the best high school children a year). The grants might be distributed by the editorial board mentioned in item three above.

5. Invitations to fSU mathematicians for half-time positions or for visits of several months every year help solve economic problems for those who can come. However, do not forget that the cost of an airplane ticket from, say, Moscow to Paris, is equal to one to two years' earnings of a full professor in Russia. Also, the most important invitations are those sent to younger people, whose salary may be five or ten times smaller.

6. Do not miss the newly organized Russian "Oberwolfach" conferences in nice places, from Moscow forests to St. Petersburg's Euler Institute, from the Volga beaches to those of the Baycal lake.

Life in Russia is still very cheap if you have dollars, and you will still meet interesting mathematicians and nice people everywhere. In the present situation, when foreign books and journals are no longer coming even to the best libraries (we have no money to pay for even those which the libraries have already received), personal communications become more and more important. You will find a real greed for mathematical news among the younger generation of Russian mathematicians. No one wants to emigrate, if there is any hope for him or her at home.

7. The Russian Academy of Sciences either has sold or is trying to sell the Nauka Publishing house to the Pleyade group, known in Russia as the owner of *Penthouse* magazine.

The rights for the translation of Russian mathematical journals into English will be transferred, according to this decision, to a new Interperiodica (private?) company, created by the academy apparatchiks together with the Pleyade group.

The first step was the decision of the President of the Academy (in the summer of 1992) that all the journals belong to the Praesidium of the Academy and not to the editorial boards or to the scientific departments of the academy, which had created the journals and had considered themselves the owners of the journals; accordingly, they signed the contracts for the translations copyrights with the AMS and other scientific publishers in the West. This strange decision to transfer ownership to the Praesidium has become possible because of the strange legal situation in Russia, where property ownership is not well defined.

The evident result of the transfer of the translations from the scientific to the commercial publishers will be the sharp decline of the scientific quality of the translations.

What is perhaps less evident to the Western reader is that the royalty money paid to the authors and to the Editorial Boards of the Russian mathematical journals for the translations was a substantial help to active researchers in mathematics.

The new arrangement will probably divert this flow to the pockets of the Academy apparatchiks who own Interperiodica.

The London Mathematical Society (LMS) succeeded in preserving the rights of the English translation of the *Uspehi Mathematicheskih Nauk* (translated by the LMS as *Russian*

Mathematical Surveys). The AMS has been less successful, and hence the other journals (*Izvestia, Math. Sbornik, Funkt. Anal. and Appl.*, etc.) are in extreme danger, since the editors of the English translations have not secured the continuation of their contracts.

The problem of preservation of the scientific quality of the translations of the Russian mathematical journals into English is of great importance. Saving at least the best journals from the sharks of commercial publishing would be a great service to fSU mathematicians.

If you wish to help fSU mathematicians by sending them books, journals, reprints, and preprints, it will be highly appreciated. You may contact, for instance, the mathematical societies, which are independent nonprofit unions of mathematicians (at Moscow, St. Petersburg, Novosibirsk, and some other larger cities of Russia and of the fSU).

Thanks to those who will one way or another try to help Russian mathematics survive.

Will mathematics survive? Report on the Zurich Congress

V.I. Arnold

Every four years mathematicians from around the world gather together at their International Congress of Mathematicians to find out who are the new champions (as in the Olympic Games or the "Hamburg accounting" described by Shklovskii).

In August 1994 the Congress took place in Zurich. (This is the third time it has been held there; the very first Congress was in Zurich, in 1897.) The Congresses are not numbered, because not everyone agrees on which past Congresses should be included.

In the years prior to the Congress, a specially selected International Program Committee (whose membership is secret until the opening of the Congress) chooses invited speakers. This year's invited talks were 16 plenary talks of one hour and 156 sectional talks of 45 minutes. There are 19 sections (mathematical logic, algebra, number theory, ... to history of mathematics and mathematical pedagogy), and the talks were given in seven halls simultaneously. Every day one could attend six talks. An invitation to speak at the Congress is considered a great honor; it can be (unfortunately?) very important for the career of a mathematician in a very strained world job market.

This year among the 16 plenary speakers 3 were from the Russian school of mathematics (in Kyoto, 1990, there had been 4 out of 15). Among the 156 sectional speakers I counted 14 from the Russian school (in Kyoto, 19 out of 139). In these counts I did not consider the present place of employment. Why our standing has dropped from 14% to 9% in four years — say, by a third — remains to be explained.

At the Congress the names of those to be honored with Fields Medals were announced: J. Bourgain (France and and USA), E. Zelmanov (Russia and USA), J.-C. Yoccoz (France), P.-L. Lions (France). These medals, awarded to mathematicians aged at most 40, are often compared to the Nobel Prizes (there is no Nobel Prize for mathemat-

V. I. Arnold

If we follow Vladimir Nabokov, who said the right way to introduce Tolstoy is just to say, *"Anna Karenina,"* we may identify *V. I. Arnold* by mentioning ABC, ADE, KAM, Liouville tori, the 13th and 16th Hilbert problems.... Then there is "strange duality" of Lobachevskii triangles, which may be the first appearance (1974) of what is now called mirror symmetry. For recent news of him, in the form of plane curves and wave fronts, see the 21st volume of *Advances in Soviet Mathematics* (American Math. Society, 1994) and the 5th volume of the University Lecture Series (ibid.).

[1] This is a translation of an article in the general-circulation magazine *Eureka*, somewhat amplified by the author.

Originally published in: *Math. Intelligencer* **17**(3), 6–10 (1995), © 1995 Springer-Verlag New York
Extended translation from the magazine *Eureka*, September 7, **14**(20) (1994)

Reprint by Springer Nature Switzerland AG 2022

ics). The comparison is undeserved: unlike the Nobel Prizes, the Fields Medals pass by many of the truly outstanding people, and in particular Russians.

To give three medals at once to representatives of the French mathematical school, and all three of them noted for the art of manipulation of inequalities, is hardly a help to the international prestige of French mathematics[2]—particularly coming at a time when the president of the International Mathematical Union (the organization naming the Fields Committee) was a well-known French analyst. It is hard to affirm a causal relation between these circumstances; that would imply a quite unlikely degree of corruption. But it is one more reminder how invalid is the comparison of Fields Medals to Nobel Prizes. The Nobel Committee asks the opinion of a much wider group of specialists than the Fields Committee does, and as a rule it does not leave itself open to suspicions (even unfounded) of the sort pointed out above. (I am afraid I heard such suspicions expressed by all too many participants at the Congress, from many countries and specialties.)

This year, many at the Congress reacted to the names of the medalists with, "Who is he?" Well, according to Plutarch, for a young person a medal is not a reward but a deposit toward future achievement. Let us hope that this year's laureates go on to achievements justifying the investment.

I counted 10 women among the invited speakers. This included—a special honor—plenary talks in the opening session (M. Ratner, "Interaction between ergodic theory, Lie groups, and number theory") and the closing session (I. Daubechies, "Wavelets and other methods of localization in phase space"). Outside of the regular program was a special "Emmy Noether Lecture" by a woman mathematician. This was delivered by Academician O.A. Ladyzhenskaya (St. Petersburg).

At the General Assembly of the International Mathematical Union (a sort of mathematicians' UN) held in Lucerne just before the Congress, the American delegation proposed to "increase the representation of women and to achieve a better balance of ethnic groups" among invited speakers. This proposal was rejected by the Assembly as a hidden insult to both women and ethnic groups. One Assembly delegate remarked, "It is strange that, contrary to their usual practice, Americans didn't mention sexual minorities."[3]

Registered Participants in the Zurich Congress: 2370 (Down from 4102 at Kyoto)

Breakdown: from		
	USA	443
	Switzerland	229
	Japan	228
	Russia	194
	Germany	191
	France	162
	Canada	82
	...	

The Assembly did adopt a resolution to make public when appointed the identity of the Chair of the Program Committee (so that this person could come under pressure). Delegates from developing countries hoped that the entire deliberations of the Program Committee and the sectional committees would be made public. Countries without an established mathematical tradition tend to be represented in the Assembly by politicians rather than mathematicians. The resolution by the Assembly carries a very real danger which could have grave consequences for the world mathematical community.

The difference between the tenth best, who will be invited, and the eleventh, who will not, is very small.

Each sectional subcommittee ("panel") is supposed to name ten or so as best among the twenty or so most active people in its area (who did not give invited talks at previous congresses). The difference between the tenth best, who will be invited, and the eleventh, who will not, is very small. An attempt to make discussion of this difference public will only add to the weight of extra-scientific considerations (representation of different countries, sexes, nationalities, etc.). The relatively few women speakers at the Zurich Congress won this honor in fair competition with men, without allowances being made.[4]

The Assembly gave special attention to the public image of mathematics.

At the beginning of this century a self-destructive democratic principle was advanced in mathematics (especially by Hilbert), according to which all axiom systems have equal right to be analyzed, and the value of a mathematical achievement is determined, not by its significance and usefulness as in other sciences, but by

[2] Here I take a view contrary to that of P. Cartier.

[3] The remark was found offensive by many. Let me try to explain why. It was heard as a witty *reductio ad absurdum*: the speaker meant to thwart efforts for participation of women and minorities by putting them in analogy with efforts for participation of homosexuals. The wit relies, then, on the listener finding absurd the notion of defending homosexuals' participation. The insult to homosexuals was incidental to the speaker's intent to laugh down the American proposal, but it hurt. —Editor's Note.

[4] Certainly. But this fair competition occurred only after vocal demands that Congresses stop underrepresenting women. See Lenore Blum, *Mathematical Intelligencer* vol. 9, no. 2 (1987), 28–32. —Editor's Note.

its difficulty alone, as in mountaineering. This principle quickly led mathematicians to break from physics and to separate from all other sciences. In the eyes of all normal people, they were transformed into a sinister priestly caste of a dying religion, like Druids, parasitic on science and technology, recruiting acolytes in the mathematical schools by Zombie-like mental subjection.

...the angel of topology and the devil of abstract algebra fight for the soul...

Bizarre questions like Fermat's problem or problems on sums of prime numbers were elevated to supposedly central problems of mathematics. ("Why *add* prime numbers?" marvelled the great physicist Lev Landau. "Prime numbers are made to be multiplied, not added!")

Unfortunately, mathematicians themselves contributed a lot to entrenching this image of their science, especially to entrenching the myth of its inaccessibility to the uninitiated.

Hermann Weyl, one of the greatest mathematicians of our times (who worked, by the way, in Zurich), said, "In these days the angel of topology and the devil of abstract algebra fight for the soul of each individual mathematical domain."[5]

In the first half of the century, the devil was winning. A special "axiomatic-bourbakist" method of exposition of mathematics was invented, to make it perfectly pure. For example, suppose we are saying that the value of a product is unaffected by the order of the factors. If we want, we can define multiplication using "rules for addition of columns." That the answer is independent of the order of multiplication can be deduced purely formally from one of these rules without knowing anything of the content of the operation of multiplication. This formal proof is urged on students by the criminal bourbakizers and algebraizers of mathematics.

If we didn't know the content of the idea of addition—if we had not first counted apples or pebbles or (with Mayakovskii) cigarette butts or locomotives—clearly we could not understand the formal proof. It is convincing only to those who have undergone a distinctive algebraic perversion of the mind, and is useless for teaching and for all applications. The grave consequences of this perversion for mathematical education in Russia and elsewhere are well known. Whole generations of mathematicians came along knowing no other style of mathematics—and of course, no other sciences. Avenging their humiliating experience in school, leaders of most countries, like the proverbial pig under the oak, planned and implemented the extermination of mathematics. According to American data, this process will take 10 to 15 years.

[5] "Invariants," *Duke Math. J.* 5 (1939), see p. 500.

Their logic is simple. England got nothing for Newton inventing the calculus; or Germany for Leibniz devising the notation the whole world uses; or France for Poincaré creating modern mathematics (topology and dynamical systems), which is indispensable in, say, radio. The American leaders, in accordance with the opinion of their voters and taxpayers, are not about to fund fundamental research (such as mathematics) unless it is proved that countries where fundamental research gets funding (Russia, France) are better off than those where it gets almost none.

Selfish calculations by the separate states lead them to scrap fundamental research needed by humanity as a whole (especially mathematics) as soon as military confrontation ends: no star wars, no supercollider, no mathematics.

The return of contemporary mathematics to the mainstream of natural science, seen everywhere over the last

There is no scientific distinction between pure and applied mathematics, just a social one.

few decades, has not yet been reflected in the conception of mathematics and mathematicians held by the "person in the street." This is true for both "pure" and "applied" mathematics.

For that matter, there isn't a scientific distinction between pure and applied mathematics so much as a social one. The "pure" mathematician is paid to do mathematics, the "applied" mathematician to solve a particular problem. If a number-theorist were paid to solve the Fermat problem, then number theory would be an applied field—like the theory of Galois fields and curves over finite fields, where research is funded by the CIA, KGB, and similar agencies for purposes of cryptography.

Columbus as he set sail was in the position of the applied mathematician: he was being paid to solve a specific problem, finding a passage to India. The discovery of the New World, however, was more analogous to pure mathematics. Coastwise navigation brought the Spanish economy much more short-term benefit than the unprofitable voyages of Columbus.

Contemporary applications of mathematics, including "computer science" and applications of computers, draw on reserves of wealth accumulated by "pure" mathematics of previous generations. In contrast to geography, discoveries on the order of Columbus's are still possible—and happen yearly.

Explaining these discoveries to the uninitiated is, to be sure, not easy. Behold the Princeton mathematician John Conway come to address this problem before an audience of three thousand in the Kongressenhaus of Zurich. He appears on the brightly lit dais in shorts, sandals, and windbreaker. "Nobody knows," he says, "how to fill our ordinary three-dimensional space as densely as possible with identical spheres. It is supposed that the best way is to pack the balls in rows and layers, in the way I'll show you now." The lecturer pulls out of his windbreaker pocket something all crumpled like a handkerchief. This turns out to be a piece of some kind of plastic that quickly uncrumples and becomes a blue ball the size of a baby's head. "Let us put next to it a few more balls," says Conway and takes about ten more out of the same pocket. He lays them adjacent on the table so they form a lattice of equilateral triangles. "Now," says the lecturer, "let us put another layer on top" — and fishes in another pocket of the windbreaker for red balls. When the third layer (of green balls, from a third pocket of the windbreaker) has been put in place, everyone clearly understands the layered packing of all of space.

"Now I don't need this ball any more," says Conway, and takes a ball from the top of his pyramid and hurls it into the hall somewhere between the twentieth and the fortieth row. "I don't need these either," and he goes on throwing colored balls to all corners of the hall. When all the balls have been thrown (and caught with a happy shout by somebody in the audience), Conway remarks, "Now I don't need the windbreaker either," and takes it off and throws it on the floor. The shorts stay on throughout the lecture.

Speakers are trying to show what great scientists they are more than to impart something to the audience.

Eccentric as it was, Conway's was one of the most understandable talks in the Congress.

The trouble is the progressive conversion of congresses into Reputation Fairs: speakers are trying to show what great scientists they are more than to impart something to the audience, and they think their purpose is served by incomprehensible lectures. (This is especially so in the section talks.)

In my opinion, the best talks at the Congress were the one by Clifford Taubes (Harvard, USA) surveying the geometry of 4-dimensional manifolds (in connection with physics of gauge fields) and the one by Jürg Fröhlich of Zurich about his recent theory of the Quantum Hall Effect.

The topology of three or four dimensions has proved to be more complicated than either the topology of curves and surfaces or that of five or more dimensions. For example, only in 4 dimensions are there "fake Euclidean spaces," topologically equivalent to ordinary space but not admitting a global smooth coordinate system.

All these fake 4-spaces have, by the way, a nice description in terms of dynamical systems: they are the orbit spaces of certain smooth vector-fields (with no zeros) in the usual Euclidean 5-space. Yet, as far as I know, no one has ever written such a vector-field explicitly. May its components be elementary functions? polynomials?

Three interesting talks were devoted to the theory of *mirror symmetry*, a striking connection, recently discovered by physicists, between apparently unrelated mathematical theories, belonging to algebraic geometry, singularity theory, topology, and combinatorics of convex polyhedra. Many assertions of this theory remain so far only conjectures (corroborated by vast experimental data and by the equality of integers with a great many digits occurring in the different theories). Yet in the talks of M. Kontsevich ("Homological algebra of mirror symmetry"), of A. B. Givental' ("Homological geometry and mirror symmetry"), and of D. R. Morrison ("Mirror symmetry and moduli spaces of conformal field theories"), one could get a relatively harmonious view of a theory to come.

In the talk by Föllmer (Bonn) on financial mathematics, it was pleasant to hear about "Russian options," introduced by Shiryaev and Shepp. I had hitherto heard only of European and American options.

I was greatly impressed by the short description of the work of Nevanlinna Prize winner A. Wigderson, given at the opening session by Yu. Matijasevich (St. Petersburg). This is about new ideas in complexity of solution of mathematical problems and applications of probabilistic ideas to proof theory. Finally, we have the possibility in rigorous mathematics of finding proofs that are correct, not with certainty, but with an extremely small probability of error (say, 10^{-500}).

Everyone knows that some problems have solutions that are easy to verify but hard to find. An example is the decomposition of an integer into prime factors. If a factor is known, then one can verify the divisibility quite fast (even if the dividend has two hundred digits and the divisor one hundred). Yet finding a factor is very difficult: one has to just go through the possibilities, and the time required is enormous (growing exponentially with the number of digits of the given number). Problems of exponential complexity are out of reach of computers in practice, and will remain so however the technology may be perfected. This makes factorizations usable in schemes for transmitting secret information over public communication channels.

However, it remains an unproved conjecture that these classes really are more than polynomially complex (though a list has been assembled of thousands of problems, each of which is known to be "hard" if any is). The new advance is the elaboration of a broad theory based on the still-unproved hypothesis of the existence of hard problems. Especially interesting is the study of randomized algorithms — algorithms including random tests — and of possible means of derandomizing them — that is, replacing their stochastic elements by pseudorandom number generators. It is shown in particular that an arbitrarily complex proof can be made verifiable arbitrarily fast.

All this activity based on an unproved conjecture (with sometimes paradoxical consequences) reminds me of the work of Lobachevskii, who constructed a beautiful theory of his geometry, undeterred by having an unproved hypothesis at the foundation. Now we know that there are two geometries, one where Lobachevskii's hypothesis is satisfied and one where it is not. They simply describe the geometry of different surfaces.

American universities boast about what famous Russian mathematicians they have rejected.

It seems doubtful that there can be a mathematics that contains exponentially hard problems (impossible to solve without combinatorial search) and another that does not. In any case, various aspects of deterministic, randomized, and derandomized algorithms provided many interesting lectures at Zurich (the section on computer science).

Most talks at the Congress, however, were like sermons. The lecturers plainly didn't expect that listeners would understand anything. Sometimes they went so far as to state obviously false theorems to the respectfully silent auditorium. The sermon mood was so pervasive that most of the introducers didn't even ask for questions at the end. And when some old-fashioned professors, like J. Moser (Director of the Mathematical Institute of ETH Zurich, the principal mathematical center in Switzerland), did urge people to ask questions, very few listeners overcame fear of exposing their ignorance sufficiently to do so.

The talks differed from sermons, however, in not being free. For those not registered as participants, the fee to attend a talk was considerable, as for a concert or a play.

I take pleasure in reporting that representatives of the Russian school generally were on the more comprehensible side. It is part of our tradition that a survey talk should emphasize new ideas and illuminating examples and not technical details.

I find rather worrisome the distinct shift in interests of our younger researchers (especially those working in the West) from directions long pursued by us to those fashionable in the USA. Such a shift of interests (doubtless related to the difficult conditions of job-hunting in American universities, some of which boast about what famous Russian mathematicians they have rejected) is inevitably negative. World leaders in one field leave it to race in a pack of jostling competitors following some other leader. Could this explain the distinct decrease in the proportion of our mathematicians among speakers at Congresses?

It is a pleasure to note also the large number of young Congress participants, including graduate students, from Russia and other countries of the former Soviet Union. Their attendance was made possible by generous support from the Swiss Organizing Committee of the Congress and the Soros Foundation.

Swiss mathematicians did everything possible to make our stay pleasant: participants were offered trips all over Switzerland (Lucerne, Interlaken, Bern, etc.), trips to the mountains (to Rigi Kulm overlooking the Vierwaldstätter See), to the Rhine waterfall (comparable to Niagara), concerts of classical and folk music. I was impressed by the small and little-known Bürlet art gallery in Zurich—Rembrandt and Franz Hals, El Greco and Goya, Canaletto and Tiepolo, Greuze and Ingres, Corot and Courbet, Cézanne, van Gogh, Matisse, Pissarro, Picasso.

After the tiring Congress, I spent a day at the home of my old friend A. Haefliger near Geneva. We climbed from 1500m to 3000m in the mountains near the Rhone valley, about halfway between the Jungfrau and the Matterhorn, and I got to swim in a glacial lake. On the return I picked mushrooms, sorrel, blueberries, and wild strawberries, and made my hosts a dinner from these gifts of nature (overcoming their doubts as to their edibility). The next day I returned to Moscow.

Steklov Mathematical Institute
ul. Vavilova 42
117966 Moscow GSP-1
Russia

Why study mathematics? What mathematicians think about it

V.I. Arnold[1]

WHY MUST WE STUDY mathematics? In 1267 this question was answered by the English philosopher Roger Bacon: "[H]e who ignores it can not know the other sciences nor the things of this world.... And what is worse, men who are ignorant of it do not perceive their own ignorance, and therefore seek no remedy for it." At this point I could have ended my lecture, but people think that maybe something has changed over the last seven centuries . . .

Let's look at more recent evidence. One of the creators of quantum mechanics, Paul Dirac, says that when you build a physical theory, no physical conceptions should be trusted. So what *is* to be trusted? According to the famous physicist, only a mathematical scheme—even if, at first glance, it isn't connected with physics.

Indeed, physics has cast aside all the purely physical conceptions prevalent at the beginning of this century, while the mathematical models incorporated into the physicists' arsenal gradually acquire physical meaning. Here the stability of mathematics clearly manifests itself.

Thus, mathematical modeling is a fruitful method of understanding in the natural sciences. We'll approach mathematical models from another angle, examining the problems of mathematical education.

Three approaches to teaching math

In Russian mathematical education (both secondary and higher) we follow the European system based on the "Bourbakization" of mathematics. (Nicolas Bourbaki is the pseudonym of a group of French mathematicians who, since 1939, have been publishing a series of books in which the principal divisions of modern mathematics were presented formally—that is, by means of the axiomatic method based on set theory.)

The formalization of mathematics leads to a certain formalization of its teaching, demonstrating the costs of Bourbakization in mathematical education. Here's a dramatic example. Second-year pupils at a French school are asked, "How much is two plus three?" The answer follows, "Since addition is commutative, it's three plus two."

A truly remarkable answer! It's absolutely correct, but the pupils didn't even think about simply adding these two numbers, because their instruction laid stress on the properties of operations. In Europe, educators had already become aware of the shortcomings of this approach and began to back off from Bourbakization.

In the last few years, Russian mathematical education had undergone Americanization. It's based on the principle: teach what's needed for practical application. So someone who doesn't think he or she will need mathematics need not study it at all. Mathematics is optional for high school juniors and seniors—for instance, one third of high school seniors don't take algebra. The effect is illustrated by the following example. A test for 14-year-old American students asked them to estimate (not compute, but just estimate) what happens to the number 120 if we take 80 percent of it. Three versions of an answer were offered: it will increase, remain the same, or decrease. The right answer was chosen by about 30% of the tested students. This means that they checked answers haphazardly. The conclusion: nobody knows anything.

The second particular feature of the American approach to teaching math is computerization. The fascination with computers by itself doesn't contribute to the development of thinking ability. Take another example from an American test.

There are 26 students in a class. They are going on a car trip. A car can take one parent and four students. How many parents should be invited to help?

[1]These are slightly abridged notes from a lecture given by Arnold, one of the leading mathematicians of our time, at the Republic Institute for Advancing the Qualifications of Educators in Moscow on April 16, 1992. The notes were submitted by Prof. Y. Fominykh of Perm.—*Ed.*

Art by Yury Vashchenko

Originally published in: *Kvant*, no. 1/2, 5–15 (1993), © 1993 Russ. Acad. Sciences
English translation: *Quantum* **5**(1), 25–31 (1994), © 1994 Springer-Verlag. All rights reserved
Reprint by Springer Nature Switzerland AG 2022

A typical answer is 65 parents. A computer gives out 26 ÷ 4 = 6.5. And a student already knows that if the solution must be an integer, one should do something with the decimal point—for instance, just throw it away.

And now look at an example from an official examination in 1992 for students:

Which of the following pairs most closely resembles the relation between angle and degree—

(a) time and hour;
(b) milk and quart;
(c) area and square inch
(and so on)?

The answer is area and square inch, because the degree is the minimal unit of angle measure, and the square inch is the minimal unit of area, whereas an hour can be divided into minutes.

The authors of this problem were obviously taught according to the American system. I'm afraid we'll soon get to this level, too.[2] It's only surprising that there are so many outstanding mathematicians and physicists in the United States.

Today our mathematical education slowly turns from the European system to the American. As always, we are late, lagging behind Europe by about 30 years, so 30 years later we should be ready to set things aright and get out of the dead-end where we'll be driven by the American educational system with its pragmatism, optionality, and mass computerization.

Our traditional mathematical education used to be at a higher level and was based on the culture of arithmetic problems. Even as recently as twenty years ago some families still had copies of the old "merchant" problems. Now it's all gone. The algebraization of the last reform of mathematical education turns students into robots. It is the arithmetic approach that demonstrates the "meatiness" of the math we teach.

Consider, for example, these problems:

(1) There are three apples. One is taken away. How many are left?

(2) How many cuts with the saw are needed to divide a log into three pieces?

(3) Boris has three more sisters than he has brothers. How many more girls than boys are in his family?

From the point of view of arithmetic these are all different problems—their content is different. And the intellectual effort needed to solve the problems is quite different, too, although the algebraic model is always the same: 3 – 1 = 2. What strikes you first of all in mathematics is the surprising universality of its models and their incomprehensible effectiveness in applications.

As the great Russian poet Vladimir Mayakovsky said: "The man who first formulated that 'two and two is four' was a great mathematician, even if he obtained this truth from adding two cigarette butts to two cigarette butts. All his followers, even though they may have added immeasurably greater things—say, locomotives to locomotives—are not mathematicians." To "count locomotives" is the American way of teaching math. It is disastrous. The example of the development of physics at the beginning of this century shows that locomotive mathematics turned out to be worse than cigarette-butt mathematics: applied mathematics couldn't keep pace with physics, while theoretical mathematics supplied everything physicists needed for further development of their science. Locomotive mathematics lags behind practice: while we teach how to calculate with abacuses, computers appear. We must teach how to think, not how to push buttons.

Admittedly, a mathematical model doesn't always give immediate practical returns. Sometimes it may prove useful only after two thousand years. That's what happened with *conic sections.*

Conic sections and the law of gravity

Conic sections were discovered in ancient Greece and described by Apollonius of Perga (265–170 B.C.) in an eight-volume treatise. But the need for this theory arose only in the 16th century, when Johannes Kepler was deriving his laws of planetary motion.[3] His teacher, Tycho Brahe, had scrupulously measured the positions of the planets of the solar system in the observatory at Uraniborg over the course of 20 years. After his teacher's death, Kepler got down to the mathematical processing of the results of these observations and found that the trajectory of Mars, for instance, is an ellipse.

An ellipse is the locus of points such that the sum of the distances from these points to two given points, called the foci, is constant. A remarkable theorem—which, unfortunately, is not proved at school—says that the section of a cone by a plane tilted at a large enough angle to its axis is an ellipse. Its proof is pretty simple (see figure 1). The two spheres inscribed in the cone and

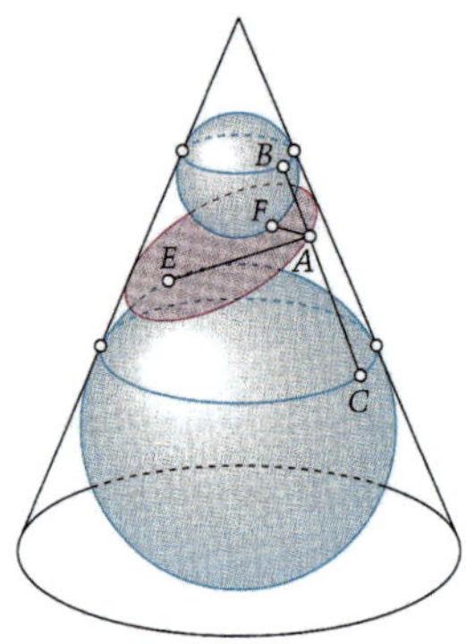

Figure 1

Ellipse with foci F and E and Dandelin spheres. Two tangents drawn to the same sphere from the same point are congruent, so FA + EA = BA + AC = BC, a constant.

[2]A New York professor, Joe Birman, explained to me that for him as an American the "correct" solution to this problem was absolutely clear. "The point is," he said, "I can imagine precisely the level of idiocy of the author of these problems."

[3]See also "An Act of Divine Providence" in the May/June 1993 issue of *Quantum* and "The Fruits of Kepler's Struggle" in the January/February 1992 issue.—*Ed.*

touching the plane (at the foci E and F of the ellipse in the section) used in this proof are called Dandelin spheres.

To understand the chain of Kepler's reasoning, we'll need a few simple facts about the geometry of the ellipse. It can be shown that the length of the *major semiaxis* of an ellipse OK (fig. 2), usually denoted a,

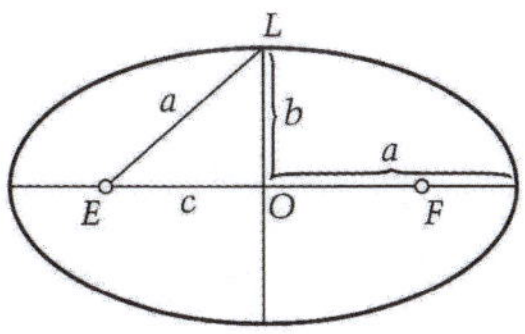

Figure 2

Foci, semiaxes, and eccentricity of an ellipse.

is equal to the length of the hypotenuse EL of the triangle with legs $b = OL$ (the *minor semiaxis*) and $c = EO$. The ratio c/a characterizes the ellipse's shape and is called its *eccentricity,* because it's proportional to the displacement of the foci from the center of the ellipse. The eccentricity is usually denoted by e.

By the Pythagorean Theorem, the ratio of the semiaxes equals $b/a = \sqrt{1-e^2} \cong 1 - e^2/2$ for small e. It follows that an ellipse with a small eccentricity is virtually indistinguishable from a circle. For instance, if $e = 0.1$, then the minor axis is shorter than the major axis by only 1/2,000 of the length of the latter. For such an ellipse with major axis 1 meter long, the minor axis will be shorter by only half a centimeter, yielding an unnoticeable difference between such an ellipse and a circle. However, the foci are 5 cm apart from the center, which is quite noticeable.

The formula $b/a = \sqrt{1-e^2} \cong 1-e^2/2$ (which means that the longer leg of a stretched right triangle is practically as long as the hypotenuse and gives a good approximation of the difference between their lengths) is one of the most remarkable facts in all of mathematics. (Unfortunately this isn't taught in school.)

For example, suppose you're

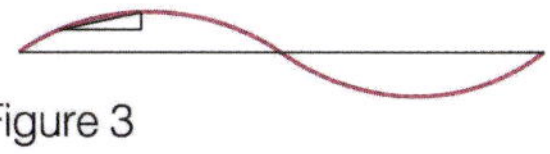

Figure 3

How much longer is a sine curve than a straight line?

coming back home along a sine curve. How much will your path be longer than the straight one (fig. 3)? The first impression (twice as long) exaggerates the length, of course. Yet it seems that the curved path will be about half again as long as the straight length. In actual fact, it's only about 20% longer. The reason is that the greater part of the sinusoid is only slightly tilted toward the axis, so the corresponding hypotenuses are barely longer than the legs.

Here is another application of this formula. The engines of the first jet planes were attached to the wings near the fuselage, so the air rushing from the engines was harmful to the tail assembly. The designers, who knew and felt the formula we've been examining, turned the engines by a small angle α (fig. 4).

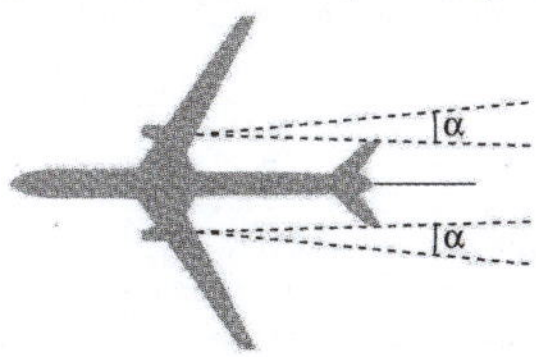

Figure 4

Saving the tail.

The tail assembly was saved (the deviation of the stream of air is proportional to α), while the net thrust remained practically the same (the loss is approximately $\alpha^2/2$, where α is measured in radians—for an angle of 3° only about 1/800 of the power is lost).

Let's turn back to Kepler. First he thought that the orbit of Mars is a circle. But the Sun happened to be offset by about 0.1 of the orbit's radius from the orbit's center. Kepler didn't stop at this result (remarkable in itself)—because he knew the theory of conic sections. Kepler knew that an ellipse with a small eccentricity looks very much like a circle, and he examined how the small remaining deviation of the orbit from a circle behaved. It's interesting that this verification was made possible only by the exceptional precision of Tycho Brahe's observations, performed with the naked eye. At the time, astronomers did not trust telescopes much, and even later, at the end of the 17th century, it still had to be proved that telescopic observations could attain as high a precision as those with the naked eye.

New physics often begins with refinement of the last significant digit of the previous theory. If Kepler had been satisfied with an eccentric circular orbit, or if Tycho Brahe's observations were less exact, the development of celestial mechanics (and, perhaps, of all of theoretical physics) could have been delayed, perhaps even for centuries.

The orbit of Mars turned out to be slightly oblate in the direction perpendicular to its diameter passing through the Sun by approximately half a percent—that is, by $e^2/2$. This led Kepler to the idea of elliptical planetary orbits.

If the theory of conic sections had not been worked out by mathematicians in advance, certain fundamental laws of nature wouldn't have been discovered in time, modern science and technology would not have arisen, and our civilization would have remained at the medieval level. Or, at the very least, the paths of history would have been totally different.

Kepler discovered the laws of planetary motion, but the fact that planets actually move in ellipses was proved by Sir Isaac Newton in his book *Mathematical Principles of Natural Philosophy* (1687), which laid the basis for all of modern theoretical physics. He derived from his law of universal gravitation the fact that planetary orbits are ellipses. Note that before Newton this problem was examined by his contemporary Robert Hooke. He studied the law of a body's motion in a gravitational field assuming that the force of gravity is in inverse proportion to the square of distance. Having approximately integrated the equation of the motion, Hooke drew orbits and discovered that they look like ellipses. His scientific honesty didn't allow him to call them ellipses, and he couldn't prove they were ellipses. So Hooke called them ellipsoids and proposed that Newton prove that Kepler's first law (that planets move in ellipses) follows from the law of inverse squares. Newton, who knew the ancient theory of conic sections very well, met this challenge by means of intricate constructions based on elementary geometry.

Later, second-order curves began to appear in scientific research more and more often. Why did this model prove so fruitful in application? Why, in particular, does the conic-section model describe planetary motion? It's a mystery. An enigma. There's no answer to this question. We believe in the power of rational science. Newton saw here a proof of God's existence: "This most graceful combination of the Sun, planets, and comets could not happen other than by the intention and the power of the mightiest and wisest creature. . . . It governs all not as the world's soul, but as master of the Universe, and by its supremacy should bear the name of God Almighty."

Modern space explorers also use the properties of conic sections when they plan the launching of satellites. So the foundation for modern physics and the scientific and technical revolution was established by a classical work by Apollonius. Yet the renowned Greek was thinking only about the beauty of the mathematical model when he investigated conic sections.

Computers, quantum mechanics, and Riemann surfaces

Another example is the story of the creation of computers. Long before the first computer came to being, its two principal mathematical components were lying in wait in mathematics: mathematical logic (specifically, Boolean algebra, created by George Boole, 1815–1864) and the schematic design of a computing machine. The first adding machine was designed by the French mathematician Blaise Pascal in 1641.

A third example is the development of wave mechanics by Erwin Schrödinger. By the time Schrödinger took up the oscillation problem, Werner Heisenberg's matrix version of quantum mechanics was already known. It wasn't clear how a discrete rather than a continuous spectrum could be obtained from the theory of waves in all of space. Schrödinger got help from the well-known German mathematician Hermann Weyl. Without Weyl's results in spectral theory on an infinite interval, we would never have heard about the famous Schrödinger equation. It's the same story again: a mathematician turns up with a theory ready to apply—that of boundary conditions at infinity—and all that was left to do was use it.

The next example is Riemann surfaces. They were introduced by the German mathematician Bernhard Riemann in the middle of the last century. These are the surfaces obtained by correspondingly cutting and gluing a number (or even an infinite number) of planes of a complex variable. Topologically, such a surface can be a sphere, a sphere with handles (fig. 5), and so on. The theory of Riemann surfaces was developed as part of the theory of functions of a complex variable. Later they proved useful in completely different problems. For instance, elliptic integrals get a simple geometric treatment on Riemann surfaces.

Karl Gustav Jacobi proved that Riemann surfaces "govern" two other problems:

1. Determining the number of ways of representing a given integer as the sum of four squares

$$N = x^2 + y^2 + z^2 + u^2;$$

2. Investigating oscillations of a pendulum, which leads to the differential equation

$$x'' = -\sin x.$$

Trig functions and counting snakes

As a fifth example, consider the so-called Bernoulli–Euler triangle:

$$\begin{array}{cccccccccccccccc}
&&&&&&&1&&&&&&&\\
&&&&&&1&&0&&&&&&\\
&&&&&0&&1&&1&&&&&\\
&&&&2&&2&&1&&0&&&&\\
&&&0&&2&&4&&5&&5&&&\\
&&16&&16&&14&&10&&5&&0&&\\
&0&&16&&32&&46&&56&&61&&61&\\
272&&272&&256&&224&&178&&122&&61&&0
\end{array}$$

It's filled in as follows. In line number 0, we write 1. Every odd line (first, third, and so on) is filled out from the right by writing in each position the sum of all the numbers in the previous line to the right of this position. Even lines are filled out in the same way but from the left.

The miracle concealed in this triangle was discovered a hundred years ago. The clue to it is given by this "simple" theorem (mathematicians

Figure 5
Riemann surfaces.

often disguise the fact that everything is quite simple):

$$\sec t + \tan t = \sum_{n=0}^{\infty} k_n \frac{t^n}{n!},$$

where the coefficients k_n are the numbers on the sloping sides of the Bernoulli–Euler triangle (k_n is the nonzero extreme number in the nth line).

It follows that the coefficients along the left slope yield the expansion into a power series of the function $\tan x$ (it's odd, and so its expansion contains only odd terms):

$$a_1 = \frac{k_1}{1!} = 1, \quad a_3 = \frac{k_3}{3!} = \frac{2}{6} = \frac{1}{3},$$

$$a_5 = \frac{k_5}{5!} = \frac{16}{5!} = \frac{2}{15}, \quad \ldots$$

and

$$\tan x = x + \frac{x^3}{3} + \frac{2x^5}{15} + \cdots.$$

Similarly, the right slope yields the expansion of the secant function.

The Bernoulli–Euler triangle delivers the topological classification of the real polynomials $x^{n+1}\ a_1 x^n + \ldots + a_{n+1}$ all of whose n critical values (local extrema) are real and different.

The graph of such a polynomial resembles a snake, so I'll call it a snake. All possible types of snakes for $n \le 4$ are shown in figure 6. Two snakes are assigned to the same type it they can be transformed into each other by a smooth change of independent and dependent variables preserving orientation.[4]

Consider, for instance, the snakes of the polynomials of degree 4 ($n = 3$). The three critical points will necessarily go in the order minimum-maximum-minimum. The topological type of a snake is determined by whether the last minimum is lower or higher than the first. So the number of types for $n = 3$ is 2.

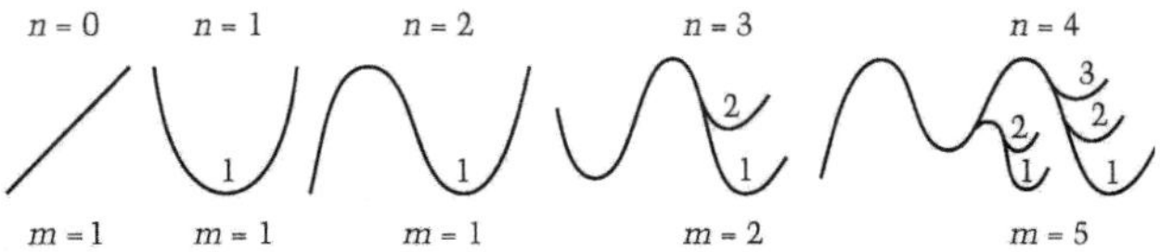

Figure 6

Classification of snakes: n is the number of bends, m is the number of snake types.

For snakes with $n = 4$ (max-min-max-min), the second maximum can be lower or higher than the first. In the first case there are two positions for the second minimum (below or above the first minimum, but always below both maxima); and in the second case three positions are possible. In all, there are five types.

If you have enough patience to draw all snakes with $n = 5$, you'll convince yourself that there are 16 of them, and for $n = 6$ there are 61 (these still lend themselves to drawing). Whenever Euler's number 61 crops up in some classification, we should look around and see if the rest of these numbers are nearby.

Now let's classify snakes by their "tails"—that is, by their rightmost critical points. Number as $1, 2, \ldots, n$ all n critical values in ascending order (from the bottom up). The number of a critical point will be called its height. The heights of the snakes' tails are indicated in figure 6. For instance, the five snakes with $n = 4$ are distributed by tail height as follows:

Height of tail	1	2	3	4
Number of snakes	2	2	1	0

Comparing the bottom line in our table with the numbers in the Bernoulli–Euler triangle (namely, the third line, counting from line 0), we see that they are the same. And, naturally, the sum $2 + 2 + 1 + 0 = 5$ (which is equal to k_4) is the total number of types of snakes with four bends.

You can make sure that the numbers of snakes with different tail heights for a certain number n of bends will always coincide with the corresponding line in the triangle (in some order). After this fact, unexpected enough in itself, has been noted (which necessarily requires some experimental work—sketching snakes), we can prove that the distribution of snakes by tail height satisfies the recursive law that defines the Bernoulli–Euler triangle.[5]

As to the analytic formula

$$K(t) = \sec t + \tan t$$

for

$$K(t) = \sum_{n=0}^{\infty} k_n \frac{t^n}{n!},$$

it can be proved as follows.

Take any snake with $n + 1$ bends, $n \ge 1$. Choose its highest local maximum and pull it upward to infinity. This will tear our snake into two shorter snakes and decrease the total

[4]That is, if the polynomials $p(x)$ and $q(x)$ that define the snakes are related by the equation $p(x) = f(q(g(x)))$, where the functions $f(x)$ and $g(x)$ have positive derivatives. You can think of these transformations as irregular shrinking and stretching of the coordinate plane along the axes without folding it or turning it over.—*Ed.*

[5]Take a snake with n bends, unbend its tail, and turn it over to get a new snake with $n - 1$ bends. Let the tail height of the first snake be h. Since the last critical point of a snake is always a minimum, the height of the next-to-last bend (a maximum) of the first snake is greater than h. Our operation turns this point into the last critical point of the second snake, so the tail height of this new snake is any number from 1 to $n - h$. This leads to the following relation for the numbers $s(n, h)$ of snakes with n bends whose tail height is h: $s(n, h) = s(n-1, 1) + s(n-1, 2) + \ldots + s(n-1, n-h)$. If we build a triangle from these numbers so that line 0 consists of the single number $s(1, 1) = 1$, and its $(n-1)$st line is $s(n, 1), s(n, 2), \ldots, s(n, n)$ for any even n, and $s(n, n), s(n, n-1), \ldots, s(n, 1)$ for any odd n, $n \ge 2$, we'll see that the above relation is just the one used to fill out the Bernoulli–Euler triangle. In particular, we see that the total number of snakes with n bends is equal to $s(n, 1) + \ldots + s(n, n) = s(n + 1, 1) = k_n$.—*Ed.*

number of critical points by one. Similarly, we can pull the lowest local minimum to $-\infty$ (and turn over the left piece to make it a regular snake).

This reduction leads to the following recursive relation for $n \geq 1$:

$$2k_{n+1} = \sum_{i=0}^{n} \binom{n}{i} k_i k_{n-1}.$$

Here the product $k_i k_{n-1}$ counts the pairs of all possible "subsnakes"; the binomial coefficient $\binom{n}{i}$ takes care of possible different mutual arrangements of the critical values in the two pieces (any i numbers chosen from $1, 2, \ldots, n$ can be the heights of the bends in, say, the left piece with respect to all the n bends in the entire snake); and the factor 2 on the left accounts for the two ways in which every $(n + 1)$th snake can be torn and so enters the sum on the right. For $n = 0$ the factor 2 must drop out (why?).

In terms of the function K, this relation can be written as the differential equation

$$2\frac{dK}{dt} = 1 + K^2$$

(see if you can verify it directly). It follows that $K(t) = \sec t + \tan t$ (this function satisfies the equation and the initial condition $K(0) = k_0 = 1$).

From optimal fishing to optimal reforms

And the last example. Let's consider a model of the variation in the number of a certain population of animals (say, the number of fish in a pond or ocean).[6] In the simplest case the situation is described by the model $x' = kx$, where $x = x(t)$ is the size of the population at time t—that is, the rate of change of the population is proportional (with the coefficient k) to the population itself. The solution to this equation is the exponential function $x(t) = x(0)e^{kt}$.

However, in actual practice the living conditions of the population worsen as x increases, and the coefficient k decreases. Taking, for instance, $k = a - bx$, we get the so-called logistic equation. In the case $a = b = 1$, its solutions tend to a stable population level $x = 1$ (fig. 7).[7]

If, in addition, a certain quota c for the capture and consumption of a part of the population is introduced, the equation becomes only a little more complicated: $x' = x - x^2 - c$. This is the simplest model of fishing.

For a quota $c < 1/4$, a stationary solution is established again (fig. 8). For $c > 1/4$, we get rapid extinction (fig. 9).

For $c = 1/4$ the stationary regime is at the level $1/2$ (fig. 10). But this state is unstable: small random deviations lead to a catastrophe—annihilation of the population. How can an optimal catch be achieved while keeping the population at a certain stable level? The answer is not to assign a fixed plan for the catch, but rather to introduce feedback—that is, a catch quota proportional to actual resources. In a model with feedback $x' = x - x^2 - kx$, the optimal value of the coefficient is $k = 1/2$. With this choice of k all solutions stabilize at the level $x_0 = 1/2$, which means that the average catch over a long period will be $kx_0 = 1/4$ (fig. 11). This is the same catch as with the highest permissible fixed plan for the catch.

Figures 7–11 present solutions to the model equation of the population change $x' = x - x^2 - c$. The red curves at left in all the figures graph the rate of change $v(x) = x - x^2 - c$ (in coordinates rotated for convenience); the blue curves at right are the solutions. The population grows at points with $v(x) > 0$, decreases for $v(x) < 0$, and is steady at points with $v(x) = 0$.

$c = 0$

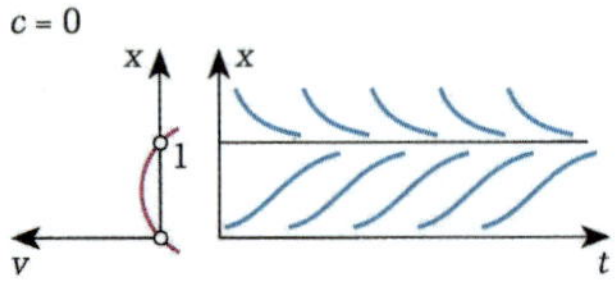

Figure 7

Stabilization of the population. All trajectories tend to the stable solution $x = 1$.

$c < 1/4$

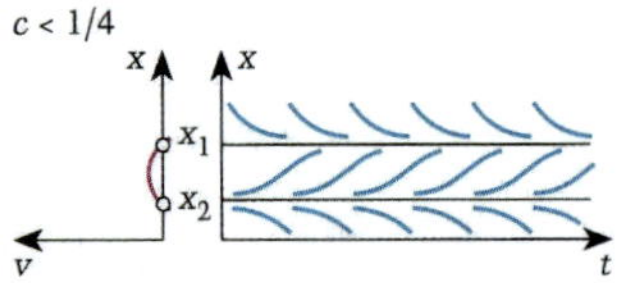

Figure 8

Model of fishing with a "moderate" catch. If the initial population $x(0) < x_2$, the population dies out; trajectories with $x(0) > x_2$ all approach $x = x_1$.

$c > 1/4$

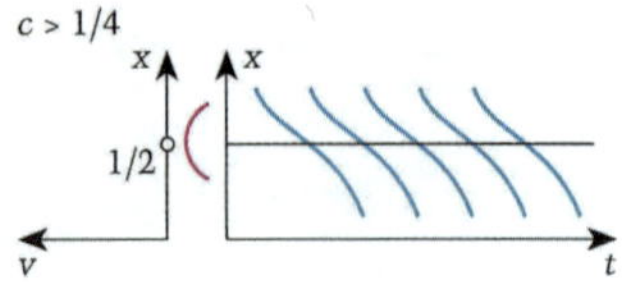

Figure 9

Overcatch. The population always dies out.

$c = 1/4$

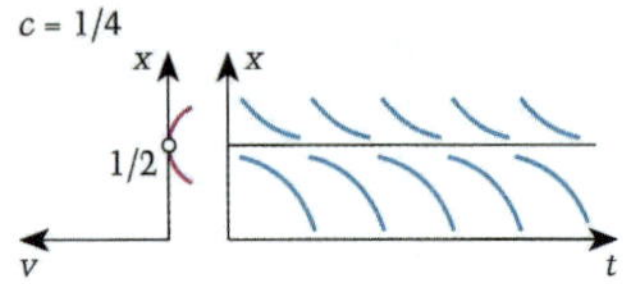

Figure 10

Destruction of the population by the optimal plan. The trajectories in the domain $x < 1/2$ lead to extinction; those with $x \geq 1/2$ tend to the steady state $x = 1/2$, but in the long run, due to small random perturbations, necessarily hit the dangerous domain as well.

$c = kx$

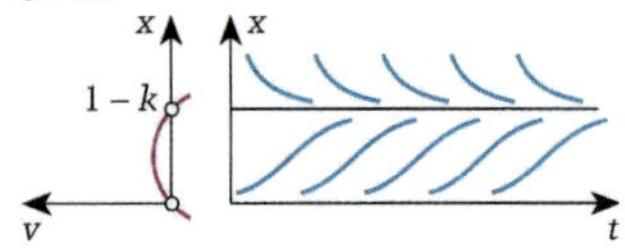

Figure 11

Stabilization of the system through feedback. (Compare with figure 7.)

[6]See also "The World according to Malthus and Volterra" in the July/August 1992 issue.—*Ed.*

[7]Don't be surprised at this strange population size: it's just a model, but a model that gives a more or less correct qualitative picture of the situation.—*Ed.*

Greater productivity in this case is impossible. But with a rigid plan the system loses its stability and is guaranteed to self-destruct, while feedback stabilizes it, and small variations in the coefficient k do not lead to disaster.

It wouldn't be bad if the persons who make crucial decisions were familiar with similar models and other rules for choosing strategic social options.

Simpler mathematical considerations—the fact that laws of nature are described by differential equations—allow us to understand certain seemingly paradoxical phenomena in our life. For several decades the state of the Russian economy has been a matter of concern for specialists: militarization, monopolism, and the general incompetence of the leadership caused the second derivative to become stably negative (that is, the rate of development steadily slowed). This didn't really scare those who didn't understand mathematics, because the first derivative was still positive (social well-being kept increasing). But mathematicians know that a permanently negative derivative, even of a higher order, ultimately makes the first derivative negative—that is, leads to a decrease in production and in the welfare of society, and this process of deterioration, when it becomes noticeable, will accelerate. Because of the inertia of the system, there are no means at all for instantaneously changing its state at this point, since changes of any kind affect only the sign of a higher derivative (for our *perestroika*, the third or even fourth derivative). Thus, the economic degradation we observe is caused by old mistakes made at the time of production growth rather than by wrong new decisions. Unfortunately, these elementary mathematical facts are too difficult to explain to a pillaged people inclined to ascribe all its hardships to failed reforms. Any reforms must certainly lead to a worsening, even if they are absolutely the correct steps to take.

Plans in this country were usually devised so as to optimize production for 20 years ("long enough for our lifetime"). It's clear to a mathematician that optimal planning of this sort must result in the complete destruction of all resources by the end of that period (otherwise, the remaining resources could have been used and, therefore, the plan wouldn't be optimal). Fortunately, the plans used to be "corrected" and were never fulfilled. However, the basic tendencies were kept, so roughly speaking, we had eaten all we had by the beginning of *perestroika*.

Attempts to create detailed "day-by-day" programs of economic reform are similar to attempts to plan the entire economy and are like trying to give minute-by-minute instructions to someone driving from Moscow to St. Petersburg: "At such-and-such a minute, turn right; at such-and-such a minute, turn left..." Success can be achieved only through feedback. That is, what's needed is not a program (trajectory) but, in mathematical terms, a vector field in the space of the system's states, a mechanism for making decisions as required by the attained state rather than the calendar date.

Some of these points should be kept in mind as well when it comes to reforming the educational system. Our examples show that "there is nothing more practical than a good theory." It's essential that educators not chase after the practical need of the moment, but rather have the long-term goals of society constantly in view.

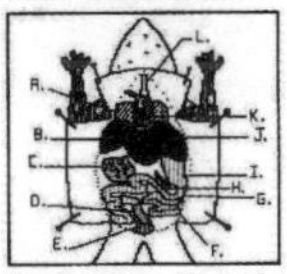

Preface to the Russian translation of "The Geometry and Physics of Knots" by M.F. Atiyah

V.I. Arnold

English originals: M.F. Atiyah. The Geometry and Physics of Knots. Lezioni Lincee. Cambridge Univ. Press, Cambridge, 1990, x + 78 pp. M.F. Atiyah. Geometry of Yang–Mills Fields. Lezioni Fermiane. Scuola Normale Superiore Pisa, Pisa, 1979, 99 pp.

Russian translation: Mir Publishing House, Moscow, 1995, 192 pp. Translators: V.N. Leksin and I.G. Shcherbak. V.I. Arnold's preface is on pp. 5–6.

Preface translated by Boris Khesin.

Fundamental laws of physics have a simple description in purely geometric terms.

This fact (remaining a mystery even today) was so striking for Newton that he regarded it as a proof of God's existence.

The entire development of theoretical physics has convincingly shown that only a consistent geometrization allows one to capture the full variety of observable phenomena. The achievements of Newton and Hamilton, Maxwell and Gibbs, Einstein and Dirac, Feynman and Yang deliver numerous and well-known examples of fruitfulness of geometric concepts in physics.

Today, however, we witness the converse process: the use in fundamental mathematics of concepts developed in theoretical physics. Free from the draining Bourbaki-type algebraic education, as well as from the necessity to prove (or at least formulate) rigorously their statements, physicists have managed to predict profound mathematical facts in topology and algebra, in number theory and algebraic geometry.

The present book by M. Atiyah, one of the greatest mathematicians of our time (master of Trinity College in Cambridge, UK, President of the Royal Society of London, a foreign member of the Russian Academy of Sciences, etc.) is a concise and informally written survey of several recent achievements in geometry based on the discovery of profound connections between topological quantum field theory (E. Witten), the theory of knot invariants (V. Jones), and the invariants of low-dimensional manifolds constructed by A. Floer and S. Donaldson.

Written with a mastership typical for M. Atiyah, this small book covering elementary basics of the new theories, along with the author's extended Fermi lecture notes on the geometry of Yang–Mills fields, which are included in the Russian edition as an appendix, allow any mathematician and theoretical physicist to get acquainted with fundamental (and hence in fact simple) ideas and notions that appeared at the border between physics and geometry. S. Donaldson writes that the best reference on the theory of the instanton and monopole

Translation of: The Geometry and Physics of Knots (Russian), Mir Publishing House, Moscow, 1995, pp. 5–6,
Translated by Boris Khesin

equations are the Fermi lectures by M. Atiyah, but they are hard to obtain.[1] You are holding in your hands just those hard-to-get Atiyah's lecture notes.

[1] See p. 101 of the paper: S.K. Donaldson. The Yang–Mills equations on Euclidean space. In: Perspectives in Mathematics. Anniversary of Oberwolfach 1984. Edited by W. Jäger, J. Moser, and R. Remmert. Birkhäuser Verlag, Basel, 1984, 93–109. The editors are grateful to Prof. S.K. Donaldson for pointing out this reference. (Editors' remark)

A comment on one of A.D. Sakharov's "Amateur Problems"

V.I. Arnold

Translation of: *Kvant*, 1991, no. 5, p. 12.
Translated by Boris Khesin.

Problem 6 (A.D. Sakharov). Here are formulations of two families of theorems related to number theory.

1) The sequences $a_n = n! + 1$, $a_n = [n \ln \ln n]! + 1$,[1] etc. (here [ ] stands for the integer part of a number) contain an infinite number of primes, since the series $\sum(1/\ln a_n)$ diverges.

2) The sequences $b_n = (n^2)!+1$, $b_n = [n(\ln n)^\alpha]!+1$,[2] etc. for which the series $\sum(1/\ln b_n)$ converges, contain a finite number of primes.

How could such a problem occur to A.D. Sakharov?

Here we present a comment by academician V.I. Arnold, a member of the editorial board of our journal.

Comment. To the best of my knowledge, the "theorems" of A.D. have not been proven. However, at the "physical level of rigor" they follow from the "randomness" of the distribution of primes among positive integers, the theorems on asymptotics of the number of primes, and the following well-known fact of probability theory.

Lemma 1. Let independent events $1, 2, \ldots$ have probabilities $p_1, p_2, \ldots$ Then the probability of occurrence of an infinite number of those events is equal to 1 if the series $p_1 + p_2 + \cdots$ diverges, and it is equal to 0 if this series converges.

Proof. The probability of occurrence of none of those events is $p = (1 - p_1)(1 - p_2) \cdots$; the probability of occurrence of only the ith event is $p\,r_i$, where $r_i = p_i/(1 - p_i)$. The probability of occurrence of events only with numbers $i, \ldots, j$ is $p\,r_i \cdots r_j$. If the series diverges, then $p = 0$, and hence for every combination of a finite number of events its probability is also equal to 0. Consequently, the probability of occurrence of a finite number of events only is 0. On the other hand, if the series converges, then the probability that only a finite number of events will occur is equal to

$$p\left(1 + \sum r_i + \sum r_i r_j + \cdots\right) = p(1 + r_1)(1 + r_2) \cdots = 1,$$

[1] $a_n = [n \ln n]! + 1$ in the original. This is obviously a misprint, cf. part 2) of Problem 6. (Editors' remark)

[2] Here $\alpha > 0$. (Editors' remark)

Translation of: *Kvant*, no. 5, p. 12 (1991),
Translated by Boris Khesin

since $1 + r_i = 1/(1 - p_i)$. The lemma is proven.

Apply Lemma 1 "without hesitation" to the (of course, not at all random) "events" "$n! + 1$ is a prime", $n = 1, 2, \ldots$ We set the "probability" p_n of such an event to be the fraction of primes among the integers from 1 to $N = n! + 1$ (i.e., the ratio of the number of primes smaller than N to N itself).

Lemma 2. The series $\sum p_n$ diverges.

Proof. The average density of the distribution of primes on the segment from 1 to N decreases inversely proportional to $\ln N$ as N increases (Hadamard's theorem),[3] $p_n \sim 1/\ln N$. Furthermore, $\ln(n!+1) \sim \ln n! \sim n \ln n$ according to Stirling's approximation, while the series $\sum 1/(n \ln n)$ is divergent. Therefore, so is the series $\sum p_n$. Lemma 2 is proven.

Assuming independence of the events "the number $n!+1$ is a prime" for different n (whatever that means) and (illegitimately) applying Lemma 1, we conclude that the "probability" that $n! + 1$ is a prime for an infinite number of n is equal to 1, since the series $\sum p_n$ diverges, according to Lemma 2.

This "physical" reasoning (i.e., one not making rigorous sense) convinces any sane person (but a mathematician) that the sequence $n! + 1$ does contain infinitely many primes.

For the sequence $n^2! + 1$, the series made of $p_n \sim 1/\ln(n^2!) \sim 1/(2n^2 \ln n)$ converges. Consequently, the probability of infiniteness of the number of primes of the form $n^2! + 1$ is equal to 0 (whatever this "probability", meaningless from a mathematician's viewpoint, stands for). A reasonable person will conclude that the set of primes of the form $n^2! + 1$ is finite.

Indeed, $4! + 1$ is divisible by 5, while $9! + 1$ is divisible by 19, and actually I do not know a single prime of the form $n^2! + 1$ beyond $1! + 1 = 2$.

On the other hand, the numbers $n! + 1$ are prime for $n = 1, 2, 3, 11$, while for $n = 4, 5, 6, 7, 8, 9, 10$ they have nontrivial divisors (respectively, $5, 11, 7, 71, 61, 19, 11$).

[3] See M.I. Bashmakov's article "On Bertrand's postulate", *Kvant*, 1971, no. 5, 4–7.

Comments on two of A.D. Sakharov's "Amateur Problems"

V.I. Arnold

Translation of: *Academician A.D. Sakharov, Scientific Works*, TsentrKom Publishing House, Moscow, 1995, pp. 509–511; see pp. 503–504 for the formulations of Problems 8 and 11.

Translated by Boris Khesin.

Problem 8 (A.D. Sakharov). Here are statements of two families of theorems related to number theory.

1) The sequences

$$a_n = n! + 1, \quad a_n = [n \cdot \ln \ln n]! + 1, \quad \text{etc.}$$

(here [ ] stands for the integer part) contain an infinite number of primes, since the series $\sum 1/\ln a_n$ diverges.

2) The sequences

$$b_n = (n^2)! + 1, \quad b_n = [n(\ln n)^\alpha]! + 1, \quad \alpha > 0, \quad \text{etc.}$$

for which the series $\sum 1/\ln b_n$ converges, contain a finite number of primes.

Give an estimate for the number $S(n)$ of primes among the first n terms of the sequence $n! + 1$.

A.D. Sakharov's comment:

For the heuristic consideration of this problem (as well as of its versions and generalizations) I employed a semi-intuitive notion of probability of being a prime number; one should also take into account that a number of the form $a_n = n! + 1$ is not, of course, divisible by any prime smaller than n. This reasoning is certainly far from the requirements of mathematical rigor. I do not know if there exists a more rigorous consideration of the problem.

Comment on Problem 8 (V.I. Arnold). To the best of my knowledge, the "theorems" of A.D. have not been proven. However, at the physical level of rigor they "follow" from the "randomness" of the distribution of primes among positive integers and the following well-known fact of probability theory:

<u>Lemma 1.</u> *Let independent events $1, 2, \ldots$ have probabilities $p_1, p_2, \ldots$ Then the probability of occurrence of an infinite number of those events is equal to 1 if the series $p_1 + p_2 + \cdots$ diverges, and it is equal to 0 if this series converges.*

Translation of: *Academician A.D. Sakharov, Scientific Works*, TsentrKom Publishing House, Moscow, 509–511 (1995),
Translated by Boris Khesin

Proof of Lemma 1. Denote the probability of occurrence of none of those events by $p = (1 - p_1)(1 - p_2)\cdots$. Then the probability of occurrence of only the ith event is $p\,r_i$, where $r_i = p_i/(1 - p_i)$. The probability of occurrence of events only with numbers $i, \ldots, j$ is equal to $p\,r_i \cdots r_j$.

If the series diverges, then $p = 0$, and hence the probability of occurrence of exactly the given finite combination of events is equal to 0 for any combination. Therefore, the total probability of occurrence of only a finite number of events is also equal to 0.

On the other hand, if the series converges, then the probability that only a finite number of events will occur is equal to

$$p\left(1 + \sum r_i + \sum r_i r_j + \cdots\right) = p(1 + r_1)(1 + r_2)\cdots = 1,$$

since $1 + r_i = 1/(1 - p_i)$. The lemma is proven.

Apply Lemma 1 without hesitation to the "events" $\{n! + 1 \text{ is a prime}\}$, $n = 1, 2, \ldots$ We set the "probability" p_n of such an event to be the fraction of primes among the integers from 1 to $N = n! + 1$ (i. e., the ratio of the number of primes smaller than N to N itself).

Lemma 2. The series $\sum p_n$ diverges.

Proof of Lemma 2. The average density of the distribution of primes on the segment from 1 to N decreases inversely proportional to $\ln N$ as N increases, by Hadamard's theorem: $p_n \sim 1/\ln N$. However, $\ln(n! + 1) \sim n \ln n$ according to Stirling's approximation, while the series $\sum 1/(n \ln n)$ diverges. Therefore, so does the series $\sum p_n$. Lemma 2 is proven.

Assuming the primality (property of being prime) of the numbers $n! + 1$ and $k! + 1$ to be independent events with probabilities p_n and p_k (whatever all that means) and (illegitimately) applying Lemma 1, we obtain that the "probability" that the set of primes of the form $n! + 1$ is infinite is equal to 1, since the series $\sum p_n$ diverges, according to Lemma 2.

This "physical" reasoning (i.e., one not making rigorous sense) convinces any sane person (but a mathematician) that the sequence $n! + 1$ does contain infinitely many primes.

For the sequence $n^2! + 1$, the series made of $p_n \sim 1/\ln(n^2! + 1) \sim 1/(n^2 \ln n^2)$ converges. Hence the "probability" of infiniteness of the set of primes of the form $n^2! + 1$ is equal to 0 (whatever this "probability", meaningless from a mathematician's viewpoint, stands for). A reasonable person will conclude that the set of primes of the form $n^2! + 1$ is finite.

Remark by the editors of the collection of Sakharov's works. The statement that "a sequence a_n contains an infinite number of primes if the series $\sum 1/\ln a_n$ diverges", as well as its converse, should be understood only "on average", in the probabilistic sense: every time one **randomly** takes a number from a certain neighborhood of a_n (rather than exactly the number a_n), while the width of this neighborhood does not exceed a by an order of magnitude. Otherwise for each of the above statements one can easily find particular counterexamples. For instance, the series $a_n = 2^n$ contains only one prime (for $n = 1$), while the series $\sum 1/\ln a_n$ diverges. Conversely, let b_n be the smallest prime from the set of primes greater than, say, 2^{n^2} and different from $b_1, b_2, \ldots, b_{n-1}$. Such a prime exists for every n, since the set of primes is infinite. Thus we have an infinite sequence of primes for which $\sum 1/\ln b_n < \infty$.

Remark by the editors of the present volume of Arnold's Collected Works. Here are other counterexamples which are closer to the sequences dealt with in Problem 8: $a_n = n!$ and b_n is the smallest prime that is greater than $n^2!$ and b_{n-1}. In fact, the statement that "a sequence a_n contains an infinite number of primes provided that the series $\sum 1/\ln a_n$ diverges" does not hold even in the probabilistic sense pointed out in the preceding paragraph. Indeed, consider the segments of positive integers

$$a_{n,1} = n! + 2, \quad a_{n,2} = n! + 3, \quad \ldots, \quad a_{n,n-1} = n! + n$$

of unboundedly increasing lengths $n - 1$ ($n \geq 2$). One can also consider narrower segments centered at $\bar{a}_n = (a_{n,1} + a_{n,n-1})/2 = n! + n/2 + 1$. The series $\sum 1/\ln \bar{a}_n$ diverges. However, **all** the numbers $a_{n,k}$ are composite since $a_{n,k} = n! + k + 1$ is divisible by $k + 1$. On the other hand, the principle "let an increasing sequence c_n of positive integers be *generically arranged with respect to the set of primes*, then this sequence contains an infinite number of primes if and only if the series $\sum 1/\ln c_n$ diverges" seems to be true (and, maybe, it can be rather easily formulated in a rigorous manner and proven). That a particular sequence is not generically arranged with respect to the set of primes is usually obvious (like in all the counterexamples above). Unfortunately, it is not clear at all how to establish (even quite non-rigorously) that a given sequence is generically arranged with respect to the set of primes. Consequently, this principle is probably not very useful in the analysis of specific individual sequences.

Problem 11 (A.D. Sakharov)**.** Here one presents a simple example of a hydrodynamic motion causing the "magneto-hydrodynamic dynamo" effect. In an incompressible conducting medium, consider a closed toroid, together with a "frozen-in" magnetic field H, of large radius R, the cross section of the toroid being S and its magnetic flux being HS (Fig. 1*a*). The toroid is stretched by a factor of 2:

$$R \to 2R, \quad S \to \tfrac{1}{2}S, \quad H \to 2H.$$

The field energy $H^2SR/4$ is increasing by a factor of 4 (Fig. 1*b*). Then the toroid is twisted as a figure-eight while preserving its cross section and the cord length (Fig. 1*c*). Then the figure-eight is folded in half into a toroid of the original size, but with a field that is twice as strong (Fig. 1*d*). The whole process is repeated unboundedly many times.

Comment on Problem 11 (V.I. Arnold)**.** Apparently this construction was described for the first time by S.I. Vainshtein and Ya.B. Zeldovich in the paper "On the origin of magnetic fields in astrophysics (Turbulent mechanisms of 'dynamo')", *Uspekhi Fiz. Nauk*, 1972, v. 106, no. 3, 431–457 (in Russian). See also the books "Turbulent Dynamo in Astrophysics" by S.I. Vainshtein, Ya.B. Zeldovich, and A.A. Ruzmaikin (Nauka, Moscow, 1980, in Russian) and "Magnetic Fields in Astrophysics" by Ya.B. Zeldovich, A.A. Ruzmaikin, and D.D. Sokoloff (Gordon and Breach, New York, 1983).

One should also mention a seemingly similar mechanism suggested by H. Alfvén in 1950,[1] which is actually not equivalent to the one proposed here. For a detailed discussion

[1] "Discussion of the origin of the terrestrial and solar magnetic fields", *Tellus*, 1950, v. 2, no. 2, 74–82. (Editors' remark)

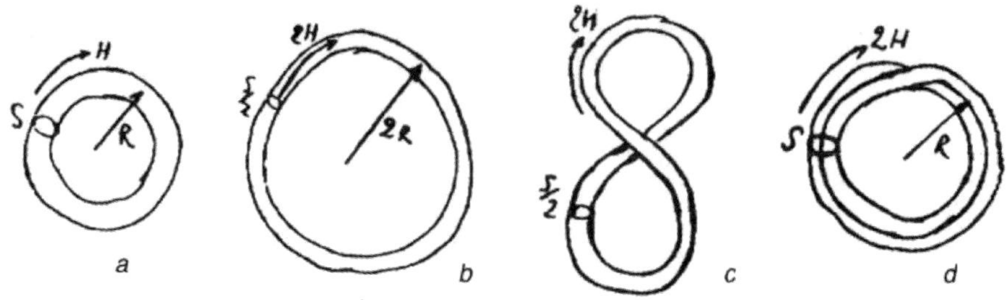

Fig. 1 Figures by A.D. Sakharov

of the difference see "Selected Works. Vol. 1. Chemical Physics and Hydrodynamics" by Ya.B. Zeldovich (Nauka, Moscow, 1984, pp. 44–45, in Russian[2]).

In mathematics a mechanism similar to the one proposed by A.D. and Ya.B. (in a *compressible* medium) was earlier considered for other purposes in the theory of dynamical systems for constructing "strange attractors", so-called "van Dantzig's solenoids" (in the thirties). There is still no rigorous mathematical foundation of the proposed mechanism (for an incompressible medium), although most experts believe in it. From the mathematical viewpoint, one would like to construct a volume-preserving diffeomorphism g (a smooth mapping with a smooth inverse) of a bounded domain V of the three-dimensional space into itself, that increases the energy of the frozen-in magnetic field exponentially with iterations of the mapping:

$$E(g_*^n V) \geq \lambda^n E(V), \quad \lambda > 1.$$

The frozenness means that the field vectors can be thought of as drawn on the medium particles and being stretched along with those particles; here the field energy is the integral of the square of the length of the field vector over the whole domain.

The mapping suggested by A.D. takes a part of the domain boundary inside and it is not a diffeomorphism, it is even not smooth on the boundary. Modifications of this mapping allow one to make it smooth, however then the stretched torus folded in half is mapped not exactly into itself, but almost into itself: a part of that sticks out beyond the boundary of the original solid torus. Because of that one can prove an exponential growth of the field under iterations not in the whole volume of the solid torus but only for some magnetic lines which constitute a set of measure zero. There is still no proof of an exponential growth of the energy under iterations of such mappings — it is a challenge for mathematicians.

[2] These two pages are a part of V.I. Arnold's review "Mathematics in the works of Ya.B. Zeldovich". (Editors' remark)

Acknowledgements

Springer-Verlag thanks the publishers of V.I. Arnold's papers for having granted permission to reprint them in these Collected Works.

The numbers behind the original sources correspond to the numbers given in the Table of Contents.

We have made every effort to identify the copyright owners for all articles included in this book in order to obtain reprint permission. Some of our requests however remained unanswered. The following list includes all sources.

Translated from Academician A.D. Sakharov, Scientific Works	27
Reprinted from Developments in Mathematics: The Moscow School	15
Reprinted from Duke Math. Journal	1
Translated from El País (Futuro)	21
Reprinted from First European Congress of Mathematics (Paris, 1992)	12
Translated from Gazette des Mathématiciens	20
Reprinted from J. Algebraic Geometry	4
Translated from Kvant	19, 26
Reprinted from Math. Intelligencer	23
Reprinted from Notices Amer. Math. Soc.	22
Translated from Passion des Formes	18
Reprinted from Proc. Steklov Inst. Math.	11
Reprinted from Quantum	24
Translated from Revista Matemática de la Universidad Complutense de Madrid	9
Reprinted from Russ. Acad. Sci. Izv. Math.	2
Reprinted from Russ. Math. Surveys	3, 13
Reprinted from Selecta Math. (N. S.)	14
Reprinted from Singularities and Bifurcations	10
Reprinted from St. Petersburg Math. J.	5, 8
Reprinted from The Floer Memorial Volume	7
Translated from The Geometry and Physics of Knots (Russian translation)	25
Reprinted from Topological Methods in Modern Mathematics	6
Translated from Topological Methods in Nonlinear Analysis	16
Reprinted from Trends and Perspectives in Applied Mathematics	17

Printed by Wilco bv, the Netherlands